S0-AMO-264

Evolved Cellular Network Planning and Optimization for UMTS and LTE

Evolved Cellular Network Planning and Optimization for UMTS and LTE

Edited by
Lingyang Song and Jia Shen

CRC Press is an imprint of the
Taylor & Francis Group, an **informa** business
AN AUERBACH BOOK

CRC Press
Taylor & Francis Group
6000 Broken Sound Parkway NW, Suite 300
Boca Raton, FL 33487-2742

Printed in the United States of America on acid-free paper
10 9 8 7 6 5 4 3 2 1

International Standard Book Number: 978-1-4398-0649-4 (Hardback)

Library of Congress Cataloging-in-Publication Data

Evolved cellular network planning and optimization for UMTS and LTE / editors,
Lingyang Song, Jia Shen.
p. cm.
"A CRC title."
Includes bibliographical references.
ISBN 978-1-4398-0649-4 (hardcover : alk. paper)
1. Cell phone systems--Planning. 2. Universal Mobile Telecommunications System.
3. Long-term evolution (Telecommunications) I. Song, Lingyang. II. Shen, Jia, 1977- III.
Title.

TK5103.485.E96 2011
621.3845'6--dc22 2010025218

Visit the Taylor & Francis Web site at
http://www.taylorandfrancis.com

and the CRC Press Web site at
http://www.crcpress.com

Contents

Contributors

Antonios Alexiou
University of Patras
Patras, Achaia, Greece

Matthew Baker
Alcatel-Lucent
Cambridge, United Kingdom

Tarek Bejaoui
Mediatron Lab
University of Carthage
Sfax, Tunisia

Christos Bouras
University of Patras and RACTI
Patras, Achaia, Greece

Hsiao Hwa Chen
Department of Engineering Science
National Cheng Kung University
Tainan City, Taiwan, Republic of China

Américo Correia
Instituto de Telecomunicações
ISCTE-IUL
Lisboa, Portugal

Rui Dinis
Instituto de Telecomunicações
FCT-UNL
Caparica, Portugal

Petar Djukic
Department of Septenes and
Computer Engineering
Carleton University
Ottawa, Canada

Avraham Freedman
NICE Systems Ltd.
Intelligence Solutions Division
Ra'anana, Israel

Mariusz Głąbowski
Poznan University of Technology
Faculty of Electronics and
Telecommunications
Chair of Communications and
Computer Networks
Poznan, Poland

Fernando Gordejuela-Sánchez
Centre for Wireless Network Design
(CWiND)
University of Bedfordshire
Luton, United Kingdom

Bin Han
Beijing University of Posts and
Telecommunications
Beijing, People's Republic of China

Xuemin Huang
NG Networks Co., Ltd.
Suzhou, People's Republic of China

Vasileios Kokkinos
University of Patras and RACTI
Patras, Achaia, Greece

Moshe Levin
NICE Systems Ltd.
Cellular Technology Department
Ra'anana, Israel

Li Li
Beijing University of Posts and
Telecommunications
Beijing, People's Republic of China

Guangyi Liu
Research Institute of China Mobile
Beijing, People's Republic of China

Anis Masmoudi
Mediatron Lab
University of Carthage
Tunisia and ISECS Institute
University of Sfax
Sfax, Tunisia

Asad Mehmood
Department of Signal Processing
School of Engineering
Blekinge Institute of Technology
Blekinge, Sweden

Abbas Mohammed
Department of Signal Processing
School of Engineering
Blekinge Institute of Technology
Blekinge, Sweden

Bilal Muhammad
Department of Electronics Engineering
IQRA University
Peshawar Campus NWFP
Peshawar, Pakistan

Nidal Nasser
Department of Computing
and Information Science University
of Guelph
Guelph, Canada

Mugen Peng
Beijing University of Posts and
Telecommunications
Beijing, People's Republic of China

Balaji Raghothamon
Airvana
Chelmsford, Massachusetts

Mahmudur Rahman
Department of Septenes and
Computer Engineering
Carleton University
Ottawa, Canada

Mohammad S. Sharawi
Electrical Engineering Department
King Fahd University of Petroleum and
Minerals (KFUPM)
Dharan, Saudi Arabia

João Silva
Instituto de Telecomunicações
ISCTE-IUL
Lisbon, Portugal

Nuno Souto
Instituto de Telecomunicações
ISCTE-IUL
Lisboa, Portugal

Anand Srinivas
Airvana
Chelmsford, Massachusetts

Maciej Stasiak
Poznan University of Technology
Faculty of Electronics and Telecommunications
Chair of Communications and Computer Networks
Poznan, Poland

Meixia Tao
Institute of Wireless Communication Technology
Department of Electronic Engineering
Shanghai Jiao Tong University
Shanghai, People's Republic of China

Jay Weitzen
Airvana and university of Massachusetts Lowell
ECE Department
Chelmsford, Massachusetts

Changqing Yang
Beijing University of Posts and Telecommunications
Beijing, People's Republic of China

Halim Yanikomeroglu
Department of Septenes and Computer Engineering
Carleton University
Ottawa, Canada

Jianhua Zhang
Wireless Technology Innovation (WTI) Institute of Beijing
University of Posts and Telecom
Beijing, People's Republic of China

Jie Zhang
Centre for Wireless Network Design (CWiND)
University of Bedfordshire
Luton, United Kingdom

Jietao Zhang
Huawei Wireless Research
Shenzhen, People's Republic of China

Xiaobo Zhang
Alcatel-Lucent Shanghai Bell
Shanghai, People's Republic of China

Piotr Zwierzykowski
Poznan University of Technology
Faculty of Electronics and Telecommunications
Chair of Communications and Computer Networks
Poznan, Poland

INTRODUCTION

I

Chapter 1

Introduction to UMTS: WCDMA, HSPA, TD-SCDMA, and LTE

Matthew Baker and Xiaobo Zhang

Contents

1.1 Progression of Mobile Communication Provision

A key aim of modern cellular communication networks is to provide high-capacity coverage over a wide area. The cellular concept was first deployed in the U.S. in 1947. By breaking the coverage area down into many small cells, the total system capacity could be substantially increased, enabling more users to be served simultaneously.

The first cellular systems avoided interference between the cells by assigning a particular operating frequency to each cell; cells in the same vicinity were assigned different frequencies. The level of inter-cell interference in such systems can be reduced by assigning more frequencies, at the expense of reduced spectral efficiency. The total number of frequencies used is termed the frequency reuse factor. A high frequency reuse factor gives good isolation between cells but makes poor use of the scarce and expensive spectrum resource. An example of a cellular network with a frequency reuse factor of 3 is shown in Figure 1.1.

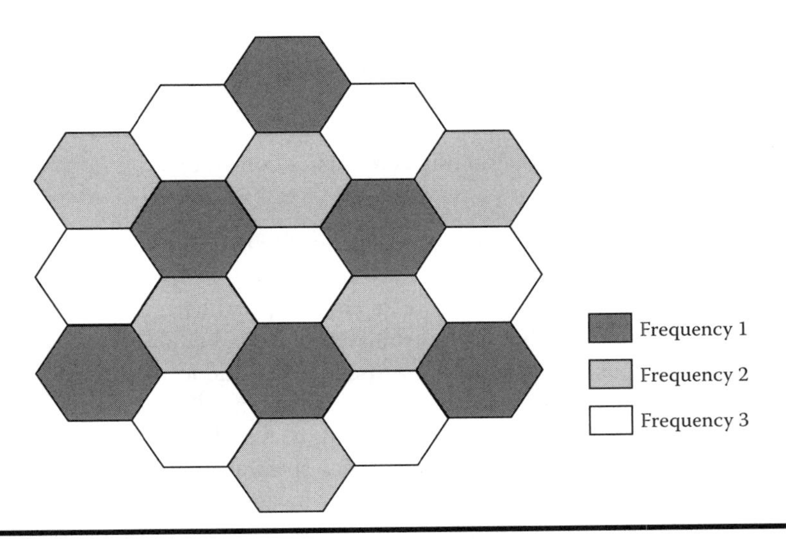

Figure 1.1 An example of a cellular communication network with frequency reuse factor 3.

The use of different frequencies in cells that are close to each other continued as the predominant cellular technique for the next four decades, up to and including the Global System for Mobile Communications (GSM), which was the first cellular system to achieve worldwide penetration, with billions of users. Such widespread deployment has led to a high level of understanding of network planning issues for GSM, in particular in relation to frequency reuse planning. Practical network deployments are never as straightforward as the simplistic example shown in Figure 1.1, and complex software tools have been developed to model propagation conditions and enable optimal frequency assignments to be achieved.

Projections of increasing demand for wide-area communications supporting new applications requiring high data rates led to the development of a new generation of cellular communication system in the late 1980s and the 1990s. These systems became known as 3rd Generation systems, aiming to fulfil the requirements set out by the International Telecommunication Union (ITU) for the so-called IMT-2000* family. Broadly speaking, such systems aimed to achieve data rates up to 2 Mbps.

The 3rd Generation system which has become dominant worldwide was developed in the 3rd Generation Partnership Project (3GPP) and is known as the Universal Mobile Telecommunication System (UMTS). 3GPP is a partnership of six regional Standards Development Organizations (SDOs) covering Europe (ETSI), Japan (ARIB and TTC), Korea (TTA), North America (ATIS), and China (CCSA).

* International Mobile Telecommunications for the year 2000.

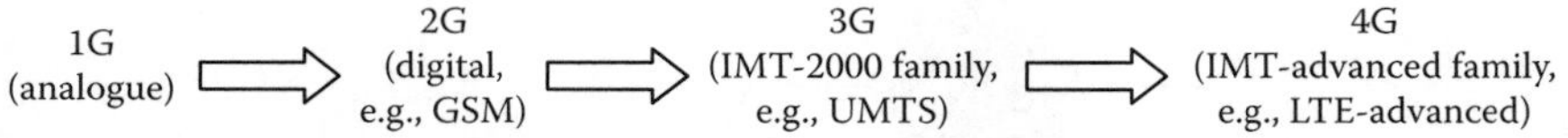

1G (analogue) ⇒ 2G (digital, e.g., GSM) ⇒ 3G (IMT-2000 family, e.g., UMTS) ⇒ 4G (IMT-advanced family, e.g., LTE-advanced)

Figure 1.2 The generations of mobile communication systems.

In contrast to the time division multiple access (TDMA) used by GSM, UMTS used a new paradigm in multiple access technology, being based on code division multiple access (CDMA) technology. CDMA technology had been known for decades from military applications, but its suitability for use in cellular systems was not demonstrated until the 1990s when it was used in the American "IS95" standard.

The use of CDMA requires a fundamental change in cellular network planning and deployment strategies, largely resulting from the fact that it enables a frequency reuse factor of 1 to be used. This can achieve high spectral efficiency but necessitates careful control of inter-cell interference. The principles of CDMA as utilized in UMTS are discussed in the following section, together with an introduction to some of the resulting network planning and deployment issues.

The subsequent sections of this chapter introduce the evolutions of UMTS which continue to be developed. First, high-speed packet access (HSPA) brings a significant shift from predominantly circuit-switched applications requiring roughly constant data rates toward packet-switched data traffic. This is accompanied by new quality of service (QoS) requirements and consequent changes for network planning.

In parallel with the widespread deployment and continuing development of HSPA, a radical new step is also available in the form of the long-term evolution (LTE) of UMTS. LTE aims to provide a further major step forward in the provision of mobile data services, and will become widely deployed in the second decade of the 21st century. LTE continues with the spectrally efficient frequency-reuse-1 of UMTS, but introduces new dimensions for optimization in the frequency and spatial domains. Like UMTS, LTE itself is progressively evolving, with the next major development being known as LTE-advanced (LTE-A), which may reasonably be said to be a 4th Generation system.

The succession of generations of mobile communication system are illustrated in Figure 1.2.

1.2 UMTS

The first release of the UMTS specifications became available in 1999 and is known as "Release 99." It provides for two modes of operation depending on the availability of suitable spectrum: the frequency-division duplex (FDD) mode, suitable for paired spectrum, uses one carrier frequency in each direction, while the time-division duplex (TDD) mode allows UMTS to be deployed in an unpaired spectrum by using different time slots for uplink and downlink transmissions on a single carrier frequency.

One of the principle differences of UMTS compared to previous cellular systems such as GSM is that it is designed to be a wideband system. In general, this means that the transmission bandwidth is greater than the coherence bandwidth of the radio channel. This is advantageous in terms of making the system more robust against multipath fading and narrowband interference. In UMTS FDD mode, this is achieved by means of a 5-MHz transmission bandwidth: Regardless of the data rate of the application, the signal bandwidth is spread to 5 MHz to make use of the full diversity of the available channel.

In Release 99, the TDD mode of UMTS also makes use of a 5-MHz carrier bandwidth, but in later releases of the specifications other bandwidths were added: the second release, known as "Release 4" (for alignment with the version numbering of the specification documents), introduced a narrower 1.6-MHz TDD carrier bandwidth, while Release 7 added a 10-MHz TDD bandwidth. The 1.6-MHz option for TDD is used for the mode of UMTS known as time division-synchronous CDMA (TD-SCDMA), which is introduced in Section 1.4.

The key features introduced in each release of the UMTS specifications are summarized in Figure 1.3.

Regardless of the duplex mode or bandwidth option deployed, UMTS is structured around a common network architecture designed to interface to the same core network (CN) as was used in the successful GSM system. The UMTS terrestrial* radio access network (UTRAN) is comprised of two nodes: the radio network controller (RNC) and the NodeB. Each RNC controls one or more NodeBs and is responsible for the control of the radio resource parameters of the cells managed by those NodeBs. This is a key difference from GSM, where the main radio resource management functions were all provided by a single radio access network node, the base transceiver station (BTS).

Each NodeB in UMTS can manage one or more cells; a common arrangement comprises three 120°-segment-shaped cells per NodeB, formed using fixed directional antennas. Higher-order sectorization may also be deployed, for example, with 6 or even 12 cells per NodeB. A deployment using three cells per NodeB is shown in Figure 1.4.

The terminals in a UMTS system are known as user equipments (UEs). At any given time, a UE may be communicating with just one cell or with several cells simultaneously; in the latter case the UE is said to be in a state known as soft handover, which is discussed in more detail in Section 1.2.1.4. In order to facilitate mobility of the UEs within the UMTS network, interfaces are provided between RNCs to enable a connection to be forwarded if the UE moves into a cell controlled by a different RNC. For a given UE, the RNC which is currently acting as the connection point to the CN is known as the serving RNC (SRNC), while any intermediate RNC is referred to as a drift RNC (DRNC). The main standardized network interfaces are

* Terrestrial as opposed to satellite.

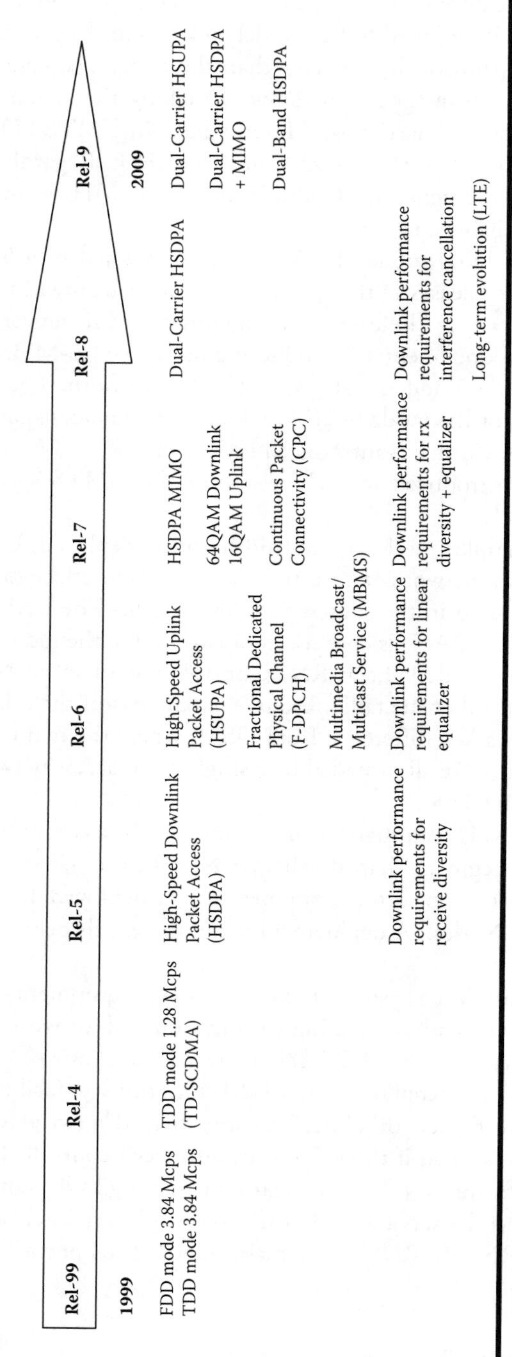

Figure 1.3 Key features of each UMTS release.

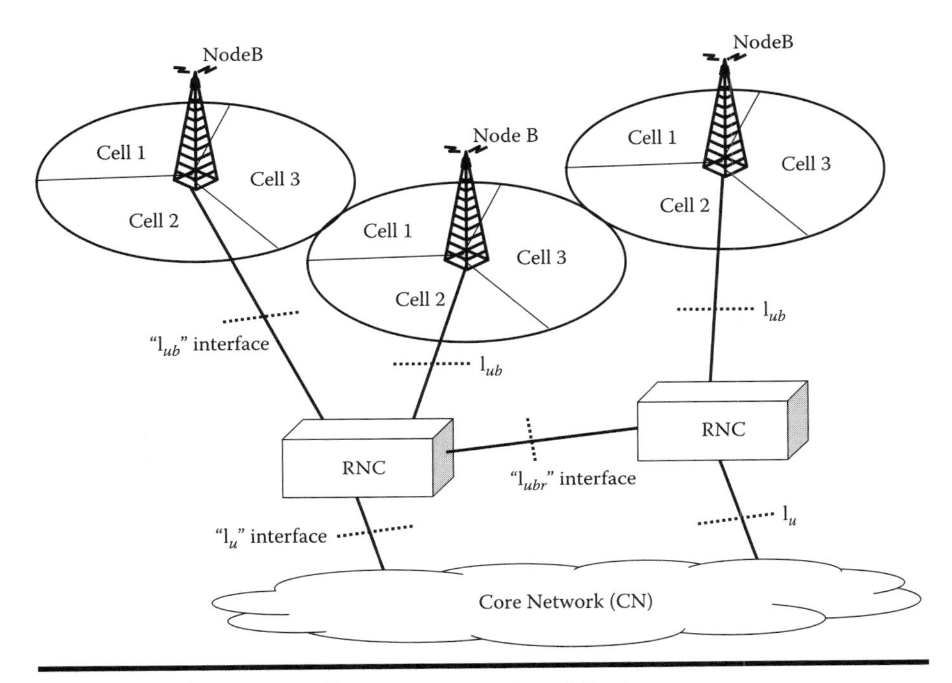

Figure 1.4 The UMTS radio access network architecture.

also shown in Figure 1.4: the "I_{ub}" interface between the NodeB and RNC, the "I_{ur}" interface between RNCs, and the "I_u" interface between the RNC and the CN.

1.2.1 Use of CDMA in UMTS

An understanding of the principles of CDMA is essential to the ability to deploy UMTS networks efficiently. In this section therefore, an introduction to CDMA in general is given, followed by an explanation of how CDMA is adapted and applied in UMTS specifically, and an overview of some particular aspects of the technology that are relevant to cellular deployment.

1.2.1.1 Principles of CDMA

The basic principle of CDMA is that different data flows are transmitted at the same frequency and time, and they are rendered separable by means of a different code sequence assigned to each data flow. This is in contrast to FDMA and TDMA, which use different frequencies and different time slots respectively to separate the transmissions of different data flows. In CDMA,* each data symbol to be transmitted is multiplied by a higher-rate sequence known as a spreading sequence, which

* We focus here on "direct sequence" CDMA as used in UMTS.

increases (spreads) the signal bandwidth to the desired transmission bandwidth. By assigning different spreading sequences to different data flows, taking care that the different spreading sequences have low cross-correlation between them, the signals can be separated at the receiver even though they all use the same spectrum.

A simple example is shown in Figure 1.5, where two data flows are transmitted from a base station, each data flow being destined for a different user and therefore being assigned a different spreading sequence. Each user's receiver, knowing its assigned spreading sequence in advance (by means of suitable configuration signalling), correlates the received signal with its spreading sequence over the duration of each data symbol, thereby recovering the transmitted data flow. This process is also known as despreading. The length of the spreading sequence is known as the spreading factor (SF); hence the rate and bandwidth of the spread signal are SF times the rate and bandwidth of the original data flow. The symbols after spreading are known as chips, and hence the rate of the spread signal is known as the chip rate. The chip rate is chosen to fit the available channel bandwidth (5 MHz in the case of UMTS FDD mode), and the SF is set for each data flow depending on the data rate, to increase the transmitted rate up to the chip rate. A low-rate data flow would therefore be assigned a high SF, and vice versa.

Ideally, the spreading sequences would be fully orthogonal to each other (as is the case with the example in Figure 1.5), thus resulting in no interference between the different data flows, but in practice this is not always possible to achieve. One reason for this is that insufficient orthogonal sequences exist of practical length; it has been shown in [1] that the full multiple access channel capacity is achieved by means of non-orthogonal sequences coupled with interference cancellation at the receiver. Moreover, orthogonal sequences often exhibit other properties which are less desirable. In particular, if it is desired to use the sequences for synchronization of the receiver, a sharp single-peaked autocorrelation function is required, with low sidelobes; orthogonal sequences often exhibit multiple peaks in their autocorrelation functions.

A further factor affecting the performance of spreading sequences is lack of time-alignment. Many types of orthogonal sequences are only orthogonal if they are time-aligned. Lack of time-alignment can occur due to the transmitters not being synchronized (in the case of the sequences being transmitted by different terminals or base stations), but also due to different propagation delays in the radio channel. The latter can also cause self-interference to occur even when only a single sequence is transmitted, and equalization is required to remove such interference.

1.2.1.2 CDMA in UMTS

In UMTS, two families of codes are used for the CDMA spreading sequences: orthogonal codes and non-orthogonal pseudo-noise (PN) codes.

The orthogonal codes are used for the spreading operation and are therefore referred to as spreading codes. They have exactly zero cross-correlation if they are

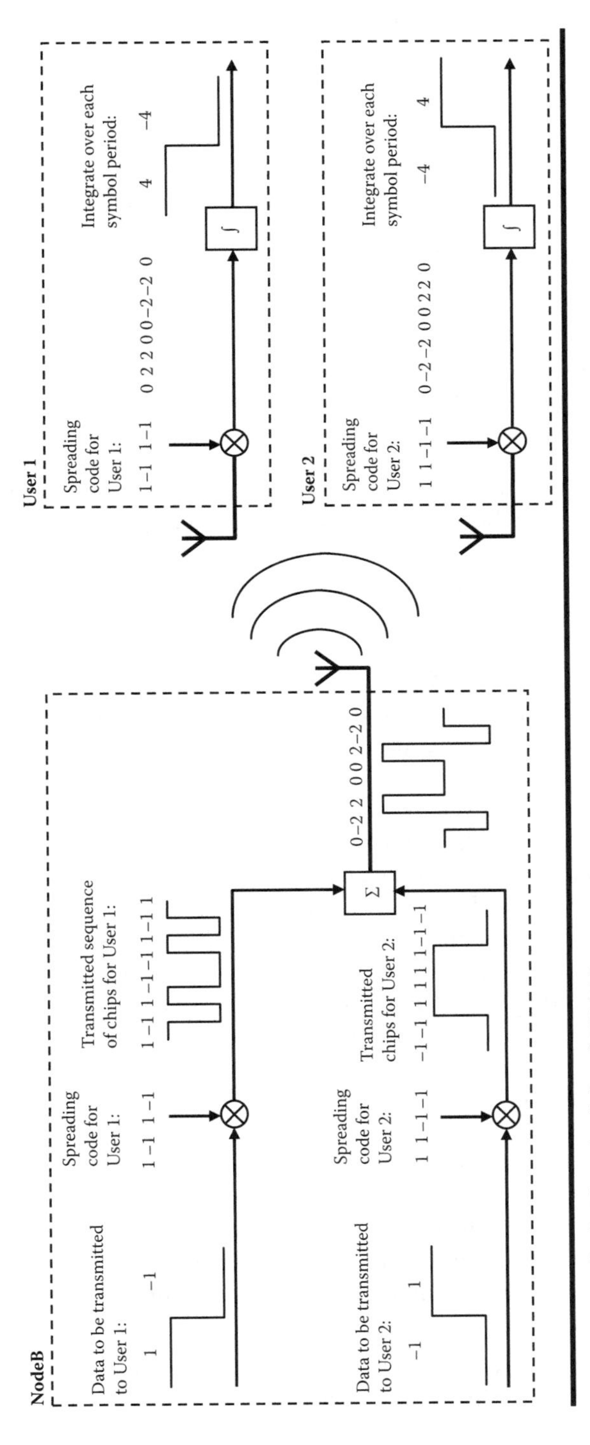

Figure 1.5 The basic principle of CDMA.

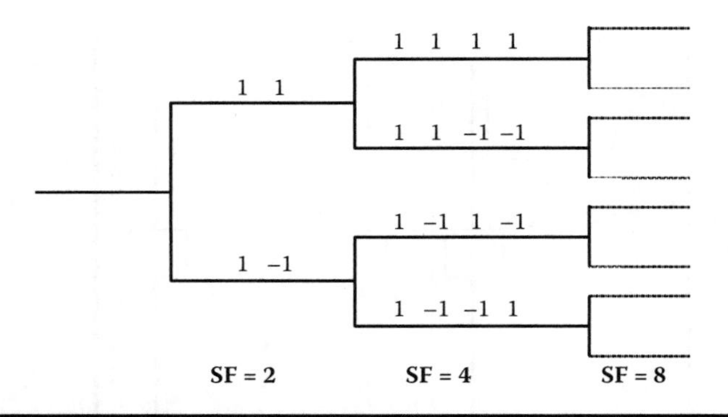

Figure 1.6 Orthogonal variable spreading factor code tree.

time-aligned, but poor and variable cross-correlation if they are not time-aligned. This means they are suitable for separating data flows (a.k.a., channels) transmitted from a single source—from a single UE in the uplink and from a single cell in the downlink—where it can be guaranteed that the transmit timing of the different channels is aligned.* The orthogonal codes are therefore also known as channelization codes.

The orthogonal codes are Walsh-Hadamard codes [2], selected in a systematic tree-like structure to enable code sequences of different lengths (i.e., different SF) to be chosen depending on the data rate. This structure is often referred to as an orthogonal variable spreading factor (OVSF) code tree [3]. It is illustrated in Figure 1.6. Any code of a given SF is not only orthogonal to any other code of the same SF in the tree, but is also orthogonal to all the codes of higher SF, which are offshoots from a different code of the same SF.

The chip rate in UMTS FDD is 3.84 Megachips per second (Mcps), which, after application of a suitable spectrum mask, fits comfortably within the 5-MHz channel bandwidth typically available for UMTS. As an example, a data channel with a symbol rate (after channel coding) of 120 kbps would use an OVSF channelization code with $SF = 3.84/0.12 = 32$ to spread the signal to this bandwidth. For TD-SCDMA, the chip rate is 1.28 Mcps, corresponding to a 1.6-MHz channel bandwidth.

The second family of codes used in UMTS are the non-orthogonal PN codes, which are known as scrambling codes. These codes do not have zero cross-correlation even if they are time-aligned with each other, but on the other hand the cross-correlation remains low regardless of the time-alignment. They are therefore well-suited to the separation of signals from different sources—from different UEs in the uplink and different cells in the downlink—by virtue of whitening the interference between them. Moreover, the autocorrelation function of these codes usually has

* Note, however, that there is still some loss of orthogonality at the receiver, due to the self-interference arising from multipath propagation delays.

only one strong peak, and they can therefore help with timing acquisition and maintenance of synchronization.

The scrambling codes are applied at the chip rate after the spreading operation. The combined spreading and scrambling operations in UMTS are illustrated in Figure 1.7 (for the case of the downlink transmissions).

The downlink scrambling codes are usually statically assigned during the network deployment. A large number of downlink scrambling codes are available in UMTS, in order to facilitate assignment without complex planning. Each cell has a primary scrambling code which must be discovered by the UE before it can access the network. To aid this discovery process, the available primary scrambling codes are grouped into 64 groups of 8. The identity of the group to which the primary scrambling code of a particular cell belongs is discovered from a synchronization channel broadcast by the cell. As part of the network planning process for UMTS, the timing of the synchronization channels must be set appropriately. This involves ensuring that cells in the same vicinity have different timings in order to enable a UE to distinguish the synchronization channels from different cells and select the strongest. The particular scrambling code used within the group is then identified from the common pilot channel (CPICH), which is also broadcast from each cell. The CPICH from every cell uses a fixed sequence defined in the UMTS specifications, spread by a specified channelization code of SF 256, and scrambled by the primary scrambling code used in the cell. A UE can therefore identify the primary scrambling code of the cell by performing eight correlations of the known CPICH sequence with the signal received. The CPICH is an important channel as it also provides the phase reference for the UE to demodulate other downlink channels transmitted by the NodeB.

One limitation of the spreading and scrambling code structure in UMTS is the limited number of orthogonal spreading codes available. In the uplink this is not a

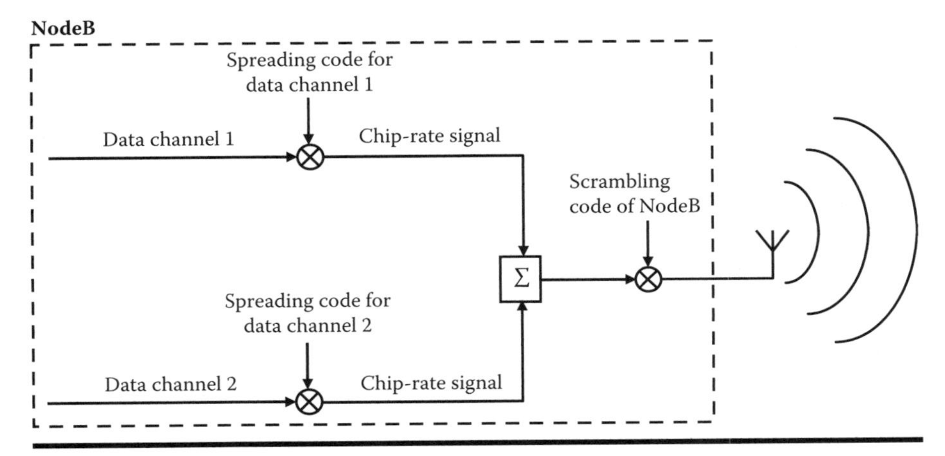

Figure 1.7 Spreading and scrambling in UMTS.

problem, because the number of channels transmitted by a single UE is smaller than the number of codes available. However, in the downlink, the number of transmitted channels in one cell is typically much larger, owing to the need to separate the transmissions to all the different UEs in the cell. One solution to this is to use one or more additional "secondary" scrambling codes in each cell; each additional secondary scrambling code enables the whole OVSF code tree to be reused, but this comes at the expense of additional intra-cell interference due to the fact that the PN scrambling codes are not orthogonal to each other. Other solutions to the downlink channelization code shortage problem in UMTS include the increased use of time-multiplexing, as discussed in more detail in Section 1.3.

In the uplink, the scrambling codes used by each UE to separate their transmissions from those of other UEs are assigned by radio resource control (RRC) signaling following an initial random access transmission by the UE.

1.2.1.3 Power Control

In a CDMA system, where the different signal transmissions potentially interfere with each other, the transmission power of each signal needs to be carefully controlled so that it arrives at the receiver with sufficient signal-to-interference ratio (SIR) to achieve the desired QoS, yet not cause excessive interference to the other signal transmissions and thereby limit the capacity of the system.

This is especially important in the uplink, where the non-orthogonal PN scrambling codes used to separate the users, and the lack of synchronization of the transmissions, result in the system capacity being limited by intra-cell interference. In the absence of power control, the differing path losses of different UEs would result in the signals transmitted by UEs close to the NodeB drowning out those from UEs at the cell edge.* Additionally, the received signal strength from a moving UE typically fluctuates rapidly due to the fast fading that arises from the constructive and destructive superposition of signals propagating by different paths in the radio channel.

Uplink power control in UMTS is designed to compensate for both the path loss and the variable fast fading. This is achieved by a closed-loop design, whereby the NodeB regularly measures the received SIR from each UE, compares it with a target level set to achieve the desired QoS, and sends transmitter power control (TPC) commands back to each UE to instruct them to raise or lower their transmission power as necessary. This operation occurs at 1500 Hz, which is sufficiently fast to counteract the fast fading for terminals moving at vehicular speeds of several tens of km/h.

In parallel with the closed-loop TPC command feedback process, an "outer" control loop also operates to ensure that the SIR target is set at an appropriate level. The outer loop operates more slowly, with the NodeB measuring the block error rate (BLER) of the received uplink data blocks, and adjusting the target SIR to ensure that a target BLER is met. The BLER is used as the primary indicator of QoS.

* This is often known as the "near-far problem."

The combined operation of the inner and outer power-control loops is illustrated in Figure 1.8.

Appropriate power control configuration is a key aspect of network optimization in UMTS and is closely related to call admission control (CAC). If too many users are admitted to a particular cell, the rise in interference that they cause to each other will force the closed-loop power control to raise the power of all the UEs. This in turn causes further interference, which may result in the power control system becoming unstable and creating a severe degradation of uplink capacity.

1.2.1.4 Soft Handover and Soft Capacity

As noted earlier, in a CDMA system like UMTS with a frequency-reuse factor of 1, a UE can receive transmissions from multiple cells simultaneously. Similarly, a UE's uplink transmissions can be received simultaneously by multiple cells. When a UE is in this state, it is said to be in soft handover. If the multiple cells are controlled by the same NodeB, it is described as softer handover, which is characterized by the TPC commands transmitted by the different cells to the UE being identical.

For the downlink transmissions in soft handover, the UE can combine the soft values of the received bits (typically in the form of log-likelihood ratios [LLRs]) from the different cells prior to decoding. In the uplink, soft combining may also be used in the case of softer handover, but where different NodeBs are involved, selection-combining is used, whereby the RNC selects decoded packets from whichever NodeB has managed to decode them successfully. The soft handover state is illustrated in Figure 1.9.

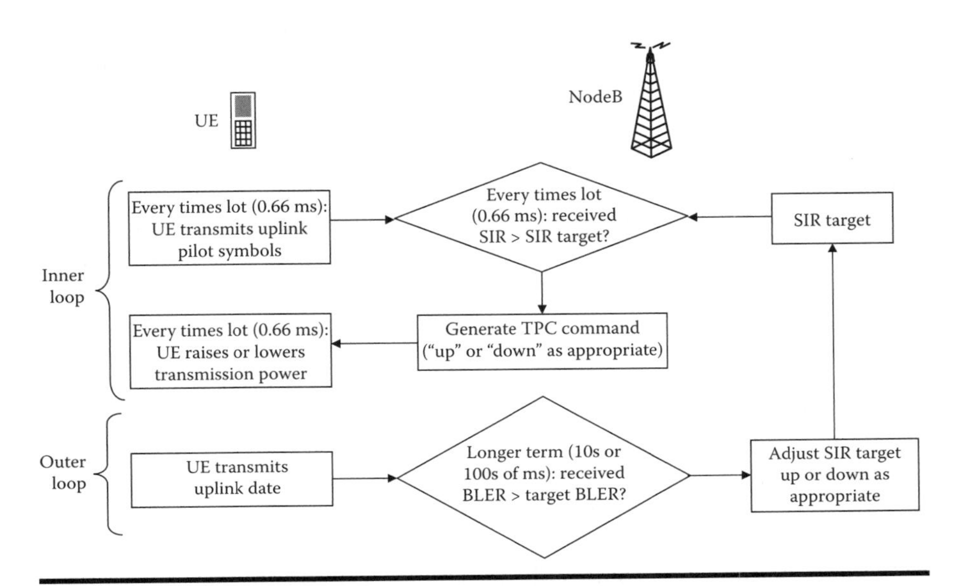

Figure 1.8 Closed-loop power control in UMTS.

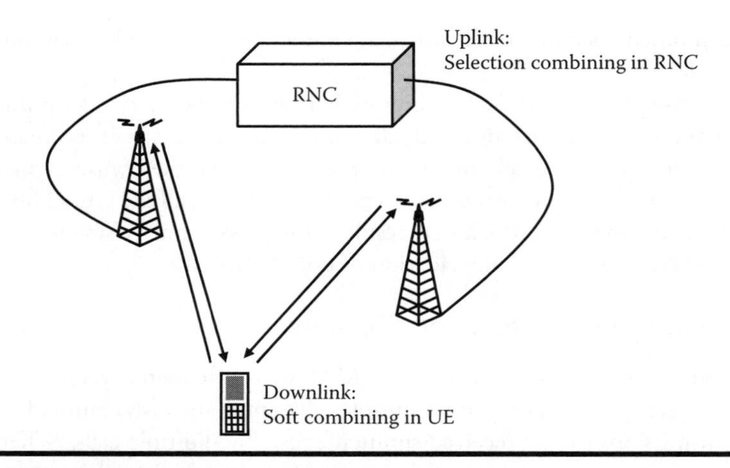

Figure 1.9 Soft handover in UMTS.

Soft handover can play an important role in increasing network capacity in UMTS, since it provides a source of diversity that allows the uplink transmission power and the transmission powers of each of the downlink cells to be significantly reduced compared to the case of single-cell transmission and reception. In typical macro-cellular UMTS network deployments, 20% to 40% of the UEs are likely to be in soft handover at any time.

The set of cells with which a UE is communicating is known as the active set. Cells are added to and removed from the active set based on measurements of the SIR of the CPICHs from the different cells. The network configures thresholds for the UE to determine when a UE should transmit a CPICH SIR measurement report to the network, and the network then uses these measurement reports to decide when to instruct the UE to add a cell to the active set or remove one from it. Some hysteresis is usually used, to avoid "ping-pong" effects, whereby a cell is repeatedly added to or removed from the active set of a UE near a cell border. On the other hand, where high-mobility UEs are involved, or in environments with dramatic discontinuities in propagation conditions (e.g., in "Manhattan" type dense urban areas), it is important that the thresholds are configured to ensure sufficiently rapid updating of the active set to avoid calls being dropped.

Unlike with multiple access schemes that are orthogonal in time or frequency, where the the capacity of each cell depends on the number of time slots or frequencies available, in a CDMA network like UMTS the quality of the links can be traded off against the number of users in the cell. If an additional user is allowed to set up a call in a cell, the existing users will experience a small rise in the interference level, but for most users this will not result in their call being dropped. Any calls which might be dropped would tend to be at the cell edge, where users would usually already be in soft-handover with another cell and can simply transfer to that cell. Thus the effective size of a cell automatically reduces as more users set up connections, and vice versa. This is known as cell breathing, and gives the network operator flexibility to manage varying densities of users.

1.2.2 Deployment Techniques in UMTS

A number of optional aspects are available in UMTS to increase capacity and/or improve QoS. Some of these are introduced briefly here.

1.2.2.1 Transmit Diversity

In the downlink, transmit diversity can be configured to improve the link quality. Two transmit diversity schemes were defined in the first release of UMTS: a space-time block code (STBC) known as space-time transmit diversity (STTD), and a closed-loop beamforming mode.

The STTD scheme uses an orthogonal coding scheme as shown in Figure 1.10 to transmit pairs of data symbols s_1 and s_2 from two antennas at the NodeB. This scheme can be shown to achieve full diversity gain when using a linear receiver [4]. However, the orthogonality of the transmissions from the two antennas is only achieved if the channel gain is constant across the two transmitted symbols of each pair, and is therefore not suitable for high-mobility scenarios. A further drawback of this scheme is that the orthogonality is also lost if the radio channel exhibits frequency selectivity [5]; the orthogonality cannot be restored by linear processing. Consequently, the usefulness of the STTD scheme is limited in practice.

In the closed-loop beamforming mode, identical data symbols are transmitted from the two NodeB antennas. Fast feedback from the UE is employed to select the optimal phase offset to be applied to the transmission from one of the NodeB antennas to steer a beam in the direction of the UE (maximizing the received signal energy by constructive superposition). Orthogonal CPICH patterns are transmitted from the two NodeB antennas without any phase offset, and the UE reports the phase offset which would maximize the received signal strength based on its measurements of the received CPICH signals. The NodeB then applies the selected phase offset to the data transmissions only. One limitation of this scheme is that the performance depends significantly on whether the UE performs hypothesis testing to confirm

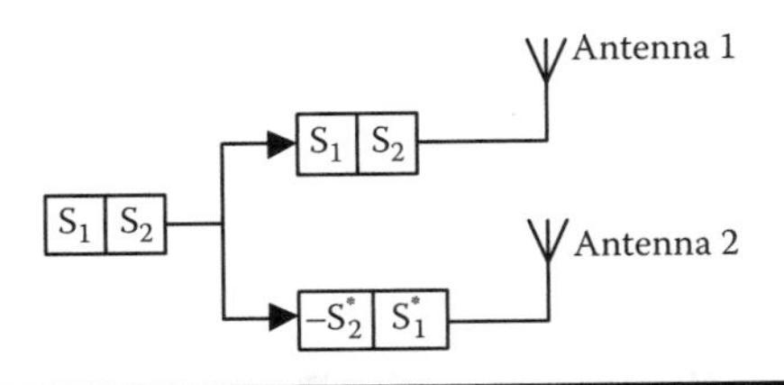

Figure 1.10 Space-time transmit diversity (STTD).

whether the NodeB actually used the same phase offset as recommended by the feedback; this in turn depends on the reliability of the feedback signaling. Some enhancements to this scheme are available in later releases of the HSPA specifications (see Section 1.3.3).

1.2.2.2 Receiver Techniques

The performance of the receiver has a significant impact on the throughput and QoS that can be achieved.

The first CDMA receivers were based on an architecture known as the "rake" receiver [6], owing to its resemblance to a garden rake (see Figure 1.11). The rake receiver is an extension of the concept of the matched filter, and is designed to combine CDMA signals received over a time-dispersive multipath channel. It consists of a tapped delay line, where the delays are set to match the time differences between the corresponding path delays. At each delay (known as a rake "finger," the received signal is correlated with the known CDMA spreading sequence, and the outputs from the different rake fingers are typically combined using the Maximal Ratio Combining (MRC) principle. This weights each path by its SNR, as well as canceling the phase rotation of the channel, resulting in a final output SNR equal to the sum of the SNRs of the signals received via each individual path. (This assumes that the interference can be modeled as AWGN.) The basic performance requirements for UMTS assume a rake receiver.

A variety of more advanced receivers are possible, at both the UE and the NodeB. A simple enhancement is the use of multiple antennas for receive diversity. MRC can be used to combine the signals from the different antennas in the same way as the signals from different rake fingers. In UMTS, enhanced performance requirements known as "Type 1" are defined for UEs implementing MRC-based receive diversity.

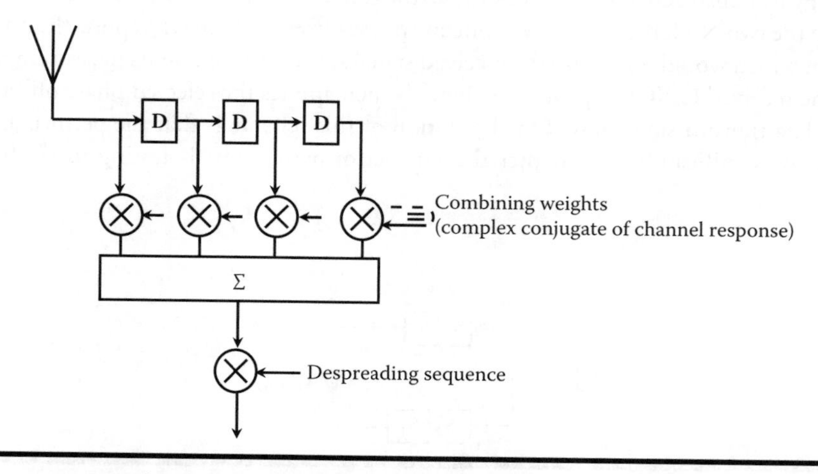

Figure 1.11 Rake receiver architecture.

Further enhancement is possible by means of linear equalization techniques. A number of categories of linear equalizer exist. Zero-forcing (ZF) equalizers aim to eliminate ISI and can give good performance in time-dispersive channels with high SINR. However, in low SINR conditions the ZF approach causes noise amplification to occur at frequencies where the channel gain is low. This problem can be addressed by using instead a minimum mean-squared error (MMSE) criterion to determine the equalizer coefficients. An MMSE equalizer aims to minimize the error in the received bits taking into account the estimated channel impulse response and noise power. Enhanced performance requirements known as "Type 2" are defined for UEs implementing linear MMSE equalization (or similar techniques). This approach may also be extended to dual-antenna receivers, in which case the UMTS performance requirements are known as "Type 3."

Multiple antennas may also be used to cancel interference, for example by considering the spatial characteristics of the received interference and adapting the antenna combining coefficients so as to set a null in the direction of the strongest interferer. For a UE with two receive antennas, this enables interference from one NodeB to be reduced in order to increase the SINR of the signal from the serving NodeB. "Type 3i" performance requirements are defined for such a case.

Yet more complex receiver architectures can use non linear techniques. For example, decision feedback equalization (DFE) takes into account previously demodulated symbols to improve the rejection of Inter-Symbol Interference (ISI) in later symbols. Alternatively, successive interference cancellation (SIC) may be used, whereby one signal (usually that with the highest SINR) is fully decoded (including channel decoding), before being reconstructed and subtracted (without noise) from the total received signal. This can dramatically increase the SINR for the next data stream and hence enhance its decoding, at the expense of significantly increased complexity and delay. SIC techniques are particularly appropriate for the uplink receiver in the NodeB, where signals from many interfering UEs have to be decoded, as well as for multi-stream (MIMO) transmissions in the downlink (see Section 1.3.3).

1.2.3 Network Planning Considerations for UMTS

Many of the aspects of UMTS and CDMA that affect network planning and optimization have been introduced earlier. In many cases, an extensive range of parameters and options are provided in the UMTS specifications to enable these aspects to be configured and tuned to maximize performance in particular scenarios. For example, for handover, thresholds may be configured to optimize the trade-off between fast handover and ping-pong behavior, depending on the environment and characteristics of each cell.

In addition, the network operator may consider other implementation-related techniques to improve performance. For example, network synchronization [whereby the NodeBs are tightly synchronized, often by means of an external time reference such as the Global Positioning System (GPS)] can help improve the performance of inter-cell interference cancellation techniques.

Additionally, the physical configuration of the NodeB antennas is an important consideration. NodeB antenna configurations should be chosen appropriately for the scenario of each cell (e.g., depending on the beamwidths required. Additionally, NodeB antenna down-tilting may be used to control downlink inter-cell interference; interference to neighboring cells can be reduced by increasing the down-tilt, at the expense of some coverage reduction in the cell in question. A specification is provided in UMTS for the control of remote electrically tilting antennas.

Other physical characteristics such as the sitting of the NodeBs (e.g., above or below rooftop level) also play an important part in the resulting propagation characteristics and network peformance.

1.3 HSPA

As introduced in Figure 1.3, the completion of the first release of the UMTS specifications was followed by extensions known as high-speed packet access (HSPA). The main stimulus for this was the rapid growth of packet data traffic, necessitating both much higher data rates and a switch from constant data-rate circuit-switched traffic (chiefly voice) toward Internet Protocol (IP)—based packet-switched traffic. The first enhancement was to the downlink, where high-speed downlink packet access (HSDPA) was introduced in Release 5 of the UMTS specifications, driven predominantly by the growth of Internet download traffic; this was followed in Release 6 by high-speed uplink packet access (HSUPA), as attention began to focus on services requiring a more symmetric uplink/downlink traffic ratio such as e-mail, file sharing (including photographs and videos), and interactive gaming.

1.3.1 Principles of HSPA

The transition to a packet-switched service model required a fundamental change of approach in the radio interface compared to Release 99, leading to improved performance. Instead of providing a constant data rate regardless of the radio propagation conditions, a packet-switched model allows the instantaneous data rate on the radio interface to vary, taking advantage of instances of good radio conditions to provide very high peak data rates, and reducing the data rate when the "cost" of transmission is higher (i.e., when the radio propagation conditions are worse so that more transmission power or more bandwidth is required to maintain the same data rate). This enables the spectral efficiency of the overall system (taking all users into account) to be increased considerably.

This approach to the utilization of the radio channel can be exploited in two ways—firstly between multiple users, by means of dynamic multiuser scheduling, and secondly on individual radio link by link adaptation.

1.3.1.1 Dynamic Multiuser Scheduling

Multiuser scheduling exploits the fact that different users in a cell experience different variations in radio propagation conditions. This is sometimes known as "multiuser diversity," and is particularly the case in mobile scenarios where fast fading is a characteristic of the propagation environment. At each transmission opportunity, a scheduler in the NodeB can use knowledge of the propagation conditions of each user to select users that maximize the instantaneous system capacity, as illustrated by a simple example in Figure 1.12. This usually requires feedback from the UEs in order to provide sufficient information to the scheduler; in HSDPA, this feedback is provided by channel quality indicator (CQI) signaling which is transmitted from the UE to the NodeB based on the UE's measurements of the CPICH.

In practice, the scheduling function cannot aim solely to maximize system capacity (i.e., the sum rate to all users); such an approach is "unfair," always selecting UEs that are situated close to the NodeB. Most network operators require a more uniform distribution of QoS provision, and typical multiuser packet schedulers will therefore take a number of factors into account. A well-known approach is the proportional fair (PF) scheduler [7, 8], which selects users on the basis of their current instantaneous channel capacity weighted by the inverse of their actual average throughput achieved in a past time window. A typical ranking for each user could thus be as follows:

$$\frac{R_k(n)}{T_k(n)} \tag{1.1}$$

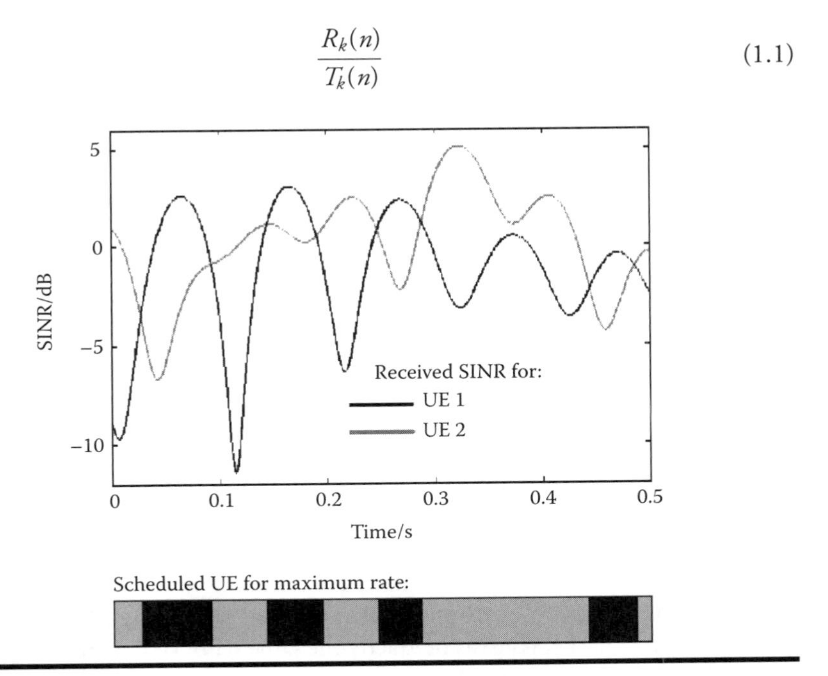

Figure 1.12 Multiuser scheduling to maximize system capacity.

where $R_k(n)$ is the predicted instantaneous achievable rate for the kth user at the nth scheduling opportunity, and $T_k(n)$ is the kth user's average past throughput, for example, calculated as $(1 - \alpha) \cdot T_k(n - 1) + \alpha R_k(n)$.

This means that a user which has not been scheduled for a long time will tend to be scheduled when its radio channel is relatively good compared to its average, rather than by comparison to the channels of other users. This enables a trade-off to be achieved between maximizing system capacity and providing a fair level of QoS to all users.

1.3.1.2 Link Adaptation

For the users that are selected by the scheduler, the signal-to-interference ratio (SIR) varies depending on the state of fast fading, shadowing, and path loss. The Release 99 approach to coping with such variations was to adapt the transmission power as described in Section 1.2.1.3. However, this does not make the most effective use of the available spectrum, as a high power is used when the channel capacity is lowest. Since the transmission resources can be reallocated to different users by multiuser scheduling, it is more efficient to reduce the data rate allocated to a given user when the channel conditions worsen, and increase the rate when they improve. This is known as link adaptation, and is done by varying the modulation order and channel coding rate, collectively referred to as the modulation and coding scheme (MCS).

In order to increase the dynamic range of the link adaptation, HSDPA introduced the possibility of 16QAM modulation in Release 5, allowing a doubling of the peak data rate in good channel conditions compared to the QPSK modulation used in Release 99. 16QAM modulation with a high code rate can therefore be used when radio conditions permit, while QPSK with a lower code rate can be applied for more robust communication in lower SIR conditions.

1.3.1.3 Hybrid ARQ

In conjunction with link adaptation, HSDPA and HSUPA introduced the concept of Hybrid Automatic Repeat reQuest (HARQ). This is a combination of forward error correction (FEC) and ARQ. It enables a higher code rate to be used for initial transmissions, which may succeed if the SIR turns out to be sufficiently high; retransmissions are used to ensure successful delivery of each packet. HARQ adapts automatically to the variations of the radio channel, and overcomes the inevitable errors in predicting the exact radio propagation conditions for each user at each scheduling instant: if the radio conditions turn out to be worse than expected, a retransmission will take place rapidly, under the control of the lowest protocol layers. This provides good resilience regardless of the detailed network configuration parameters.

The simplest form of HARQ is known as chase combining [9], where each retransmission of a packet consists of exactly the same set of systematic and parity bits as the initial transmission, and the receiver combines the soft values of each bit (typically by adding the log-likelihood ratios) before reattempting the decoding. HSDPA and HSUPA support a more advanced version of HARQ known as incremental redundancy (IR), where retransmissions may comprise different sets of systematic and parity bits from the initial transmission. This offers improved performance by providing additional coding gain as each new retransmission is received and the overall code rate of the combination reduces. IR also allows the total number of bits in a retransmission to be different from the initial transmission.

Chase combining and IR are illustrated in Figure 1.13.

1.3.1.4 Short Subframe Length

In order to support dynamic multiuser scheduling, link adaptation, and HARQ, HSPA operates using a much shorter unit of transmission time than Release 99—reduced to 2 ms from a minimum of 10 ms in Release 99. This means that scheduling decisions can be updated rapidly, the MCS can be adapted to follow fast changes in radio channel conditions, and HARQ retransmissions can take place without causing an intolerable delay for the application.

Importantly, these new functions are brought under the control of a single serving NodeB for each UE, instead of being controled by higher-layer signaling from the RNC, as is the case in Release 99. This helps reduce latency. These features also mean that the role of soft handover in HSPA is reduced compared to Release 99, since in the downlink it would be difficult for two NodeBs to make simultaneous decisions on scheduling and link adaptation based on radio channel conditions.

The overall result is that, compared to Release 99, multiple access in HSPA makes reduced use of CDMA and moves closer toward a TDMA-like structure, with a smaller number of users being selected for higher-rate transmission in each subframe. This has the added benefit of reducing inter-user interference. This change of principle is illustrated in Figure 1.14.

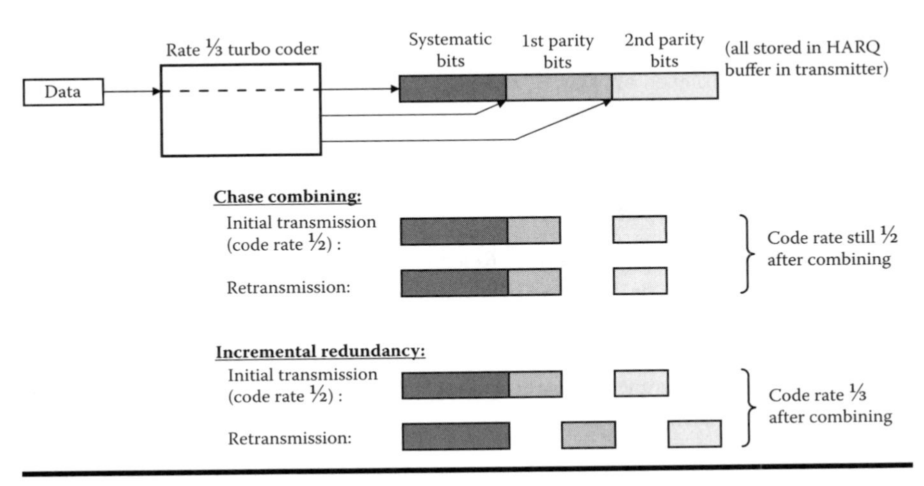

Figure 1.13 HARQ schemes.

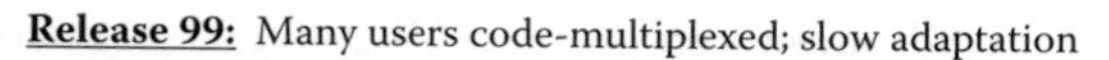

Release 99: Many users code-multiplexed; slow adaptation

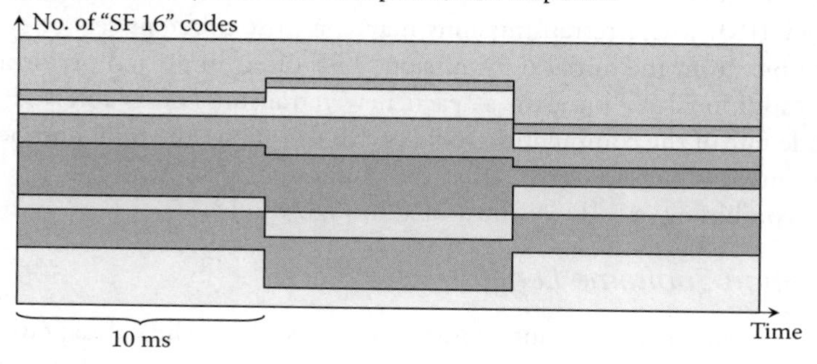

HSPA: Fewer users code-multiplexed; fast adaptation

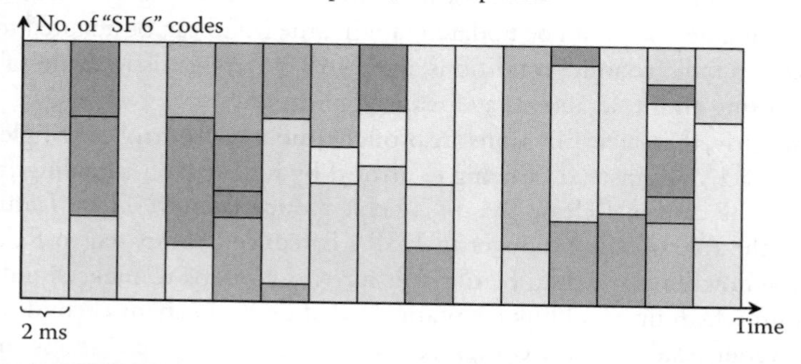

Figure 1.14 Change of emphasis from CDMA toward TDMA as Release 99 evolves to HSPA.

This evolution in the direction of TDMA is also observed in the downlink control signaling from Release 6 onwards, where TPC commands to multiple UEs may be time-multiplexed on a single channelization code,* thus helping mitigate the downlink code shortage problem mentioned in Section 1.2.1.2.

1.3.2 MBMS for HSPA

In addition to HSDPA and HSUPA, Release 6 of the UMTS specifications introduces a feature known as multimedia broadcast/multicast service (MBMS). This is designed to offer an efficient mode of data delivery to large numbers of UEs without needing to set up individual data connections and physical channels for each. The overall efficiency of broadcast and multicast data delivery can therefore be increased using MBMS.

* This time-multiplexed control channel is known as the Fractional Dedicated Physical Channel.

In order to further improve efficiency, MBMS data may be transmitted from multiple cells and combined in the UE, giving macro-diversity. In Release 7, this is further enhanced by enabling the different cells to be configured to transmit using the same scrambling code, thereby enabling the data to be combined in the equalizer of the UE receiver before decoding.

1.3.3 HSPA Evolution

As demand for higher data rates and improved QoS continues to increase, UMTS has continued to evolve beyond the initial releases of HSPA.

Three main directions can be observed for this evolution:

- Higher-order modulation
- Advanced multiple-antenna techniques
- Multi-carrier operation

Higher-order modulation enables high peak data rates to be achieved in scenarios with high SIR, such as in indoor hotspot cells. 64QAM was introduced for the downlink in Release 7, and 16QAM for the uplink in Release 8.

Multiple-input multiple-output (MIMO) antenna operation was added to HSDPA in Release 7, making HSDPA the first standardized cellular system to support the transmission of multiple data streams to each UE by means of multiple antennas at each end of the radio link. MIMO aims to exploit spatial multiplexing gain by making positive use of the multiple propagation paths to separate different data streams transmitted simultaneously using the same frequency and code. The MIMO solution adopted for HSDPA uses beamforming from two antennas at the NodeB to generate two spatially orthogonal beams to the UE, each carrying an independent data stream, as illustrated in Figure 1.15. The system relies on scattering in the radio channel to enable both beams to be received at the UE.

The same MIMO scheme can also provide a robust transmit-diversity mode, and from Release 9 onwards this is available even for UEs that do not support the high data rate dual-stream MIMO scheme.

Multi-carrier operation enables network operators to offer higher data rates irrespective of the SIR, as well as allowing more efficient use of diverse spectrum allocations. In Release 8, dual-carrier HSDPA was introduced, whereby two adjacent 5-MHz radio channels can be used simultaneously to a single UE. This can also give a small improvement in performance compared to using two carriers independently, as the scheduler can take the qualities of both carriers into account.

HSUPA operation on two adjacent carriers was introduced in Release 9, as was non adjacent dual-carrier HSDPA, allowing operators to make use of licences for operating carriers in different parts of the radio spectrum.

Future evolutions of HSPA may extend to operation with more than two simultaneous carriers in later releases. Dual and multi-carrier operation will pose new

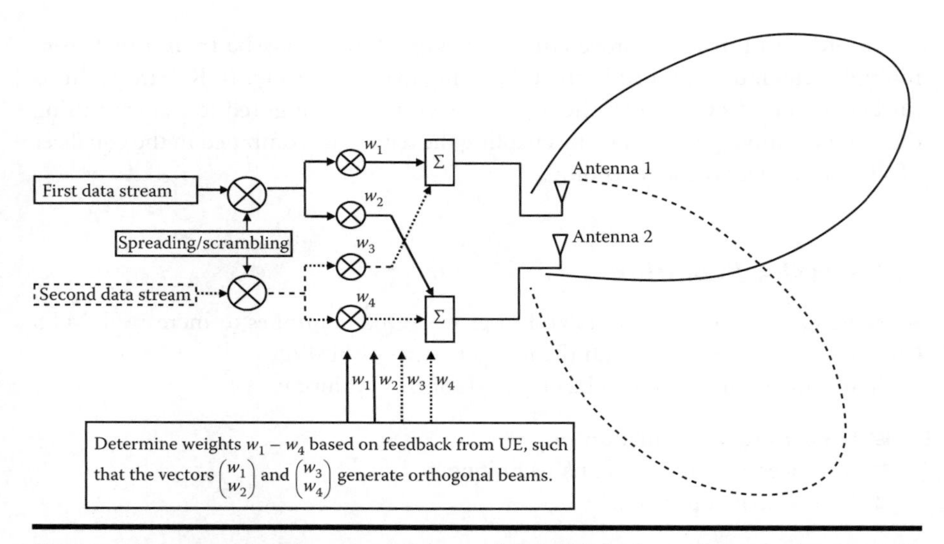

Figure 1.15 The dual-beam MIMO spatial-multiplexing scheme of HSDPA in Release 7.

challenges for network planning and optimization for HSPA, as it becomes necessary to take into account the planning of different frequencies and combinations of aggregated carriers in addition to the original aspects of UMTS network planning. In some cases, this may even extend to the different aggregated carriers having significantly different coverage areas if they are in different bands.

1.4 TD-SCDMA

1.4.1 Historical Perspective of TD-SCDMA

Time division-synchronous code division multiple access (TD-SCDMA) was submitted to the ITU by CATT (China Academy of Telecommunication Technology) as one of the candidate 3G standards for IMT2000 in 1998. It was accepted by the ITU in May 2000. Subsequently, in 2001, the TD-SCDMA concept was accepted by 3GPP and included in Release 4 of the UMTS specifications.

Compared to WCDMA and CDMA2000, TD-SCDMA was relatively immature in the early years. The industrialization of TD-SCDMA did not progress smoothly until 2006, when TD-SCDMA was accepted as a Chinese national communication industry standard by the Chinese Information Industry department. With the granting of a 3G licence in China specifically for TD-SCDMA in 2007, the industrialization of TD-SCDMA then began to grow rapidly.

1.4.1.1 TD-SCDMA Standardization in 3GPP and CCSA

Following TD-SCDMA's integration into Release 4 of the 3GPP UMTS specifications, subsequent versions of TD-SCDMA were standardized in 3GPP in Release 5 onwards, broadly aligned with the standardization of the FDD mode of UMTS.

TD-SCDMA is also standardized by CCSA (China Communications Standards Association). As a member of 3GPP, CCSA generally follows the released TD-SCDMA versions in 3GPP, integrating features from the 3GPP specifications into the CCSA standard. However, some new features were introduced first into the CCSA standard, and then injected into 3GPP—for example, a multi-carrier version of TD-SCDMA was standardized first in CCSA around the Release 5 timeframe of 3GPP, and included into Release 7 in 3GPP. The approximate relationship between the versions of TD-SCDMA released by 3GPP and CCSA is shown in Figure 1.16.

1.4.2 Deployment of TD-SCDMA

From 2005 to 2008, several TD-SCDMA test networks were deployed in China, Korea, and Europe, as a result of which the TD-SCDMA industry gained significant experience in network layout, and production maturity also improved.

The first TD-SCDMA commercial licence was granted to CMCC (China Mobile Communication Company) in January 2009. By the end of May 2009, there were 39,000 base stations deployed in 38 cities in China, and 0.85 million subscribers. Current deployment plans for TD-SCDMA include 85,000 base stations covering 238 Chinese cities, aiming for 10 million subscribers before the end of 2009.

1.4.3 Key TD-SCDMA-Specific Technologies

TD-SCDMA has several different features from UMTS FDD, including a narrower bandwidth and hence lower data rate, the possibility to support asymmetric

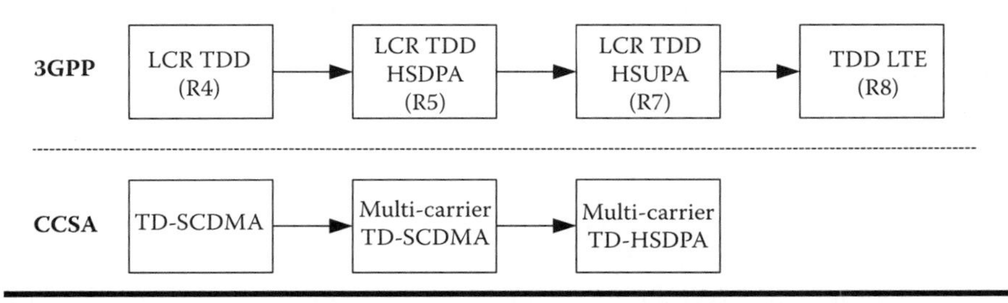

Figure 1.16 TD-SCDMA development in 3GPP and CCSA.

downlink/uplink (DL/UL) allocation of transmission resources, and, being a TDD system, the potential to benefit from transmit/receive channel reciprocity. Therefore, several TD-SCDMA-specific technologies have been developed to exploit these features.

1.4.3.1 Time Synchronization

TD-SCDMA is a synchronized system that requires strict synchronization (1/8 chip granularity) between the demodulators in the UE and NodeB. This enables complete orthogonality of all delays of the orthogonal spread spectrum codes during despreading, thus avoiding multiple-access interference. This helps overcome the interference limitation of asynchronous CDMA technology caused by the lack of time-alignment of the codes at the receiver; the TD-SCDMA system capacity and spectral efficiency are therefore improved.

TD-SCDMA utilizes two special time slots for open-loop DL and UL synchronization, respectively; these are referred to as DwPTS (downlink pilot time slot) and UpPTS (uplink pilot time slot). The UE first achieves DL synchronization via correlation detection on the DwPTS, and then transmits a random access channel (RACH) sequence on the UpPTS so theNodeB can calibrate the UL transmission timing. According to the received timing of the RACH preamble, the NodeB can adjust the UE's UL transmission timing by sending a DL control signal. During the ensuing communication, downlink control signals known as SS (synchronization shifting) are employed to maintain closed-loop UL synchronization. This synchronization mechanism enables several key TD-SCDMA technologies, such as smart antennas and joint detection. These are discussed in the following sections.

1.4.3.2 Smart Antennas

A typical smart antenna system is composed of an antenna array with M antenna elements and M coherent RF transceivers. Following the RF part, a baseband digital signal processing (DSP) module exploits the spatial dimension by adaptively forming multiple RF beams. Figure 1.17 shows a typical smart antenna system.

Based on the received signals from the different UEs, the smart antenna generates suitable phase differences between the elements of the antenna array to produce multiple transmission beams. Each beam is directed toward a specific terminal and moves with it automatically, thus reducing co-channel interference and improving downlink capacity. This spatial beamforming can result in a significant reduction of the total RF transmission power without reducing the cell coverage. Similarly on the receiving side, a smart antenna can greatly improve the reception sensitivity and reduce co-channel interference from UEs in different locations.

Theoretically, the larger the M, the higher the beamforming gain. On the other hand, too many antenna elements will result in unacceptable complexity and cost. As a compromise, TD-SCDMA typically employs eight antenna elements (although

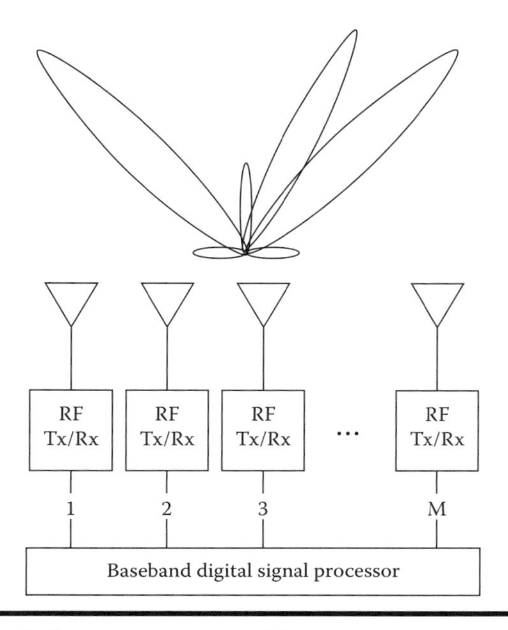

Figure 1.17 A typical smart antenna architecture.

four are sometimes used). The performance gain of smart antennas may be influenced by several factors such as multipath propagation and the Doppler frequency. Smart antennas are therefore usually combined with some interference elimination techniques such as joint detection (JD). It should also be noted that smart antennas are applicable for the data channels but not for broadcast or common control channels, where UE-specific beams cannot be formed.

1.4.3.3 Joint Detection

CDMA-based systems suffer from multipath inter-symbol interference (ISI) and multiple access interference (MAI). These sources of interference destroy the orthogonality of the different code channels, and hence reduce the system capacity. Simple detectors such as the rake receiver are sub optimal because they only consider one user's signal and do not take into account the interference from other users in the system.

The JD algorithms on the other hand are designed to process the signals of all users as useful signals, making use of the NodeB's knowledge of the spreading codes, amplitudes, and timing of each signal, to reduce the multipath and multiple-access interference. Combined with smart antenna technology, joint detection technology can achieve better results.

JD can achieve significant performance gains compared to a single-user receiver. However, JD schemes are complex and computationally intensive (complexity grows exponentially as the number of users increases) because most of the operations are matrix- and vector-based operations. TD-SCDMA, however, largely avoids this problem by limiting the number of users in a given time slot to 16, using a maximum spreading factor of 16. This means that the number of users that need to be processed in parallel is manageable, and furthermore these users are synchronized. This results in a joint detector of reasonable complexity that can easily be implemented in today's parallel computing architectures.

1.4.3.4 Baton Handover

Baton handover (BH) is a handover approach that can be regarded as being intermediate between hard handover and soft handover. Based on the UE positioning technologies available in TD-SCDMA (i.e., smart antennas and UL timing synchronization), the NodeB estimates whether a UE has entered a handover region or not. During configured handover measurement periods, the UE then obtains the system information of the target cell [e.g., scrambling code, transmission timing, and power of the primary common control physical channel (P-CCPCH)] by reading the relevant messages on the serving cell's broadcast channel (BCH) or forward access channel (FACH). From this information, the UE can accurately deduce the appropriate uplink transmission timing and power, based on the received timing and power; this is known as the pre-synchronization procedure. In this way, BH reduces handover time, improves the handover success rate and reduces the call drop rate.

During the handover execution process, the UE establishes a link with the target cell and releases the link with the serving cell almost at the same time. The whole process is like a relay race in field sports—hence the name baton handover.

1.4.3.5 Multi-carrier TD-SCDMA and TD-SCDMA HSDPA

The Release 4 version of TD-SCDMA features a 1.6-MHz channel bandwidth and 1.28-Mcps chip rate. In order to utilize the radio spectrum in a more flexible way, and to achieve higher peak data rates, a multi-carrier feature was introduced into TD-SCDMA by CCSA and later by 3GPP.

One primary carrier and several secondary carriers are configured for one cell, and all the carriers employ the same scrambling code and midamble sequence so as to reduce the measurement complexity for the UE.* The common control channels are usually configured on only the primary carrier so as to save radio resources and reduce the required transmission power.

---

* Midambles are used in UMTS TDD mode, including TD-SCDMA, for the purpose of channel estimation and synchronization in the absence of the continuous CPICH that is provided in FDD mode.

The multi-carrier version of TD-SCDMA can achieve a theoretical peak downlink data rate of *N* times the single-carrier TD-SCDMA HSDPA peak rate of 2.8 Mbps (where *N* is the number of carriers).

Moreover, the multi-carrier version of TD-SCDMA supports more flexible radio resource allocation than the original single-carrier version. For example, adjacent cells can allocate different carriers as the primary carrier so as to avoid inter-cell interference on channels such as the broadcast channel, which cannot benefit from smart antenna techniques.

1.5 LTE and Beyond

The transition from circuit-switched mobile service provision to packet-switched is completed with the advent of the long-term evolution (LTE) of UMTS. Further growth in demand for packet data services, fueled by the arrival of mobile terminals with much more advanced capabilities for images, audio, video, e-mail, and office applications, led to the need for a further radical step in radio access network design.

1.5.1 Context of LTE

3GPP took the first steps toward LTE at the end of 2004, when the industry came together to make proposals for the requirements and suitable technologies for the new system. In order to maximize its longevity, it was decided to embrace the opportunity to design a completely new radio access network architecture and radio interface, without being constrained by attempting to retain backward compatibility with the UMTS radio access network.

This meant that LTE was able to take advantage of the possibility of using much wider channel bandwidths (up to 20 MHz), partly facilitated by the allocation in 2007 of large new spectrum bands by the ITU for global use by "IMT"-designated systems, as well as exploiting advances in theoretical and practical understanding and processing capabilities.

Targets were set for LTE to support at least 100 Mbps in the downlink and 50 Mbps in the uplink, with average and cell-edge spectral efficiencies in the range two to four times those provided by Release 6 HSPA.

1.5.2 Principles of LTE

In addition to a 20-MHz carrier bandwidth, some fundamental technologies of LTE can be identified as follows:

- New multiple-access schemes, based on multi-carrier technology
- Advanced multi-antenna technology
- Fully packet-switched radio interface

- Flat network architecture
- Common design for operation in a paired and unpaired spectrum

Each of these is introduced briefly in the following subsections. For a thorough explanation of LTE, the reader is referred to [10].

1.5.2.1 Multi-Carrier Multiple Access

In place of the CDMA multiple access scheme used in UMTS, which suffer from interference arising from non-orthogonality, LTE has adopted orthogonal multi-carrier multiple access schemes: orthogonal frequency division multiple access (OFDMA) in the downlink, and single-carrier frequency division multiple access (SC-FDMA) in the uplink.

OFDMA breaks the wideband transmitted signal down into a large number of narrowband subcarriers. These are closely spaced such that they are orthogonal to each other in the frequency domain, as shown in Figure 1.18, resulting in a high spectral efficiency. Different groups of subcarriers can be allocated to transmissions for different users.

As OFDMA uses multiple subcarriers in parallel, the symbol rate on each subcarrier is low compared to the total combined data rate. This means that the symbol duration is long, so that the delay spread that arises from multipath propagation can be contained within a guard period occupying only a small proportion of each symbol duration. In order to maintain orthogonality of the subcarriers, the guard period is generated as a cyclic prefix (CP), by repeating some samples from the end of each symbol at the beginning. This is illustrated in Figure 1.19. This enables the degradation that arises from inter-symbol interference (ISI) to be avoided, provided that the propagation delay spread is less than the CP length.

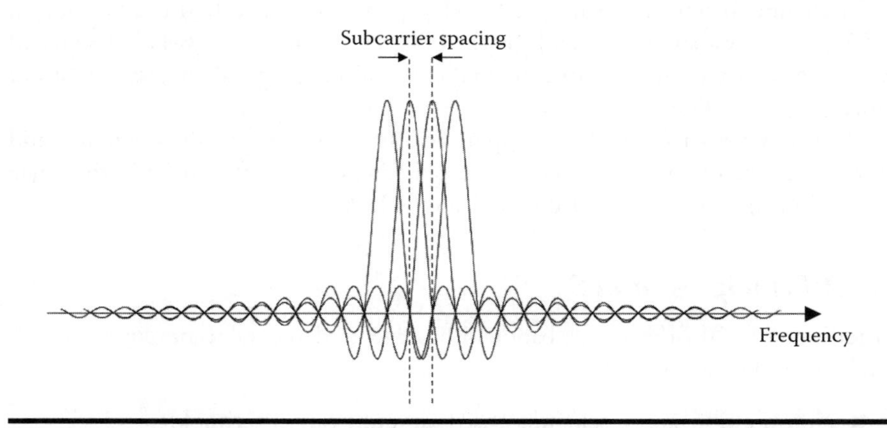

Figure 1.18 The frequency spectra of all the OFDMA subcarriers are orthogonal to each other.

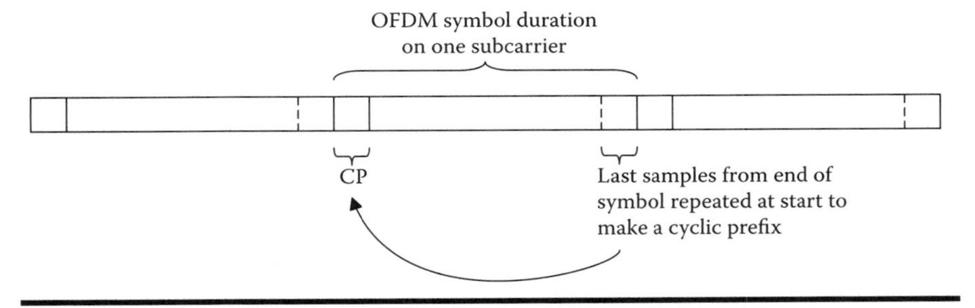

Figure 1.19 Guard period generated as a cyclic prefix to eliminate ISI.

The subcarrier spacing is chosen as a trade-off between resilience against delay spread and resilience against frequency shifts such as Doppler shifts caused by mobility and receiver non-idealities. A small subcarrier spacing enables a large number of subcarriers to be used in a given spectrum allocation, thereby enabling a long CP to be used without it representing a high overhead as a proportion of the symbol duration. However, a small subcarrier spacing is more sensitive to inter-carrier interference (ICI). As a compromise between these factors, LTE uses a fixed subcarrier spacing of 15 kHz, regardless of the system bandwidth. For operation in a wide range of different bandwidths, LTE has the flexibility of using different numbers of subcarriers.

The basic structure of an OFDMA transmitter is shown in Figure 1.20. After serial-to-parallel conversion of the data stream to be transmitted (to map it onto the multiple subcarriers), the orthogonal multi-carrier signal can be generated in a very straightforward manner by means of an inverse Fourier transform. This can be implemented as an inverse fast Fourier transform (IFFT), further reducing the complexity.

The OFDMA receiver is very low in complexity, as it can be simply implemented using a fast Fourier transform (FFT). As each subcarrier has a narrowband spectrum (i.e., bandwidth less than the coherence bandwidth of the radio channel),

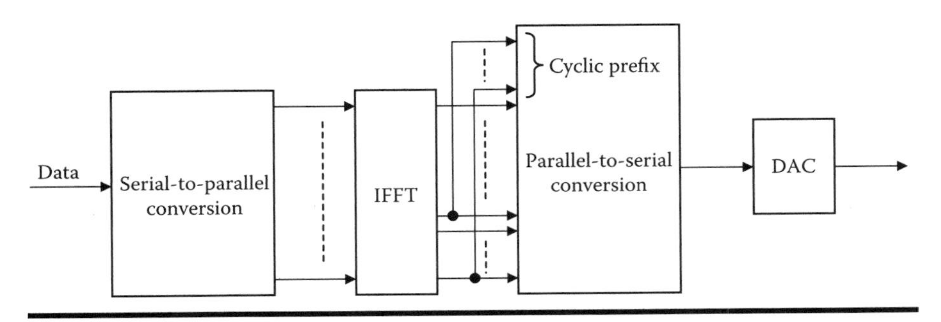

Figure 1.20 OFDMA transmitter structure.

the channel gain can be compensated in the receiver by a single multiplication operation. Thus, frequency-domain equalization is very straightforward for OFDMA, and the receivers have low complexity.

The low-complexity receiver operation makes OFDMA ideally suited to downlink transmissions, where the cost of the receiver in the mobile terminals is crucial. However, there is one drawback of OFDMA that makes it less suitable for the uplink, namely that the peak-to-average power ratio (PAPR) of the transmitted signal is high. The SC-FDMA scheme adopted for the LTE uplink is a variation of OFDMA but with an initial precoding stage using a discrete Fourier transform (DFT), which results in each subcarrier carrying a linear combination of data symbols instead of each data symbol being mapped to a separate subcarrier. This results in a single-carrier waveform that exhibits a significantly lower PAPR than OFDMA.

The structure of an SC-FDMA transmitter is illustrated in Figure 1.21. The use of SC-FDMA for the LTE uplink results in a high degree of commonality between the downlink and uplink signal structures, including the 15-kHz subcarrier spacing.

1.5.2.2 Multi-Antenna Technology

As explained in [9], multiple antennas can provide three kinds of gain, depending on the scenario and configuration: diversity gain, array gain, and spatial multiplexing gain.

Antenna diversity provides resilience against multipath fading and improves coverage, particularly for control and broadcast channels for which techniques based on UE-specific beamforming cannot be used.

LTE provides schemes for downlink transmit diversity for up to four antennas at the eNodeB. With two transmit antennas at the eNodeB, a space-frequency block code (SFBC) is used, which operates in a similar way to the STBC code available in UMTS Release 99, except that symbols are treated in pairs across pairs of subcarriers in the frequency domain instead of in-pairs in the time domain. With four transmit antennas at the eNodeB, the antennas are treated in pairs, each using independent SFBC schemes in a frequency-switched transmit diversity (FSTD) scheme.

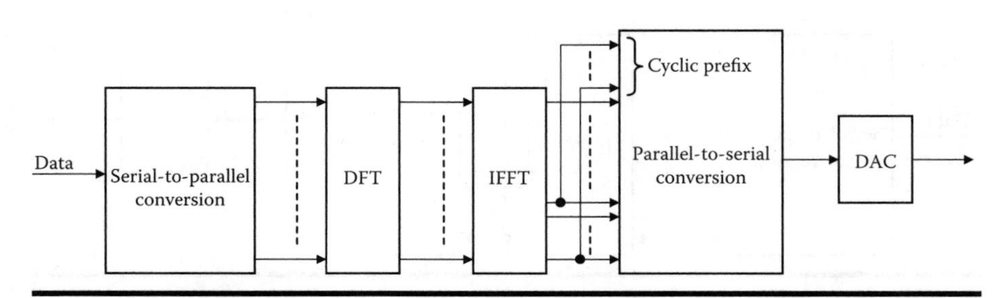

Figure 1.21 SC-FDMA transmitter structure.

In the uplink, the first release of LTE assumes that, for reasons of cost, UEs would possess only a single power amplifier and would therefore be capable of transmitting from only one antenna at a time. However, LTE does enable some benefit to be derived from uplink transmit diversity by means of antenna switching, whereby the UE changes the antenna from which it transmits from time to time.

Receive diversity also provides robustness against multipath fading. In order to assist network planning, especially for coverage of the broadcast channel, it is important that network operators can assume a certain baseline level of performance for all UEs. The specified LTE performance requirements therefore assume that all UEs have at least two receive antennas, from which the signals are combined using a technique like maximal ratio combining (MRC).

Array gain makes use of phase adjustments at each of the antennas to cause constructive superposition to direct a beam in the desired direction for a particular UE, as illustrated in Figure 1.22. This can give a significant increase in signal power at the receiver. Similar techniques can also be used to reduce interference at a receiver, by setting a null in the beam pattern in the direction of a strong interfering signal—for example, in the direction of a dominant interfering cell in the downlink, or in the direction of a high-power UE in a neighboring cell in the case of the uplink.

Two modes of multi-antenna transmission are provided in the LTE downlink which exploit array gain. The first is a closed-loop mode in which the UE uses the cell's broadcast reference signals to derive an estimate of the downlink channel response in order to feed back a recommendation of the optimal precoding vector (selected from a specified codebook of vectors) to be applied at the transmitter to

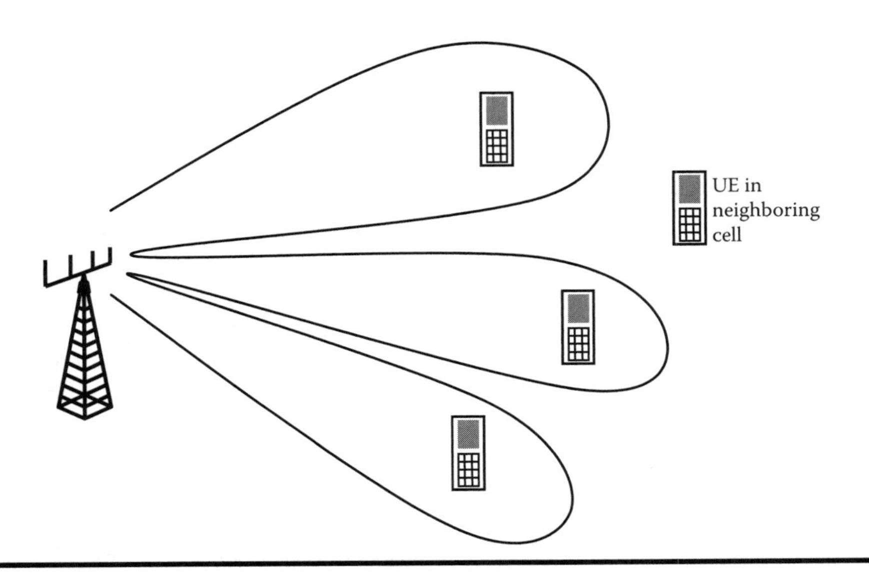

Figure 1.22 A beam pattern from an antenna array.

maximize the received signal power. The second mode is open-loop, in the sense that no explicit feedback is sent from the UEs to guide the transmitter precoding at the eNodeB. Instead, it is assumed that the eNodeB has a correlated array of antennas (e.g., eight closely spaced elements) and that the uplink and downlink channels are therefore spatially correlated. The eNodeB then estimates the angle of arrival of uplink transmissions and forms a transmission beam in the same direction. The precoding that is required in this latter case is not constrained by a finite codebook, and UE-specific reference signals are therefore precoded with the same precoding vector as is used for the data and transmitted together with the data to provide the UE with the phase reference for demodulating the data.

Spatial multiplexing makes use of multiple antennas at each end of the radio link to transmit multiple data streams using the same time and frequency resources. This relies on there being sufficient uncorrelated paths through the radio channel to be able to separate the parallel data streams at the receiver by means of different combinations of the signals from the receive antennas. Each such combination of antennas is typically referred to as a spatial layer.

The first release of LTE provides for up to four spatial layers to be transmitted simultaneously to a UE (although only the highest category of UEs has the capability to support more than two layers). The spatial multiplexing is an extension of the closed-loop beamforming mode described earlier. Based on measurements of the downlink reference signals, the UE identifies the preferred rank (number of layers) for the downlink transmission, and feeds back a recommendation of a corresponding precoding matrix. In suitable propagation conditions, this provides the possibility for the peak data rate to be increased by a factor of up to four compared to single-layer transmission.

1.5.2.3 Packet-Switched Radio Interface

Following the same route as HSPA, LTE benefits from packet switching at the radio interface by means of dynamic scheduling, link adaptation, and HARQ. Unlike UMTS, the LTE physical layer makes no use of constant-rate power-controlled channels; the data rate and frequency-domain transmission resources are adapted on the basis of multiuser scheduling considerations and radio channel quality.

Dynamic scheduling in the frequency domain is facilitated by the multi-carrier transmission schemes in uplink and downlink which enable different groups of subcarriers to be allocated to different UEs depending on the radio channel conditions in different parts of the carrier bandwidth. This is a new degree of freedom for data scheduling which was not available in UMTS.

The only exception to dynamic scheduling in LTE is known as "semi-persistent scheduling," whereby a particular set of subcarriers is allocated to a UE at regular intervals with a particular modulation and coding scheme. This is particularly suitable for services like voice over IP (VoIP). However, even in this case, HARQ retransmissions are scheduled dynamically, allowing link adaptation to be used.

1.5.2.4 Flat Network Architecture

LTE uses not only a new radio interface but also a new network architecture, illustrated in Figure 1.23. In order to facilitate dynamic scheduling and mobility, the LTE radio access network consists of a single type of node, the eNodeB, in place of the NodeB and RNC of UMTS. Nearly all the functions of the UMTS RNC are the responsibility of the eNodeB in LTE, including radio resource management (RRM) and scheduling.

This helps ensure that LTE has a very low latency, both for packet transmission and for connection setup.

The eNodeBs are interconnected by a standardized interface known as the X2 interface, across which both control signaling and data can be transferred. The connection between the eNodeB and the core network [known for LTE as the evolved packet core (EPC)] is provided by the S_1 interface to the packet data network gateway (PDN-GW) node.

1.5.2.5 Evolved MBMS

LTE also provides support for an enhanced version of MBMS from Release 9 onwards, known as "evolved MBMS" or simply "eMBMS." eMBMS in LTE exploits the OFDM signal of the LTE downlink to give high spectral efficiency when multiple cells transmit the eMBMS data in a mode known as multimedia broadcast single frequency network (MBSFN) operation.

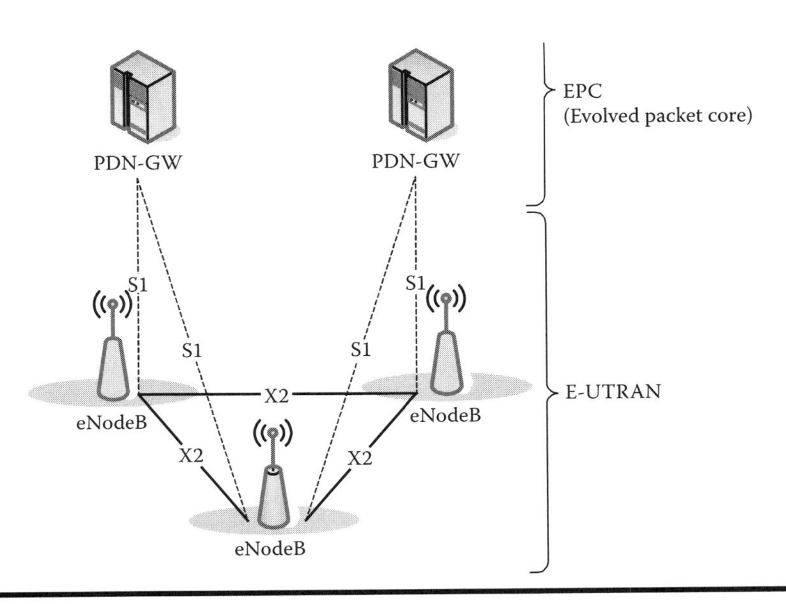

Figure 1.23 The LTE radio access network uses a flat architecture consisting of a network of interconnected eNodeBs.

In MBSFN operation, MBMS data is transmitted simultaneously from multiple cells. A UE receiver will therefore receive multiple versions of the signal with different delays due to the different propagation delays from the different cells. Provided that the transmissions from the multiple cells are sufficiently tightly synchronized for each to arrive at the UE within the cyclic prefix at the start of each OFDM symbol, there will be no intersymbol interference (ISI). This makes the MBSFN transmission appear like a transmission from a single large cell. This gives a significant increase in the received SINR, as inter-cell interference is translated into useful signal energy for the eMBMS data reception.

1.5.3 Network Planning Considerations for LTE

As a result of the radically new radio access network in LTE, there have been many different considerations for network planning compared to previous systems such as UMTS.

1.5.3.1 Interference Management

An important aspect is the control of inter-cell interference, since LTE is designed to operate with a frequency reuse factor of unity. Cell-specific scrambling is used to whiten inter-cell interference. However, further measures may be employed to mitigate inter-cell interference.

The ability of LTE to support frequency-domain scheduling (as introduced in Section 1.5.2.1) is an important enabler for interference management. In particular, neighboring cells may cooperate to avoid scheduling transmissions for cell-edge users in the same time/frequency resource blocks in adjacent cells, and some standardized signaling is provided to support this.

For the downlink, signaling can be exchanged between eNodeBs over the X2 interface so that one eNodeB can inform the neighboring eNodeBs whether it is planning to keep the transmit power for each group of 12 subcarriers below a certain upper limit. This enables the neighboring cells to take into account the expected level of interference on each group of subcarriers when scheduling transmissions to UEs in their own cells in, for example, avoiding scheduling transmissions to cell-edge UEs on those subcarriers.

For the uplink, eNodeBs can exchange a reactive "overload indicator" (OI) and a proactive "high interference indicator" (HII). The OI indicates whether an eNodeB has detected a high level of interference on certain groups of subcarriers. The HII can be used by an eNodeB to inform its neighboring eNodeBs that it intends to schedule uplink transmissions by cell-edge UEs on certain groups of subcarriers, and therefore that high interference might occur on those subcarriers. Neighboring cells may then take this information into consideration in scheduling their own users.

In addition to making use of frequency-domain scheduling to manage inter-cell interference in the uplink, the eNodeB can also control the degree to which each UE compensates for the path loss when setting its uplink transmission power.

This is known as fractional power control and can be used to maximize system capacity by trading off fairness for cell-edge UEs against the inter-cell interference generated toward other cells.

These aspects of interference management are important factors to take into account when optimizing LTE network deployment.

1.5.3.2 Other Aspects of Network Planning

Other aspects related to LTE network planning that must be taken into account chiefly relate to the configuration of cell identities and sequences used in neighboring cells.

LTE provides 504 distinct cell IDs, grouped into three groups of 168; each group has a corresponding primary synchronization sequence (PSS), which is used by the UE for the first stage of cell identification and synchronization. These cell group IDs would normally be allocated to adjacent cells, such as to three 120° cells of an eNodeB; arranging a suitable allocation of the cell group IDs in relation to the cells of neighboring cells may be more complex in practical network deployments. The cell IDs within the groups are conveyed by the secondary synchronization sequences (SSSs) which must also be allocated to the cells.

LTE also provides the possibility to plan the allocation of sequences for the uplink reference signals. This enables groups of reference signal sequences to be assigned to UEs in neighboring cells such that they have low cross-correlation with each other. Alternatively, reference sequence group hopping may be configured to avoid the necessity for planning, so that even if UEs in neighboring cells end up using reference signals with high cross-correlation, the resulting inter-cell interference should not persist in consecutive subframes.

1.5.3.3 Network Self-Optimization

LTE provides some network self-optimization tools which can be used to optimize some aspects of the network configuration automatically. These tools are collectively known as self-optimizing networks (SON), and they may include:

- Coverage and capacity optimization
- Energy-saving features for the eNodeBs
- Inter-cell interference coordination
- Automatic configuration of Cell ID
- Optimization of mobility and handover, including, for example, detection of handovers that are too early, too late, or to a cell that is not the most appropriate, reducing the occurrence of unnecessary handovers, and optimization of cell reselection parameters
- Load balancing
- RACH optimization
- Automatic identification of the neighboring cells for each cell

Some degree of automatic configuration of such features can reduce the effort required in manual planning and optimization of LTE networks. However, this facility comes at the expense of increased complexity, usually in the eNodeBs. Many of the SON aspects listed here will be further developed in subsequent releases of the LTE specifications.

1.5.4 Future Development of LTE

Further developments of the LTE specifications are continuing to follow, providing further-enhanced capacity and data rates.

Some extensions are provided in Release 9, including support for UE positioning and enhanced beamforming. As mentioned earlier, a much more major step known as LTE-advanced (LTE-A) will follow in Release 10, as the LTE system is developed to meet the demands set by the ITU for the IMT-advanced family of systems, and beyond.

While the first version of LTE (Release 8) can already satisfy many of these demands in terms of spectral efficiency, the peak data rate requirements and some particular deployment scenarios require that significant new features are introduced into LTE. These changes will be introduced in a backward-compatible way, allowing a smooth migration from networks built on Release 8 toward later releases, with full compatibility with legacy Release 8 terminals.

One prominent feature of LTE-A is bandwidth aggregation, whereby multiple carriers at different frequencies, and possibly of different bandwidths, are used in conjunction with each other in a single cell, as illustrated schematically in Figure 1.24. This will enable network operators to make more effective use of diverse spectrum allocations, as well as multiplying the available data rates by the number of aggregated carriers. In total, up to 100 MHz would be accessible for a single LTE-A link in each direction.

The LTE-A will also feature enhancements to multi-antenna transmission schemes. Instead of supporting a maximum of four layers in the downlnk and one in the uplink, LTE-A will support up to eight layers in the downlink (for antenna configurations up to 8×8) and four layers in the uplink, enabling peak data rates in excess of 1 Gbps in the downlink and 500 Mbps in the uplink to be achieved when propagation conditions allow.

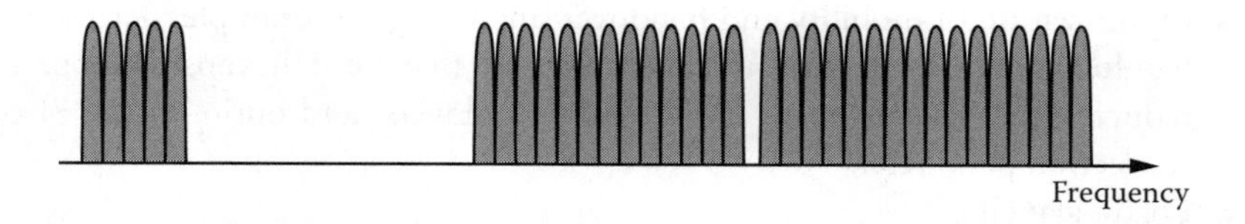

Figure 1.24 Carrier aggregation for LTE-Advanced.

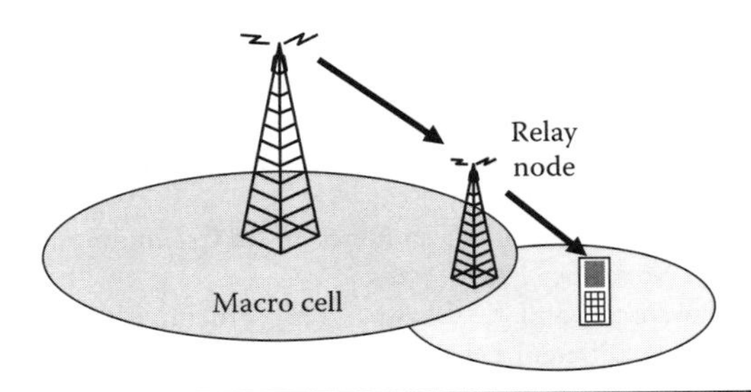

Figure 1.25 Relaying for LTE-advanced.

Enhancements to multiuser MIMO, as well as cooperative multi-Point (CoMP) transmission in which multiple cells transmit simultaneously to a given UE, are also under consideration for LTE-A.

Yet a further aspect of LTE-A is the introduction of relay nodes (RNs), which can forward data from an eNodeB to a UE using the same spectrum and radio interface between the eNodeB and the RN as between the RN and the UE. RNs can in some scenarios be useful for filling coverage gaps, extending the coverage of rural cells, and potentially in some cases for capacity enhancement. A typical configuration is shown in Figure 1.25.

All of these aspects will bring new possibilities and challenges for LTE network deployment and optimization.

1.6 Network Planning and Optimization

The remainder of this book is devoted to an explanation of techniques for network deployment planning and optimization for the evolved systems introduced in this chapter.

All network deployment is ultimately constrained by the radio propagation environment, and an understanding of propagation scenarios is therefore essential to efficient network planning. Chapter 2 therefore explains the main approaches to radio propagation modeling that are applicable to cellular systems.

Chapter 3 considers some techniques for assessing the performance of deployed networks and identifying where improvements are needed.

Chapters 4 to 10 consider in detail network planning aspects for UMTS, including WCDMA, TD-SCDMA, HSPA, and MBMS. Aspects discussed include coverage, radio resource management, and interference management. Both macrocellular and small-cell deployments are considered.

Finally, Chapters 11 to 16 then consider network planning for LTE and possible future evolutions of LTE.

References

[1] A. J. Viterbi, "Very Low Rate Convolutional Codes for Maximum Theoretical Performance of Spread-Spectrum Multiple-Access Channels," *IEEE J. Select. Areas Commun.*, Vol. 8, No. 4, pp. 641–649, May 1990.

[2] H. F. Harmuth, "Applications of Walsh Functions in Communications," *IEEE Spectrum*, Vol. 6, pp. 82–91, November 1969.

[3] F. Adachi, M. Sawahashi, and K. Okawa, "Tree-Structured Generation of Orthogonal Spreading Codes with Different Lengths for Forward Link of DS-CDMA Mobile Radio," *Electronics Letters*, Vol. 33, pp. 27–28, January 1997.

[4] S. M. Alamouti, "A Simple Transmit Diversity Technique for Wireless Communications," *IEEE Journal on Selected Areas in Communications*, 16 (8): pp. 1451–1458.

[5] E. Lindskog and A. Paulraj, "A Transmit Diversity Scheme for Channels with Intersymbol Interference," *IEEE International Conference on Communications*, pp. 307–311, June 2000.

[6] R. Price and P. Green, "A Communication Technique for Multipath Channels," *Proc. IRE*, Vol. 46, pp. 555–570, Mar 1958.

[7] P. Viswanath, D. N. C. Tse, and R. Laroia, "Opportunistic Beamforming Using Dumb Antennas," *IEEE Transactions on Information Theory*, Vol. 48, No. 6, pp. 1277–1294, June 2002.

[8] T. E. Kolding, "Link and System Performance Aspects of Proportional Fair Scheduling in WCDMA/HSDPA," *Proc. 58th IEEE VTC*, Vol. 3, pp. 1717–1722, Orlando, FL, 2003.

[9] D. Chase, "Code Combining – A Maximum-Likelihood Decoding Approach for Combining an Arbitrary Number of Noisy Packets," *IEEE Trans. Comm.* Vol. 33, No. 5, pp. 385–393, May 1985.

[10] S. Sesia, I. Toufik, and M. P. J. Baker (eds.), "LTE—The UMTS Long Term Evolution: From Theory to Practice," Wiley, 2009.

Chapter 2

Overview of Wireless Channel Models for UMTS and LTE

Abbas Mohammed and Asad Mehmood

Contents

2.1 Introduction

Designing, analyzing, and deploying communication systems requires the efficient utilization of available resources for reliable transfer of information between two parties. However, in practical systems some amount of unpredictability is tolerated in order to achieve better consumption of available resources. Thus the performance of signal processing algorithms, transceiver designs, etc., for a communication system are highly dependent on the propagation environment.

A correct knowledge and modeling of the propagation channels is a central prerequisite for the analysis and design of the long-term evolution (LTE) at both the link level and the system level and also for the LTE specifications for the mobile terminal and the base station performance requirements, radio resource managements to ensure that the resources are used in an efficient way, and in RF system scenarios to derive the requirements and in system concept evolution. The use of multiple transmit/receive antenna techniques is an important feature of LTE. Multiple antenna techniques used in LTE (e.g., in spatial diversity), take the advantage of multipath dispersion to increase the capacity. However this requires that the spatial correlation between antenna elements should be low, which is difficult to obtain in practical systems. When assessing multiple antenna techniques, it is important that relevant features (e.g., spatial correlation) of the channel are modeled in an efficient way. Therefore, standard MIMO channel models also have great significance in the design and analysis of an LTE system.

This chapter gives an overview of standard channel models for Universal Mobile Telecommunication Systems (UMTS) and the upcoming LTE. The emphasis is on some general channel models used during the evolution of UMTS and LTE, and specific channel models for LTE as well. The chapter is organized as follows: Section 2.2 describes the basics of multipath channel modeling. In Section 2.3, different approaches for developing generic channel models are discussed, which are used to build standard channel models for LTE. Section 2.4 describes standard channel models for UMTS and LTE and Section 2.5 gives an overview of recently developed MIMO channel models for LTE. The chapter ends with concluding remarks.

2.2 Multipath Propagation Channels

The difficulties in modeling a wireless channel are due to the complex and varied propagation environments. A transmitted signal arrives at the receiver through different propagation mechanisms shown in Figure 2.1: the line-of-sight (LOS) or free space propagation, scattering, or dispersion due to contact with objects with irregular surfaces or shapes, diffraction due to bending of signals around obstacles, reflection by the objects with sizes that are large compared to the wavelength of the electromagnetic wave, transmission through irregular objects, and so on. The presence of multipaths (multiple scattered paths) with different delays and attenuations gives rise to highly complex multipath propagation channels. It is significant here to note that the level of information about the environment a channel must provide is highly dependent on the category of communication system under assessment. To predict the performance of single-sensor narrowband receivers, classical channel models, which provide information about signal power level distributions and Doppler shifts of the received signals, may be satisfactory [1, 2]. The advanced technologies (e.g., LTE) built on the typical understanding of Doppler spread and fading; however, also incorporate new concepts such as time delay spread, direction of departures, direction of arrivals and adaptive array antenna geometry, and multiple antenna implementations at both sides of the wireless link, or multiple-input multiple-output (MIMO) [3–9].

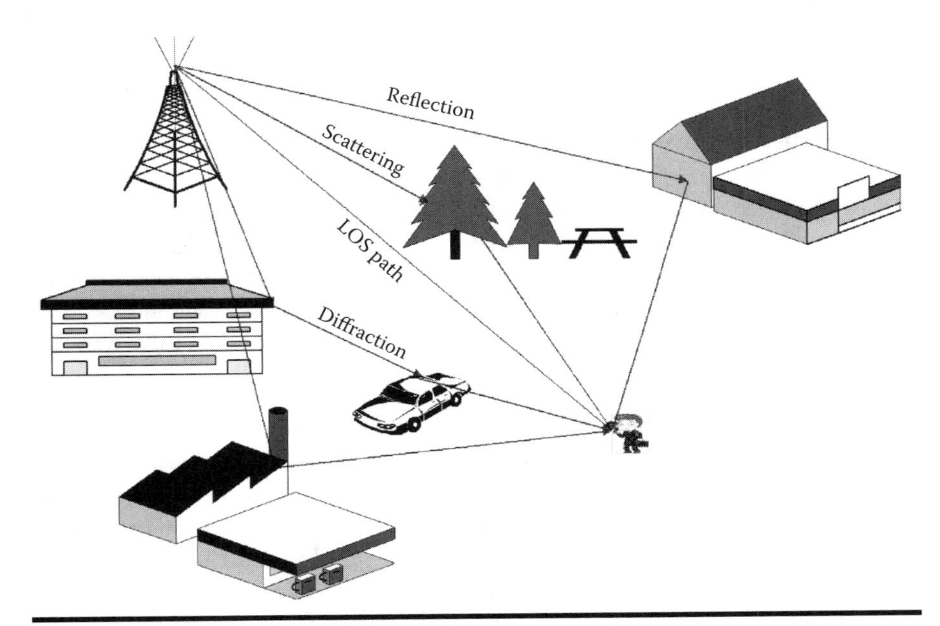

Figure 2.1 Radio propagation environment. [From 3GPP Technical Specification 25.996, "Spatial Channel Model for Multiple Input Multiple Output (MIMO) Simulations (Release 8),"V8.0.0, 12-2008. http://www.3gpp.org.] © 2008. 3GPP™.

The degradation in the received signal level due to multipath effects can be classified into large-scale path loss components, medium-scale slow varying components with log-normal distribution, and small-scale fast fading components with Rayleigh or Rician distribution depending on the absence or presence of LOS components between the transmitter and receiver [2, 10]. Thus, a three-stage propagation model can be used to describe a wireless cellular environment. The three stages are:

1. **Large-scale propagation model**: This model is used to characterize the received signal strength by averaging the amplitude or power level of the received signal over large transmitter–receiver separation distances in the range of hundredths or thousandths of a wavelength. The large-scale models are often derived from measured data. However, semi-empirical models are employed in smaller areas to achieve higher accuracy. For this purpose, theoretical models are used, which are then fitted to measured data to obtain a desired model for a particular propagation scenario.
2. **Medium-scale propagation model**: This model determines the gradual variations of the local mean amplitude or the local mean power of the received signal over a time-variant multipath channel when the mobile station moves over distances larger than a few tens or hundreds of a wavelength. Some existing components will disappear while new components will appear. It is observed that variations of the local mean power of the received signal follow log-normal distribution, which is called slow fading or shadowing. The shadowing is caused by obstructions like trees and foliage. The mean and standard deviations of the received power are determined from large-scale propagation models in the environment of interest.
3. **Small-scale propagation model**: This model is used to characterize the rapid variations of the received signal strength due to changes in phases when a mobile terminal moves over small distances on the order of a few wavelengths or over short time durations on the order of seconds. Since the mean power remains constant over these small distances, small-scale fading can be considered as superimposed on large-scale fading for large-scale models. The most common description of small-scale fading is by means of the Rayleigh distribution.

Multipath signals arrive at the receiver with different propagation path lengths, called multipath taps, and different time delays. The multipath signals with different phases sum constructively or destructively at the receiver, giving rise to time varying multipath taps. The power distribution of channel taps is described by a distribution function depending on the propagation environment. The most severe multipath channel is the Rayleigh fading channel in which there is no line-of-sight path and the channel taps are independent. In the case of the Rician fading channel, the fading dips are low due to the presence of line-of-sight components in addition to the dispersed paths.

The behavior of a multipath channel needs to be characterized in order to model the channel. The concepts of Doppler spread, coherence time, and delay spread and coherence bandwidth are used to describe various aspects of the multipath channel. The maximum value of delay spread gives the delay difference between the first and the last channel tap in the power delay profile. The coherence bandwidth is the inverse of the delay spread. If the coherence bandwidth is greater than the transmission bandwidth, then the frequency components of the signal will undergo frequency flat fading. A frequency selective fading results if the coherence bandwidth is less than the transmission bandwidth. The Doppler spread arises due to motion of the mobile station and gives a maximum range of Doppler shifts. If there is only one path from the mobile terminal to the base station, then the Doppler spread will be zero with a simple shift of carrier frequency. The inverse of the Doppler spread gives coherence time of the channel during which the channel statistics do not change significantly.

2.3 Modeling Methods for Multipath Channel Modeling

New channel models can be developed in two stages [5]. First, setting up a framework for a generic channel model and identifying a set of parameters that needs to be determined for the description of the channel. Second, conducting measurement campaigns and extracting numerical values of parameters and their statistical distributions. In the first stage, different methods can be employed (e.g., the deterministic methods are based on measured impulse responses and ray tracing algorithms); the stochastic approaches include tapped delay line models, frequency domain modeling, and geometry-based stochastic modeling. In this section, a brief overview of different modeling methodologies for the first stage is described.

2.3.1 Deterministic Channel Computation

Deterministic propagation modeling is intended to reproduce the actual radio propagation process for a given environment. These methods are suitable for environments where the radio waves interact with fairly simple geometric obstacles such as streets and buildings. The geometric and electromagnetic characteristics of the corresponding environment and of the radio links can be stored in files (environment data) and the corresponding propagation process can be modeled through analytical formulas or computer programs. Deterministic channel models are potentially accurate and meaningful. Due to the high accuracy and adherence to the real propagation process, deterministic models can be advantageous to use in situations when time is not sufficient to conduct measurements or some specific cases that are difficult to measure in the real world.

Although electromagnetic models such as finite difference in time domain (FDTD) and methods of moments (MoM) can be helpful in the study of near field problems, these models are not used for field prediction at radio frequencies because of small wavelengths with respect to the dimensional scale of the environment. The most appropriate deterministic method for radio propagation, at least in urban area, is the ray-tracing model [11].

In a wireless channel, multipath propagation at higher frequencies can be modeled by applying geometrical optics (GO) theory. This theory is based on ray approximation, which is suitable when the wavelength is significantly small compared to the dimensions of the obstacles. Under this assumption, the electromagnetic waves following multiple paths can be expressed as a set of rays or beams where each ray represents a piecewise linear path connecting two terminals. In the ray-tracing method, the position of the transmitter and the receiver is specified initially and then all the possible rays (paths) between the transmitter and the receiver can be determined by applying geometrical optics rules and geometric considerations. The rays can be characterized from the propagation environment by their amplitude, Doppler shifts, delays, angle of departure, angle of arrival, and polarization. Once the complete information (database) about the environment is known, including the positions of the transmitter and the receiver, then by applying the fundamental laws of electromagnetic waves propagation, channel properties can be derived from the positions of the scatterers. If instead of rays, beams of finite transverse dimension are used, then the resulting model is called beam launching or ray splitting. The beam launching models are suitable for large areas and permit faster field strength prediction. On the other hand, the ray-tracing method is more suitable for point-to-point field prediction and gives accurate results as compared to the beam launching method. High computational burdens and a difficulty to maintain suitable and detailed databases are the main drawbacks of ray-based models.

2.3.2 Geometry-based Stochastic Channel Models

Geometry-based stochastic channel modeling approach also uses the ray-based modeling concept, which is a commonly used approach in directional channel modeling for performance assessment of systems including adaptive antenna arrays. In deterministic geometrical modeling approaches (e.g., ray tracing in previous subsection), the locations of scatterers are prescribed deterministically from a database. On the other hand, in geometric-based stochastic channel models (GSCM) the locations of the scatterers are chosen in a stochastic manner with a certain probability distribution of multipath delays and where the direction of departures and arrivals is determined by the ray-based approach. The distribution of scatterers depends on the environments. In indoor environments, the scatterers are located around both ends of the wireless link. In the case of highly mounted antennas, a scatterers' presence only around the mobile terminal is considered. Each scatterer can be characterized by its own direction of arrival, direction of departure and propagation delay using the ray-based approach. When the channel parameters of each ray (i.e., complex amplitudes, Doppler frequency, delays, direction of departures and arrivals) are determined, the channel behavior can be characterized in multidimensions.

The GSCM approach has some advantages [12, 13]. The approach used by GSCM is more practical, and channel parameters can be obtained through simple geometric considerations. Many effects like delay drifts, direction of arrivals, and small-scale fading by superposition of individual scatterers, are implicitly reproduced. All the information lies in the distribution of the scatterers, which do not make the model complex. The transmitter, receiver and scatterer locations, appearance/disappearance of propagation paths, and shadowing can be determined in a straightforward way. Different versions of GSCM are described in [11].

2.3.3 Non-geometrical Stochastic Channel Models

In non-geometrical stochastic channel models, all parameters such as the location of the scatterers, direction of departures, direction of arrivals and so on, describing the paths between the transmitter and the receiver are determined statistically, without referencing the geometry of the physical environment. There are two non-geometrical stochastic channel models in the literature [11]: the extended Saleh-Valenzuela model and the Zwick model. The Zwick model considers the multipath components (MPCs) individually, while the extended Saleh-Valenzuela model uses the cluster of the MPCs. The details of these models are explained in [10].

2.3.4 Correlation-based Channel Models

Correlation-based channel models are simple and have a low computational complexity compared to geometric-based channel models. These models are used to describe the correlation properties between all transmit/receive antenna pairs and are useful in the study of correlation impacts on any performance parameter of the system. Spatial correlation can be defined explicitly by the spatial correlation matrix. In ray-based modeling, correlation is present in the channel matrix implicitly. For zero mean complex circularly symmetric Gaussian channels, the channel correlation matrix for the description of the MIMO channel behavior is defined in [14].

$$R_{\text{full}} = E[Vec(H)Vec(H)^H] \tag{2.1}$$

In the preceding equation, $Vec(\cdot)$ stacks all elements of the matrix H into a large vector, H is the channel matrix for single tap delay, and $(\cdot)^H$ is the Hermitian transpose. The channel correlation matrix R is different for each channel tap. MIMO channel spatial correlation properties are captured by the matrix R at both ends of a wireless link. The correlation matrix R based on the Rayleigh fading channel is defined in [15] as

$$R = Vec^{-1}(R^{1/2}g) \tag{2.2}$$

where, g is a circularly symmetric Gaussian vector having zero mean and unit variance, and Vec^{-1} is the inverse vectorization operation. To simulate the Ricean fading K-factor, the LOS signal is included in the signal.

The spatial correlation matrices can be derived from ray-based models, channel matrices based on measurements, or from analytical calculations. The most popular correlation-based model is the Kronecker model, which is computationally simpler than the full correlation matrix R_{full}. This model requires that the correlation matrix at the receiver be independent of the direction of transmission. In this case, channel matrices are obtained [15] using,

$$H = R_{TX}^{1/2} G R_{RX}^{1/2} \tag{2.3}$$

where, G is the i.i.d. (independent identically distributed) complex Gaussian matrix, R_{TX} and R_{RX} are correlation matrices of the transmitter and the receiver, respectively.

The main advantage of correlation-based channel models is that these models are simple and have low computational complexity. On the other hand, these models cannot be generalized simply to other configurations since spatial correlations depend on antenna configurations, and so new correlation coefficients are required for each configuration.

2.4 Standard Channel Models

When designing an LTE system, different requirements are considered (e.g., UE and BS performance requirements, radio resource management requirements, RF system scenarios) to derive the requirements. The standard channel models play a vital role in the assessment of these requirements. In the following section, some standard channel models are discussed that are used in the design and evolution of the UMTS-LTE system.

2.4.1 COST Channel Models

COST stands for the "European Co-operation in the Field of Scientific and Technical Research." Several COST efforts were dedicated to the field of wireless communications, especially radio propagation modeling; COST 207 for the development of Second Generation of Mobile Communications (GSM), COST 231 for GSM extension and Third Generation (UMTS) systems, COST 259 "Flexible personalized wireless communications (1996–2000)" and COST 273 "Toward mobile broadband multimedia networks (2001–2005)." These projects developed channel models based on extensive measurement campaigns, including directional characteristics of radio propagation (COST 259 and COST 273) in macro, micro, and picocells, and are appropriate for simulations with smart antennas and MIMO systems (MIMO models are discussed in Section 2.5). These channel models form the basis of ITU standards for channel models of beyond 3G systems (e.g., LTE).

2.4.1.1 COST 259 Directional Channel Model

COST 259 was a European Research initiative in the field of "Flexible Personalized Wireless Communications," which encompassed representatives of the key manufacturers, many network operators, and universities. One of the contributions of COST 259 was to propose set standards to overcome the limitations of channel models developed in the past [16, 17].

The COST 259 directional channel model was originally developed for simulations of systems with multiple antennas either at the base-station or mobile terminal (i.e., MISO systems). This channel model is developed for 13 different types of environments, covering macro-, micro-, and picocells, which are given in Table 2.1 [16]. Each environment is illustrated by a set of external parameters (e.g., radio frequency, MS and BS heights, BS position) and global parameters, which are probability density functions describing a specific environment (e.g., scatterers are characterized by Poisson distribution), defining the propagation characteristics as a whole.

A layered approach, which distinguishes between the external (fixed) small-scale effects and the large-scale effects, allows well-organized parameterization. It is categorized into the following subsequent layers:

- The upper layer describes different propagation scenarios that represent a group of environments with similar propagation characteristics.
- The middle layer deals with non stationary large-scale effects. These effects include angular spreads, delay spreads, shadowing, and the appearance/disappearance of far-scattering clusters. These effects are described by their

Table 2.1 Different Types of Propagation Environments for Macro-, Micro-, and Picocells

Macrocell	*Microcell*	*Picocell*
General typical urban (GTU)	General urban microcell (GUM)	General office LOS (GOL)
General rural area (GRA)	General urban bad microcell (GUBM)	General office NLOS (GON)
General bad urban (GBA)	General open place (GOP)	General corridor LOS (GCL)
General hilly terrain (GHT)	General open place NLOS (GPN)	General corridor NLOS (GCN)
		General factory hall (GFH)

corresponding probability density functions with different parameters for different propagation environments. These parameters may be modeled as correlated log-normally distributed random variables.

- The lower layer deals with small-scale fading effects caused by interference due to rapid fluctuations of amplitudes and phases in multipath components. The statistics for small-scale fading are obtained from large-scale fading effects.

In a specific scenario, clusters of scatters are distributed at random fixed places in the coverage area according to a specific probability density function. The clusters of scatterers are characterized by the angular spreads and the RMS delay spreads, which are correlated random variables and are obtained deterministically from the positions of the MS and the BS. The intra-cluster variations are modeled stochastically. Each scatterer is described by a random complex coefficient that follows Gaussian distribution.

In spite of its general applicability, COST 259 has some limitations which restrict its applicability. First, the scatterers are assumed to be stationary, so the channel variations originate only due to the MS movement. Secondly, a rich scattering environment is required to describe the envelope of delay attenuation as complex Gaussian, which is the case in this model. However, this assumption is not supported by some environments of the channel model, which is a common assumption for all other channel models.

2.4.1.2 COST 273 Channel Model

Despite efforts made in the previous COST projects, difficulties arose in designing channel models for MIMO systems (MIMO models are discussed in Section 2.5). The 3GPP made efforts to develop a channel model that is used for third-generation cellular systems (UMTS). The description of this channel model is limited to a particular set of environments with specific parameters. This model is specific from a simulation point of view but restricts its general applicability. Thus, COST 273 decided to develop a new channel model for MIMO systems.

The channel model can be considered as an extension of the COST 259 directional channel model [18, 19]. The main difference between this model and the COST 259 directional channel is that it uses the same generic channel model for all environments while the COST 259 model uses different models for macro-, micro-, and picocells. In addition, some new environments are defined to incorporate new MIMO applications (e.g., fixed wireless access scenarios and peer-to-peer). Two types of parameters are used to describe each environment.

1. **External parameters**: These parameters describe the environment and keep fixed values during a simulation run (e.g., carrier frequency, base station and mobile station antenna heights, building heights, antenna orientations, antenna scenarios, path loss models).

2. **Global parameters**: Global or stochastic parameters are a set of probability density functions and a set of statistical moments describing a specific environment (e.g., the number of scatterers is characterized by Poisson distribution).

The COST 273 channel model includes three types of scatterer clusters, local clusters around the BS or the MS, single interaction clusters, and twin clusters to model the concept of multiple interactions. A cluster is divided into two representations, one as seen by the BS and the other corresponding to the MS side. The advantage of splitting up a cluster in two is that the angular distributions of energy at the BS and the MS can be modeled independently based on the marginal densities of the angular spectra of corresponding clusters.

Each ray radiated from the transmitter is scattered by a scatterer in a cluster and it reaches the receiver after bouncing at the corresponding scatterer of the twin cluster. The twin clusters are linked through the stochastic cluster link delay concept. The link delay guarantees realistic path delays (e.g., obtained from measurements), while the position of the cluster is determined by the angular statistics of the cluster. All scatterers inside a cluster have the same link delay [6].

The mean angles and delays of the clusters are modeled by geometric considerations, and the small-scale fading and intra-cluster spreads can be modeled by either the tapped delay line approach or by the geometrical representation. The total impulse response can be written as the sum of the clusters' double directional impulse responses, which is as follows [6]:

$$P(\tau, \theta_{BS}, \varphi_{BS}, \theta_{MT}, \varphi_{MT}) = P_\tau(\tau)\, P_\theta^{BS}(\theta_{BS})\, P_\varphi^{BS}(\varphi_{BS})\, P_\theta^{MT}(\theta_{MT})\, P_\varphi^{MT}(\varphi_{MT}) \tag{2.4}$$

In the preceding equation, τ is the delay, θ_{BS}, φ_{BS}, θ_{MT}, φ_{MT} are the respective azimuth and elevation spreads at the BS and the MS, respectively. The model assumes that azimuth spreads, elevation spreads (i.e., angular spreads), and delay spreads in a cluster are independent on a per-cluster basis. However, as a whole there can be significant coupling between DoDs and DoAs.

2.4.2 ITU Channel Models

The International Mobile Telecommunications (IMT-2000) was an initiative of the International Telecommunication Union (ITU) for the evolution of European Telecommunications Standardization Institute (ETSI) standards for second-generation mobile systems (GSM) to third-generation Universal Mobile Telecommunications Systems (UMTS). ITU standard channel models were used for the development of the 3G IMT-2000 group of radio access systems [20, 21]. The aim of these channel models is to develop standards that help system designers and network planners with system designs and performance verification. Instead of defining propagation models for all possible environments, ITU proposes a set of test environments in [21] that adequately spans all possible operating environments and types of user mobility. The proposed ITU test environments may not resemble the actual mobile's user operating environments, but they give a very good overview of how a mobile user performs in different operating environments. The complete description of all possible scenarios can be found in [22].

1. **Indoor office test environment:** The indoor office test environments, where both the base station and users are located indoors, are characterized by small cells and low transmit powers. Path losses and shadowing effects are due to scattering and the attenuation by floors, walls, and metallic structures such as partitions and filing cabinets. Fading can follow Rayleigh or Rician distribution depending upon the location of the user. Indoor channel models based on ITU recommendations are used for modeling indoor scenarios. The average powers and the relative delays of taps for ITU channel models in indoor scenarios are given in Table 2.2 [21].
2. **Outdoor to indoor and pedestrian test environment:** For outdoor to indoor and pedestrian environments, base stations with low antenna heights are situated outdoors, while pedestrian users are to be found inside buildings and residences. Path loss rules of R^{-2} to R^{-6} can be applied for different ranges (e.g., LOS on a canyon-like street rule where there is Fresnel zone clearance to the region where there is no longer Fresnel zone clearance [21]). Shadowing, caused by hindrance from trees and foliage, follows log-normal distribution and results in the received signal power variations with standard deviations of 12 dB for indoor and 10 dB for outdoor environments, respectively. The building penetration loss average is 12 dB, with a standard-deviation of 8 dB. ITU recommends that in modeling microcells, the outdoor to indoor and

Table 2.2 Average Powers and Relative Delays for ITU Indoor Office Test Environment

	Channel A		*Channel B*		
Tap No.	*Relative Delay (ns)*	*Average Power (dB)*	*Relative Delay (ns)*	*Average Power (dB)*	*Doppler Spectrum*
1	0	0	0	0	Classical
2	50	−3	100	−3.6	Classical
3	110	−10	200	−7.2	Classical
4	170	−18	300	−10.8	Classical
5	290	−26	500	−18	Classical
6	310	−32	700	−25.2	Classical

Table 2.3 Average Powers and Relative Delays for ITU Indoor to Outdoor and Pedestrian Test Environment

Tap No.	Channel A		Channel B		Doppler Spectrum
	Relative Delay (ns)	Average Power (dB)	Relative Delay (ns)	Average Power (dB)	
1	0	0	0	0	Classical
2	110	−9.7	200	−0.9	Classical
3	190	−19.2	800	−4.9	Classical
4	410	−22.8	1200	−8	Classical
5	NA	NA	2300	−7.8	Classical
6	NA	NA	3700	−23.9	Classical

pedestrian models are to be used to represent multipath conditions. The average powers and the relative delays for the taps of multipath channels based on ITU recommendations are given in Table 2.3 [21].

3. **Vehicular test environment:** This type environment is categorized by large macrocells with higher capacity, limited spectrum, and a large transmit power. A path loss exponent of 4 and log-normal shadow fading with a 10 dB standard deviation are suitable in urban and suburban areas. In rural areas, path loss may be lower than previous, while in mountainous areas, if the BS location is suitably selected to avoid path blockages, a path loss attenuation exponent closer to two may be appropriate. The vehicular models (Table 2.4 [21]) are used to model multipath propagations in macrocells regardless of whether the user is inside the car or not.
4. **Mixed test environment:** This type of environment takes account of environments [e.g., a vehicular environment (macrocells) and outdoor-to-indoor test environment (microcells) in the same geographical area]. In this test environment, fast-moving terminals are connected to macrocells and slow-moving terminals (pedestrians) are associated with microcells to achieve higher capacity. For example, a dense urban environment may be modeled as consisting of 30% of the pedestrian channel model at a speed of 50 km/h and 70% of pedestrian channel model at a speed of 3 km/h. Likewise, other environments (e.g., suburban or rural environments) may be modeled as percentage mixtures of ITU channel models at various speeds.

To assess these propagation environments, reference models for each operating environment have been given both on system level calculations and link level software simulations. The key parameters to describe each propagation model are time delay

Table 2.4 Average Powers and Relative Delays for ITU Vehicular Test Environment

	Channel A		*Channel B*		
Tap No.	*Relative Delay (ns)*	*Average Power (dB)*	*Relative Delay (ns)*	*Average Power (dB)*	*Doppler Spectrum*
1	0	0	0	−2.5	Classical
2	310	−1	300	0	Classical
3	710	−9	8900	−12.8	Classical
4	1090	−10	12900	−10	Classical
5	1730	−15	17100	−25.2	Classical
6	2510	−20	20000	−16	Classical

spread, its structure and statistical variability, overall path loss prediction, including path loss, excess path loss, shadowing, maximum Doppler shifts, and operating radio frequency [22].

2.4.3 Extended ITU Models

The analysis done by ITU-R showed that evolution of 3G systems to future generation networks will require technology changes on large scale while new quality of service (QoS) requirements will require increased transmission bandwidth. Thus, LTE channel models require more bandwidth as compared to UMTS channel models to account for the fact that channel impulses are associated to the delay resolution of the receiver [23]. The LTE channel models developed by 3GPP are based on the existing 3GPP channel models and ITU channel models. The extended ITU models for LTE were given the names Extended Pedestrian A (EPA), Extended Vehicular A (EVA), and Extended Typical Urban (ETU), which do not actually represent urban environments because of small cell sizes with large delay spreads of up to 5 ps. Another proposal to categorize these models in terms of delay spreads are representing these with low delay spread (LD), medium delay spread (MD), and high delay spread (HD), respectively. The low delay spread models are used to represent indoor environments and small cell sizes, while medium delay spread and high delay spread are used to model urban environments with large cell sizes. The high delay spread models are according to the typical urban GSM model [24]. The resulting model parameters, number of taps, RMS delay spread, and maximum excess tap delay are shown in Table 2.5, and RMS delay spreads values for tap delay line models are given in Table 2.6 [24].

Table 2.5 Power Delay Profiles for Extended ITU Models

	EPA Model		*EVA Model*		*ETU Model*	
Tap No.	*Excess Delay (ns)*	*Average Power (dB)*	*Excess Delay (ns)*	*Average Power (dB)*	*Excess Delay (ns)*	*Average Power (dB)*
1	0	0	0	0	0	−1
2	30	−1	30	−1.5	50	−1
3	70	−2	150	−1.4	120	−1
4	80	−3	310	−3.6	200	0
5	110	−8	370	−0.6	230	0
6	190	−17.2	710	−9.1	500	0
7	410	−20.8	1090	−7	1600	−3
8			1730	−12	2300	−5
9			2510	−16.9	5000	−7

The Doppler frequencies for these LTE channel models are defined on a basis similar to what was used for UTERA. Just as the three channel models are classified on the basis of low, medium, and large delay spreads, a similar approach is adopted to define Doppler frequencies (low, medium, and high) for the Doppler environments. The Doppler frequencies for LTE channel models with low, medium, and high Doppler conditions are 5 Hz, 70 Hz, and 900 Hz, respectively [24]. The delay spreads and the Doppler frequencies provide a framework from which possible scenarios for the operating environment can be selected. The following combinations of delay spread and Doppler spread are proposed in [25]: extended pedestrian A 5 Hz, extended vehicular A 5 Hz, extended vehicular A 70 Hz, and extended typical urban 70 Hz.

Table 2.6 Summary of Delay Profiles for LTE Channel Models

Model	*Channel Taps*	*Delay Spread (rs)*	*Max. Access Tap Delay (span)*
Extended pedestrian A model (EPA)	7	45	410 ns
Extended vehicular A model (EPA)	9	357	2510 ns
Extended typical urban model (ETU)	9	991	5000 ns

The propagation scenarios for LTE with speeds from 120 km/h to 350 km/h are also defined in [25, 26] (e.g., the high-speed train scenario at speed 300 km/h and 350 km/h). The maximum carrier frequency over all frequency bands is $f_c =$ 2690 MHz and the Doppler shift at speed $v =$ 350 km/h is 900 Hz.

2.4.4 3GPP Channel Model

Since 1998, ETSI's standardization of third-generation mobile systems has been carried out in the 3rd Generation Partnership Project (3GPP). The 3GPP standards for radio propagation in UMTS are mainly based on ITU models. These standards stipulate certain performance tests for both mobile stations and base stations under different propagating conditions covering additive white Gaussian noise and multipath fading environments. As an example, the 3GPP propagation models used for performance evaluation of different multipath environments are shown in Table 2.7 [27]. All taps have the classical Doppler spectrum [28]. These models are used as reference for the specification and testing of both uplink and downlink data channels with defined parameters and data rates of 64 kbps, 144 kbps, 384 kbps, and 2048 kbps.

Case 1 in Table 2.7 is a single tap heavily faded model and is almost identical to the ITU pedestrian S model at speed 3 km/h. Case 2 and case 4 are similar (mobile speed of 3 km/h), but the former has less fading and a higher multipath diversity. Case 3 has four channel taps and a mobile terminal speed of 120 km/h, which is similar to the vehicular A channel. Case 5 is similar to case 1, but the mobile terminal speed is 50 km/h. Case 6 and case 3 have the same number of taps, but the UE speed is 250 km/h in case 6. The signals arriving at the receiver are considered independent (i.e., no correlation among the received signals), a case appropriate for these channel models.

2.5 MIMO Channel Models

The spatial characteristics of a radio channel have a significant effect on the performance of multiple-input multiple-output (MIMO) systems. The MIMO techniques take the advantage of multipath effects in the form of spatial diversity to significantly improve SNR by combining the outputs of de-correlated antenna arrays with low mutual fading correlation. The other technique to improve the gain of a system using multi-antenna arrays is spatial multiplexing, which creates multiple parallel channels between the transmitter and receiver sides. Multiple antennas at the transmitter and/or the receiver side can be used to shape the overall beam in the direction of a specific user to maximize the gain. This technique is called beamforming. The large MIMO gains can be achieved by low spatial correlation. The antenna separation, in terms of the wavelength of the operating frequency, has a significant impact on the spatial correlation. To achieve a low fading correlation, the antenna separation

Table 2.7 Average Path Powers and Relative delays for 3GPP Multipath Channel Models for specific cases

Case 1 Speed 3 km/h		*Case 2 Speed 3 km/h*		*Case 3 Speed 120 km/h*		*Case 4 Speed 3 km/h*		*Case 5 Speed 50 km/h*		*Case 6 Speed 250 km/h*	
Relative Delay (ns)	*Avg Power (dB)*	*Relative Delay (ns)*	*Avg Power (dB)*	*Relative Delay (ns)*	*Avg Power (dB)*	*Relative Delay (ns)*	*Avg Power (dB)*	*Relative Delay (ns)*	*Avg Power (dB)*	*Relative Delay (ns)*	*Avg Power (dB)*
0	0	0	0	0	0	0	0	0	0	0	0
976	−10	976	0	260	−3	976	0	976	−10	260	−3
		20000	0	521	−6					521	−6
				781	−9					781	−9

should be large. The small sizes of wireless devices restrict large antenna separation, depending upon the wavelength of the operating frequency. An alternative solution to achieve low correlation is to use antenna arrays with cross-polarizations (i.e., antenna arrays with polarizations in orthogonal or near orthogonal orientations (discussed in Section 2.5.4)).

Different channel modeling approaches (see Section 2.3) are used to develop MIMO channel models for LTE. The 3GPP/3GPP2 spatial channel and its extension SCME model, described in Sections 2.5.1 and 2.5.2, respectively, are ray- or geometric-based stochastic channel models, while ITU models, described in Sections 2.4.2 and 2.4.4, are correlation-based channel models. Section 2.5.3 details the IST-WINNER channel model which is also the geometric-based stochastic channel model. Section 2.5.4 describes the concepts of polarized antenna arrays and in subsections 2.5.4.1 and 2.5.4.2, an overview of the 3GPP-polarized SCM model and the theoretical MIMO channel model is presented using polarized antenna arrays. Section 2.5.5 deals with the LTE evolution channel, and in Section 2.5.6, a comparison of standard MIMO channels is made.

2.5.1 3GPP Spatial Channel Model

The standardization bodies for third-generation cellular systems (3GPP and 3GPP2) jointly established a double directional geometry-based spatial channel model (SCM) for modeling, analysis and evaluation of MIMO concepts in outdoor environments for a system with bandwidth 5 MHz and 2 GHz frequency band (e.g., HSDPA, LTE) [29]. The modeling formation is similar to the COST 259 directional channel model but with some differences. First, it is not a continuous model but advises a precise discrete implementation (i.e., DOAs and DoDs are supposed to be fixed), and second, the movement of the mobile terminal is not continuous on a large scale within a cell; the model illustrates diverse positions of the mobile terminal within a cell. The SCM model describes two types of models: a calibration-level model based on tapped delay line approach taps is described in both the delay and angular domains, and a geometric-based (ray-tracing) stochastic model for system level simulations. The spatial description of the channel is illustrated in terms of directional distribution of multipath components and the angular energy distribution at both the mobile station and the base station.

The calibration model is used to give only one snapshot of the channel behavior [29]. This model is not intended for performance assessment of the behavior of systems or algorithms; its purpose is to only verify the accuracy of simulation implementations. The calibration model, described by 3GPP/3GPP2, can be seen as a spatial extension of ITU-R channel models [21], where the channel characteristics are described as tap delay lines. The taps are independently fading with different delays and each tap is illustrated by its own mean angular spread, power azimuth spectrum (Laplacian or uniform), and mean direction at both the BS and the MS. These parameters are fixed so the channel characterizes stationary channel conditions. The Doppler spectrum is characterized by introducing the direction of travel and speed of the MS.

The SCM model used for the comparison and performance evolution of MIMO systems is called the system level simulation model. The system level simulations typically consist of multiple UE, base stations, and multiple cells/sectors. Performance metrics such as power delay profiles, angle spreads, and throughput are generated over a large number of drops D—a "drop" is specified as a simulation run for given cells/sectors. These channel observation periods (i.e., drops) are significantly separated in space or time and channel large-scale parameters remain constant within a drop but the channel suffers from fast fading according to the movement of the mobile terminals, which varies randomly. The base station can schedule its transmission according to the channel state information provided by the mobile terminals. The spatial parameters are described according to the geometrical framework shown in Figure 2.2 [29].

The SCM model distinguishes between three propagation environments: urban macrocell, suburban macrocell, and urban microcell. The overall methodology is similar for these environments but there are some optional features in the basic propagation scenarios; far scatterer clusters in case of bad urban environment, modified angular distribution at the MS in the urban canyon model representing dense urban areas for both the urban macro and urban micro scenarios, and a LOS component in the urban microcellular case based on the Ricean K-factor. In addition, the spatial parameters like delay spread, angular spread, and so forth, are different for each of these environments.

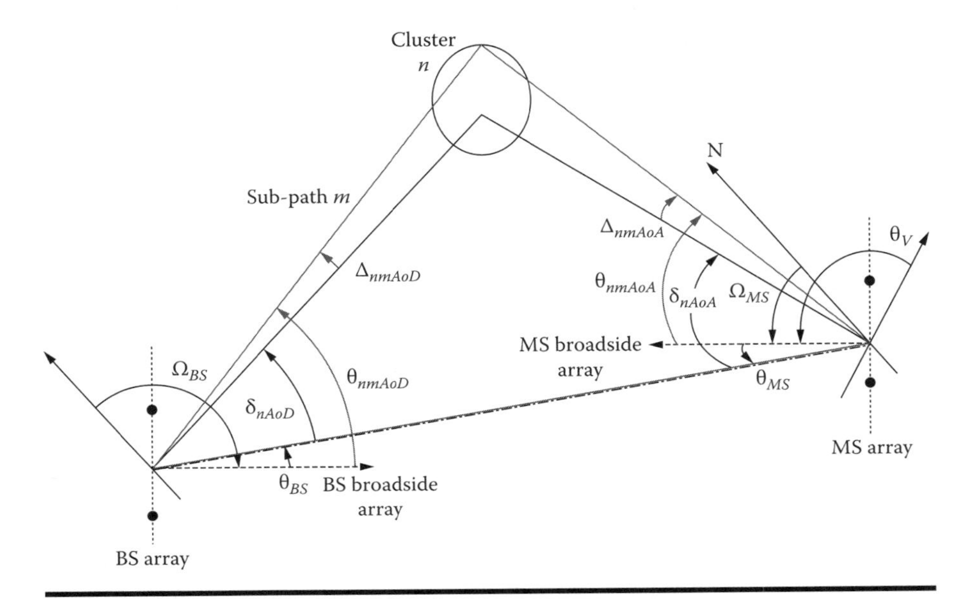

Figure 2.2 Geometry of BS and MS angular parameters.

The simulation model is a geometry-based stochastic model in which the movement of the mobile terminal within a given cell and the orientations of antenna arrays chosen at random are modeled geometrically. This channel model is based on ITU-R models described in Section 2.4, so the number of propagation paths with different delays is six for each environment. The paths are described by their mean angles and delays, which are correlated random variables with normal or log-normal probability density function [30].

Each path arrives at the BS or at the MS with angular dispersion. This dispersion is modeled by representing each path by a number of sub-paths with the same delays but different DoDs and DoAs dispersed around the mean angles with different fixed offsets $\Delta\phi_i$, where i represents the number of sub-paths. In all environments, angular dispersion for each path is composed of 20 sub-paths. The per-path angular dispersion at both the UE and the base station is described by Laplacian distribution, which is obtained by giving 20 sub-paths the same power, and fixed azimuth directions with respect to the nominal direction of the corresponding path. Addition of the different sub-paths gives Rayleigh or Rice fading.

The angle spread, delay spread, and shadow fading are correlated random variables. Path losses for the environments are determined by the COST 231-Hata model for urban and suburban macrocells and the COST 231-Walfish Ikegami model for microcells.

The SCM model was designed for different antenna radiation patterns, antenna orientations, and geometry to be applied. For example, antenna patterns and antenna spacing at the base station can be varied using antenna patterns for three-sector, six-sector cells, or omni-directional pattern and inter-element spacing of 0.5, 4, and 10 wavelengths. The composite angle spreads, delay spreads, and shadow fading which can be correlated random variables depending on the employed scenario, are applied to all sectors or antennas of the given base. When all parameters and antenna effects are specified, we can extract analytical formulation from the physical model. During each drop, a different correlation matrix is obtained for the analytical model. Table 2.8 shows the main parameters for this channel model.

2.5.2 Extended 3GPP Spatial Channel Model

The channel models play a vital role for the performance evolution and comparison of communication systems. With the advancement of communication technologies, these models need to be refined to incorporate challenging advanced communication algorithms. The SCM model, described in Section 2.5.1, operates in the 2-GHz frequency band and supports bandwidth up to 5 MHz. An extension to the 3GPP/3GPP2 model was made and used within the European IST-WINNER project. This is known as the SCME (SCM-Extension) model [31, 32]. The main contribution of this spatial channel model is broadening the channel bandwidth from 5 MHz to 100 MHz in the 2- and 5-GHz frequency bands. The focus was to extend the model in such a way that it remains backward-compatible with the conceptual

Table 2.8 The Main Parameters of 3GPP Spatial Channel Model Realizations

Parameter	*Suburban Macro*	*Urban Macro*	*Urban Micro*
No. of paths (N)	6	6	6
No. of sub-paths per path (M)	20	20	20
Mean angle spread at BS	5°	8°, 15°	19°
Per-path angle spread at BS (Fixed)	2°	2°	5° (LOS and NLOS)
Mean angle spread at MS	68°	68°	68°
Per-path angle spread at MS (Fixed)	35°	35°	35°
Mean total delay spread (r)	0.17 μs	0.65 μs	0.25 μs
Std. deviation for log-normal shadowing	8 dB	8 dB	NLOS:10 dB, LOS:4 dB
Path loss model (dB)	$31.5 + 35 \log_{10}(d)$	$34.5 + 35 \log_{10}(d)$	NLOS:$34.53+38 \log_{10}(d)$ LOS:$30.18 + 26 \log_{10}(d)$

approach of the 3GPP/3GPP2 SCM model. The extension is based on the shortcomings of the existing SCM model (i.e., large bandwidth support, no LOS component in case of macrocells, and short-term time variations in system level model).

To extend the model, the bandwidth extension is done in such a way that it remains compatible with the original 5-MHz bandwidth, by introducing the concept of the intra-cluster delay spread. The idea was initially proposed by Saleh and Valenzuela for indoor channel modeling. The idea of the intra-cluster delay has also been employed for outdoor scenarios in COST 259. The 20 sub-paths of a path are divided into subsets, called mid-paths. These mid-paths define the intra-cluster delay spread and have different delays and power offsets relative to the original path. Each mid-path consisting of a number of sub-paths acts as a single tap (delay resolvable component). Grouping together a number of sub-paths makes the fading distribution of that tap approximately Rayleigh distributed. The angle spreads (AS) assigned to the mid-paths are optimized in such a way that the angular spread of all mid-paths combined is minimized. The resulting SCME impulse response has a good approximation to the respective SCM impulse response. Due to bandwidth extension, the number of tap delays increases from 6 to 18 or 20 depending upon the propagation scenario. Table 2.9 [31] shows mid-paths powers and delays for SCME.

Table 2.9 The SCME Mid-Path Power-Delay Parameters

Scenario	*Suburban Macro Urban Micro*		*Urban Micro*	
No. of Mid-paths per Path	3		4	
	Power	Relative Delay (ns)	Power	Relative Delay (ns)
1	10/20	0 ns	6/20	0 ns
2	6/20	7 ns	6/20	5.8 ns
3	4/20	26.5 ns	4/20	13.5 ns
4	-	-	4/20	27.6 ns

One other contribution of SCME is the evolution of spatio-temporal parameters for fixed tap delay line (TDL) models called cluster delay-line models. The model parameters: power, delays, and angles of departures and arrivals are assigned fixed values, illustrating all MIMO propagation parameters. The tap-delay line model is similar to the SCM link level model; however, it can be closely approximated to the SCM system level model, which is optimized for small frequency autocorrelations.

The path loss for the SCME model in the 5 GHz band is proposed on the basis of path loss models used in the SCM model with an offset of 8 dB to the 2-GHz path loss model. The COST 231 Walfish Ikegami model is selected as the standard path loss model for all scenarios.

The SCME model offers a number of optional features which can be employed depending upon the specific simulation purpose. In the SCM model, the LOS option is for urban micro only. The SCME model also incorporates the K-factor option (i.e., the LOS option for urban and suburban macro scenarios by assigning the same parameters to both scenarios). The SCME also features the time evolution of system level parameters (i.e., it introduces the optional drifting of the path delays and angles of arrivals and departures). In the SCM model, all the propagation parameters stay fixed and independent during the observation periods, which are significantly separated from each other in space or time. This approach is also followed in the SCME model; the length of these intervals is extended by adding the short-term time unpredictability of some channel parameters within the drops. The channel parameters stay independent between the drops. Because of fixed geometry assumption, the sub-path delays and scatterer angles do not change at the BS, but due to movement of the MS, these parameters vary during a drop as seen from the MS. The drifting of these parameters is intended to the testing of beamforming algorithms. Another optional feature of the SCME model is drifting of shadow fading, which is modeled by the exponentially shaped spatial autocorrelation function, which shows that correlation of shadow fading decreases exponentially with distance.

2.5.3 WINNER Channel Model

The European WINNER (Wireless World Initiative New Radio) project is responsible for developing new radio concepts for Beyond-3G systems (e.g., LTE) using a frequency bandwidth of 100 MHz and a radio frequency lying between 2 and 6 GHz in spectrum. The latest developments in MIMO channel modeling are made within WINNER WPs (Work Packages). The WINNER models used the GSCM principle and generic approach for all scenarios with the same generic structure. Generic multilink double-directional models are developed for system level simulations while cluster delay line (CDL) models with reduced statistical variability of small-scale parameters are used for calibration and comparison purposes. Extensive measurement campaigns conducted by five partners in different European countries provide the background for the parameterization of various scenarios.

Initially there was no broadly accepted channel model suitable for WINNER system parameters. In the beginning, the 3GPP/3GPP2 SCM model was selected for outdoor simulations. Due to the limited frequency applicability range and narrow bandwidth, some modifications were made to cope with more advanced simulations. However these initial models were not adequate for advanced level simulations. The main requirements were the proper categorization of spatial properties for MIMO support, consistency in space, time, and frequency (e.g., inherent association between Doppler and angle spreads); a set of possible channels and some limited randomized channels; statistical variability of bulk parameters, and extended polarization support. Consequently, new WINNER models had to be developed. The WINNER channel models were developed in two phases of the IST-WINNER project [33].

In the first stage, the so-called WINNER generic channel model for immediately required propagation scenarios with a limited number of parameters was created based on channel measurements at 2 and 5 GHz. The stochastic channel modeling approach provides unlimited double-directional channel realizations. This generic channel model is a ray-based multilink double-directional model, which is scalable, antenna independent, and capable of channel modeling for MIMO systems. Channel characterization parameters (e.g., delay spreads, angle spreads, and power delay profiles), cross-polarization, shadow fading, and path loss extracted from measurements for the scenarios of interest and the respective statistical distributions can be integrated into the generic model. The following scenarios are of interest in phase 1: indoor typical urban microcell, suburban macrocell, typical urban macrocell, rural macrocell, and stationary feeder link.

In the second stage [34], the WINNER-I channels models were upgraded and new multidimensional channels were developed based on the measurement campaigns. More parameters were included and the frequency range was increased to cover the 2 to 6 GHz spectrum. The numbers of scenarios are increased to 13 based on the feedback from other work packages. The propagation scenarios of concern are indoor office, indoor to outdoor, outdoor to indoor, large indoor hall, urban micro cell, bad urban microcell, stationary feeder, suburban macrocell, urban macrocell, rural macrocell, and rural moving networks. Table 2.10 shows the specific scenarios according to the environments [i.e., for wide area (WA), metropolitan area (MA), and local area (LA) environments] [34].

Measurement campaigns showed that the differences between indoor-to-outdoor and outdoor-to-indoor scenarios are negligible; therefore, these scenarios are merged together. These geometric-based stochastic channel models use a generic channel modeling approach, which means that the number of antennas, antenna configurations and geometry, and antenna beam patterns can be changed without varying the basic propagation model. The new features of the second stage include representation of the elevation of rays, LOS components taken as random variables, and moving scatterers in fixed links. This method facilitates the same channel data in different system level and link level simulations [34].

Path loss models for various propagation scenarios are also developed on the basis of measurement campaigns conducted within the WINNER and from open literature. The general structure of the path loss model is of the form [15]:

$$P_l = A\log_{10}(d[m]) + B + C\log_{10}\left(f_c\left[\frac{GHz}{5.0}\right]\right) + X \tag{2.5}$$

The free space path loss is of the form:

$$P_l = 20\log_{10}(d) + 46.4 + 20\log_{10}\left(\frac{f_c}{5.0}\right) \tag{2.6}$$

where f_c is the carrier frequency and d is the separation between the transmitter and the receiver. The parameters A, B, and C are respectively the path loss exponent, intercept, and path loss frequency dependence. The parameter X is optional for specific cases. Details about these parameters are given in [15].

2.5.4 Multi-polarized MIMO Channel Models

The multiple-input multiple-output systems using arrays of spatially separated antennas at both ends show a dramatic increase in capacity by exploiting the multipath effects. However, in a LOS scenario, the MIMO systems show reduced performance since the LOS components overpower the multipath components in the received signal. The exploitation of polarization dimension results in the improved performance of MIMO systems. Indeed the orthogonal polarization ideally offers a complete channel separation, with a full de-correlation between the transmitter and receiver sides [35]. With spatially separated and cross-polarized antenna arrays, both the polarization diversity and polarization multiplexing can be achieved (e.g., two dual-polarized spatially separated arrays form four antenna arrays. This concept can be extended for $n_r \times n_r$ MIMO systems with the assumption that antenna arrays consist of $n_t/2$ and $n_r/2$ dual-polarized sub arrays.

Table 2.10 Selected Propagation Environments of the IST-WINNER Channel Model

Scenario	Definition	LOS/NLOS	Environment	Frequency (GHz)	Mobile Velocity (km/h)	Notes
A1	Indoor/residential	LOS/NLOS	LA	2–6	0–5	
A2	Indoor to outdoor	NLOS	LA	2–6	0–5	AP inside UT outside. Outdoor, environment urban
B1	Typical urban microcell	LOS/NLOS	LA, MA	2–6	0–70	
B2	Bad urban microcell	NLOS	MA	2–6	0–70	Same as B1 with long delays
B3	Large indoor hall	LOS/NLOS	LA	2–6	0–5	
B4	Outdoor to indoor microcell	NLOS	MA	2–6	0–5	Outdoor typical urban B1. indoor A1
B5	LOS stationary feeder	LOS	MA	2–6	0	
C1	Suburban	LOS/NLOS	WA	2–6	0–120	
C2	Typical urban macrocell	LOS/NLOS	WA/MA	2–6	0–120	
C3	Macrocell bad urban macrocell	NLOS	WA/MA	2–6	0–70	Same as C2 with long delays
C4	Outdoor to indoor macrocell	NLOS	MA	2–6	0–5	Outdoor typical urban C2-Indoor A1
D1	Rural macrocell	LOS/NLOS	WA	2–6	0–200	
D2	1)Moving networks-BS-MRS rural 2)Moving networks-MRS-MS rural	LOS LOS/OLOS/NLOS	WA LA	2–6 2–6	0–350 0–5	Large doppler variability Same as A1 NLOS

The cross-polar transmissions (e.g., from a horizontally polarized transmit antenna to a vertically polarized receive antenna) should be zero. But in real propagation scenarios there is always some depolarization due to the following reasons: linearly polarized antennas have nonzero patterns for cross-polarized fields. Therefore, signals arriving at the, say, vertically polarized antenna from a horizontally polarized antenna will not be zero. Also, due to the multipath scattering effects (i.e., diffuse scattering, diffraction, reflection, and so on), the polarization of the incident electromagnetic wave at the receiver may change [35]. In the following subsections, the two MIMO channel models employing the concept of polarization arrays are described briefly.

2.5.4.1 3GPP Polarized Spatial Channel Model

The antenna separation has a significant impact on the spatial correlation (i.e., the larger the separation between the antenna arrays, the lower the spatial correlation, and vice versa). The large MIMO gains can be achieved with low spatial correlation. Since multiple antennas on the handheld devices require spacing much less than half of the wavelength of the carrier frequency, polarized arrays are likely to be the primary choice to implement multiple antennas [36].

The 3GPP/3GPP2 spatial channel model described in Section 2.5.1 has several optional features. One of the optional features is the use of multi-polarized antenna arrays at both the transmitter and receiver sides. All other features, like the angular spread, delay spread, DOAs, AOAs, power delay profiles, path loss modeling, movement of the mobile terminal to model Doppler spreads, the "drop" concept, the number of propagation environments and the procedures to characterize them, are the same as for the SCM model. The sub-paths in the case of multi-polarized arrays are to be determined as follows.

As described in Section 2.5.1, the multipath signals arrive at the BS or at the MS with angular dispersion; this dispersion is modeled by expressing each path (total paths are six) by a number of sub-paths with the same delays but different DoDs and DoAs distributed around the mean angles. In all environments, angular dispersion for each path is composed of 20 sub-paths. In case of polarized arrays, to consider the effects of signal leakage into the cross-polarized antenna orientations due to scattering, additional M sub-paths at the BS and M sub-paths at the MS are created. The angle of departures and angle of arrivals for these sub-paths are calculated in the same way as in the case of the co-polarized SCM model, which follows Laplacian or uniform distribution. The phase offsets for the cross-polarized elements are also determined. The phase $\varphi_{(n,m)}^{(x,y)}$ is the phase offset of the *mth* sub-path of the *nth* path between x-component (vertical-polarized or horizontal-polarized) of the MS antenna array and y-component (vertical-polarized or horizontal-polarized) of the BS antenna array. The phase offsets $\varphi_{(n,m)}^{(x,x)}$, $\varphi_{(n,m)}^{(y,x)}$, and $\varphi_{(n,m)}^{(y,y)}$ uniformly distributed in $(0, 360^\circ)$, are also determined—the fading is seen independent between orthogonal polarizations, therefore, sub-path phases are modeled randomly. The propagation characteristics of horizontal-to-horizontal paths are equivalent to those of vertical-to-vertical paths. The co-polarized and cross-polarized sub-paths are decomposed into vertical and horizontal components based on the co-polarized and cross-polarized orientations.

The leakage of power P2 of each sub-path in the horizontal direction is set relative to the power P1 of each sub-path in the vertical direction according to the cross-polarization discrimination relation (XPD) ratio (i.e., XPD = P1/P2). For all scenarios, the XPD is calculated from the following distribution [29]:

$$\mathrm{P2} = \mathrm{P1} - \mathrm{A} - \mathrm{B} \times \eta(0, 1) \tag{2.7}$$

The term $\eta(0, 1)$ is a Gaussian random number with zero mean and unit variance. P1 and P2 are the respective powers of sub-paths in the horizontal or vertical directions; A and B are the relative mean path powers and standard deviation of cross-polarization discrimination variation, respectively. The coupled powers of the V-H (vertical-horizontal) and H-V (horizontal-vertical) XPD are the same by symmetry.

At the receiving antennas, the horizontal and vertical components are decomposed into components that are co-polarized with the receiving antennas and are added. The bulk parameters, path losses, and log-normal shadow fading are calculated in the same way as for the SCM model. Based on these calculations, a number of independent channel realizations are found using the drop concept.

2.5.4.2 Theoretical Polarized MIMO Channel Model

Widely used channel models such as the 3GPP/3GPP2 spatial channel model described in the earlier subsection, have limited support for cross-polarization MIMO channels in LOS scenarios. A modified SCM for cross-polarized MIMO channels for both line-of-sight and non-LOS (NLOS) scenarios are found in [37]. This generic theoretical cross-polarized MIMO channel model is derived using the co-polarized MIMO channel model for both LOS and NLOS scenarios. This model shows that cross-polarization discrimination (XPD) depends on the polarization mismatch between the transmit and the receive antenna pairs and antenna patterns. Another feature of this model is that it can be fitted to the SCM model to predict the dual-polarized MIMO channels for LOS and NLOS scenarios.

For LOS case, the channel model for co-polarized antenna elements is derived first and is then extended by element-wise multiplication with a matrix having the polarization mismatch loss between the antenna pairs at both ends and the effect of the azimuth direction of the mobile terminal as well.

The channel matrix for the line-of-sight case can be written as [37]:

$$H_{LOS} = \begin{bmatrix} H_{N\times M} \cdot A^{VV}_{N\times M} & H_{N\times M} \cdot A^{VH}_{N\times M} \\ H_{N\times M} \cdot A^{HV}_{N\times M} & H_{N\times M} \cdot A^{HH}_{N\times M} \end{bmatrix} \tag{2.8}$$

Equation 2.8 can be written in matrix form as:

$$H_{LOS} = H_{2N\times 2M} \bullet A_{2N\times 2M} \tag{2.9}$$

where the dot between the two matrices denotes an element-wise multiplication. $H_{2N\times 2M}$ is the LOS channel matrix for the 2M-transmitter and the 2N-receiver co-polarized MIMO system. The matrix $A_{2N\times 2M}$ illustrates the polarization mismatch between the transmit and receive antenna elements. When antennas at both the ends are strictly aligned, cross-polar transmissions are considered zero, but in reality there is always some depolarization. To represent this, the polarization rotation angle is introduced in the model. Since the normal vectors of the transmitting and receiving antennas do not lie along the LOS path, in this model azimuthal displacements of the antenna pairs are also taken into account. To accomplish this, a displacement angle factor is multiplied to the $A^{VH}_{N\times M}$ and $A^{HH}_{N\times M}$ of the polarization mismatch matrix [37]. This concept can be extended to the $M \times N$ MIMO channel to obtain the arbitrarily polarized MIMO channel.

In the case of NLOS, the signals transmitted with horizontal or vertical polarization are not zero at the cross-polarized receiver. The model describes the polarization rotation is distributed within $(0, 2\pi)$. The model for NLOS scenarios models the amplitude and phase changes during multipath effects and also the polarization rotation angle for each path, respectively. The elements of the channel matrix are the sum of all multipath fields for the corresponding antenna pairs. It is shown that in rich scattering environments, elements of the channel matrix can be modeled as Gaussian random variables. Thus, in a rich scattering environment, there is not much impact by polarization on the channel statistics. This theoretical model can be integrated into the SCM model. The details are found in [37].

2.5.5 MIMO Channel Model for LTE Evolution

Initially, the SCME channel model, discussed in Section 2.5.2, was taken as the standard model for the design of LTE, BS and UE performance requirements, radio resource management requirements, and RF designing to derive requirements without applying the full complexity of the channel model in [38]. The full implementation of the SCME channel requires a significant amount of simulation time, and its backward compatibility with the SCM puts limitations on its performance. Therefore, some simplifications were made in the SCME channel model to obtain a MIMO tap delay line model for design purposes of LTE. The modifications were performed using the following three steps.

First, the statistical variability of the SCME model is removed by defining the fixed tap delay line models. The fixed tap delay line models are defined for four environments given in Table 2.10. The optional feature, polarization, of the SCM and SCME models, which gives independence between the antennas, are also included in the simplified models. These models are also intended to be used in the system level simulations.

In the second step, the antenna configurations at the MS and BS are defined. At the BS, two spatially separated dual polarized (+45/−45) antenna elements are taken with three-sector or six-sector antenna patterns according to the calibration model in the SCM model. The radiation pattern for the three-sector or six-sector antenna is as follows:

$$A(\theta) = -\min[12(\theta/\theta_{3dB})^2, A_m] \quad \text{where} \quad -180^\circ \leq \theta \leq 180^\circ$$

For a three-sector antenna pattern: $\theta_{3dB} = 70^\circ$, $A_m = 20$ dB and maximum gain: 14 dBi

For a six-sector antenna pattern: $\theta_{3dB} = 35^\circ$, $A_m = 23$ dB, maximum gain: 17 dBi

The separation between antenna elements is chosen to be 0.5λ or 4λ with polarization assumed to be unchanged over all AoDs. The azimuthal directions of the BS antennas are set so that the angle of departure of the first tap occurs at +20° in all scenarios.

Two types of MS scenarios are considered: a laptop with two dual-polarized spatially separated antennas (vertical–horizontal) and a handset with two orthogonally polarized (vertical–horizontal) antennas as shown in Figure 2.3 [38]. The antenna pattern shapes are the same as in the case of the BS but with a wider beamwidth

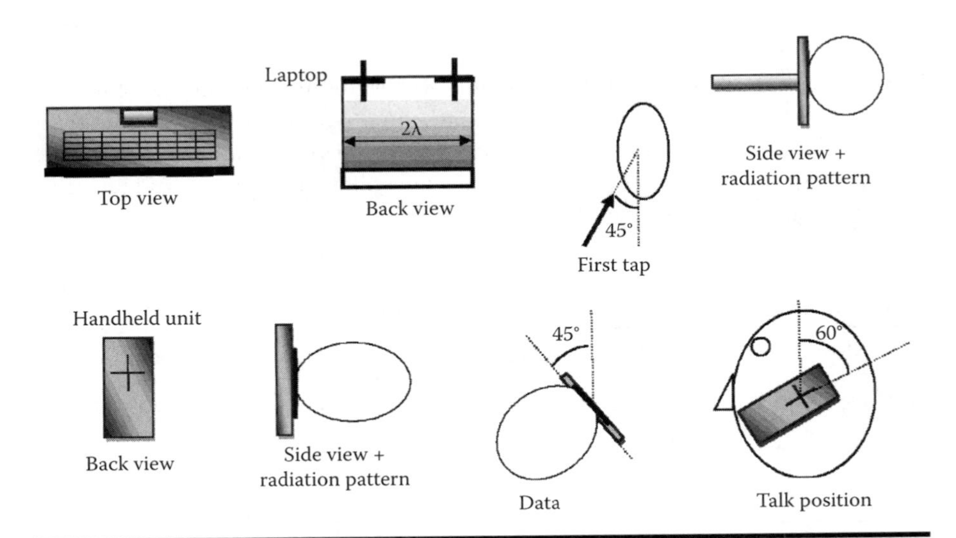

Figure 2.3 Antenna positions for two scenarios, using handset and laptop.

and different side lobe levels. The polarizations are assumed to be pure horizontal and vertical in all directions with an antenna nominal position. The handset can be in the talk or web browsing position. In talk position, the lobe is in the horizontal direction and the handset is turned 60° (polarizations are also rotated). In the data position, the MS is at 45° such that the lobe has its maximum partially downwards.

The azimuthal directions of the MS antennas are adjusted such that the angle of arrival of the first tap occurs at +45° in all scenarios. The parameter values for antenna patterns are:

Handheld, talk position: $\theta_{3dB} = 120°$, $A_m = 15$ dB, maximum gain: vertical: 3 dBi, horizontal: 0 dBi

Handheld, data position: $\theta_{3dB} = 120°$, $A_m = 5$ dB, maximum gain: vertical: 3 dBi, horizontal: 0 dBi

Laptop: $\theta_{3dB} = 90°$, $A_m = 10$ dB, maximum gain: 7 dBi, spatial separation: 2λ

In the third step, using the angular and polarization conditions with the antenna configurations described earlier, correlation matrices per channel tap are calculated for the LTE evolution model. The polarization covariance matrices are determined instead of the correlation of the polarization combinations to account for the power imbalances between different combinations of antenna polarizations and between the MS antennas. The Kronecker product of the BS and the MS spatial correlation matrices and the polarization covariance matrix is used to obtain the total per tap covariance matrix,

$$R_{tap} = p_{tap} \cdot g_{BS,tap} \cdot g_{MS,tap} \cdot A \otimes \Gamma \otimes B \tag{2.10}$$

where p_{tap} represents the relative power of the tap, $g_{BS,tap}$ is the BS antenna gain, and $g_{MS,tap}$ is the gain at the MS antenna. A and B are the correlation matrices of the BS and the MS, respectively, and Γ represents the polarization covariance matrix. The channel models, the propagation environments and the BS and MS arrangements are given in Table 2.11. [38].

Table 2.11 The Channel Models and Propagation Environments for the BS and the MS Arrangement

Model	*Propagation Environment*	*BS Arrangement*	*MS Arrangement*
SCM-A	Suburban macro	3-sector, 0.5λ spacing	Handset, talk position
SCM-B	Urban macro (Low spread)	6-sector, 0.5λ spacing	Handset, data position
SCM-C	Urban macro (High spread)	3-sector, 4λ spacing	Laptop
SCM-D	Urban micro	6-sector, 4λ spacing	Laptop

The two models SCM-C and SCM-D are used for evaluating laptops with two receiving antennas. In these models, channel realizations are calculated with one of the two dual-polarized antennas.

2.5.6 Comparison of SCM, SCME, WINNER, and LTE Evolution Models

A comparison of the SCM, its extension SCME, and the WINNER model is made in terms of small-scale and large-scale effects in the MIMO radio channel in [39]. All three models use the same generic ray- or geometric-based stochastic approach. The delay spreads, angular spreads, departure and arrival angles are used to characterize these channel models, which are chosen randomly from the appropriate probability distributions. The values of these parameters are different in different simulation runs for a single user or in multi-simulation runs for different users. The channel models are compared in [39] in terms of system bandwidth, the correlation between large-scale parameters at system level, antenna arrays and polarization, drop concept, cluster delay spreads, cluster angular spreads, and complexity issues.

The channel model for LTE evolution, discussed in Section 2.5.5, is a simplified version of SCME analogous to link level calibration models. The procedure of finding spatial correlation matrices in the SCM/SCME and the WINNER is different from that of the LTE evolution model. The spatial correlation for the SCM/SCME and the WINNER is defined by the angular geometry and nominal direction of the sub-paths of each delay tap while the LTE evolution model describes deterministic correlation values for different clusters. Table 2.12 describes the main features and Table 2.13 shows key parameter values of these models [39].

Table 2.12 Different Attributes of SCM, SCME, WINNER, and LTE Evolution Models

Attribute	*SCM*	*SCME*	*WINNER*	*LTE Evolution*
Indoor scenarios	No	No	Yes	No
BW >20 MHz	No	Yes	Yes	Yes
Indoor-to-outdoor and outdoor-to-indoor scenarios	No	No	Yes	No
Elevation angle AoD/AoA	No	No	Yes	No
Intra-cluster delay spreads	No	Yes	Yes	Yes
Cross-correlation between LSPs*	No	No	Yes	No
Time evolution of model parameters	No	Yes	Yes	No

*Large-scale parameters

Table 2.13 Key Parameters of SCM, SCME, WINNER, and LTE Evolution Models

Parameter	*SCM*	*SCME*	*WINNER*	*LTE Evolution*
No. of scenarios	3	3	12	4
No. of taps	6	18–24	4–24	18
Maximum bandwidth (MHz)	5	100*	100**	20
Carrier frequency (GHz)	2	2–6	2–6	-
No. of clusters	6	6	4–20	6
No. of mid-paths per cluster	1	3–4	1–3	3
No. of sub-paths per cluster	20	20	20	-

*Artificial bandwidth from 5 MHz bandwidth
**Based on 100 MHz measurements

2.6 Channel Modeling Role in Cell Planning and Optimization

Designing a cellular network is a challenging task that requires the use of available resources in an efficient way to allow networks to accommodate as many users as possible. One of the main purposes in the design of a cellular network is to meet the traffic demands of all potential users with an adequate quality of service (QoS) and with an acceptable percentage of blocked calls in the network [40]. Standard channel models assist network planners and system designers to meet these objectives. Therefore, realistic spatial and temporal radio propagation models are a critical part of any radio system design or network planning and optimization process. To design wireless networks with optimal user and frequency allocation and cellular planning, important statistical characteristics (e.g., path loss, log-normal shadowing, small-scale, or fast fading) must be predicted [41]. Detailed cell planning also includes coverage, spectral efficiency, capacity and parameter planning, parameters related to power control, neighbor cells, signaling, radio resource management, and so on. On the basis of these characteristics, cellular network designers are able to predict strict link budgets to generate cellular maps of areas of service with desired QoS and to optimize the data transfer within each radio communication channel [41, 42].

There are two ways to design a radio network [42]: the network planners can either produce their own propagation models for cell planning of different environments in the given area, or existing channel models are used that are generic in nature. Although the use of standard channel models is economical from a cost and time perspective, these models are usually not employed directly. The reason is that standard models are developed by taking particular cities into account. So changes must be made through drive tests called correction factors.

The use of different channel models in cellular planning and optimization have trade-offs between them. The use of deterministic channel models based on the uniform theory of diffraction (UTD) and the geometrical theory of diffraction (GTD) has been a widely adapted deterministic technique for radio propagation modeling in indoor, outdoor, and rural environments [40, 43]. The combined use of deterministic channel models, ray-tracing acceleration techniques, and topographical/morphological databases give more accurate results for microcellular networks planning and design. However, these techniques are computationally complex, requiring large amounts of data and computer memory for simulations.

When designing the interfaces between picocells and macrocells or between distant picocells, a better approach to work out propagation losses is to apply empirical or semi empirical methods. This is because in modeling these environments, deterministic methods require large amounts of computational resources and the improvement in precision is insignificant. These channel models can provide enough information necessary in the network design process in the case of macrocells and microcells. The deterministic channel models are adequate for microcells and picocells independently, but their implementation in macrocell designs makes them less favorable due to large computation times.

2.7 Conclusion

This chapter presents an overview of the important features of wireless channel modeling and standard channel models for UMTS and LTE communication systems. Standard channel models play a vital role in the design and performance assessment of advanced transceivers techniques and smart antennas employed to establish reliable communication links in mobile communication systems such as UMTS and LTE.

From different channel modeling approaches, it is evident that there is no standalone master method to obtain radio channels with desired characteristics. There is always a trade-off between complexity and accuracy in modeling a radio channel. The channel models SCM and SCME, based on 3GPP specifications, are geometry-based stochastic models, suitable for LTE and give more accurate results for real propagation scenarios. However, these models are complex to some extent and require long simulation times. On the other hand, extended ITU models are correlation-based and show low computational complexity in the performance assessment of systems involving advanced antenna array concepts.

The state-of-art channel models such as WINNER models provide additional features for state-of-the-art communication systems like UMTS and LTE. An efficient modeling of the spatial characteristics in MIMO systems is necessary to determine the improved transmission techniques for LTE. The spatial characteristics in multi-antenna channel modeling, including polarization effects, are expected to be crucial in the coming LTE standards and future communication systems. Thus, new and improved channel models are necessary to evaluate the parameters and performance of these future systems.

References

[1] P. A. Bello, "Characterization of Randomly Time-Variant Linear Channels," *IEEE Trans. Comm. Systems*, CS-11, 360, 1963.

[2] M. Ibnkahla, (ed.), *Signal Processing for Mobile Communications*, CRC Press: New York, 2005.

[3] H. Holma and A. Toskala, (eds.), *LTE for UMTS: OFDMA and SC-FDMA Base Band Radio Access*, John Wiley & Sons Ltd, 2009.

[4] S. R. Saunders and A. A. Zavala, *Antennas and Propagation for Wireless Communication Systems*, John Wiley & Sons Ltd, The Atrium, Southern Gate, Chichester, West Sussex, PO19 8SQ, England, 2007.

[5] A. F. Molisch, "A Generic Model for MIMO Wireless Propagation Channels in Macro- and Micro Cells," IEEE Trans. Signal Processing, vol. 52, no. 1, 2004.

[6] M. Steinbauer, "The Radio Propagation Channel-A Non-directional, Directional, and Double-Directional Point-of-View," Ph.D. dissertation, Vienna University of Technology, Vienna, Austria, 2001.

[7] M. Steinbauer, A. F. Molisch, and E. Bonek, "The Double Directional Radio Channel," *IEEE Antennas and Propagation Magazine*, vol. 43, no. 4, pp. 51–63, 2001.

[8] J. K. Cavers, *Mobile Channel Characteristics*, Kluwer Academic Publishers, New York, 2002.

[9] J. Maciej, Nawrocki, M. Dohler, and A. H. Aghvami, (ed.), *Understanding UMTS Radio Network Modeling, Planning, and Automated Optimization*, John Wiley & Sons, The Atrium, Southern Gate, Chichester, West Sussex PO19 8SQ, England, 2006.

[10] T. Rappaport, *Wireless Communications, Principles and Practice*, Prentice-Hall: Englewood Cliffs, NJ, 1996.

[11] P. Almers, E. Bonek, A. Burr, N. Czink, M. Debbah, V. Degli-Esposti, H. Hofstetter, P. Kyosti, D. Laurenson, G. Matz, A. F. Molisch, C. Oestges, and H. Ozcelik, "Survey of Channel and Radio Propagation Models for Wireless MIMO Systems," *EURASIP Journal on Wireless Communications and Networking* (special issue on space-time channel modeling for wireless communications), 2007.

[12] G. D. Durgin, "Theory of Stochastic Local Area CHANNEL Modelling for Wireless Communications," Doctor of Philosophy in Electrical Engineering, Blacksburg, Virginia, December, 2006.

[13] A. F. Molisch, A. Kuchar, J. Laurila, K. Hugl, and R. Schmalenberger, "Geometry-based Directional Model for Mobile Radio Channels: Principles and Implementation," *European Transactions on Telecommunications*, vol. 14, no. 4, pp. 351–359, 2003.

[14] J. D. Parsons, *The Mobile Radio Propagation Channel, 2nd Edition*, John Wiley & Sons Ltd, 2000.

[15] J. Meinilä, T. Jämsä, P. Kyösti, D. Laselva, H. El-Sallabi, J. Salo, C. Schneider, and D. Baum, "IST-2003-507581 WINNER: Determination of Propagation Scenarios," D5.2 v1.0, 2004.

[16] A. F. Molisch, H. Asplund, R. Heddergott, M. Steinbauer, and T. Zwick, "The COST 259 Directional-Channel Model A-I: Overview and Methodology, *IEEE Transactions on Wireless Communications*, 12, pp. 3421–3433, 2006.

[17] L. M. Correia, (ed.), *Wireless Flexible Personalised Communications (COST 259 Final Report)*, John Wiley & Sons, Chichester, UK, 2001.
[18] L. Correia, (ed.), *Mobile Broadband Multimedia Networks*, John Wiley & Sons, New York, 2006.
[19] I. D. Sirkova, "Overview of COST 273 Part I: Propagation Modelling and Channel Characterization," Sofia, Bulgaria 29 June – 1 July 2006. http://www.lx.it.pt/cost273.
[20] S. Sesia, I. Toufik, and M. Baker, *LTE – The UMTS Long Term Evolution*, John Wiley & Sons Ltd, 2009.
[21] ITU-R M.1225 International Telecommunication Union, "Guidelines for Evaluation of Radio Transmission Technologies for IMT-2000," 1997.
[22] ITU-R M.1034 International Telecommunication Union, "Requirements for the Radio Interface(s) for International Mobile Telecommunications-2000 (IMT-2000)," 1997.
[23] T. B. Sørensen, P. E. Mogensen, and F. Frederiksen, "Extension of the ITU Channel Models for Wideband (OFDM) Systems," in *Proc. IEEE Vehicular Technology Conf.*, Dallas, Sept. 2005.
[24] 3GPP Technical Specification 36.803, "User Equipment (EU) Transmission and Reception," Release 8, V0.3.0, May 2007.
[25] Ericsson, Nokia, Motorola, and Rohde and Schwarz, "R4-070572: Proposal for LTE Channel Models," www.3gpp.org, 3GPP TSG RAN WG4, meeting 43, Kobe, Japan, May 2007.
[26] Ericsson, "R4-070994—LTE Channel Models: High-speed Scenario," TSG-RAN Working Group 4 (Radio) meeting #43 bis Orlando, FL, June 25–29 2007.
[27] 3GPP Technical Specification 25.101, "UE Radio Transmission and Reception (FDD)," Release 5, V5.2.0, 2002-3.
[28] 3GPP Technical Specification, 25.943, v4.2.0, "Deployment Aspects," (Release 4), www.3gpp.org.
[29] 3GPP Technical Specification 25.996, "Spatial Channel Model for Multiple Input Multiple Output (MIMO) Simulations (Release 8)," V8.0.0, 12-2008. http://www.3gpp.org.
[30] A. Algans, K. I. Pedersen, *Associate Member, IEEE*, and Preben Elgaard Mogensen, *Member, IEEE*, "Experimental Analysis of the Joint Statistical Properties of Azimuth Spread, Delay Spread, and Shadow Fading," *IEEE Journal on Selected Areas in Communications*, vol. 20, no. 3, April 2002.
[31] D. S. Baum, J. Salo, G. Del Galdo, M. Milojevic, P. Kyösti, and J. Hansen, "An Interim Channel Model for Beyond-3G Systems," in *Proc. IEEE Vehicular Technology Conference*, Stockholm, May 2005.
[32] Elektrobit, Nokia, Siemens, Philips, Alcatel, Telefonica, Lucent, and Ericsson, "R4-050854: Spatial Radio Channel Models for Systems Beyond 3G," 3GPP TSG RAN WG4 Meeting 36, London, UK, August 29–September 2, 2005.
[33] Commission of the European Communities, "IST-WINNER Project," http://www.ist-winner.org.
[34] Commission of the European Communities, IST-WINNER II Project Deliverable D1.1.2, V1.2 "WINNER II Channel Models," http://www.ist-winner.org.
[35] C. Oestages and B. Clerckx, "MIMO Wireless Communications: From Real World Propagation to Space-code Design," AP.
[36] T. Hult, A. Mohammed, Z. Yang, and D. Grace, "Performance of a Multiple HAP System Employing Multiple Polarization," Invited Paper, Special Issue of *Springer Wireless Personal Communications Journal*, vol. 52, Issue 1, January 2010.
[37] L. Jiang, L. Thiele, and V. Jungnickel, "On the Modeling of Polarized MIMO Channel," Fraunhofer Institute for Telecommunications, Heinrich-Hertz-Institute Einsteinufer 37, D-10587, Berlin, Germany.

[38] Ericsson, "R4-060101: LTE Channel Models for Concept Evaluation in RAN1," TSG-RAN Working Group 4 (Radio) meeting 38 Denver, CO, February 13–17, 2006.

[39] M. Narandzic, C. Schneider, R. Thomä, T. Jämsä, P. Kyösti, and X. Zhao, "Comparison of SCM, SCME, and WINNER Channel Models," in *Proc. IEEE Vehicular Technology Conference*, Dublin, April 2007.

[40] M. F. Câatedra and J. P.-Arriaga, *Cell Planning for Wireless Communications Artech House Mobile Communications Library*, Artech House, Inc., 1999.

[41] N. Blaunstein and C. Christodoulou, *Radio Propagation and Adaptive Antennas for Wireless Communication Links*, John Wiley & Sons, Inc., 2007.

[42] A. R. Mishra, *Fundamentals of Cellular Network Planning and Optimization 2G/2.5G/3G ... Evolution to 4G*, John Wiley & Sons Ltd, The Atrium, Southern Gate, Chichester West Sussex PO19 8SQ, England, 2004.

[43] C. Smith and C. Gervelis, *Cellular System Design and Optimization*, McGraw-Hill Professional, 1996.

Chapter 3

Virtual Drive Test

Avraham Freedman and Moshe Levin

Contents

3.1 Introduction

The cellular telephone has had a major effect on everybody's life. Since its early days, we have witnessed a larger and larger penetration of the mobile telephone service and it has been quickly adapted as a major means of personal communications. The cellular telephone has continuously evolved over the relatively short years of its existence. From the prestigious car phone of earlier days, use of wireless communications has become ubiquitous and personal. Indoor usage is significantly larger than outdoor usage. It keeps evolving as the current trend is for the data traffic, of which the usage is constantly increasing, and the transition to all-IP (Internet Protocol) networks is the key for the next-generation networks [1,2]. User demand for higher data rates does not come without its toll. Higher data rates essentially lead to higher bandwidth systems, and the scarcity of spectrum resources makes it necessary to resort to other means to satisfy the higher-capacity demands. These may include adding base stations, increasing the spectral efficiency of the physical layer by various means such as additional antenna [multiple-input multiple-output (MIMO)] systems, using diversity and spatial multiplexing to raise the reliability of the wireless channel, and to increase its capacity. One of the most important means used by operators is sophisticated network optimization techniques, which enable adapting the network better to the needs and demands of the users and better exploiting existing resources.

The problem of adaptation and optimization is a major one that accompanies the cellular network operator throughout the network lifetime. The operator relies on a variety of information sources, starting from the basic geographic information, market information, network management reports, and customers' complaints. Some of the most important sources of information are the measurements taken by the terminals. In this chapter, we introduce the concept of the virtual drive test (VDT), which combines event and other terminal reports together with a location engine for the purpose of network optimization. The idea of using the measurements for network optimization has been introduced in the past [3], in the broader context of self-optimizing networks, which envisions the network using measurements, and expert systems performing self-configuration and self-organization.

The strength of the concept stems from the fact that mapping an event to a geographical location provides significant insight into the network performance. Mapping enables the actual geographic real time display of parameters such as pilot chip energy (E_c), pilot chip energy to interference spectral density ratio (E_c/I_o), and pilot pollution for the Universal Mobile Telecommunication System (UMTS). Similar parameters for the the long-term evolution (LTE) system or the wireless interoperability for microwave access (WiMAX) system are the received signal strength (RSS) and signal to interference and noise ratio (SINR) as a function of load. The resulting data can be used for network optimization as well, namely tuning the multitude of system parameters to improve the system performance based on some key performance indicators (KPIs). An example of such an optimization project is described next. In this case, the optimization process improved the dropped call rate from 0.66% down to 0.55%. The total traffic supported by the network increased after the optimization by 10%.

In the first section, we survey the basic methods used by operators to predict, measure, and monitor the network. The second section introduces the VDT concept and explains the geolocation principles on which it is based. We also present the concept of a full 3D coverage database used as an integral part of the VDT. The third section describes methods for network optimization using the virtual drive tests and the results of such an optimization are demonstrated for an existing cellular network. The fourth section elaborates on the potential of VDT for LTE and future-generation systems, and discusses some implementation issues involved. The chapter concludes with a summary and a discussion of open issues.

3.2 Prediction and Measurement of System Coverage

Measurements and network monitoring are essential for the operation of any cellular radio network. A lot of effort is made by network operators to maintain current, comprehensive, and accurate databases of their networks. This database, together with new traffic demand data, is used by the operators to improve their networks and adjust their operation.

Drive and walk tests are the most common measurement campaigns used by operators [4, 5]. The measurements are aimed to provide an accurate and updated view of the system coverage and interference scenarios; however, in contrast to the actual indoor-dominated usage patterns, most of the drive test data are taken from roads and streets. Thus, information about coverage indoor and, even more importantly, in higher floors is lacking.

The coverage of a wireless system is defined as geographic area where the system can provide its service with the acceptable quality at the required availability. This geographic area is traditionally defined as a two-dimensional area covering the region where the system operates. In a computerized network planning tool, this area is partitioned into a set of bins.

Obviously, a 2D area representation is not sufficient to describe the actual service conditions experienced by the users within the region, as it does not represent reliably the different operation conditions of users in the same bin. Terminals on the top floors of a multistorey building do not experience the same reception conditions as the ones at the first floor, not to mention the basement. In order to account for build-up areas, a 3D database, for which different bins are allocated to different floors in a building, is much more appropriate. For a bin located within a building, predicting the signal strength involves a calculation of the signal strength as a function of higher antenna height and the building penetration loss. Different bins can also be allocated to roads, in which the reception conditions might be quite different from areas next to them, where the terminals are stationary or move slowly. This 3D coverage is the basis for the virtual drive test concept to be described next. The reception condition in each bin is determined according to the received signal strength at that location from the serving cell or cells, the level of interference from other cells or external sources, and the receiver performance at the specific channel conditions at that area. For a given system and geographical conditions, the main parameters to be estimated are the signal strength of all the signals arriving at a certain point. System performance is further determined as well by the traffic density and demand. Operators use three sources of information to estimate signal strength, and hence deduce the signal to interference and noise ratio:

- Predictions, based on empirical or deterministic propagation models
- Measurements by special drive test or walk test equipment
- Reports from the network, from various interface probes or from the management system

3.2.1 Accurate Propagation Modeling: Cons and Pros

Prediction of the received signal strength at a given point within a geographical area is made by use of propagation modeling. While it is theoretically possible to calculate the electromagnetic field strength at any point in space using Maxwell equations, the amount of information and the level of details needed for that is so large that it is practically impossible.

In order to predict the signal strength, one uses a model that would simplify the calculation. A large variety of models exist. Those models use only limited information of the environment, and they basically differ by the type and level of the information used. As the models are not accurate, a measure of uncertainty has to be added to the prediction. Another factor of uncertainty is the signal fading, which stems from the fact that the user may be moving and the environment may change. Thus, in addition to the propagation model, a fading model should be defined to account for the signal level uncertainty. Comprehensive discussion on the subject can be found in many textbooks, among them [6–8].

The models are classified into physical models and empirical models. Empirical models, like the Okumura-Hata [9, 10] or COST [11] models are based upon measurement campaigns wherein the path loss is measured in a variety of conditions. Okumura's measurements, for instance, included a set of path loss measurements in the 150-MHz to 3000-MHz frequency bands, limited to ranges between 1 km and 20 km, and antenna heights between 30 m and 200 m at the base station and between 1 m and 10 m at the terminal station. The measurements were performed in various areas to test the path loss estimation in different environments, such as rural, suburban, urban, and dense urban areas. The model provides the median of the measurements made at a given set of conditions, together with uncertainty that was found to be log-normally distributed with a standard deviation of 5 to 8 dB. The empirical models are not specific as they do not refer to a specific propagation path. As such, they do not require knowledge of the terrain, buildings, and other specific geographical information. Another significant drawback is that they are limited to the range of parameters and conditions in which the measurements were taken. Their main usage is for performance estimation (e.g., during the system design phase), and not for planning, monitoring, or optimization.

A physical model is based on a physical simplified representation of the environment and it calculates the path loss for this simplified environment. The model that takes into account most of the effects is the ray-tracing model. In this model, the energy emitted by the transmitter antenna to all possible directions is modeled as rays, each presenting a part of the spherical wavefront. The model traces each ray and determines the physical phenomena that affect it as it propagates. Database accuracy (especially ground slopes and land use data) and the extensive computations needed for a comprehensive analysis are major drawbacks for this model, limiting its use to small urban areas.

Other physical models go further toward simplification. The knife-edge model, for instance, replaces each obstacle between transmitter and receiver with a very thin, but infinitely wide screen. The model then calculates the diffraction of the electromagnetic wave over such a screen. Physical models require specific geographical information, such as terrain, land use, and building contour and height, and provide a path loss estimation specific to a given location. Unlike the empirical models, they are not limited to a given range of parameters.

Signal strength prediction is a very flexible and convenient means for network monitoring and optimization. In fact, prediction is the only way to estimate the performance of a green field deployment. One of the major advantages of predictions is the ability to use it to study the impact of system modifications. For example, the impact of rotating an antenna can be very easily simulated on a computerized planning tool, using prediction. However launching a measurement campaign or a drive test to measure that impact is not so simple. However prediction has its deficiencies. It is, as explained earlier, inherently inaccurate and thus can not be totally relied upon for absolute performance estimation. Actual measurements, such as drive tests, are required.

3.2.2 Drive and Walk Tests: Limits and Constraints

Measurements campaigns, such as drive and walk tests, are a very good source of information about signal strength values. In these campaigns, a specific instrument, typically a sensitive multichannel or scanning terminal receiver, equipped with GPS, is used. This instrument is capable of measuring instantaneously the strengths of a number of base stations, identifying the source, and providing a log of the measurements, together with a time stamp and a location. Such instruments are typically mounted on vehicles and are driven around in the analysis area. The data is later processed by relevant network planning and optimization tools. A carry-on version of those instruments enable a researcher to perform a walk test, wherein the instruments can be carried in areas not accessible to vehicles.

Drive and walk tests provide accurate and reliable data that reflect the actual reception conditions in the analyzed area. The accuracy of such measurements can be as good as 1 to 2 dB, provided they are well taken, averaging out the fading effects. Some guidelines and analysis for the measurement error in a drive test can be found in [12]. It should also be noted that GPS readings may also suffer from artifacts such as multipath geometric dilution of precision due to unfavorable satellite constellations, and so on, that would reduce the accuracy of the measurement location. However, the main disadvantages, except for the cost involved, are the latency in obtaining drive test results, and the fact they are limited to roads and public places and cannot provide estimates for the higher layers of the 3D coverage database. Considering the trend for indoor usage, this actually means a lack of capability to be relevant for most of the traffic.

3.2.3 Other Methods for Database Updates

Network reports may include reports from the various network elements, terminals, base stations, radio network controllers, and so on, or external systems such as network management systems and probes capturing the management and control traffic over the various interfaces. Those reports provide updated and full information, reflecting the actual operation of the network. However, they are typically aggregate of measurements taken from a cell level and above, making it quite difficult to pinpoint a specific problem.

On the other hand, some of the network reports include terminal reports. Terminal reports, as collected by the radio network controller or the management system are a valuable source of measurements that can fill in the gaps of drive and walk tests. Among the most significant reports are those made by the terminals in response to some events, such as dropped calls or failed connections. Those reports are important since, once enhanced by a location engine that attributes geographic coordinates to each call, they provide a focused view on problematic areas and enable problem solution and network performance improvement. The usage of terminal reports provides in-situ performance monitoring, as opposed to the "in-vitro" nature of drive testing.

The mobile station of a cellular network is in fact a sensor by its own right. It measures the surrounding base stations' (BS) power, and the distance to them. It also measures the spatial and temporal signature of the signals arriving from each BS. Some of the terminals, equipped with GPS (global positioning system), directly measure their positions. The BS is making such measurements as well, on the uplink channel. Those measurements are used mainly for proper operation of the physical layer, or for handover control, and are not always reported back to the network.

Those measurements, when available, can be used to determine the terminal position, and thus the report can be turned into a measurement point—this is the basis of the virtual drive test (VDT).

3.3 Virtual Drive Test

3.3.1 Virtual Drive Test Concept

The VDT is composed of the following phases:

- Acquiring terminal reports from the network
- Using the terminal report for positioning of the terminal at the time of the report
- Mapping the terminal measurements at the time of the report
- Using the reported information for network optimization

3.3.2 Network Report Acquisition

Report acquisition highly depends on the system and its architecture and may require some dedicated hardware. In GSM (global system mobile) or UMTS systems, a probe may be attached to the A/Abis interfaces or Iub interface connecting the BSC (base station controller) or RNC (radio network controller) to the MSC (moile switching center), whereas in next-generation systems, such as LTE [13, 14] and WiMAX [15, 16], the architecture is basically flat and the interfaces between the base stations (eNodeB or BS) and the network are more distributed in nature. Thus, a central point where the required information is concentrated does not necessarily exist. However, those standards are still in development and a solution to this problem might be available.

Terminal reports of mobile scanning measurements are made quite frequently (every 480 ms, to be exact) in the case of a GSM terminal. In UMTS, however, reports are limited to events, like call start, dropped call, inter-RAT handover, and so on, which effectively reduce the rate to about once in six seconds. While the GSM terminal reports an average over six RSS readings from neighboring cells, in UMTS, reports average between two and three readings. On the other hand, the delay measurements produced by UMTS are more accurate.

3.3.3 Event and User Location Principles

Terminal positioning is the key technology that leverages the terminal reports into a viable source for VDT. Numerous techniques exist for that, mainly developed for emergency service requirements and location-based services. This has been a rich research and development area, for which [17–20] are only examples. In this section, we give a general description of the principles, properties, and deficiencies of each method. It can be stated that none of those methods provides a comprehensive solution on its own. A combination of methods can compensate for the deficiencies of each and provide a satisfactory solution.

3.3.3.1 Satellite-based Positioning [21]

Satellite-based positions, based on the GPS, or in the future the Galileo system, were developed by the Unites States government (or the European Union in the case of Galileo) to provide a global positioning service based on a constellation of satellites orbiting the earth in known trajectories. Those satellites transmit well-defined signals, synchronized to an atomic clock. The whole system is maintained and monitored by the government to ensure synchronized transmissions and precise trajectories. The GPS receiver receives signals from at least four satellites, measures the time shift of each reception, and solves its positions on the globe longitude, latitude, and altitude together with the bias of its clock compared to the system. The GPS system is designed to provide very good accuracy—on the order of magnitude of only tens of meters. The receiver is simple and is easily integrable with the cellular phone, providing the user with an inherent location-based services capability. Still, GPS-aided location suffers some problems:

- Lack of coverage indoor.
- A long time for the first fix, due to the need to download the satellite's databases (ephemeris and almanac). These databases include the description

of the satellite trajectories needed for calculation and are downloaded from the satellites as auxiliary information.

- Loss of precision due to multipath problems and the fact that the satellite's receiver may not be ideally located, could mean the resulting location determination may be erroneous (geometrical depletion of precision effect).
- Not all terminals are equipped with GPS, which reduces the number of measurements and consequently the quality of network optimization based on it.

The network may assist the GPS-equipped terminals in several ways. For example, it can transmit the ephemeris almanac data from the base station, thus shortening the time for first fix. GPS signals can also be transmitted by indoor base stations, mimicking the GPS operation and enabling indoor location. An important technique, which can be utilized by a cellular network is differential GPS, in which a fixed and known receiver is used to correct for system biases in a certain area. The corrections are applied to terminals, thus improving their accuracy.

Despite the problems mentioned here, one can certainly state that GPS-based location, especially if enhanced by assisted GPS (AGPS), can provide a very good basis for optimization.

3.3.3.2 Time of Arrival-based Positioning [22]

The terminal and base station are capable of determining the propagation delay between them. This measurement is essential for all cellular systems, as of second generation. This propagation delay can be transformed into range. The intersections of at least three circles of such ranges from four base stations is a good unambiguous estimation of the terminal location. This is the general time of arrival (TOA) method. The TOA is usually measured by a terminal to its serving cell. The TOA of a terminal to a base station other than its serving cell is not always available. Rather, the difference of the time of arrival between the serving cell and other cells is. In this case, the loci of all points for which the difference of the time of arrival between two cells is known, is a hyperbole. The intersection of three hyperboles, or two hyperboles and a circle of absolute time of arrival, provides the location estimate.

TOA-based techniques are also susceptible to problems:

- Limited measurement resolution, which translates to low measurement accuracy
- Multipath, which distorts the range measurement
- Synchronization problems, especially between base stations, which adds an inherent bias to the propagation delay measurement

It should be noted that there are different types of measurements which result in quite different levels of accuracy. For example, a UMTS terminal reports, at the beginning of each call, a propagation delay measurement result. Figure 3.1 shows the distance, as estimated by a UMTS terminal using that measurement, as a function of the actual range. It can be observed that a certain report value may result from a

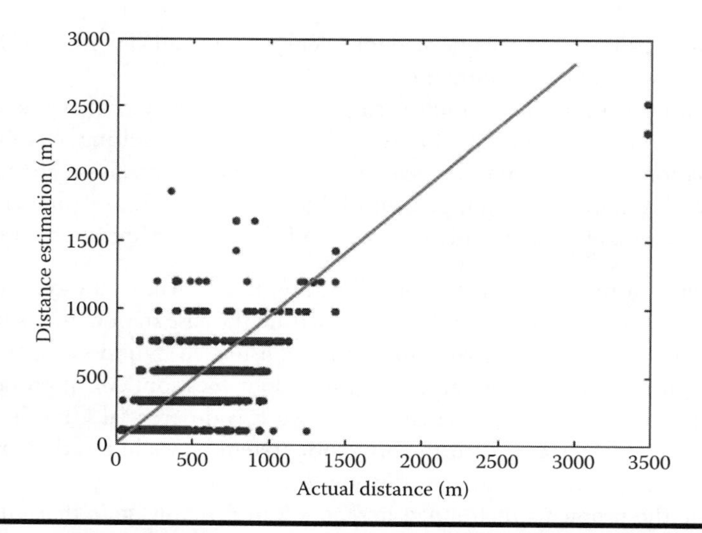

Figure 3.1 Measured distance estimation of UMTS terminals.

very large range of actual measurements. This very real picture shows that a single measurement made by a terminal is not very reliable. On the other hand, the round-trip time (RTT) measurement, standardized especially for location-based services is made with an accuracy of 1/16 of a chip, which is translated to about 5-m resolution, so accuracy is mainly a result of propagation effects. For the purpose of network optimization it is important to note that for complete statistics it is necessary to locate terminals even without dedicated measurements.

Another factor that should be taken into account is the ability of a terminal to measure the delay to a number of base stations, so that a location fix can be made. Figure 3.2 depicts the probability that a given number of cells were reported by a GSM or UMTS terminal. The graph was made using statistics of a large number (a few million) of actual terminal reports. It should be noted that for UMTS, the number of measured cells is higher than the active set size. This is the hearability problem of distant base station signals. It can be solved by adding idle periods, with the penalty of reducing the time allocated to payload communication. Similar problems may be encountered in LTE and WiMAX systems, since both, as is the case for CDMA-2000 and UMTS systems, are expected to operate in a single frequency environment.

3.3.3.3 Angle of Arrival-based Positioning [23]

A receiver equipped with an array of antennas can use it to determine the angle of arrival of the impinging signals, using a variety of direction-finding techniques. In an open space, the intersecting vectors from at least two receivers can be used to determine the transmitter location. For a long time, the main drawback of those

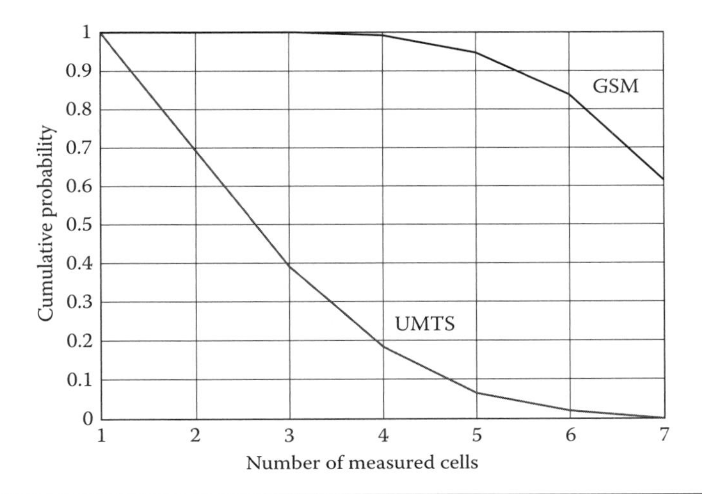

Figure 3.2 Cumulative probability of the number of measured cells.

techniques has been the fact that additional antennas and signal processing equipment are required to be installed at the base station, which made this solution quite costly. In the next-generation systems, such as LTE and WiMAX, it is envisioned that the base stations, as well as the terminal stations will be equipped with antenna arrays to make use of the MIMO spatial multiplexing techniques. Those arrays can well be used for the purpose of location. However, some problems exist in the implementation of those techniques. Low measurement resolution, due to a low number of antennas and a relatively small span of the antenna arrays, leads to low measurement accuracy. Multipath effects result in a distributed direction of arrival for a single signal. Furthermore, if elevation information is needed to locate terminals at high areas and high floors, additional antenna array and more complex processing is needed. The additional elevation array does not contribute to enhanced spatial multiplexing capability, and thus it is not cost effective.

Presently, angle measurements is not used as a primary technique for location. However, those techniques might develop to be very useful, as they are enhanced with geographical information and advanced angle determination techniques that make use of the signal properties as well as high-resolution angular spectrum estimation.

3.3.3.4 Positioning based on Radio Signal Strength Measurements [24, 25]

The cellular terminal constantly performs signal strength measurements of its serving cell and its neighbors. This is an essential part of its operation. Those measurements can be used to pinpoint its location. The principle of operation is by comparing the measured result to the signal strength predictions, as described in the previous section. Alternatively, a database of measured results resulting from drive tests or virtual drive tests can be created instead of predictions, thus providing a better reference for the RSS at any point. The received signal strengths as measured by the terminal at given locations creates a "fingerprint" that can be compared to the fingerprint of the measured or predicted results. Figure 3.3 shows the fingerprint of a mobile unit at a location, which includes the set of RSS measured to all the neighboring cells. Prediction-based location has the advantage that no additional equipment or procedures are needed. Another advantage is its ubiquity—the RSS measurements are available indoor and outdoor at lower and upper floors, and may be the only means of distinguishing between measurements made at different heights.

Obviously, RSS-based location suffers from the inherent inaccuracy of the prediction process, which can be improved as measurements are accumulated and replace the predictions. It also suffers from fading and measurement inaccuracy. Missing measurements, which can result for various reasons, such as fading, base station malfunction, base station load, and more, can skew the location determination a great deal.

3.3.3.5 Location Based on Channel Pattern Matching

The fingerprinting method used for RSS-based positioning belongs to a family of pattern matching techniques. More elaborate pattern matching can be made using the large extent of physical parameters measured by the mobile station as well as the base station. These include the power delay profile (PDP) and the frequency response, related to it. For multi-antenna systems, the power angular profile (PAP)

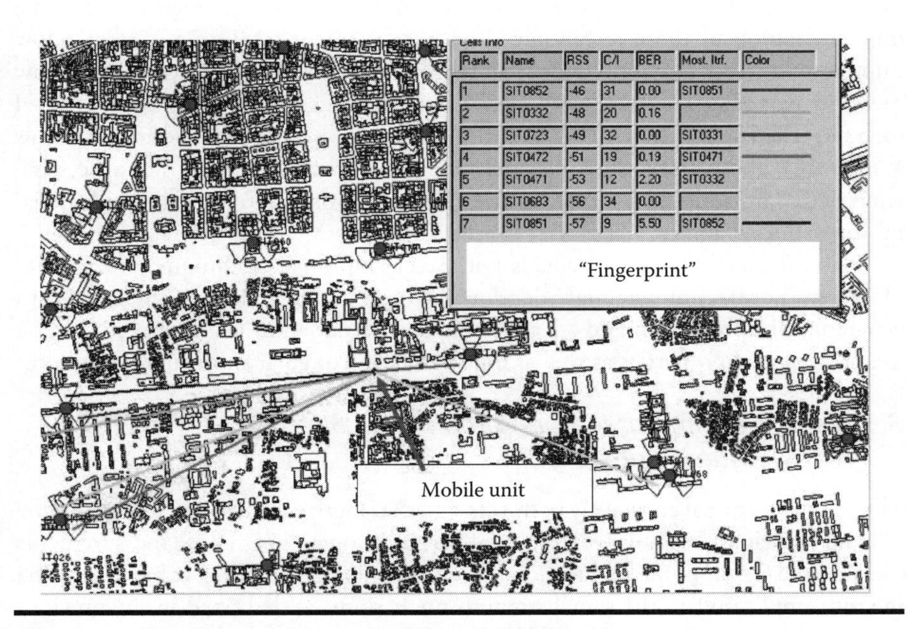

Rank	Name	RSS	C/I	BER	Most. Itf.	Color
1	SIT0852	-46	31	0.00	SIT0851	
2	SIT0332	-48	20	0.16		
3	SIT0723	-49	32	0.00	SIT0331	
4	SIT0472	-51	19	0.19	SIT0471	
5	SIT0471	-53	12	2.20	SIT0332	
6	SIT0683	-56	34	0.00		
7	SIT0851	-57	9	5.50	SIT0852	

Figure 3.3 RSS fingerprint of a mobile unit at a location.

is also measured. All those measurements provide a lot of information about the terminal environment, which can be deduced by matching the measured pattern to a pattern calculated or measured previously. We shall not elaborate further on those issues, but we will surely see their implementation in the future.

3.3.3.6 Location Based on Cell ID and Sector Azimuth

The most elementary location method is the base station identification number, known as the cell ID. The cell ID is transmitted by the cell and reported by the terminal in every report. Once the cell ID is known, the general geographical location can be easily identified, which is quite sufficient for a large number of the location-based applications (e.g., location-based advertising). The cell ID, together with the propagation delay information and the sector antenna direction, can narrow down the possible area where the terminal could be located. Figure 3.4 shows an example of such an area bounded by the sector width in angle and the propagation delay limit. As one can see in the figure, this area is quite large and does not provide an accurate location. It should be noted that the cell coverage area is far from being contiguous. Figure 3.5 demonstrates that. Each color in that figure indicates the coverage area of another cell. Shadowing and street canyons make the coverage areas of neighboring cells interlace. It should also be emphasized that for higher floors the coverage might be quite different. Thus, a cell ID indication does not necessarily imply a bounded contiguous area in which the terminal can be found.

3.3.3.7 Ways to Enhance Location Ability

It is possible to use external sources of information in order to enhance the location ability. A terminal reporting a large number of handovers is indicative of a vehicle.

Figure 3.4 Area determined by cell ID plus propagation delay.

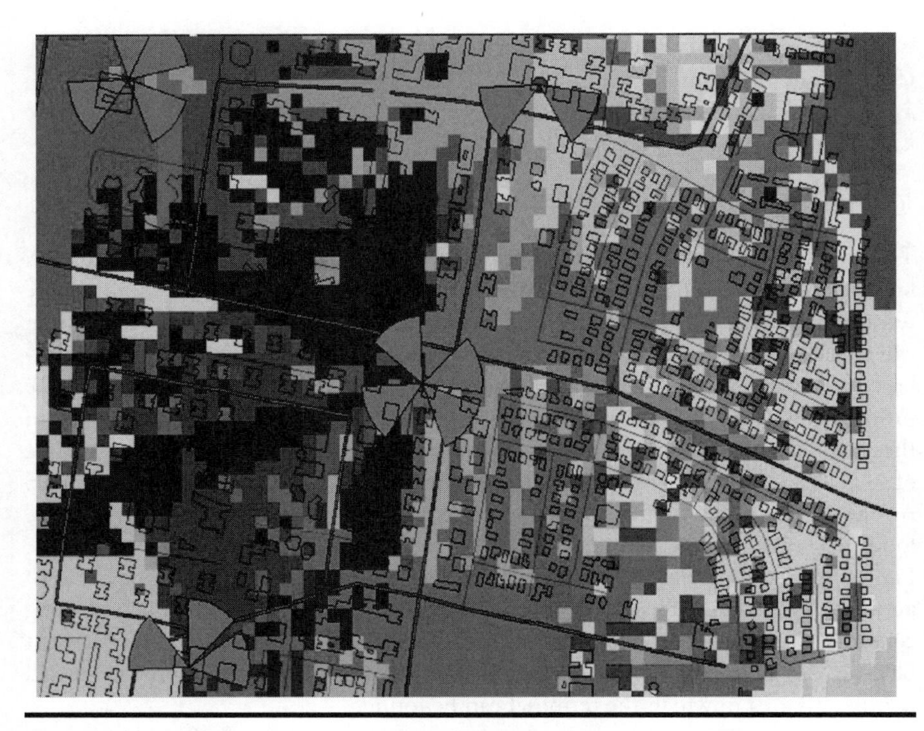

Figure 3.5 Coverage area of cellular cells in a typical urban area.

Road data can be used in order to associate the vehicle to a road, and then smooth the measurements by tracking it along that road.

As mentioned earlier, one of the key factors to a successful virtual drive test is the ability to locate terminals indoor. A terminal reporting an indoor base station (a picocell or a femtocell) as the serving cell is indicative of an indoor terminal, with a location in the vicinity of that cell. The fact that the coverage of such a cell is small provides quite an accurate estimate of its location. Sophisticated location algorithm, using the 3D coverage database, can position a terminal within a floor in a building, as shown in Figure 3.6. In this figure, the location of a specific terminal is within a building. The darkest color shade represents the highest probability of location. Some indoor picocell deployments also include a distributed antenna system (DAS). The DAS makes it more difficult to identify the floor, as the coverage of all the floors is uniform. External signals, from macrocells, that might be picked up by a terminal may serve as an indication of a low or high floor location.

As in building terminals served by outdoor base stations, a trial to test the accuracy of locations based on RSS measurements was performed. Location estimations were made for 21,276 calls, with a mean error of 97 m. This is a result of non-calibrated 3D predictions. A histogram of the result is shown in Figure 3.7. The abscissa is the error in location, compared to the actual location of the terminal making the call, while

Figure 3.6 Terminal report location.

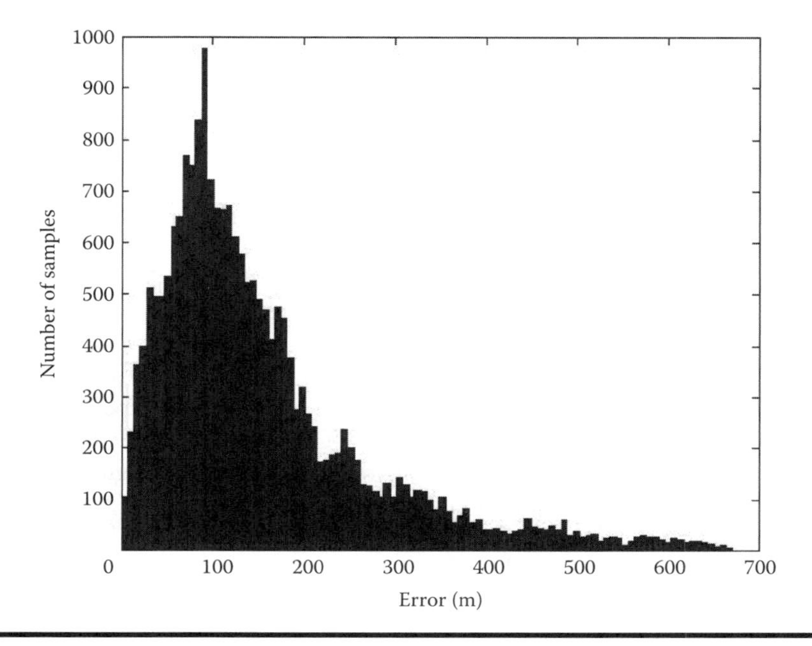

Figure 3.7 Error histogram of indoor location.

the number of calls within each location error bin is shown on the y-axis. Although the error seems to be quite a large one, compared to the average size of buildings, it is quite a good result considering that those are non-calibrated measurements, taken at relatively low SNR conditions.

3.4 Mapping the Measured Information

The first step of any treatment is diagnostics. Similarly, mapping the VDT results provides insight into the system performance and may direct the RF engineer to identify problematic regions. Such mapping may include traffic and service maps, coverage maps (such as E_c, E_c/I_o for a UMTS system), coverage problems (e.g., pilot pollution) and special event maps (call drops, Inter RAT handovers), including the reasons for those events.

Two types of maps can be generated based on the VDT results. The first is the image map type, which is a map depicting aggregated performance per bin (mean, median, minimum, or maximum). The advantage of such a type of maps is that those maps provide a quick and clear view of the area of interest and focus the engineer on problematic spots. The second one is a point map type—such maps generate a GIS database, where each VDT measurement is represented by a symbol, and which contains all the relevant information recorded for that measurement. Thus, by clicking that point the engineer can get this information and perform a drill-down analysis, required to detect the potential problem (e.g., a polluting cell that is responsible for drops in an area).

3.4.1 Map Types

In this section, a few of the large variety of maps will be demonstrated. Naturally, space is limited only to a small number of examples.

3.4.1.1 Traffic and Services Maps

The goal of the network is to serve its customers, and the first step would be to find out where they are, or rather what is the user density. Mapping the reported events is a very good indication of that. The events to be mapped can also be filtered according to service type, or height within a building, for indoor calls, etc. Figure 3.8 shows an example of such a map, showing high floors (20 m to 100 m) HSDPA traffic for the Port Authority building and its surroundings in New York City. The buildings are colored by the absolute number of calls recorded in the area, as indicated by the legend. Note the very high number of such calls in the Port Authority building itself, relative to the high-rises around it.

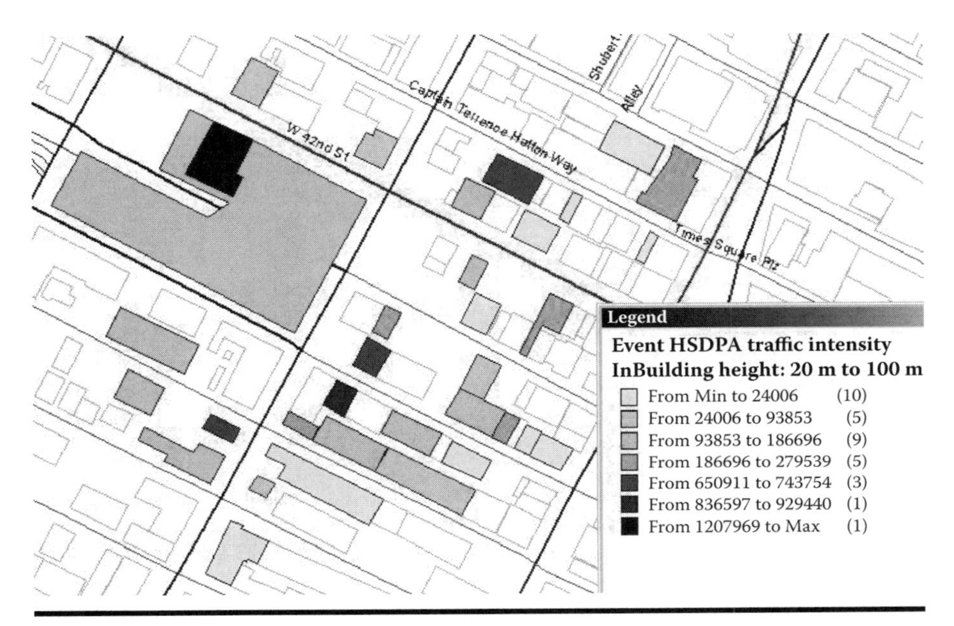

Figure 3.8 In-building HSDPA traffic, NY.

3.4.1.2 Coverage Maps

Coverage is, of course, the means to provide service to the user. An image map displaying the received pilot strength of a UMTS system, expressed as the chip energy (E_c), is shown in Figure 3.9. Each point in the map is colored according to the measured chip energy.

3.4.1.3 Special Events Maps

The terminal can be programmed to send a report as a response to some abnormal events that may occur. In case of a UMTS system, those events can be a detection of a call drop, pilot pollution, handover to another radio access technology (known as inter RAT handover) or similar events. A powerful tool for diagnostics is the mapping of those events. Figure 3.10 is an example of such a map. It provides the percentage of dropped calls per bin. An area where this value is too high certainly indicates a problem in the network. Another example is given in Figure 3.11. The map shows the estimated location of each terminal report, or event, filtered to show only those calls that were transferred from UMTS to a GSM technology. Each event is colored according to the E_c/I_o level. Each point is presented by a symbol, representing the identification of the terminal as done by the tool. This can be an indoor terminal, outdoor terminal, or a vehicle on the move. An example of the possible drill-down results is given by the window next to the map.

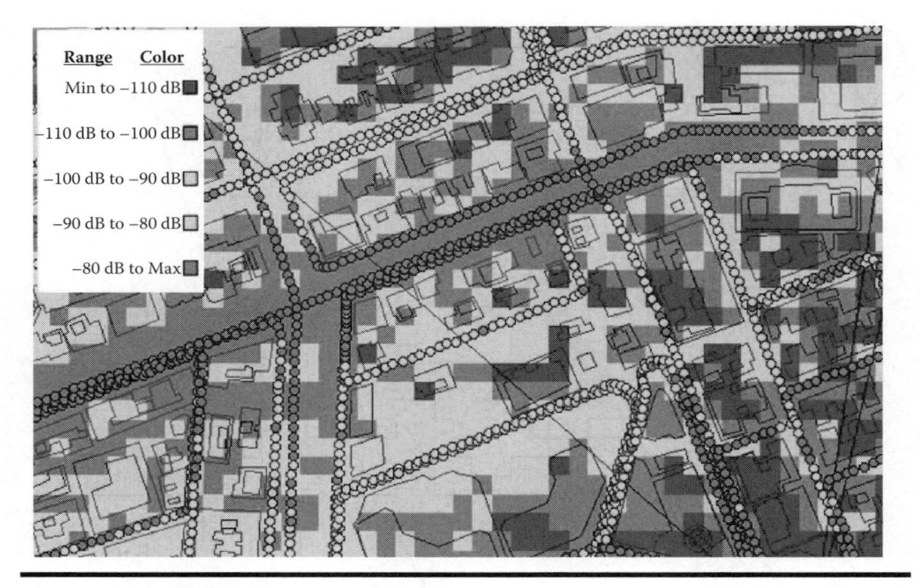

Figure 3.9 Chip energy(E_c) coverage map.

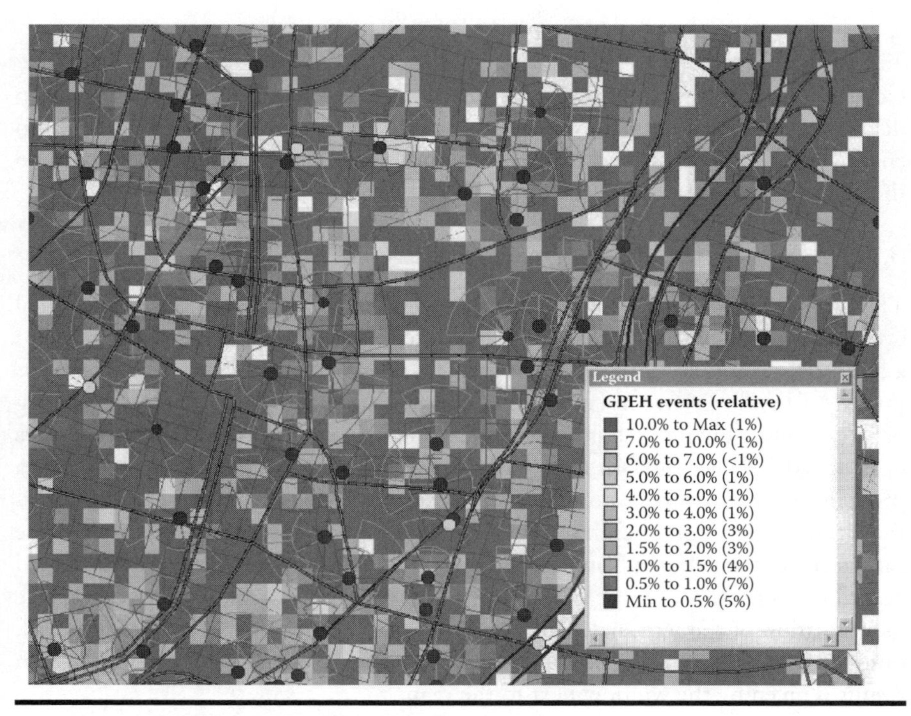

Figure 3.10 Percentage of dropped calls per bin.

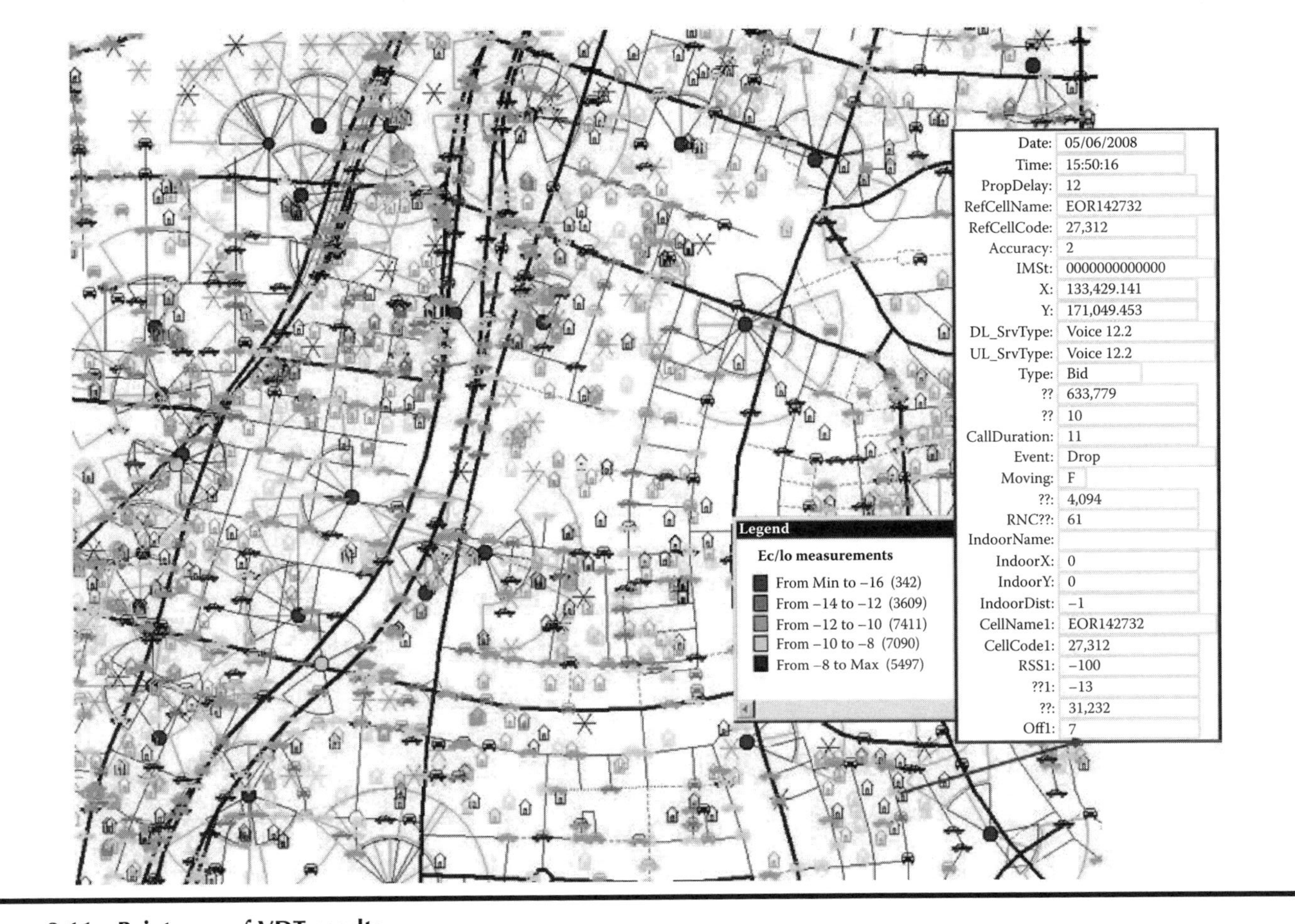
Date: 05/06/2008
Time: 15:50:16
PropDelay: 12
RefCellName: EOR142732
RefCellCode: 27,312
Accuracy: 2
IMSt: 0000000000000
X: 133,429.141
Y: 171,049.453
DL_SrvType: Voice 12.2
UL_SrvType: Voice 12.2
Type: Bid
?? 633,779
?? 10
CallDuration: 11
Event: Drop
Moving: F
??: 4,094
RNC??: 61
IndoorName:
IndoorX: 0
IndoorY: 0
IndoorDist: −1
CellName1: EOR142732
CellCode1: 27,312
RSS1: −100
??1: −13
??: 31,232
Off1: 7
Legend
Ec/lo measurements
From Min to −16 (342)
From −14 to −12 (3609)
From −12 to −10 (7411)
From −10 to −8 (7090)
From −8 to Max (5497)

Figure 3.11 Point map of VDT results.

3.5 Network Optimization Using VDT

Once terminal reports are mapped, the VDT can be used for optimization. Optimization, as defined in [4], involves monitoring, verifying, and improving of the radio network. While monitoring and verification may include network consistency checks and a coverage database update, network improvement involves tuning a set of parameters such that the network performance, as evaluated by the KPIs, improves.

The optimization problem is not a simple one, and involves tuning a very large set of parameters, such as base station locations, frequency and code planning, cell configuration, antenna direction, power control, handover parameters, and more. The tuning of the parameters should be made in order to increase one or a combination of KPIs, which could be average SINR, maximum available rate, or even financial cost. The KPIs are measured or estimated over the optimization basis, which can be a set of measurements or predictions. A large body of work has been done in order to formalize the optimization problem or present tractable heuristics to solve it. Commercially available tools present other solutions, typically a proprietary heuristics, that solve partial problems. The reader may refer to [26–32] and references therein for further discussions on network optimization problems and techniques. Other chapters in the book also refer to this problem.

3.5.1 Optimizing for Measurements and Predictions—A Unified Approach

The VDT measurements can be used as the basis for the optimization, by which the KPIs can be assessed. However, by using a well-chosen combination of predictions and measurements superior performance can be achieved.

Measurements-based elements provide:

- Ultimate accuracy of inter-cell interference. This is probably the most important feature of measurements-based optimization. In UMTS, WiMAX, and LTE networks, reducing the radio interference from other cells is the main optimization step. In order to perform that, we have to estimate the relative power of the two cells (the reference and interference). If this is done based on cells predictions, the resulting power difference can be very inaccurate and therefore dictate irrelevant optimization recommendations. On the other hand, measured data have accurate knowledge on the relative power measurement, as experienced by the user, thus providing an excellent basis for the optimization engine.
- Accurate location of calls and traffic distribution, which is crucial for the effectiveness and correctness of the optimization command.
- Event-driven optimization (i.e., attaching to each analysis point a weight which is dictated by the event that generated it). This way the cellular operator can

perform optimization that is aimed to reduce the drop rate in an area or to be engaged in a more subtle operation like handovers from UMTS to GSM.

Prediction-based elements are necessary to complete the weak points of the measured data:

- Those elements represent and reflect the impact of areas not covered by the network. This is the key contribution of those elements to the optimization process.
- They emphasize the importance of roads (relative to their weight in the total traffic). The relative importance of a drop while driving is significantly higher compared to such a drop indoor. Cellular operators learned from using mobile measurements for GSM optimization that using measured data alone degrades the overall performance on roads. One way to mitigate this effect is to use propagation prediction-based elements that increase the significance of the roads coverage.
- Predictions can be used to detect areas where measurements cannot be properly made (e.g., scrambling code collision in UMTS or subcarrier collisions in LTE and WiMAX). This issue is not as clear, but still very important.

The combination of the measurements, be they a virtual or real drive test, and predictions, can be done using a simple filtering technique, weighing the prediction with measurement results by the expected accuracy of each technique.

3.5.2 Optimization Procedures

In order to improve system performance, many parameters must be tuned. In this section some optimization procedures are presented.

3.5.2.1 Network Consistency Analysis

The cellular network is a complex system, made up of a very large number of components and elements. Base station deployment might be erroneous and so is the record of the deployment details within the database describing the network. Wrong cell location, wrong sector azimuth, and cross feeders (connection of one sector radio equipment to the antenna of another sector by mistake) are typical examples of such problems. The measurements as acquired during real and virtual drive tests are an excellent source of information for such errors. Consider for example a drive test taken in Tel-Aviv. In Figure 3.12, each cell in the network is presented as a point on a graph showing the average location error of the cell as a function of the number of measurements made in the cell. In that graph of the cell, the identification number of which is 30382, stands out as a cell with a relatively low number of measurements and a large average location error. This indicates that the coordinates of that cell are wrong.

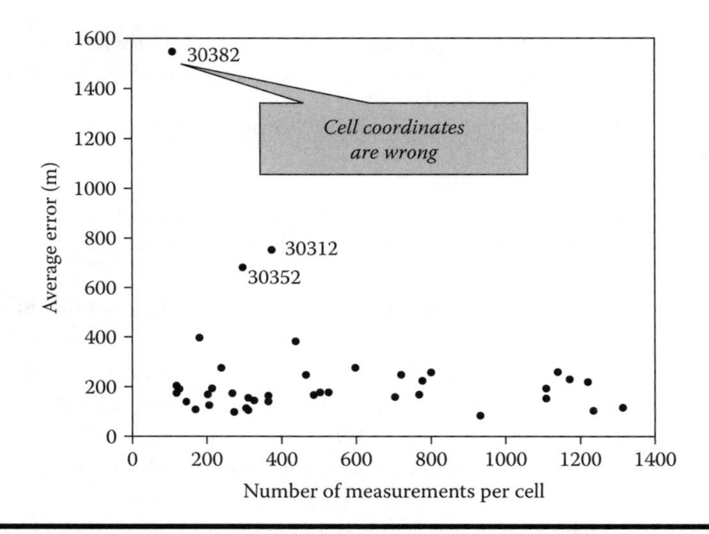

Figure 3.12 Average error versus number of measurements per cell, Tel-Aviv.

3.5.2.2 Neighbor Planning

One of the most sensitive processes that take place in a cellular network is the handover process, by which a mobile switches its serving cell (or cells). The starting point of the handover process is the neighbor list transmitted by a cell, indicating to all the terminals it serves which cells can be candidates to handover to. List optimization means trimming out cells to which the probability of successful handover is low and populating it with cells with which the serving cell has a large overlapping coverage area. Thus, mobiles and network resources are not wasted in futile scans and fruitless handover attempts.

The optimization process is based on assessment of the overlap between two cells. This overlap can be studied both by predictions and by measurements and reports of the mobile terminals. Proper weighting should be given to predictions versus measurements, and also to the different types of traffic encountered. For example, a bin representing a road segment may carry relatively very little traffic; however, it is important that the correct neighbors are available to the terminals traveling at that road. It should also be noted that, in principle, neighbor relation is unidirectional, namely if cell A includes cell B in its neighbor list, it does not imply that cell B would contain cell A in its list. A typical example is the case where cell B is down a one-way street from cell A. As long as predictions are not used for optimization, the actual geographical positioning of each measurement is not so crucial. However, when predictions and geographical information must be integrated into the process, the positioning becomes important and the full power of the VDT concept comes into play. Using VDT enables you to use the neighbor planning process even for predicted points, thus enabling you to propose new neighbors even in cases where the relation was not defined a-priori, meaning actual measurements of the neighbors are missing.

3.5.2.3 Frequency Planning, Scrambling, and Permutation Codes

The frequency channels, scrambling codes, and permutation codes are instruments by which interference from one cell can be isolated from the other. Of course each type of system, be it GSM, UMTS, LTE, or WiMAX, uses each of those instruments differently. While frequency planning is almost the only means of interference isolation in GSM, UMTS, and other CDMA systems use scrambling codes. In LTE and WiMAX, fractional frequency reuse has been adapted, in which the channels are partitioned into non-interfering segments that have to be allocated as well. Another mechanism found in WiMAX, similar to the frequency hopping mechanism of GSM, is the interference averaging mechanism. In WiMAX this is done by using different permutations of the OFDMA subcarriers. The permutation base that governs those permutations must be controlled as well.

Optimization of those parameters is based on the concept of the impact or interference matrix. An entry in this matrix describes the impact of interference of one cell on the other. The impact can be described in terms of the area loss or traffic loss at the victim cell as a function of the frequency channel offset (or code offset, in other cases). The loss can be assessed using measurements results, predictions, or, as described earlier, a combination of the measurements and predictions. Once a combination is used, geographical positioning of the measurements is necessary. The planning algorithm then works to find a frequency and code plan that would reduce the total impact of interference to minimum.

3.5.2.4 Cell Configuration

By "cell configuration" we mean the set of parameters referring to the antennas and radio deployment. Those include the number of sectors, antenna types, antenna direction in azimuth and tilt and antenna heights, as well as the cell transmission power. Tuning each of those parameters is a trade-off between the coverage of a serving cell and the interference to other cells. Tilting an antenna down, for instance, reduces the interference to far-away cells but reduces the coverage in the high floors of a nearby building.

For this optimization, positioning the measurements is essential even if the optimization is based on measurements alone. The effect of a configuration change, such as tilting the antenna, may depend very strongly on the location of the terminal and whether 2D or 3D localization is performed. The issue is demonstrated in Figure 3.13. The figure illustrates a case where an outdoor base station covers a building. On the left hand side of the figure, tilting down the serving cell antenna may improve the reception of a mobile on the first floor or on the ground outside of the building; however, it may cause a loss of coverage of the high floor terminal. If the optimization had been based on drive test alone, the impact of the lost coverage

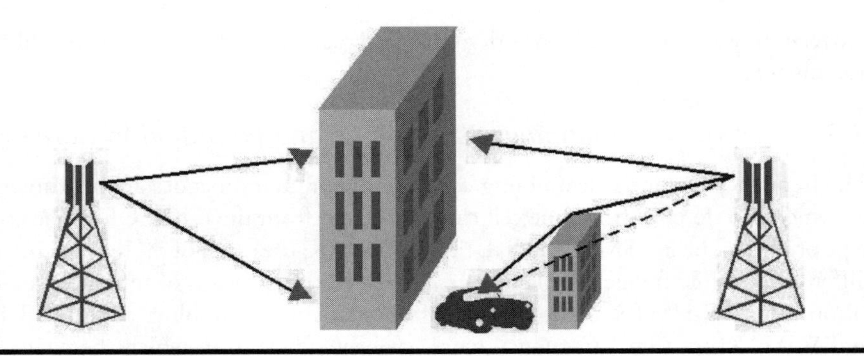

Figure 3.13 Effect of antenna tilt on high floor coverage.

would not have been taken into account since drive tests are limited to ground level only. The right-hand side of Figure 3.13 illustrates the case where antenna down tilt would not improve the reception condition of the terminal at the street, as the path is actually a diffraction path across a building. To capture such a case, a 3D propagation model should be used. The model, calibrated by the VDT or actual drive test result, would better predict this situation.

3.5.2.5 Load Balancing

The load of the cells, in terms of the number of calls, or amount of data traffic served by the cell, can be studied straightforwardly from the cell report, even without using specific terminal measurement reports. Using cell reports, load balancing can be controlled by means of handover thresholds or access control. However, in many cases a simple rotation of the base station sites could balance the load without tweaking with parameters that might affect other base stations. For this purpose, the positioning of terminals is, again, essential.

3.5.3 Requirements of Location and RSS Measurement Accuracies

As explained earlier, the location measurement process is not accurate, nor is the RSS measurement process. The impact of those errors on the VDT optimization results is not straightforward. While a single measurement may be erroneous, the overall effect is typically a result of averaging a large number of measurements, so the effect of random errors cancels out. Furthermore, the parameters being optimized are not always so sensitive to individual location or RSS measurement errors, so the overall impact on network performance is marginal. No major network change will be performed without gathering sufficient statistics such that the confidence on the conclusions is high.

3.5.4 Demonstration of an Optimization Project

An optimization project, based on the VDT principles was performed in Singapore. The network was a UMTS network of Singapore's main business district. It involved 58 outdoor cells and 71 indoor cells. No drive test results were available. Four days worth of statistics was gathered. Approximately 320,000 calls were located and used as analysis points for the optimization. An example of the mapping of the VDT results is given in Figure 3.14, which shows the mapping of drop events. Each located event is depicted as a symbol. The height of the terminal above ground is given as a label next to each symbol. A drop in a moving car can be easily spotted, as the event is positioned on a road, with a height of 1 m. Other drops were positioned indoor with different heights. One of these is marked in Figure 3.14. This terminal was located at a height of 21 m above ground, as demonstrated in the subwindow within Figure 3.14.

The cell configuration optimization process resulted in 48 proposed changes in 27 sectors, out of which there were:

- 22 azimuth changes
- 18 electrical tilt changes
- 8 mechanical tilt changes

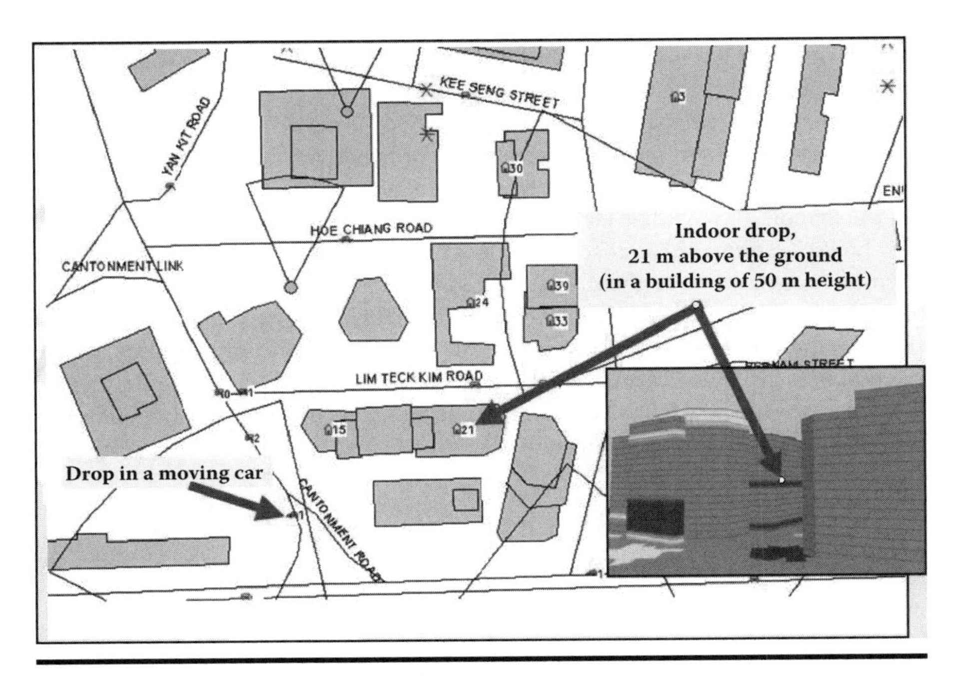

Figure 3.14 3D drop mapping—Singapore.

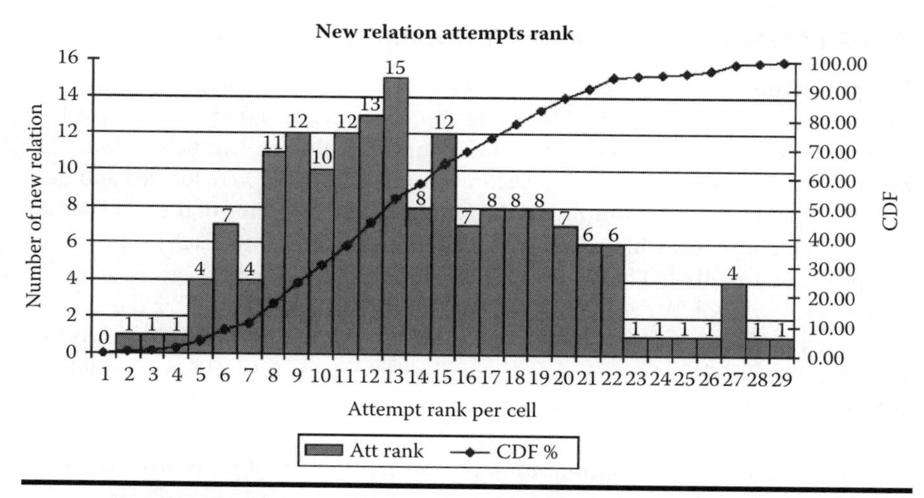

Figure 3.15 Neighbor planning result—histogram by the handover attempt rank per new relation.

Neighbor planning yielded 182 new handover neighbor relations, and proposed the deletion of 72 existing relations, as no handover attempt took place between the neighbors sharing that relation. Figure 3.15 displays a histogram showing the result of the neighbor relation optimization process. The ordinate of the histogram is the rank per attempts of the new proposed relations, and the coordinate is the number of proposed relations that achieved this rank. For example, one of the proposed relations had a rank of 2, namely the number of handover attempts made between the cell in question to the proposed neighbor, which is the second largest of all this cell's neighbors. Twenty five percent of the proposed relations were among the nine top choices, in terms of handover attempts. All the proposed relations had handover attempts associated with them. All these data show that the proposed neighbor planning improves the existing network considerably. The key performance index (KPI) for the optimization process was the call drop rate and the traffic volume. Before the process, the network had a call drop rate of 0.66% and after the optimization, the rate went down to 0.55% on a 16% improvement. In addition, the traffic rate went from 69,300 calls per hour up to 76,300 calls per hour—a 10% increase.

3.6 VDT for LTE and Fourth-Generation Systems

As the demand for mobile services grows and technology evolves, new systems are introduced to the market. At the time this book was written, the WiMAX system [15,16] and the long-term evolution (LTE) system [13,14] had made their first steps into the market. Future fourth-generation systems are being developed under the ITU initiative of IMT-advanced, according to the roadmap and vision as described in [2] and the requirements of the new-generation systems as specified in [33]. Although the detailed specifications of IMT-advanced systems have not been made yet, some of their main features are known. The effect and implications of the main features of those systems over the VDT concept will be given next.

3.6.1 Advanced System Main Features

3.6.1.1 Flat All-IP Architecture

The architecture of both LTE and WiMAX systems is, in principle, a flat all-IP architecture, and this is expected to be the type of architecture adopted for IMT-Advanced as well. Unlike the architectures of third- and second-generation cellular systems, a flat all-IP architecture is geared toward data communication and is based, as much as possible, on Internet protocols. There is no controlling entity such as the base station controller or radio network controller found in GSM or UMTS, but rather the control of the base station may also be a distributed entity.

3.6.1.2 Wideband MIMO-OFDM Air Interface

The need for services ranging from Voice over IP to video streaming, gaming, and other high-throughput consuming applications requires that next-generation systems support a higher datarate with flexibility and adaptability enough to provide for the need of a variety of users. The physical layers employ wideband channels, using orthogonal frequency division multiple access (OFDMA), which provides it with a high resilience to multipath, scalability, and flexibility in resource assignments.

An important technology integrated into the physical layers of all of these technologies is that of MIMO. The name encompasses a variety of techniques that make use of multiple antennas in the transmitter, receiver, or both. Those techniques include a receive or transmit diversity to overcome adverse fading caused by multipath, beamforming that enables focusing the transmitted or received energy to wanted directions and avoid radiation to or from unwanted ones, and spatial multiplexing that makes use of the rich scattering environment creating by that multipath in order to increase throughput. The systems are to be adaptive and select the appropriate technique (or combination of techniques) since the environment and reception conditions are not always adequate for each.

3.6.1.3 Usage of Multi-Carriers

For a high data rate, high bandwidth is needed. Even with the additional spectral efficiency offered by MIMO and OFDMA techniques, wider spectrum allocations would be necessary. Spectrum is always a problem for wireless allocations being that it is scarce and, because of the necessity to support legacy systems, it is often noncontiguous. The advanced systems are designed to be capable of operating simultaneously over a number of frequency channels, even if they are not adjacent.

This feature is the key to the flexibility needed to comply with different regulatory regimes and the requirements present in different markets. It also provides trunking efficiency—as a larger bandwidth is available to a large number of users, high multiplexing gain is achieved.

3.6.1.4 Massive Usage of Picocells, Femtocells, and Relays

The cellular concept, reusing available resources in small cells, is probably the most important factor in making the cellular system as ubiquitous and widely used as it is today. The reuse principle provides the cellular systems with increased capacity and enables them to support a large number of users. Even with the increased spectral efficiency of modern systems, a further increase in the overall capacity is expected to come from yet another step in that direction. This means that advanced systems should be made with an even larger number of smaller cells, each covering a relatively small area. These include picocells, which are small cells covering indoor public areas or high-density locations, such as shopping malls, train stations, etc., and femtocells, which are even smaller devices, covering a single residence and supporting a very small number of subscribers. The femtocells, expected to be owned and even installed by the end user, represent a real revolution in the concept of wireless cellular network architecture and operation concepts. However small, a femtocell relieves the network from load that might require much higher resources if served directly from a macrocell.

Another network element that may find larger usage in advanced networks is the relay. Simple relays can be found in present-day networks to provide coverage solutions in areas where a full base-station installation is not cost-effective. Advanced relays can serve as capacity enhancement, as well as coverage enhancement devices. By improving the reception conditions of a user group, they make it possible to serve them with less resources than would have been required from a direct link to the base station, and without the need for a full-fledged base station to be installed.

3.6.1.5 Cooperation among Network Elements

The cellular network has always been more than a group of cells working independently of each other. Cooperation between cells is necessary for handover. The introduction of soft handover made the necessary cooperation even more intense. However, advanced systems are planned to take this cooperation even further, turning the set of base stations into a large virtual array, thus enabling spatial multiplexing in both the downlink and uplink directions. Coordinated scheduling, interference mitigation, etc., are additional techniques that can enhance system capacity with different degrees of cooperation. The requirements on backhaul and communication protocols between base stations are obviously quite large.

Femtocells add another dimension to the problem. The location of a femtocell is not known a-priori and as a result some of the information needed to configure it is not always available. These include the operating frequency channels, neighbor lists, and other operational parameters. As femtocells are to be self-installable, it is necessary that they include the capability to configure themselves. A network supporting such a capability is a self-organization network (SON) or self-optimization network, and may require further cooperation between the base stations.

3.6.2 VDT Concept in Advanced Systems

The first implication of the physical layer enhancements described in the previous section is the fact that the terminal is becoming a more accurate position sensor. The wide bandwidth of operation enables more accurate range determination, and the wider use of multiple antenna systems imply that spatial information is more readily available. Received signal strength (RSS) measurements, taken over a large number of carrier channels will be more stable as multipath effects are averaged out.

The smaller cells will also have a considerable impact. The fact that the cell coverage is small implies that the location itself can be determined more accurately. Note that femtocells, of which the location is not known, cannot be used for mobile station location determination. On the other hand, the VDT concept can be used for the purpose of locating femtocells, based on its measurements (performed during its initialization phase), and the measurements of mobile terminals that measure it. Thus the location can be determined coarsely and then refined in an iterative process. When the location is established, the femtocell readings can be used to pinpoint the location of the terminals it serves.

For VDT, the physical layer information measured by the terminal and base stations have to be collected and processed in order to provide a location measurement. As the network architecture is flat and distributed, there will not necessarily be a single point of interface (as the BSC in GSM or RNC in a UMTS system). To implement the VDT concept, a distributed architecture, like that of the system itself, should be implemented. This architecture, may rely on agents residing in the various network elements to collect the results and send them over the IP interfaces to the processing entity. The flat all-IP architecture is an enabler for such an implementation which makes the information accessible to even a larger variety of applications.

As cellular systems evolve, the number of base stations increases, cooperation among them grows, and capabilities develop, it is quite likely that the VDT concept will find itself part of the intelligence embedded within the base stations and other network elements. Thus, as part of the initial configuration, the intelligent base station shall be uploaded with the geographic information of its deployment area, signal predictions within its neighborhood, and the neighboring network topologies. This information can be used by the base station to locate the terminals it serves geographically, and then use that information, as described earlier for the global virtual drive test, to update the coverage database and optimize the network. Optimization could be achieved either by fine-tuning the various operation parameters of the base station cell itself, by updating the neighbor lists or by cooperating with neighboring base stations for interference mitigation, by common scheduling, or by providing service simultaneously with neighboring cells to some terminals.

3.7 Open Issues

The concept of VDT is indeed a powerful tool for optimization, still there are some open issues in its full implementation and usage. We will start from the most practical and continue with the more theoretical ones.

- **Measurement Availability**: Recording and gathering of the required information is not yet fully standardized and fully supported by all manufacturers. Especially, for systems of which the architecture is distributed, there is no single point of interface in which the information can be directly acquired. Standardization and practices should be established such that the application would be fully supported.
- **Automatic Filtering**: The amount of data recorded is huge and needs to be even larger if more information is to be gathered to enable more accurate location. Automatic filtering needs to be developed to enable the procedure focused on the most relevant measurements.
- **Location Accuracy**: This is the main precondition for a successful VDT to take place. More studies need to be done in order to learn what would be an acceptable accuracy for VDT. Sophisticated techniques are needed to provide location at the required accuracy. Those techniques should probably include the fusion of information from various sources, including physical measurements and geographical information.
- **Optimization Techniques**: The most challenging tasks are probably in the field of network optimization. The challenge starts with the study of the proper weights between predictions and measurements to form the basis of optimization. It continues with definitions of optimization criteria, which need to take into account a set of KPIs. Finally, an efficient algorithm that could find a viable solution to this multivariate problem would be certainly a step forward. It should be emphasized that such solutions should be reliable and implementable. The most difficult task is to every so often convince the network engineer that the proposed modifications to the network would improve its performance, rather than cause unexpected problems.
- **Online Optimization**: Once a reliable solution can be found, namely a solution that could be implemented without negative effects on system performance, it can be deployed automatically online. This would be a first step toward a self-healing and self-organizing network.

3.8 Summary

The virtual drive test, a concept that combines terminal reports with terminal positioning techniques, is a powerful tool for network optimization. The VDT provides timely and ubiquitous information about the network users, and thus it presents a real and living picture of network behavior and outlines problems that might exist.

The VDT process relies on accurate positioning of terminal reports. Such reports include the measurements made by the mobile terminal during various events. Terminal location can be based upon the cell ID, propagation delay measurements, and signal strength measurements. These data are typically the results of measurements performed by the terminals for normal operation; however, the reports do not always contain all the information. The reports should be frequent enough and contain the necessary data for localization. It is expected that various measurements made by antenna arrays, both at the mobile and at the base station, will be added to provide an even better means of location. Furthermore, a large and increasing number of terminals are equipped with GPS and thus are capable of providing accurate location information. Because cellular communication is ubiquitous and personal, a large part of the traffic takes place indoor. The location capability of a VDT engine should include the ability to perform indoor location.

VDT can augment the results obtained by predictions and drive tests, and for a mature system, it can replace them both. However, it is recommended that a combination of measured and predicted data is used for optimization in order to take advantage of both. Network optimization performed for a real network has shown that the VDT can indeed be used for that purpose. The process resulted in a significant improvement of the network. VDT can be viewed as a first step toward self-healing and self-organizing networks. It provides the necessary feedback branch between terminal measurements and network optimization modules. Once the optimization procedures are integrated online as part of the network management and control system, the network can have a self-optimization capability, and adapt better to changing demands and operating conditions. Undoubtedly, in future networks such as LTE, the VDT concept will be a large part of monitoring, maintaining, and optimizing future networks.

References

[1] D. Klaus (ed.), "Technologies for the Wireless Future (WWRF)," John Wiley and Sons, Chichester UK, 2008.

[2] ITU-R Rec. M.1645, "Framework and Overall Objectives of the Future Development of IMT-2000 and Systems Beyond IMT-2000."

[3] D. Blechschmidt, "Self-Optimized Networks—Making Use of Experts' Knowledge and Approved Planning," TM Forum, Nice May 2003 http://www.tmforum.org/MobilityMOB/1245/home.html

[4] A. J. Mishra, "Fundamentals of Cellular Netwrok Planning and Optimisation," John Wiley and Sons, Chichester UK, 2004.

[5] J. Laiho, A. Wacker, and T. Novosad, "Radio Network Planning and Optimisation for UMTS," John Wiley and Sons, Chichester UK, 2nd ed. 2006.

[6] T. S. Rappaport, "Wireless Communication, Principles, and Practices," Prentice Hall, 2002.

[7] A. F. Molisch, "Wireless Communications," John Wiley and Sons, Chichester UK, 2005.

[8] N. Blaunstein, "Radio Propagation in Cellular Networks," Artech House, 2000.

[9] M. Hata, "Empirical Formula for Propagation Loss in Land-Mobile Radio Services," *IEEE Transactions on Vehicular Technology*, 29(3), 1980.

[10] Y. Okumura et al., "Field Strength and its Variability in VHF and UHF Land-Mobile Radio Service," Review of the Electrical Engineering Communication Laboratory, 16(9–10), 1968.
[11] COST 231 Final Report, "Digital Mobile Radio Toward Future Generation Systems," COST Telecom Secretariat, Brussels.
[12] J. Zhang, D. Yang, "A Quantitative Error Analysis for Mobile Network Planning, Proceedings of the International Conference on Communication Technologies," ICCT-2003, Beijing China, vol. 1, pp. 115–117, April 2003.
[13] 3GPP TS 36.201: "Evolved Terrestrial Radio Access (E-UTRA); LTE Physical Layer - General Description, V8.2.0," December 2008.
[14] 3GPP TS 36.401: "Evolved Terrestrial Radio Access (E-UTRA); Architecture Description, V8.4.0," December 2008.
[15] IEEE P802.16-2009: "IEEE Standard for Local and Metropolitan Networks. Part 16: Air Interface for Fixed and Mobile Broadband Wireless Access Systems," January 2009.
[16] WiMAX Forum: "WiMAX Forum Network Architecture Release 1 v1.2," January 2008.
[17] H. L. Bertoni and J. W. Suh, "Simulation of Location Accuracies Obtainable from Different Methods," 62nd IEEE Vehicular Technology Conference, VTC-2005- Fall. 2005, vol. 4, pp. 2196–2200, 25–28 Sept., 2005.
[18] Han-Yu Lin, Shih-Cheng Chen, Ding-Bing Lin, and Hsin-Piao Lin, "Multidimensional Scaling Algorithm for Mobile Location Based on Hybrid SADOA/TOA Measurement," *IEEE Wireless Communications and Networking Conference, 2008*, WCNC 2008, pp. 3015–3020, April 2008.
[19] S. Mingyang, T. Xiaofeng, Z. Qing, and H. Xiao, "A Believable Factor and Kalman Filtering-based Mobile Localization Algorithm," *4th IEEE International Conference on Circuits and Systems for Communications, 2008.* ICCSC 2008. pp. 723–727, 26–28, May 2008.
[20] J. M. Borokowski, "Performance of CellID+RTT Hybrid Positioning Methods for UMTS," M.Sc. Thesis, Tampere University of Technology, January, 2004.
[21] E. Kapplan, "Understanding GPS: Principles and Applications," Artech House, 1996.
[22] M. Pent, M. A. Spirito, and E. Turco, "Method for Positioning GSM Mobile Stations Using Absolute Time-Delay Measurements," *Electronics Letters*, vol. 33, issue 24, pp. 2019–2020, Nov. 1997.
[23] J. C. Liberti, Jr. and T. S. Rappaport, Smart Antennas for Wireless Communications: IS-95 and Third Generation CDMA Applications, Prentice Hall, 1999.
[24] M. Hata, "Mobile Location Using Signal Strength Measurements in a Cellular System," *IEEE Transactions on Vehicular Technology*, vol VT-29, no. 2, pp. 245–252, May 1980.
[25] K. Y. Kabalan and J. L. Mounsef, "Mobile Location in GSM Using Signal Strength Technique," Proceedings of the 2003 10th IEEE International Conference on Electronics, Circuits and Systems, 2003. ICECS 2003, vol. 1, pp. 196–199, Dec. 2003.
[26] M. J. Nawrocki, H. Aghvami, and M. Dohler, "Understanding UMTS Radio Network Modeling, Planning, and Automated Optimisation: Theory and Practice," John Wiley and Sons, Chichester, UK, 2006.
[27] J. Jiang and A. Capone, "3G/4G/WLAN/WMAN Planning and Optimization," *IEEE Wireless Communication*, vol. 13, no. 6, pp. 6–7, December 2006.
[28] A. Eisenblatter and H. F. Geerd, "Wireless network design: solution-oriented modeling and mathematical optimization," *IEEE Wireless Communication*, vol. 13, no. 6, pp. 8–14, December 2006.
[29] I. Siomina, P. Varbrand, and D. Yuan, "Automated Optimization of Service Coverage and Base Station Antenna Configuration in UMTS Networks," *IEEE Wireless Communication*, vol. 13, no. 6, pp. 16–25, December 2006.

[30] F. Ricciato, "Traffic Monitoring and Analysis for the Optimization of a 3G Network," *IEEE Wireless Communication*, December 2006, pp. 42–49.

[31] R. Chavez-Santiago, A. Raymond, and V. Lyandres. "Enhanced Efficiency and Frequency Assignment by Optimizing the Base Stations Location in a Mobile Radio Network," *Wireless Networks Online First*, January 2007.

[32] F. Gordejuela-Sánchez, A. Jüttner, and J. Zhang, "A Multiobjective Optimization Framework for IEEE 802.16e Network Design and Performance Analysis," *IEEE Journal on Selected Areas in Communication*, vol. 27. no. 2, February 2009, pp. 202–216.

[33] ITU-R Report. M.2133: "Requirements, Evaluation Criteria, and Submission Templates for the Development of IMT-Advanced," (2008).

3G PLANNING AND OPTIMIZATION

Chapter 4

WCDMA Planning and Optimization

Xuemin Huang and Meixia Tao

Contents

4.1 Introduction

This section introduces the impacts and targets of network planning. By reviewing the most significant challenges such as traffic demand and QoS (Quality of Service) criteria, the key WCDMA (wideband code division multiple access) network planning process is explained.

4.1.1 Quality, Capacity, and Economic Issues of Network Design

The increasing demand for mobile communications leads mobile service providers to look for ways to improve the quality of service and to support increasing numbers of users in their systems. Since the amount of frequency spectrum available for mobile communications is very limited, efficient use of the frequency resource is needed. Currently, cellular system design is challenged by the need for a better quality of service and the need for serving an increased number of subscribers. Network planning is becoming a key issue in the current scenario, with exceedingly high growth rates in many countries that force operators to reconfigure their networks virtually on a monthly basis. Therefore, the search for intelligent techniques, which may considerably alleviate planning efforts (and associated costs), becomes extremely important for operators in a competitive market.

Cellular network planning is a very complex task, as many aspects must be taken into account, including the topography, morphology, traffic distribution, existing infrastructure, and so on. Things become more complicated because a handful of constraints are involved, such as the system capacity, service quality, frequency bandwidth, and coordination requirements. Nowadays, it is the network planner's task to manually place BSs and to specify their parameters based on personal experience and intuition. These manual processes have to go through a number of iterations before achieving satisfactory performance and do not necessarily guarantee an optimum solution. It could work well when the demand for mobile services was low. However, the explosive growth in the service demand has led to a need for an increase in cell density. This in turn has resulted in greater network complexity, making it extremely difficult to design a high-quality network manually [1].

Furthermore, WCDMA technology is the dominant solution for the third generation (3G) cellular systems. It has been adopted by most countries deploying the UMTS (Universal Mobile Telecommunication System) networks [2]. Similar to other technologies, the deployment of WCDMA networks poses the problem to select antenna locations and configurations with respect to contradictory goals: low costs versus high performance. A key to successful planning is the fast and accurate assessment of network performance in terms of the coverage, capacity, and QoS [3]. This also makes the conventional design methods insufficient for planning mobile networks in the future. Thus, more advanced and intelligent network planning tools are required. A promising planning tool should be able to aid the human planner by automating the design processes [4, 5].

4.1.2 Radio Planning Objective

The task of radio planning is to define a set of site locations and respective NodeB configurations that addresses the coverage and capacity figures derived from dimensioning. The starting point in radio planning are the outputs from the dimensioning, especially the calculated site densities in each clutter type. The site count derived in radio planning often differs from the site count derived from dimensioning since the actual site coverage may differ significantly from the assumed empirical model(s). There is always a risk that the planned site count may exceed the estimated site count from dimensioning. As a result, several planning iterations are needed to reach a reliable figure.

One problem with radio planning is dealing with site density. Firstly, higher site density poses more difficulty in finding suitable candidates. This is true in all clutter types. In dense areas, most suitable sites are already overcrowded with 2G (the second generation mobile system) antennas. This will likely put the WCDMA antennas in less ideal positions. Secondly, there is a tendency that the candidate sites will not have comparable heights. This is a major drawback in radio planning since large differences in heights can distort the site dominance areas and cell ranges. The third problem is the bandwidth constraint which may require tighter frequency reuse. In this case, the radio plan must be as close to the ideal network to achieve good performance.

Radio network planning normally follows the dimensioning exercise. Sometimes the dimensioning process includes a rough plan to justify the site count and coverage level using some commonly accepted propagation model and generic WCDMA system modules in the planning tool. In the actual planning phase, a number of inputs are needed in order to improve the quality and accuracy of the radio plan. Depending on the selected planning tool to use, a number of inputs may be required to be fully utilized by the tool. For example, it is assumed that the following items are already well considered:

- Propagation characteristics of various areas (propagation models tuned)
- Required inputs defined (clutter maps, terrain maps, building data, etc.)
- Traffic and demographic information (i.e., per clutter type)
- WCDMA RF equipment parameters defined (antennas, RF features, etc.)
- Options for NodeB configuration (sectorized, omni, multichannel)

Two important decisions with regards to radio planning have to be considered prior to the actual planning exercise. First, the level of accuracy when it comes to coverage and capacity needs to be considered and this highly depends on the accuracy of the propagation model in the planning tool. Second, the planner needs to decide how much RF optimization will be undertaken during the planning phase. This is only possible if the planning tool together with the planning parameters and equipments models is accurate enough. It is often the case where optimization is neglected during the planning process. Post-planning optimization exercises are often costly and produce only minor improvements. It is often limited to antenna adjustments (tilting and azimuth changes).

A number of features are useful when selecting a planning tool, such as:

- Support of GSM/UMTS co-planning
- Optimal site selection—when existing or candidate sites are provided
- Support of mixed and multiple propagation models
- Support of model tuning and user-defined models
- Support of WCDMA planning features, like pilot planning and code planning
- Optimal antenna configurations (e.g., downtilting)
- Definition of mixed traffic scenarios
- Network performance simulation including semi-dynamic simulation and full dynamic simulation

A number of commercial planning tools are available in the market and some are widely used by network operators. The major factor that determines the usability of the tool is the accuracy of the RF modeling, such as propagation models, antenna configurations, interference prediction, frequency allocation, and channel models. Planning tools with WCDMA traffic models for capacity planning are advantageous.

4.1.3 WCDMA Network Planning Process

WCDMA radio planning involves a number of steps ranging from tool setup to site survey. The process is similar to any wireless network. What differs between WCDMA and other technologies are the actual site configuration, KPIs (key performance indicators), and the propagation environment since WCDMA may support mobile and fixed users where the latter may employ directional/rooftop antennas.

The final radio plan defines the site locations and their respective configuration. The configuration involves antenna height, number of sectors, assigned frequencies or major channel groups, types of antennas, azimuth and downtilt, equipment type, and RF power. The final plan will be tested against various KPI requirements, mainly coverage criteria and capacity (or signal quality). Figure 4.1 can be used as a guide in developing a planning process. The planning process also largely depends on the planning tool used.

The planning process in Figure 4.1 includes a drive test and verifications after the site survey. This procedure is not mandatory for all sites if the site count is

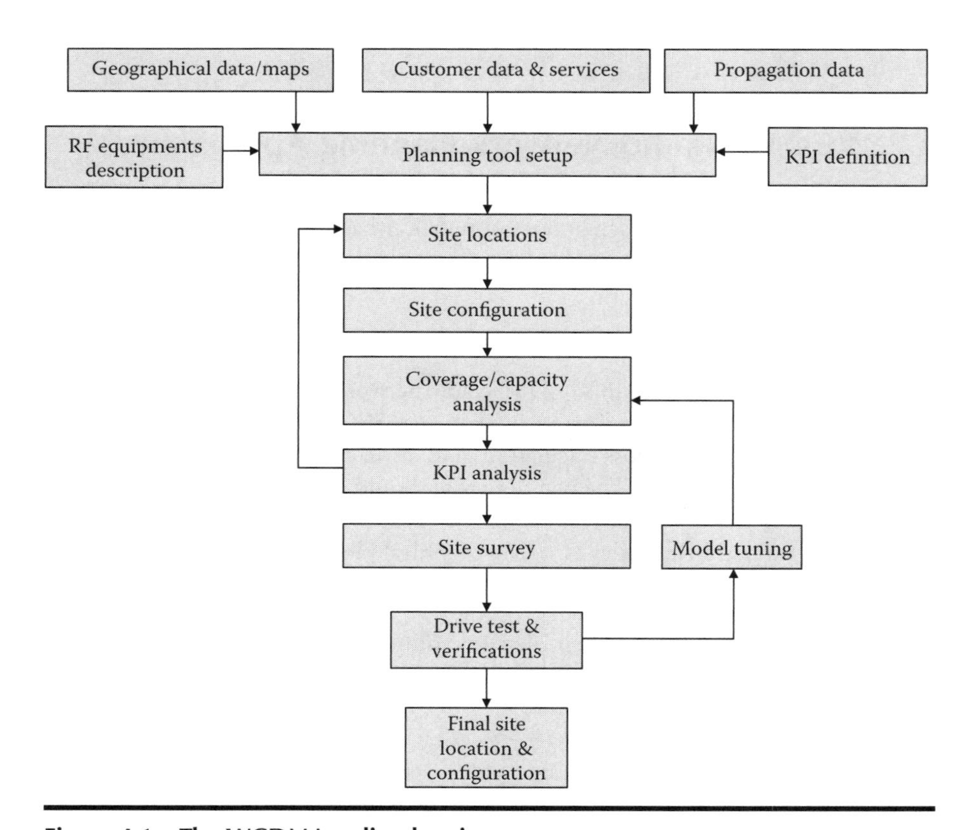

Figure 4.1 The WCDMA radio planning process.

too many. Usually, the site survey and KPI analysis give an indication of which areas are expected to have poor RF quality and which sites are involved. This is usually done when the candidate site(s) are not located in ideal locations or if the site survey finds some discrepancies with the candidate(s).

4.1.4 Challenges in WCDMA Network Planning

The differences between WCDMA and GSM radio planning are discussed in [6]. WCDMA radio offers significant processing gain due to spreading. These features are only exploited when the signal quality demands more processing. To support high data rates, the radio plan must offer very good E_b/N_o even with very big noise rise.

Another consideration in the case of WCDMA planning is the high E_b/N_o requirements to support high data rates. Although a site is expected to support high data rates for mobiles closer to it, E_b/N_o values >20 dB are only possible in the absence of interference. This requires accurate modeling of the propagation and RF equipments.

4.2 WCDMA Radio Network Planning Approaches

There are two fundamental approaches to WCDMA radio network planning. These are the path loss-based approach and the simulation-based approach. The path loss-based approach is the simplest and has been adopted by the majority of 3G operators.

4.2.1 Path loss-based Approach

The path loss-based approach to WCDMA radio network planning can be completed using a 2G (typically GSM) radio network planning tool. The planning tool must be capable of completing path loss calculations and displaying areas where specific path loss thresholds are exceeded. The planning tool should also be capable of displaying best server areas and, optionally, it should be capable of displaying downlink C/I. In each case, numerical statistics as well as graphical plots should be generated. The inputs to the planning tool for the path loss based approach are:

- 3G site candidates with their physical configuration (antenna type, antenna height, antenna tilt, antenna azimuth, feeder type, and feeder length)
- Propagation model
- Digital terrain map
- Link budget signal strength thresholds

The 3G site candidates should be those that have been selected according to the criteria specified in Section 4.2.3. They may be 2G sites that are being reused for 3G

Table 4.1 Example Translation of a 3G Link Budget Result to Planning Tool Signal Strength Threshold

Link budget result for maximum allowed path loss	140 dB
Downlink transmit power configured in the planning tool	33 dBm
NodeB antenna gain assumed in the link budgets	18 dBi
Feeder loss assumed in the link budgets	2 dB
Planning tool signal strength threshold	−91 dBm

or they may be greenfield sites that are being introduced for 3G. The propagation model should be tuned from measurements according to the recommendations in Section 4.2.4. Propagation model tuning should account for inaccuracies in the digital terrain map (e.g., areas shown to be rural clutter type, which are actually suburban clutter type). The link budget signal strength thresholds should be based upon a set of 3G service and CPICH link budgets. Guidance for WCDMA link budgets is provided in Section 4.2.5.

The WCDMA link budget results must be adjusted prior to being used within the planning tool. Link budget results are generated in terms of the maximum allowed path loss, whereas the majority of planning tools display contours of signal strength. This means that a relatively arbitrary NodeB transmit power must be selected, and then a signal strength threshold computed by subtracting the link budget maximum allowed path loss. The NodeB antenna gain and feeder loss must also be taken into account. It is common for the arbitrary transmit power to be selected to equal the CPICH transmit power. This means that signal strengths computed by the planning tool can be interpreted as CPICH RSCP. An example of the translation from a link budget maximum allowed path loss figure to a planning tool signal strength threshold is presented in Table 4.1.

In this example, it is assumed that the maximum allowed path loss from the WCDMA link budgets is 140 dB. This figure may have originated from the uplink service link budget, the downlink service link budget, or the downlink CPICH link budget (i.e., whichever link budget generated the lowest maximum allowed path loss). The planning tool is used to display contours of downlink signal strength irrespective of whether the corresponding path loss originated from an uplink or downlink link budget (i.e., signal strength is being used to provide an indication of path loss). Differences between uplink and downlink path loss (resulting from the 190-MHz frequency division duplex spacing) should be accounted for during the link budget analysis. This is described in Section 4.2.5.

The antenna gain used in Table 4.1 should equal the NodeB antenna gain assumed in the link budgets. It is likely that an actual radio network includes a range of different antenna types, each with a different antenna gain. This does not have an impact upon the results as long as the planning tool is configured with the actual antenna types. For example, if the link budgets are based upon an antenna gain of 18 dBi, whereas a specific node B has an actual antenna gain of 16 dBi, then the maximum allowed path loss resulting from the link budget will be 2 dB more relaxed than it should be. However, applying an 18 dBi antenna gain in Table 4.1 means that the signal strength threshold is 2 dB more difficult to achieve and the two effects cancel one another.

The feeder loss used in Table 4.1 should equal the feeder loss assumed in the limiting link budget (i.e., uplink service, downlink service, or downlink CPICH). If the uplink service link budget is the limiting link budget and mast head amplifiers (MHAs) are present, it is more accurate to account for the benefit of using the MHA and to configure a value of 0 dB in both Table 4.1 and in the planning tool. This assumes that the MHA provides a benefit that is exactly equal to the feeder loss. This approximation introduces inaccuracies for node B that have very short or very long feeder lengths. However, there is often a requirement to use the planning tool as a database for site-specific feeder loss values. This means that it is not possible to enter feeder loss values of 0 dB for all sites. Instead, the actual feeder loss values are entered in the planning tool and the feeder loss value assumed in the link budget (excluding the benefit of using an MHA) is applied in Table 4.1. If the downlink service link budget or the downlink CPICH link budget is the limiting link budget, then the actual feeder loss values should be entered in the planning tool and the feeder loss value assumed in the downlink link budget should be applied in Table 4.1.

Link budgets should be completed on a per-service and per-clutter type basis. Link budgets differ across services primarily due to differences in the E_b/N_o requirements and processing gains. Link budgets differ across clutter types primarily due to differences in building penetration losses and slow fading standard deviations. This means a relatively large number of signal strength contours could be displayed. It is common to reduce the number of contours that are displayed to help simplify their visualization and interpretation. The number of contours can be reduced such that only those corresponding to the most demanding service are plotted. Figure 4.2 illustrates an example of plotting the coverage contours only for the most demanding service.

The most demanding service could be the 64/128-kbps PS data service. This does not necessarily mean that the operator does not wish to offer the 64/384-kbps PS data service. The radio network could be planned according to the 64/128-kbps PS data service and then the 64/384-kbps PS data service offered on a best effort basis. In the case of Figure 4.2, a different coverage contour is plotted for each clutter type. Coverage should be checked by selecting the contour corresponding to the underlying clutter type.

The path loss-based approach to 3G radio network planning should include an analysis of the best server areas. This helps ensure good dominance and a relatively even distribution of network loading. Best server areas should be contiguous and should not be fragmented. Noncontiguous best server areas indicate that there is likely to be relatively poor dominance and increased levels of inter-cell interference.

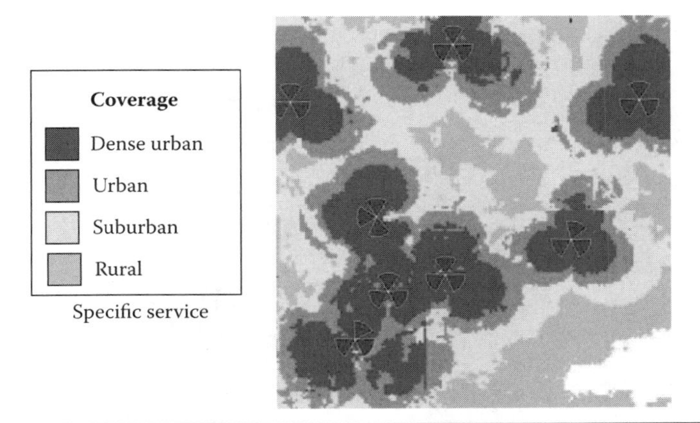

Figure 4.2 Example coverage plot from the path loss-based approach to 3G radio network planning when applying clutter dependent signal strength thresholds for a specific service.

In general, neighboring best server areas should be of approximately equal size. If there is a known traffic hotspot, then a NodeB should be located as close as possible to that hotspot and the dominance area can be smaller. Figure 4.3 illustrates an example best server plot.

This example is applicable to a population of CEC NodeB. If the radio network included ROC NodeB then, depending upon how the radio network planning tool is configured, there could be a single best server area per node B rather than a single best server area per sector.

Some operators include a downlink C/I analysis as part of their path loss-based approach to 3G radio network planning. A C/I analysis provides an indication of cell isolation and inter-cell interference. 2G radio network planning tools can be used to complete a C/I analysis by assigning a single RF carrier to all cells. Large negative values of C/I can be interpreted as areas of poor dominance where the CPICH E_c/I_o is likely to be poor. A typical threshold for the minimum allowed downlink C/I is −6 dB. This corresponds to incurring interference, which is four times stronger than the desired signal (e.g., this would occur if a UE received five signals of equal strength). A C/I analysis does not account for soft handover between cells, but this is analogous to actual CPICH signals, which are not combined while a UE is in soft handover. 3G simulations can be used to validate that a C/I analysis is generating meaningful results. Figure 4.4 illustrates an example downlink C/I plot.

In this example, there are a few relatively small locations where the downlink C/I is poor (i.e., below a value of −6 dB). The planner would have to decide whether or not these locations are important. If they are in areas where there is unlikely to be any traffic, then they can be left.

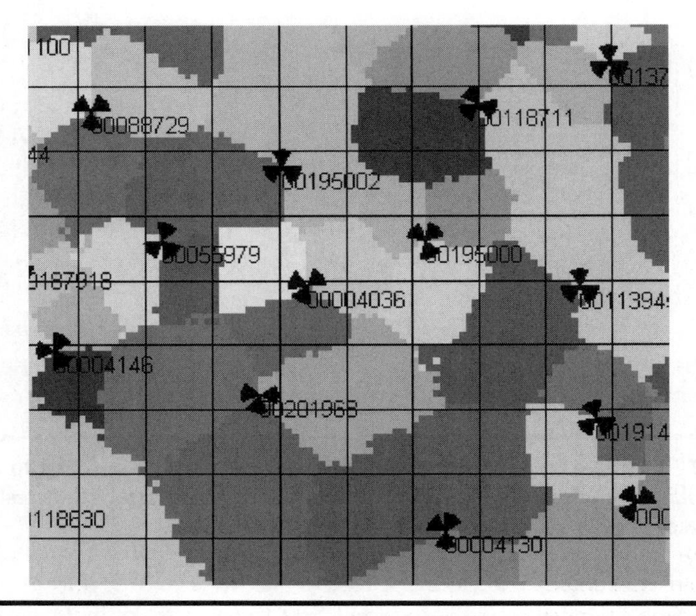

Figure 4.3 Example best server plot from the path loss-based approach to 3G radio network planning.

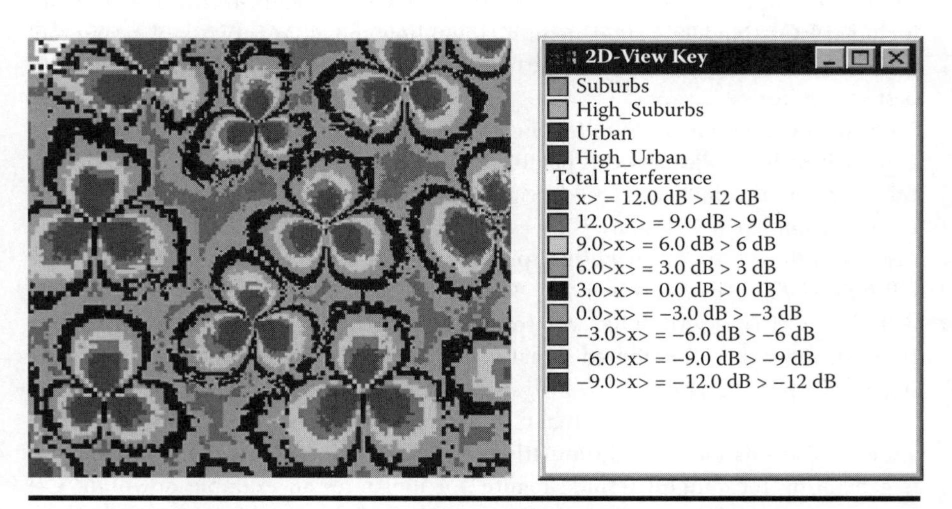

Figure 4.4 Example C/I analysis plot from the path loss-based approach to 3G radio network planning.

A C/I analysis can also be used subsequent to radio network planning when defining cluster boundaries during the prelaunch optimization phase. A C/I analysis can be used to help identify clusters that are relatively well isolated from one another. This helps ensure that RF optimization within one cluster has a minimal impact upon the neighboring clusters.

4.2.2 Simulation-based Approach

The simulation-based approach to radio network planning requires the use of a WCDMA radio network planning tool. The majority of WCDMA radio network planning tools make use of Monte Carlo simulations. Monte Carlo simulations are static rather than dynamic. This means that system performance is evaluated by considering many independent instants (snapshots) in time. A dynamic simulation evaluates performance by considering a series of consecutive instants in time. In general, dynamic simulations are more time-consuming than static simulations. In the case of static simulations, the population of UEs are redistributed across the simulation area for every simulation snapshot. For each snapshot, the uplink and downlink transmit power requirements are computed-based upon link loss, C/I requirement, and the level of interference. UE that are not able to achieve their C/I requirements are categorized as being in outage. Outage may also be caused by factors such as inadequate baseband processing resources or reaching the maximum allowed increase in uplink interference. By considering a large number of instants in the time, the simulation is able to provide an indication of the probability of certain events occurring (e.g., the probability that a UE will be able to establish a connection at a specific location). The simulation is also able to provide an indication of average performance metrics such as cell throughput and downlink transmit power. Detailed information regarding 3G simulations is provided in [7].

The inputs required for the 3G simulation-based approach to radio network planning are:

- 3G site candidates with their physical configuration (antenna type, antenna height, antenna tilt, antenna azimuth, feeder type, and feeder length)
- Propagation model
- Digital terrain map
- 3G parameter assumptions
- 3G traffic profile

The first three inputs are the same as those used for the path loss-based approach to 3G radio network planning. The 3G parameter assumptions overlap with those used to generate link budgets. 3G simulation tools usually require more parameters than a link budget (e.g., the number of hardware channels available at a NodeB, downlink orthogonality, soft handover addition, and drop windows). Some link budget parameters have a different definition in the planning tool. For example, the soft handover gain that appears within a link budget includes the diversity benefits for both fast and slow fading. The soft handover gain that appears within a 3G simulation tool usually includes only the diversity benefit for fast fading. This is because 3G simulation tools usually model slow fading explicitly and thus the soft handover gain is already included in the system modeling. The 3G traffic profile is relatively difficult to define. It requires a specification of the services used and also the extent to which they are used. In addition, it requires a specification of the geographic distribution of UE. Most 3G radio network planning tools allow UE to be distributed within polygons, along vectors, or based upon the clutter type. Some tools also allow the import of traffic maps which can be generated outside the planning tool. These traffic maps can be based upon national census population statistics if the movement of UE from home locations to work locations during the busy hour is accounted for. Alternatively, traffic maps can be generated from Erlang maps that have been recorded from an existing 2G network. This approach assumes that the distribution of 3G traffic will be approximately the same as the distribution of 2G traffic.

The simulation-based approach to radio network planning is more time consuming than the path loss-based approach. The results take more time to generate and more time to interpret. The time required to generate simulation results depends upon the size of the geographic area being modeled and also the quantity of traffic loading the network. Each simulation snapshot takes longer to complete when there are large quantities of traffic. However, there is a requirement for more simulation snapshots when there is only a small quantity of traffic. More simulation snapshots are required to ensure that the simulation results have converged. Convergence of the simulation results should always be checked toward the end of a simulation. This can be achieved by ensuring that the results are not changing significantly after successive snapshots. The majority of 3G simulation tools include functionality for a passive scan terminal. A passive scan terminal increases the rate at which graphical results are generated by the simulation tool but decreases the rate at which numerical results are generated. The passive scan terminal operates by evaluating system performance at every pixel within the simulation area at the end of each simulation snapshot. This increases the rate at which graphical results are generated because without the passive scan terminal, results can only be generated for the pixels where UE have been distributed.

The main benefit of completing 3G simulations is the relatively large quantity of information generated. This information is beneficial if it is interpreted correctly. Although the scenario is less realistic, it can be easier to interpret results if a single service is simulated at a time. The main results from a 3G simulation are typically:

- Service coverage
- System capacity
- Soft handover overhead
- Inter-cell interference
- Uplink and downlink transmit powers

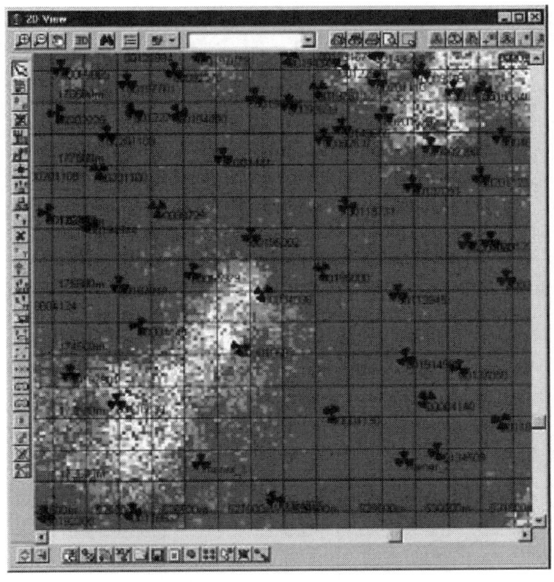

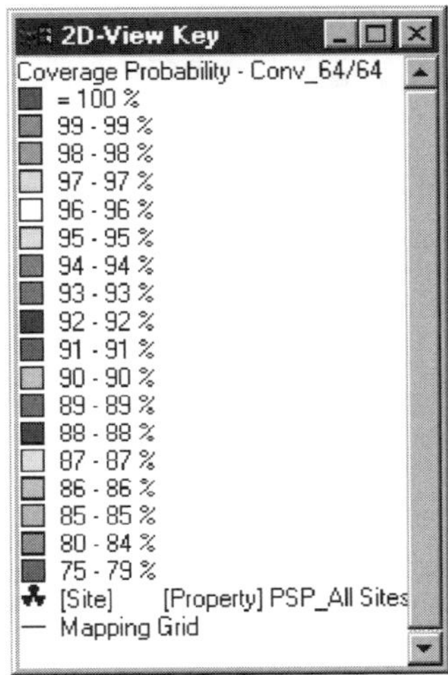

Figure 4.5 Example coverage plot for the 64/64-kbps CS data service.

- Uplink and downlink interference floors
- Connection establishment failure mechanisms

It is normal to be able to visualize each of these results using graphical plots and also to be able to study them using numerical reports. Coverage plots are usually generated on a per-service basis. An example coverage plot for the 64/64-kbps CS data service is illustrated in Figure 4.5.

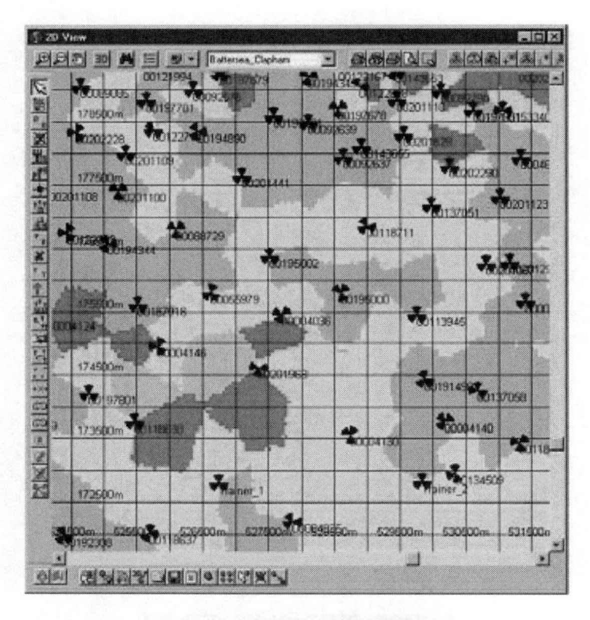

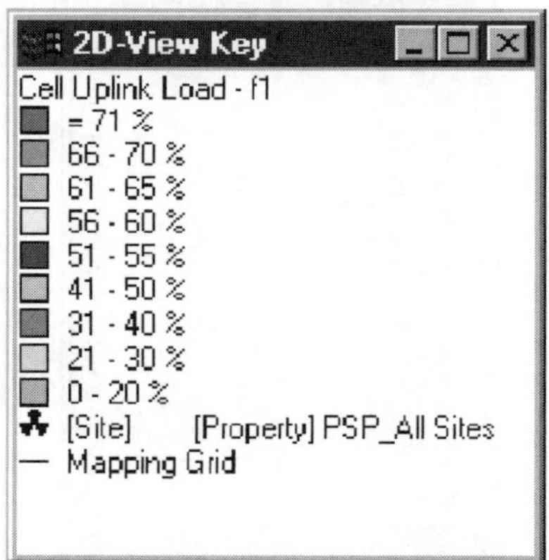

Figure 4.6 Example plot of uplink load.

Service coverage plots illustrate locations where UE were able to achieve their uplink and downlink C/I requirements. Areas where there is no coverage correspond to locations where UE could not achieve their C/I requirement as a result of either inadequate transmit power or one of the other failure mechanisms (e.g., the maximum allowed increase in uplink interference has been reached). An example plot of uplink load is illustrated in Figure 4.6. This plot provides an indication of the increase in uplink interference floor at each cell and can be used to identify cells that have reached their maximum allowed increase.

Cell Blocking Report On Service Conv_64/64							
Cell Identity	Mean Number of Failures	Probability of No Primary Channel Failure (%)	Probability of Channel Limit Failure (%)	Probability of Low Ec/Io Failure (%)	Probability of Uplink Eb/No Failure (%)	Probability of Downlink Eb/No Failure (%)	Probability of Noise Rise Failure (%)
00143655A	0.00	0.00	0.00	0.00	0.00	0.00	0.00
00143655B	0.00	0.00	0.00	0.00	0.00	0.00	0.00
00143655C	0.00	0.00	0.00	0.00	0.00	0.00	0.00
00201441A	0.00	0.00	0.00	0.00	0.00	0.00	0.00
00201441B	0.00	0.00	0.00	0.00	0.00	0.00	0.00
00201441C	0.00	0.00	0.00	0.00	0.00	0.00	0.00
00187918A	0.00	0.00	0.00	0.00	0.00	0.00	0.00
00187918B	0.00	0.00	0.00	0.00	0.00	0.00	0.00
00187918C	0.00	0.00	0.00	0.00	0.00	0.00	0.00

Figure 4.7 Example cell blocking report.

In the case where connections fail, an indication of the failure mechanism is provided in the cell blocking report. An example of this report is illustrated in Figure 4.7.

The relatively large quantity of time required to complete and analyze 3G simulations means that they are usually used for focused studies rather than wide area radio network planning. Focused studies may be used to evaluate the capacity of a section of the network or they may be used to estimate the soft handover overhead or the level of inter-cell interference. 3G simulations may also be used to help validate the path loss-based approach to 3G radio network planning.

4.2.3 Site Selection

Sites are expensive long-term investments for the operator. If sites are selected that have a poor location or a poor set of characteristics, then system performance is likely to be poor irrespective of any subsequent RF and parameter optimization. A good site location should maximize coverage across the intended area while limiting interference into neighboring areas.

Site selection criteria can be divided into two categories. The first category includes criteria that determine whether or not a site should be considered for inclusion within the 3G radio network plan. The second category includes criteria that can be used to prioritize those sites being considered for inclusion within the 3G radio network plan. In general, it is difficult to acquire sites and there will not be many sites to prioritize between. If any of the criteria belonging to the first category are not satisfied, then the site should be excluded from further consideration unless there are no alternatives and the benefit of introducing the site is believed to justify its cost. Table 4.2 presents the minimum set of criteria that should be used to determine whether or not a site should be considered for inclusion within the 3G radio network plan. These criteria should be evaluated after a site visit and not only from the information available within a radio network planning tool.

Assuming that a site satisfies the criteria in Table 4.2, then it may be considered for inclusion within the 3G radio network plan. Table 4.3 presents a set of criteria which

Table 4.2 Site Selection Criteria Used to Determine Whether Or Not a Site Should be Considered for Inclusion Within the 3G Radio Network Plan

1	Does the site allow the main beam of each proposed antenna to have good visibility of the surrounding terrain without any high obstacles blocking the view?	Yes/No
2	Can the main beam of each antenna be positioned such that it does not cross the main beam of another antenna?	Yes/No
3	Can the main beam of each antenna be positioned such that they are not shadowed by the building or structure upon which they are secured?	Yes/No
4	Can each antenna be mounted above the rooftops of the neighboring buildings without being excessively above them? Typically < 10 m above the neighboring rooftops.	Yes/No/ Not Applic.
5	Do neighboring cells have antenna heights which are within 15 m of the proposed antenna heights?	Yes/No
6	Are neighboring cells of similar size?	Yes/No
7	Is the site unlikely to be very dominant and unlikely to cause significant interference to neighboring cells?	Yes/No
8	Is the best server area of the site unlikely to be fragmented?	Yes/No
9	If the proposed site is a rooftop site, is there sufficient space for the appropriate antenna mountings to ensure that there is adequate clearance from the rooftop?	Yes/No/ Not Applic.
10	Is the site safe from new neighboring buildings which may be constructed in the future and which may block the main beam of an antenna?	Yes/No
11	Are the cabling distances between the NodeB cabinet and the antennas reasonable?	Yes/No
12	Is there access to leased lines or microwave links for transmission purposes?	Yes/No
13	Is there availability of the NodeB power supply requirements?	Yes/No
14	Is there space to accommodate the NodeB equipment?	Yes/No
15	Are rental costs acceptable?	Yes/No
16	Is there reasonable access to the site?	Yes/No

Table 4.3 Site Selection Criteria Used to Prioritize Between Sites Which are Being Considered for Inclusion Within the 3G Radio Network Plan

1	Is the site an existing GSM site?	Yes/No
2	Do antenna locations allow for changes in azimuth?	Yes/No
3	Do antenna locations allow for changes in height?	Yes/No
4	Do antenna locations allow sufficient isolation from other antennas (e.g., GSM antennas)?	Yes/No
5	Is the site away from environmentally protected or historic areas?	Yes/No
6	Is the site unlikely to require any special permits?	Yes/No
7	Is the site unlikely to cause public disapproval?	Yes/No
8	Does the site form a regular pattern with its neighbors?	Yes/No
9	Is the site close to where the traffic is expected?	Yes/No
10	Is the site capable of accommodating potential capacity upgrades?	Yes/No

may be used to prioritize between sites being considered for inclusion within the 3G radio network plan. Similar to the first set of criteria, these should be evaluated from a site visit and not from the information available within a radio network planning tool. Sites that result in the greatest number of "yes" responses should be selected to be built.

Site visits represent a significant part of the site selection process. They should be used to collect all of the information required to evaluate the suitability of a site as well as additional information for radio and transmission planning, information for site build and installation, and information for site design. Site visits are relatively expensive and time consuming and should be planned carefully. Planners completing site visits should, as a minimum, take with them:

- A paper map of the area
- A paper diagram of the building
- A coverage plot from the planning tool
- A best server plot from the planning tool
- A GPS receiver
- Binoculars and compass
- A digital camera
- An altimeter
- A tape measure or other measuring device
- Safety equipment if necessary

The paper map of the area should be used to mark the proposed antenna azimuths and general antenna locations. The paper diagram of the building should be used to make a more detailed record of the proposed antenna locations and the positions from where any photos are taken. The digital camera should be used to obtain panoramic views from the proposed antenna locations. Photos should also be taken of the neighborhood and surrounding environment. The location and visibility of neighboring sites should be recorded as should potential gaps in coverage. The different possibilities for antenna mounting should be noted and feeder length requirements estimated. The GPS receiver should be used to determine an exact set of coordinates for the site as well as the height above sea and ground level. The property address and owner should be recorded as should the possibilities for site access.

4.2.4 Propagation Modeling

Accurate propagation modeling is fundamental to both the 3G simulation-based approach and the path loss-based approach to 3G radio network planning. The key inputs to propagation modeling are the propagation model itself, the digital terrain map (DTM) and the site configuration data. Radio network planning is often completed using a set of propagation models rather than a single propagation model. Different models may be tuned for specific antenna heights or specific cell ranges. Digital terrain maps often become out-of-date and subsequently become a source of error. The site configuration data relies upon the accuracy of the antenna gain pattern as well as the antenna height, azimuth, and tilt data. Operators may have existing propagation models for their 2G radio networks. They may wish to use the same models for 3G radio network planning. In this case, their models should be compared with typical models to help identify any significant differences. It is reasonable to assume that a propagation model that has been used to plan a DCS 1800 network can also be applied when planning a 3G network.

When a project requires the definition of a new propagation model, the following tasks should be completed:

- DTM specification, purchase, and validation
- Selection of propagation model type based upon requirements
- Planning and completion of drive survey
- RF measurement post-processing and propagation model calibration
- Propagation model validation and continuous auditing

DTM requirements should be specified in terms of resolution, format, and the number of clutter categories. The resolution should be relatively high for urban and suburban areas but can be reduced for rural areas. It is typical to use a 20-m resolution for urban and suburban areas, whereas a 50-m resolution may be used for rural areas. If the resolution is too low, then the accuracy of the map and the subsequent propagation modeling becomes poor. If the resolution is too high, then the DTM is likely to be expensive and the computer processing requirement may become excessive. Resolutions higher than 20 m may be appropriate for dense urban areas where cell ranges are particularly small. If microcells with below rooftop antennas are to be planned, then it may be necessary to purchase a map that includes building vectors. Building vectors may have either two or three dimensions. The format of the DTM should be specified to match the format used by the radio network planning tool. If the required format is not available, then the purchased map will require post-processing prior to importing. The appropriate number of clutter categories depends upon the geographic area. It is typical to make use of about ten categories. Some planning tools may have a maximum number of categories that can be imported. In this case, the DTM can be post-processed to merge similar categories. If the number of categories is large, then the propagation tuning exercise becomes more difficult. Once the DTM has been purchased and imported into the radio network planning tool, a set of checks should be completed to help validate its accuracy. Clutter types and vectors should be compared with those indicated on paper maps. Likewise, the ground height data should be compared with that indicated on paper maps.

Once the digital terrain map has been validated, then an initial selection of propagation models should be made. It may be necessary to change this selection if it is subsequently found to be inappropriate during the model tuning task. The requirement for a set of different propagation models should be evaluated. In general, the types of propagation models that can be selected are limited to those supported by the radio network planning tool. Most planning tools provide the Okumura-Hata model and Walfisch-Ikegami model. The main differences between the applicability of the Okumura-Hata model and the Walfisch-Ikegami model are presented in Table 4.4.

A detailed specification for each of these models is available in [7]. A number of model types can be chosen at this stage with the final selection being made during

Table 4.4 Applicability of the Okumura-Hata and Walfisch-Ikegami Propagation Models

	Okumura-Hata	*Walfisch-Ikegami*
Frequency range	150 MHz to 1.0 GHz 1.5 to 2.0 GHz	800 MHz to 2.0 GHz
NodeB antenna height	30 to 200 m above rooftop	4 to 50 m above rooftop
UE antenna height	1 to 10 m	1 to 3 m
Range	1 to 20 km	30 m to 6 km
Applicable to	Macrocells	Microcells

propagation model tuning. The final number of models should be kept relatively small to help ensure the radio network planning process remains practical. If there are sites with significantly different cell ranges, antenna heights, or radio environments, then it may be necessary to define separate propagation models.

Once the approximate number and type of propagation model have been identified, then an RF measurement campaign should be planned to tune each model. Tuning a propagation model involves comparing measured and predicted data while adjusting the model input parameters to achieve a minimum error. The RF measurements may be recorded either from existing DCS 1800 sites or from specific test transmitters radiating continuous wave (CW) signals within the 3G downlink frequency band. It is preferable, although more expensive, to use specific test transmitters. At least eight sites are required for each propagation model. A further two sites should be measured for subsequent model validation. If test transmitters are being used, then some of the test sites should be measured with multiple antenna heights. The test sites and their antenna locations should be representative of candidate 3G sites. The test sites should be distributed across the area where the propagation model will be applied. This may involve taking measurements in a number of different areas. In the case of rooftop sites, antenna clearance angles should be checked in the same way as they are checked during the normal site design activity. Likewise, antennas should not be obstructed. Each test site should be visited before the drive route is planned to help ensure its suitability. If directional antennas are used for the test sites, the drive route should remain within the 3 dB beamwidth of the antenna. Propagation model tuning is less likely to be accurate outside this beamwidth due to the increased variance of the antenna gain pattern. The drive route should include a balance of line-of-sight and non-line-of-sight locations. The route should also provide measurements that are relatively evenly distributed in terms of distance from the test site. Each clutter type should have a minimum of 400 measurement samples after binning. There can be an overlap between routes which are used for different test sites although the entire route should not be the same. Routes should not be planned across sections of elevated road because these locations are likely to cause a discrepancy between the DTM ground height and the actual measurement height. Likewise, routes should not be planned through tunnels or cuttings. Once the drive routes have been planned, the measurements should be recorded. Differential GPS with dead reckoning should be used whenever possible. Dead reckoning helps to maintain location accuracy when GPS satellite visibility is lost.

Once the measurement data has been recorded it should be filtered, binned, and analyzed. Erroneous data should be removed, as should data which have a signal strength below −110 dBm and are less likely to be accurate. If drive routes include elevated sections of road, tunnels, or cuttings, then these measurements should also be removed. Data should be binned according to distance and using a grid that has the same resolution and reference as the DTM. The binned measurements should be analyzed in terms of plotting the number of data points per clutter type and plotting the number of data points as a function of signal strength and distance from the test site. It may also be necessary to convert the measurement data into a format that the radio network planning tool can import for tuning.

Propagation model tuning involves minimizing the standard deviation of the error between the predicted propagation loss and the measured propagation loss, while maintaining a mean error that is close to 0 dB. There is no single correct way of calibrating a propagation model. Some radio network planning tools offer functionality for automatically tuning a propagation model. The generic propagation model tuning process is iterative and involves the following steps:

- Complete a set of propagation predictions.
- Quantify the mean and standard deviation of the error between the predicted and measured data.
- Adjust one input parameter belonging to the propagation model.
- Recalculate the propagation predictions and determine whether or not the results have improved.
- If the results have improved, keep the change and repeat, otherwise undo the change and repeat.
- Repeat until the mean and standard deviation of the error cannot be reduced any further.

A detailed explanation of each of these steps is available in [7]. This document provides instructions for tuning each type of propagation model.

Once the propagation models have been tuned, they should be validated using some measurements not used for tuning. The measurements should be loaded into the tuning tool and used to determine a mean and standard deviation for the error between the measured and predicted data. The mean and standard deviation or the error should be similar to that obtained at the end of the tuning procedure. Once the propagation models have been finalized and used, they should be audited on a regular basis to ensure they remain valid. Measurements from the prelaunch optimization activity may be used to refine the propagation models. This may be done on a cluster-by-cluster basis, or if the resources are available, then it could be done on a cell-by-cell basis. Some planning tools allow the import of prelaunch or post-launch optimization drive test data for comparison against predictions.

4.2.5 Link Budgets

Link budgets are fundamental to understanding system behavior and performance. It is important that generic link budgets are refined on a per-project basis to help ensure input assumptions are aligned with implementation. Link budgets are essential when adopting the path loss-based approach to 3G radio network planning. In this case, the link budget results are used to define signal strength thresholds, which are applied throughout the radio network planning process.

4.2.5.1 Uplink Service DPCH

The generic uplink service link budgets for a suburban area are presented in Table 4.5. A column is included to specify which parameters are directly dependent upon the performance of the RAN.

The uplink bit rates are limited to those supported by the network and by the terminals. Separate link budgets should be completed for CS data and PS data services. The maximum transmit powers are defined by the capability of the terminals. Terminals typically have capabilities of either 21 dBm or 24 dBm. In practice, both transmit power capabilities are likely to be present in a network. Applying a figure of 21 dBm represents a worst-case assumption. Antenna gain and body loss assumptions are relatively open to discussion and agreement on a per-project basis. It may be appropriate to assign higher antenna gains to data services if the terminals are different to those used for the speech service (e.g., a data card). Body loss is usually reduced for data services where the terminal is typically held away from the body. The first interim result from the link budget is the uplink EIRP.

The processing gain is defined by the ratio of the chip rate to the service bit rate expressed in logarithmic units:

$$ProcessingGain = 10 \times \text{LOG}\frac{ChipRate}{BitRate}$$

Uplink E_b/N_o assumptions reflect the performance of the NodeB receiver and may vary between vendors. E_b/N_o figures should be applied for a specific propagation channel, BLER target, and physical layer configuration. The set of uplink E_b/N_o figures for use within link budgets and radio network planning tools are presented in [8]. These E_b/N_o figures have values that are less than those within the RNC database. In addition, the definition of E_b/N_o for the RNC database does not include the DPCCH overhead. The RNC database E_b/N_o values are currently hidden parameters which cannot be edited. They have been defined to be relatively high to provide a margin when completing admission control and computing the initial uplink SIR target and SIR target range. The C/I requirement for a specific service is defined by the difference between the E_b/N_o requirement and the processing gain:

$$ServiceCIRequirement = E_b/N_o - ProcessingGain$$

The target uplink load should be agreed to on a per-project basis and is typically defined on a per-clutter basis. Urban areas are usually assumed to have higher target loads. This results in a higher uplink capacity and smaller cell range. The rise over thermal noise is computed directly from the target load:

$$RiseOverThermalNoise = -10 \times \text{LOG}(1 - TargetLoad)$$

Table 4.5 Generic Uplink Service Link Budgets

Service Type	*RAN Specific*	*Speech*	*CS Data*	*PS Data*
Uplink bit rate (kbps)	No	12.2	64	64
Maximum transmit power (dBm)	UE dependent	21.0	21.0	21.0
Terminal antenna gain (dBi)	UE dependent	0.0	2.0	2.0
Body loss (dB)	No	3.0	0.0	0.0
Transmit EIRP (dBm)	UE dependent	18.0	23.0	23.0
Chip rate (Mcps)	No	3.84	3.84	3.84
Processing gain (dB)	No	25.0	17.8	17.8
Required E_b/N_o (dB)	Yes	4.4	2.0	2.0
Target uplink load (%)	No	50	50	50
Rise over thermal noise (dB)	No	3.0	3.0	3.0
Thermal noise power (dBm)	No	–108.0	–108.0	–108.0
Receiver noise figure (dB)	Yes	3.0	3.0	3.0
Interference floor (dBm)	No	–102.0	–102.0	–102.0
Receiver sensitivity (dBm)	Yes	–122.6	–117.8	–117.8
NodeB antenna gain (dBi)	No	18.5	18.5	18.5
Cable loss (dB)	No	2.0	2.0	2.0
Benefit of using MHA (dB)	No	2.0	2.0	2.0
Fast-fading margin (dB)	Yes	1.8	1.8	1.8
Soft handover gain (dB)	Yes	2.0	2.0	2.0
Building penetration loss (dB)	No	12.0	12.0	12.0
Indoor location probability (%)	No	90	90	90
Indoor standard deviation (dB)	No	10	10	10
Slow fading margin (dB)	No	7.8	7.8	7.8
Isotropic power required (dBm)	Yes	–121.5	–116.7	–116.7
Allowed propagation loss (dB)	Yes	139.5	139.7	139.7

The thermal noise power is defined by the assumed temperature of the NodeB receiver.

$$\begin{aligned} ThermalNoisePower &= 10 \times \text{LOG}(kTB) \\ &= 10 \times \text{LOG}(1.4 \times 10^{-23} \times 290 \times 3.84 \times 10^{6}) \\ &= -138 \text{ dBW} = -108 \text{ dBm} \end{aligned}$$

The noise figure assumption reflects the performance of the NodeB receiver and may vary between vendors. A typical figure for a NodeB is 3 dB. The second interim result from the link budget is the receiver sensitivity. This is computed by summing the service C/I requirement with the composite interference floor:

$$\begin{aligned} ReceiverSensitivity = \; & ServiceCIrequirement + ThermalNoisePower \\ & + RiseOverThermalNoise + ReceiverNF \end{aligned}$$

The NodeB antenna gain should be representative of the antenna type being deployed. A real network is likely to include a range of antenna types with a range of gains, but a figure of 18.5 dBi is typical for a three-sector site. Antenna gains decrease as the horizontal and vertical beamwidths increase and the antenna becomes less directional.

The cable loss assumption has relatively little importance in the uplink link budget when mast head amplifiers are included. The benefit of using mast head amplifiers is typically assumed to equal the cable loss. A more precise calculation can be completed using the Friis equation. The actual benefit of using an MHA is usually greater than the feeder loss. The benefit of using an MHA is less than the feeder loss when the feeder loss is relatively large. In the case of an MHA that has a noise figure of 2 dB and a gain of 12 dB, the benefit of using the MHA becomes less than the value of the feeder loss when the feeder loss is greater than 5 dB.

The set of fast fade margins are presented in [9]. These figures are theoretical and have been derived from simulations. The fast fade margin is dependant upon the UE speed. At lower UE speeds, the uplink inner loop power control is able to track the fast fading and so there is a requirement for a fast fade margin (i.e., the peak UE transmit power is greater than the average UE transmit power for a specific average link loss to the NodeB). At high UE speeds (greater than approximately 50 km/hr), the fast fades are experienced too rapidly for the inner loop power control to track. This causes an increase in the uplink E_b/N_o requirement, but a decrease in the fast fade margin. Higher UE speeds are also associated with an increase in the performance of channel coding and interleaving. Channel coding performs best when bit errors are randomly and uniformly distributed throughout the data. In practice, fades cause bursts of errors and so interleaving is used to distribute the errors in a relatively random manner. At low UE speeds, fades tend to be wide and the corresponding bursts of errors also tend to be wide. Wide bursts of errors are more difficult for interleaving to cope with. Interleaving is completed over the duration of the transmission time interval and a large transmission time interval could be used to improve the performance of channel coding and interleaving but at the expense of an increased delay and buffering requirement.

The set of soft handover gains are presented in [2]. These figures are theoretical and have been derived from simulations, although field trials have been completed to help validate them. The term "soft handover gain" is often defined in different ways. In general, three types of soft handover gain exist: reduction in E_b/N_o requirement as a result of greater diversity in the RAKE receiver; reduction in the fast fade margin as a result of not having to track all of the fast fades; and a reduction in the slow fade margin as a result of not having to track all of the slow fades. The reduction in the E_b/N_o requirement is assumed to be 0 dB in the uplink direction because there is already a significant quantity of diversity in the RAKE receiver (assuming dual branch receive diversity is being used). The soft handover gain appearing in the uplink link budget is the sum of the reductions in the fast and slow fade margins. This is different than the soft handover gain used for 3G simulations in a radio network planning tool because radio network planning tools typically model slow fading explicitly and thus do not require an input parameter to define the associated soft handover gain. In the case of 3G simulations, the uplink soft handover gain is simply the reduction in the fast fade margin as a result of not having to track all of the fast fades.

Building penetration loss assumptions are usually made on a per-clutter basis [10]. These figures have a relatively large uncertainty and have a significant impact upon the final link budget result. They should be defined on a per-project basis and should be agreed to with the operator. Similarly, the indoor location probability and indoor standard deviation have a significant impact upon the final link budget result. These figures should also be agreed upon with the operator. The indoor standard deviation includes the standard deviation associated with slow fading and the standard deviation associated with the uncertainty in the building penetration loss figure. Some recommendations for building penetration loss, indoor location probability, and indoor standard deviation are presented in Table 4.6.

The indoor location probability and indoor standard deviation are used to compute the slow fading margin. The slope required as an input to this function is

Table 4.6 Building Penetration Losses and Slow Fading Characteristics

	Dense Urban	*Urban*	*Suburban*	*Rural*
Building penetration loss (dB)	not defined	16	12	10
Indoor location probability (%)	not defined	90	90	90
Indoor standard deviation (dB)	not defined	12	10	9
Slow fading margin (dB)	not defined	10.1	7.8	6.6

typically assigned a value of 3.4. The third interim result from the link budget is the isotropic power requirement. This is computed by combining the receiver sensitivity with the subsequent set of gains and margins:

$$\begin{aligned} IsotropicPowerReq = {} & ReceiverSensitivity - AntennaGain + CableLoss \\ & - MHABenefit + FastFadeMargin - SoftHandoverGain \\ & + BuildingPenetrationLoss + SlowFadeMargin \end{aligned}$$

The final link budget result represents the maximum allowed uplink propagation loss to achieve the indoor location probability at the assumed uplink load. This is computed by subtracting the third interim result from the first interim result:

$$AllowedPropagationLoss = TransmitEIRP - IsotropicPowerReq$$

The uplink service allowed propagation loss should be compared with the downlink service and downlink CPICH allowed propagation losses. The comparison should account for the frequency difference between the uplink and downlink frequency bands. The downlink frequency is 190 MHz greater than the uplink frequency and so has a greater propagation loss associated with it. The frequency dependant term in the Okumura-Hata path loss equation is given by:

$$FrequencyDependantLoss = 33.9 \times \mathrm{LOG}(Frequency\_MHz)$$

This equation indicates that the downlink propagation loss over a specific distance is 1.4 dB greater than the corresponding uplink propagation loss. Coverage plots in radio network planning tools are usually generated by computing path loss from the downlink frequency. This means that the uplink maximum allowed path loss figures should be increased by 1.4 dB prior to their use within the radio network planning process. This figure should also be taken into account when comparing the allowed uplink and downlink propagation losses.

4.2.5.2 Downlink Service DPCH

The generic downlink service link budgets for a suburban area are presented in Table 4.7. A column is included to specify which parameters are directly dependant upon the performance of the RAN.

The downlink bit rates are limited to those supported by the network and by the terminals. Similar to the uplink, separate link budgets should be completed for CS data and PS data services. The maximum downlink transmit powers are defined by the RNC admission control functionality. This functionality will vary between vendors. The admission control determines the maximum downlink transmit power for real-time services according to:

$$MaxRTDLPower = \mathrm{Min}(43 + PtxDPCH\mathrm{max}, MaxDLCalculated)$$

Table 4.7 Generic Downlink Service Link Budgets

Service Type	*RAN Specific*	*Speech*	*CS Data*	*PS Data*		
Downlink bit rate (kbps)	No	12.2	64	64	128	384
Maximum transmit power (dBm)	Yes	34.2	37.2	37.2	40.0	40.0
Cable loss (dB)	No	2.0	2.0	2.0	2.0	2.0
MHA insertion loss (dB)	Yes	0.5	0.5	0.5	0.5	0.5
NodeB antenna gain (dBi)	No	18.5	18.5	18.5	18.5	18.5
Transmit EIRP (dBm)	Yes	50.2	53.2	53.2	56.0	56.0
Processing gain (dB)	No	25.0	17.8	17.8	14.8	10.0
Required E_b/N_o (dB)	UE dependent	7.9	5.3	5.0	4.7	4.8
Target loading (%)	No	80	80	80	80	80
Rise over thermal noise (dB)	No	7.0	7.0	7.0	7.0	7.0
Thermal noise power (dBm)	No	−108.0	−108.0	−108.0	−108.0	−108.0
Receiver noise figure (dB)	UE dependent	8.0	8.0	8.0	8.0	8.0

(Continued)

Table 4.7 Generic Downlink Service Link Budgets (Continued)

Service Type	*RAN Specific*	*Speech*	*CS Data*	*PS Data*		
Interference floor (dBm)	No	–93.0	–93.0	–93.0	–93.0	–93.0
Receiver sensitivity (dBm)	UE dependent	–110.1	–105.5	–105.8	–103.1	–98.2
Terminal antenna gain (dBi)	UE dependent	0.0	2.0	2.0	2.0	2.0
Body loss (dB)	No	3.0	0.0	0.0	0.0	0.0
Fast fading margin (dB)	UE dependent	0.0	0.0	0.0	0.0	0.0
Soft handover gain (dB)	UE dependent	2.0	2.0	2.0	2.0	2.0
MDC gain (dB)	UE dependent	1.2	1.2	1.2	1.2	1.2
Building penetration loss (dB)	No	12.0	12.0	12.0	12.0	12.0
Indoor location probability (%)	No	90	90	90	90	90
Indoor standard deviation (dB)	No	10	10	10	10	10
Slow fading margin (dB)	No	7.8	7.8	7.8	7.8	7.8
Isotropic power required (dBm)	Yes	–90.5	–90.9	–91.2	–88.5	–83.6
Allowed propagation loss (dB)	Yes	140.7	144.1	144.4	144.5	139.6

and for non-real-time services according to:

$$MaxNRTDLPower = \text{Min}(43 + PtxDPCH\text{max}, MaxDLCalculated, PtxDLabsMax)$$

PtxDPCHmax and PtxDLabsMax are the RNC parameters presented in Table 4.5. MaxDLCalculated is computed from the expression:

$$MaxDLCalculated = PTxPrimaryCPICH - CPICHtoRefRABOffset + 10 \times \text{LOG}\left[\frac{\left(10^{\frac{E_b/N_o\,_{SRB}}{10}} \times BR_{SRB}\right) + \left(10^{\frac{E_b/N_o\,_{Service}}{10}} \times BR_{Service}\right)}{\left(10^{\frac{E_b/N_o\,_{Ref}}{10}} \times BR_{Ref}\right)}\right]$$

The E_b/N_o figures that appear in this equation are those from the RNC database (i.e., not those from the link budget). In the case of the speech service, the E_b/N_o figure used in this equation is that for the class A bits. The E_b/N_o figures for the class B and class C bits are not used in this equation. The bit rate used for the SRB should include the layer 2 overhead (i.e., a bit rate of 3.7 kbps rather than 3.4 kbps, or 14.8 kbps rather than 13.6 kbps). The bit rate used for the service and the reference service should not include the layer 2 overhead (e.g., 64 kbps rather than 67.2 kbps). The reference service should be specified as the 12.2-kbps speech service. The RNC database parameter DLreferenceTargetBLER can be configured with any value because the look-up table within the RNC includes only a single E_b/N_o figure for all target BLER figures.

The cable loss should be the same as that assumed for the uplink service link budget. In this case, it is more important because it does not have the benefit of the MHA to compensate for it. A real network is likely to include a range of cable losses but a figure of 2 dB is typical. The MHA insertion loss should also be included as an additional passive loss. The NodeB antenna gain should be the same as that assumed for the uplink. The first interim result from the link budget is the downlink EIRP.

Similar to the uplink, the processing gain is defined by the ratio of the chip rate to the service bit rate expressed in logarithmic units. Downlink E_b/N_o assumptions reflect the performance of the UE receiver and may vary between terminals. The set of E_b/N_o figures can be found in [2]. Similar to the uplink, these E_b/N_o figures are different to those found in the RNC database. The target loading should be agreed to on a per-project basis and is typically defined on a per-clutter basis. The target loading should result in a rise over thermal noise which is representative of the downlink RSSI experienced at the cell edge. This is typically −90 dBm in an interference-limited scenario. The relationship between the target loading and the rise over thermal noise is the same as that specified for the uplink. In this context, the target load does not indicate the percentage of the downlink transmit power being used. It represents the downlink load that would be calculated from the downlink load equation for a specific number of connections. Thermal noise is calculated in the same way as for the uplink. The noise figure assumption reflects the performance of the UE receiver and may vary between terminals. A typical figure for a UE is 8 dB. The second interim result from the link budget is the receiver sensitivity. This is computed by summing the service C/I requirement with the composite interference floor (equations are presented throughout the uplink link budget description).

The terminal antenna gain and body loss assumptions should reflect the values assumed for the uplink link budget. The fast fade margin in the downlink direction is assumed to be 0 dB. This value is based upon the argument that inner loop power control has less dynamic range and less tracking to perform in the downlink direction. In general, this is true because when the UE is relatively close to a NodeB, then both the desired signal and the interference originate from the same point and experience the same fading dynamics (i.e., the signal-to-noise ratio remains approximately constant because the desired signal and interference fade in time with one another). However, when a UE is located at the cell edge then the desired signal and interference originate from more than a single point and they do not experience the same fading dynamics. Operators are likely to argue that there should be a finite downlink fast fade margin. It is acceptable to use a finite downlink fast fade margin if it is agreed to with the operator.

The set of soft handover gains can be found in [2]. Similar to the uplink, three types of soft handover gains exist: reduction in the E_b/N_o requirement as a result of greater diversity in the RAKE receiver; reduction in the fast fade margin as a result of not having to track all of the fast fades; and a reduction in the slow fade margin as a result of not having to track all of the slow fades. In the case of the downlink, it is not usual to have receiver diversity. This means that there is more likely to be a soft handover gain that reduces the downlink E_b/N_o requirement. This soft handover gain is termed the macro diversity combination (MDC) gain within the link budget. Similar to the uplink, the entry in the link budget termed "soft handover gain" is the sum of the reductions in the fast and slow fade margins. Again, this is different than the soft handover gain used for 3G simulations in a radio network planning tool because radio network planning tools typically model slow fading explicitly and thus do not require an input parameter to define the associated soft handover gain.

If the frequency dependence of the building penetration loss is ignored, then its value should be the same as that assumed for the uplink. The indoor location probability, indoor standard deviation and slow fading margin should also reflect the values assumed for the uplink. The third interim result from the link budget is the isotropic power requirement. This is computed by combining the receiver sensitivity with the subsequent set of gains and margins.

The final link budget result represents the maximum allowed downlink propagation loss to achieve the indoor location probability at the assumed downlink load. This is computed by subtracting the third interim result from the first interim result:

$$AllowedPropagationLoss = TransmitEIRP - IsotropicPowerReq$$

The downlink service allowed propagation loss should be compared with the uplink service and downlink CPICH allowed propagation losses.

4.2.5.3 Downlink CPICH

The generic downlink CPICH link budget for a suburban area is presented in Table 4.8. A column is included to specify which parameters are directly dependant upon the performance of the RAN.

The CPICH link budget is similar to the downlink service link budget. In this case, the transmit power is defined by the PtxPrimaryCPICH RNC database parameter. The downlink service E_b/N_o requirement and processing gain are replaced by the CPICH E_c/I_o requirement. This is equivalent to the downlink service C/I requirement. The terminal antenna gain is assumed to be 0 dB. This represents a worst-case assumption. The CPICH is not combined during soft handover and so there are no soft handover gains for the CPICH link budget. The downlink CPICH allowed propagation loss should be compared with the uplink service and downlink service allowed propagation losses.

4.2.5.4 Evaluation and Validation

Once link budgets have been generated for the uplink services, the downlink services and the downlink CPICH, the results should be compared to determine which is the limiting link. Table 4.9 presents a summary of the generic link budgets.

The generic link budgets indicate that the CPICH is the limiting link. However, networks are not usually planned according to coverage thresholds defined by the CPICH. Networks are usually planned according to coverage thresholds defined by either the uplink or downlink service DPCH link budgets. As a result, the CPICH coverage probability will be slightly less. The uplink link budget results have been increased by 1.4 dB to compensate for the lower propagation loss experienced in the uplink frequency band relative to the downlink frequency band. In general, this increase should always be applied, but it is particularly necessary when the link budget results are being used to compute signal strength thresholds for the path loss-based approach to 3G radio network planning (assuming that the radio network planning tool is configured to complete its path loss predictions with a downlink frequency).

Link budgets should be validated once the first phases of the radio network have been deployed. The first task is to ensure that the actual limiting link is the same as that indicated by the link budgets (e.g., the link budgets may indicate that the uplink service coverage is the limiting link, whereas field measurements may

Table 4.8 Generic Downlink CPICH Link Budget

Service Type	*RAN Specific*	*CPICH*
Maximum transmit power (dBm)	Yes	33.0
Cable loss (dB)	No	2.0
MHA insertion loss (dBi)	Yes	0.5
NodeB antenna gain (dBi)	No	18.5
Transmit EIRP (dBm)	Yes	49.0
Required E_c/I_o (dB)	UE dependant	–15
Target loading (%)	No	80
Rise over thermal noise (dB)	No	7.0
Thermal noise power (dBm)	No	–108.0
Receiver noise figure (dB)	UE dependant	8.0
Interference floor (dBm)	No	–93.0
Receiver sensitivity (dBm)	UE dependant	–108.0
Terminal antenna gain (dBi)	UE dependant	0.0
Body loss (dB)	No	3.0
Fast fading margin (dB)	No	0.0
Building penetration loss (dB)	No	12.0
Indoor location probability (%)	No	90
Indoor standard deviation (dB)	No	10
Slow fading margin (dB)	No	7.8
Isotropic power required (dBm)	Yes	–85.2
Allowed propagation loss (dB)	Yes	134.2

indicate that the downlink service coverage is the limiting link). The limiting link can be identified relatively easily from UE drive test data. If the UE is approaching its maximum transmit power capability at the time of a connection failure, then the uplink is limiting. If the UE is requesting large quantities of additional downlink power at the time of a connection failure, then the downlink is limiting. If the CPICH quality is very poor at the time of a connection failure, then the CPICH is limiting. In practice, the links should be relatively well balanced. If the network

Table 4.9 Comparison of Link Budget Results

Service Type	*Speech*	*CS Data*	*PS Data*		
Bit rate (kbps)	12.2	64	64	128	384
Uplink allowed propagation loss (original) (dB)	139.5	139.7	139.7	—	—
Uplink allowed propagation loss (adjusted for downlink band) (dB)	140.9	141.1	141.1	—	—
Downlink allowed propagation loss (dB)	140.7	144.1	144.4	144.5	139.6
CPICH allowed propagation loss (dB)			134.2		

has not been prelaunch optimized, it may be difficult to complete a meaningful evaluation of the link budgets. For example, missing neighbors will cause high levels of downlink interference and are likely to make the downlink link budgets appear pessimistic.

Once the limiting link has been validated, there are two subsequent approaches to validating the link budgets. The first approach is to validate the link budget result as a whole without validating the individual input assumptions. This involves post-processing drive test data to extract the path loss values at which connection failures occurred. The link budget results should be adjusted prior to any comparison to account for the drive test data being collected outdoors rather than indoors. The slow fade margin should also be removed to account for the UE experiencing actual slow fading during a drive test. The second approach to validating the link budget is to validate the individual input assumptions. Some input assumptions can be validated relatively easily, but others are more difficult. For example, the E_b/N_o requirement assumptions can be validated relatively easily but building penetration loss assumptions are difficult to validate.

4.2.6 Downlink Common Channel Powers

The downlink common channels are essential to the operation of the radio network. They allow a UE to synchronize to a specific cell, read system information, complete the random access procedure, receive paging messages, transfer signaling, and transfer user data.

3GPP TS25.211 [11] defines the timing relationships between each of the downlink common channels. This timing information should be used when computing either the average or the peak common channel transmit powers.

3GPP TS25.211 and TS25.213 specify that the CPICH is a stream of 1s that is spread with channelization code (256, 0). The value 256 represents the spreading factor and the value 0 represents the position within the code tree. Channelization code (256, 0) is a series of 256 1s and so the channelization code has no impact upon the original stream of 1s. TS25.213 also specifies that the CPICH is scrambled using the primary scrambling code of the cell and TS25.211 specifies that the CPICH has an activity factor of 100%.

3GPP TS25.211 and TS25.213 specify that the P-SCH is a fixed sequence of 256 chips transmitted during the first 10% of every downlink slot. The same sequence is used by every cell belonging to every operator. 3GPP TS25.211 and TS25.213 also specify that the S-SCH is a sequence of 15 blocks of 256 chips. The precise sequence depends upon which of the 64 scrambling code groups has been assigned to the cell. The 15 blocks of 256 chips occupy the first 10% of each slot over every 15-slot radio frame.

3GPP TS25.331 defines the system information content of the BCH transport channel, which is encapsulated by the P-CCPCH. TS25.213 specifies that the P-CCPCH is spread with channelization code (256, 1) and TS25.211 specifies that the P-CCPCH is scrambled using the primary scrambling code assigned to the cell. TS25.211 specifies that the P-CCPCH is broadcast during the last 90% of every slot. The P-CCPCH does not include any TFCI bits or any pilot bits.

3GPP TS25.331 defines the paging content of the PCH transport channel and signaling content of the control plane FACH transport channel which are both encapsulated by the S-CCPCH. The content of the user plane FACH transport channel is outside the scope of the RAN. The channelization code used to spread the S-CCPCH is not standardized and can be selected by the vendor. TS25.211 specifies that when an S-CCPCH encapsulates a PCH transport channel, it must be scrambled using the primary scrambling code. Otherwise, it may be scrambled using either the primary or a secondary scrambling code. Usually, only primary scrambling codes for the S-CCPCH are used. The transmission of either one or two S-CCPCH is supported. The second S-CCPCH is only required if the capacity of a single S-CCPCH is insufficient.

If one S-CCPCH is broadcast, then channelization code (64, 1) is assigned. TS25.211 specifies the set of allowed slot formats for the S-CCPCH. When one S-CCPCH is broadcast, then slot format 8 is applied. This slot format includes 8 TFCI bits, 72 data bits, and 0 pilot bits within each slot. The TFCI bits are broadcast irrespective of whether or not there is any data to transmit. This means the S-CCPCH always has a finite activity. The TFCI bits are transmitted with a greater power than the data bits. The transmit power of the data bits is defined by the RNC database parameter PtxSCCPCH1 whereas the transmit power of the TFCI bits is defined by the parameter PO1_60. If the S-CCPCH is assumed to have a transmit power equal to the value of PtxSCCPCH1, then the higher transmit power of the TFCI bits can be modeled by increasing the activity factor appropriately. If no data bits are transmitted, then the S-CCPCH has an effective activity of 25%, whereas if data bits are transmitted continuously, then the S-CCPCH has an effective activity of 115% (based upon the default value of 4 dB for PO1_60).

When one S-CCPCH is broadcast, the FACH-c can be sent simultaneously with the PCH (when the FACH-c uses an intermediate bit rate of 16.8 kbps rather than the maximum bit rate of 33.6 kbps), but the FACH-u cannot be sent in the same TTI as the PCH or in the same TTI as the FACH-c. The MAC layer within the RNC is responsible for managing the priority of the PCH, FACH-c, and FACH-u (i.e., the RNC decides which of these messages are sent first on the S-CCPCH). The priority order is PCH (highest priority), FACH-c, and then FACH-u (lowest priority). This means that FACH-u transport blocks can only be sent when there are no PCH or FACH-c messages to send. The PCH transport channel always has priority, so whenever there are PCH messages to send, the RNC schedules up to one 80-bit paging message (containing a single paging record) per 10-ms TTI. If there are simultaneous PCH and FACH-c messages to send, then the RNC can schedule both during the same time period, but the FACH-c has to make use of the transport format based upon the intermediate bit rate of 16.8 kbps. Scheduling of PCH messages for UE in RRC IDLE and CELL_PCH also needs to account for the DRX cycle length and the associated paging occasions used by the UE. UE in RRC IDLE and CELL_PCH will only listen to the PICH once per DRX cycle. The S-CCPCH frame, which includes the PCH message associated with a specific PICH frame, occurs 7680 chips after the PICH. With our current implementation, there should only ever be one positive paging indication within a PICH frame because currently it is only possible to send one paging record within the associated S-CCPCH frame. If the paging channel is experiencing congestion, it may be necessary to reduce the size of the location areas. A check should also be made to determine whether or not CELL_PCH is enabled. The use of CELL_PCH can significantly reduce the paging load because paging messages are directed to a single cell rather than an entire location area.

If two S-CCPCH are broadcast, the channelization codes with spreading factors 64 and 256 are assigned for the FACH and PCH, respectively. The first S-CCPCH, which encapsulates the control plane FACH and the user plane FACH, makes use of slot format 8. This slot format includes 8 TFCI bits, 72 data bits, and 0 pilot bits within each slot. The TFCI bits are broadcast irrespective of whether or not there is any data to transmit. The activity for this S-CCPCH can be modeled in the same way as for the case when a single S-CCPCH is broadcast. The second S-CCPCH that encapsulates the PCH makes use of slot format 0. This slot format includes 0 TFCI bits, 20 data bits, and 0 pilot bits. This S-CCPCH has an activity defined only by the activity of the data bits, and a transmit power defined by the RNC database parameter PtxSCCPCH2. The PCH capacity of using two S-CCPCH is the same as using one S-CCPCH. In both cases, a maximum of one paging message can be sent per 10-ms TTI.

3GPP TS25.211 specifies that the PICH is a sequence of 288 bits that occupy the first 96% of every 10-ms radio frame. TS25.211 also specifies that the spreading

Table 4.10 Relationship Between the Number of Paging Indicators and the Number of Bits Per Paging Indicator

Number of Paging Indicators per PICH Frame	*Number of Paging Groups per PICH Frame*	*Number of bits per Paging Indicator*	*Total Number of bits per PICH Frame*
18	18	16	288
36	36	8	288
72	72	4	288
144	144	2	288

factor for the PICH is 256. The precise channelization code is not standardized and can be selected by the vendor. For example, Nokia uses channelization code (256, 3). TS25.213 specifies that the PICH is always scrambled using the primary scrambling code assigned to the cell. The 288 bits that form the PICH frame are a concatenation of paging indicators. There may be 18, 36, 72, or 144 paging indicators per PICH frame. This is controlled by the RNC database parameter PI_amount. Table 4.10 presents the relationship between the number of paging indicators per PICH frame and the number of bits available to each paging indicator.

Decreasing the number of paging indicators per PICH frame means each paging indicator has a greater number of bits. This increases the level of redundancy in the paging indicator and thus increases its reliability for a fixed transmit power. If the number of paging indicators per PICH frame is decreased, it should be possible to decrease the transmit power of the PICH and maintain the same reliability. However, the default transmit power of the PICH is 8 dB lower than that of the CPICH and the S-CCPCH, so any reductions in the PICH transmit power have little impact upon the total NodeB transmit power. Increasing the number of paging indicators per PICH frame means the UE population can be divided into more paging groups and thus there are less UE within each paging group. This provides the benefit of reducing the frequency with which UE is instructed to read the PCH transport channel on the S-CCPCH (i.e., it reduces UE power consumption). The default number of paging indicators per PICH frame is 72 because the potential reduction in UE power consumption is more significant than the reduction in total downlink transmit power that could be achieved by decreasing the number of paging indicators. In the case of a completely unloaded cell with the default common channel powers, the total downlink transmit power would decrease by 0.2 dB if the PICH power was decreased by 3 dB. This decrease would be smaller if the cell was loaded.

3GPP TS25.211 specifies that the AICH is a chip sequence that can occupy up to 80% of each 20-ms PRACH frame. TS25.211 also specifies that the spreading factor for the AICH is 256. The precise channelization code is not standardized and

Table 4.11 Calculation of the Average Downlink Transmit Power for the Common Channels

Service Type	*Default Power*	*Minimum Activity*	*Minimum Average Power*	*Maximum Activity*	*Maximum Average Power*
CPICH	33 dBm	100%	33 dBm	100%	33 dBm
P-SCH	30 dBm	10%	20 dBm	10%	20 dBm
S-SCH	30 dBm	10%	20 dBm	10%	20 dBm
P-CCPCH	28 dBm	90%	27.5 dBm	90%	27.5 dBm
S-CCPCH	33 dBm	25%	27 dBm	115%	33.6 dBm
PICH	25 dBm	96%	24.8 dBm	96%	24.8 dBm
AICH	25 dBm	0%	—	80%	25 dBm
Total	—	—	35.5 dBm 3.5 W	—	37.5 dBm 5.6 W

can be selected by the vendor. For example, Nokia uses channelization code (256, 2). TS25.213 specifies that the AICH is always scrambled using the primary scrambling code assigned to the cell.

Table 4.11 presents the minimum and maximum average downlink transmit powers for the common channels. The difference between the minimum and maximum average transmit powers is determined by the activities of the S-CCPCH and the AICH.

The activity of the S-CCPCH has a significant impact upon the total common channel downlink transmit power. These figures define the upper limit for the CPICH E_c/I_o. The figures in Table 4.11 indicate that when a UE is in an area of good dominance, it would have a maximum possible CPICH E_c/I_o of −2.5 dB. This figure would be reduced if the cell had some activity, if the UE moved into an area where it incurred inter-cell interference or if the UE moved into an area that was thermal noise limited.

It may be necessary to adjust the common channel powers to suit a specific scenario. For example, if a CEC NodeB is facing a ROC NodeB, then the common channel powers belonging to the CEC NodeB can be reduced by typically 3 dB. This helps improve the balance of soft handover radio links when a UE is simultaneously connected to both a CEC NodeB and a ROC NodeB. The ROC NodeB includes a 5.6 dB splitter in the downlink direction and so the CPICH power is significantly less than that for a CEC NodeB. Reducing the power at the CEC NodeB helps improve the balance.

Table 4.12 presents the peak downlink transmit powers required by the common channels at the start and end of a radio frame time slot.

Table 4.12 Calculation of the Peak Downlink Transmit Power for the Common Channels

Service Type	*Active at the Start of a Slot*	*Default Power*	*Active at the End of a Slot*	*Default Power*
CPICH	Yes	33 dBm	Yes	33 dBm
P-SCH	Yes	30 dBm	No	—
S-SCH	Yes	30 dBm	No	—
P-CCPCH	No	—	Yes	28 dBm
S-CCPCH	Yes (TFCI bits)	37 dBm	Yes (data bits)	33 dBm
PICH	Yes	25 dBm	Yes	25 dBm
AICH	Yes	25 dBm	Yes	25 dBm
Total	—	39.8 dBm 9.6 W	—	37.2 dBm 5.3 W

The instantaneous downlink transmit power requirement for the common channels at the start of a time slot is relatively high. This is caused by the S-CCPCH TFCI bits coinciding with the CPICH, the P-SCH, and the S-SCH. The frame structure of the S-CCPCH is offset in time relative to that of the CPICH, the P-SCH, and the S-SCH. For example, Nokia's implementation uses an offset of exactly three time slots, which means the start of each slot still coincides. This relatively high power is required during the first 10% of each time slot. The RNC does not have visibility of this peak transmit power. The RNC is only informed of the average downlink transmit power. This figure is averaged over the duration of a radio resource indication period, which is typically 20 radio frames.

In general, simulations and field trials should be used to determine appropriate common channel transmit powers. However, a sanity check can be completed by comparing the approximate link budget of the common channels with the link budget of an example downlink service. Table 4.13 presents the parameters used to compute the maximum downlink transmit power for the 64-kbps data service. The resulting maximum downlink transmit power is also included.

This transmit power can be compared with the transmit power assigned to each of the common channels. The difference should be equal to the difference in the C/I requirements. The C/I requirement is a function of the processing gain and the E_b/N_o requirement. The E_b/N_o requirements of the downlink common channels are not well known and have been excluded from this analysis. Excluding them introduces an uncertainty in the order of 1 dB. Table 4.14 presents a comparison of the processing gain differences and the transmit power differences for the P-CCPCH and S-CCPCH.

Table 4.13 The Maximum Downlink Transmit Power for the 64-kbps Data Service

PtxPrimaryCPICH	*33 dBm*
CPICH to RefRAB Offset	2 dB
User plane bit rate	64 kbps
User plane RNC E_b/N_o	4.5 dB
Signaling radio bearer (SRB) bit rate	3.7 kbps
Signaling radio bearer (SRB) RNC E_b/N_o	8.0 dB
Reference service bit rate	12.2 kbps
Reference service E_b/N_o	8.0 dB
Maximum downlink transmit power	35.2 dBm

If the downlink common channels have the same coverage as the downlink 64-kbps data service, then the processing gain delta should equal the transmit power delta. These figures indicate that the P-CCPCH and S-CCPCH have similar coverage to the 64-kbps downlink data service when the default RNC database is used. The calculations could be repeated for other downlink services and similar results should be obtained (i.e., the maximum downlink transmit power is computed such that each downlink service has approximately the same coverage).

4.2.7 Downlink Transmit Power Calculations

A radio link represents the physical connection across the air interface between a NodeB and a UE. Admission control functionality within the RNC is responsible for calculating the maximum, minimum, and initial downlink transmit power for each radio link. Calculations are completed for each of the following scenarios:

- A UE moves from RRC Idle into CELL_DCH to establish a new dedicated radio link.

Table 4.14 Comparison of the P-CCPCH and S-CCPCH Link Budgets with the 64-kbps Data Service Link Budget

	Bit Rate	*Default Transmit Power*	*Processing Gain Delta*	*Transmit Power Delta*	*Link Budget Delta*
P-CCPCH	12.3 kbps	28 dBm	7.2 dB	7.2 dB	0.0 dB
S-CCPCH	33.6 kbps	33 dBm	2.8 dB	2.2 dB	0.6 dB

- A UE moves from CELL_FACH to CELL_DCH to establish a new dedicated radio link.
- The configuration of an existing radio link is changed while in CELL_DCH.
- A soft handover radio link addition is triggered while in CELL_DCH.

4.2.7.1 UE Moves from RRC Idle to CELL_DCH

When a UE moves from RRC Idle to CELL_DCH, it establishes a new radio link. The new radio link encapsulates four stand-alone signaling radio bearers (stand-alone SRB), which are multiplexed into a single DCH transport channel and a single DPCH physical channel. Admission control calculates the maximum, minimum, and initial downlink transmit powers for the new radio link. The NodeB is informed of the results using an NBAP: Radio Link Setup Request message. A stand-alone SRB can be configured to have a bit rate of either 3.4 kbps or 13.6 kbps. These bit rates correspond to transmission time intervals (TTIs) of 40 ms and 10 ms, respectively. Admission control uses a different set of equations for each bit rate. The maximum downlink transmit power is computed using the equations:

$$StandaloneSRBMaxDLPower = \mathrm{Min}(Ptxmax + PtxDPCHmax,\ MaxDLCalculated)$$

where,

$Ptxmax$ = 43 dBm for a 20-W cell and $Ptxmax$ = 46 dBm for a 40-W cell.

$$MaxDLCalculated(3.4\,\mathrm{kbps}) = PTxPrimaryCPICH - CPICHtoRefRABOffset + 10 \times \mathrm{LOG}\left[\frac{\left(10^{\frac{E_b/N_{o\,SRB}}{10}} \times 3.7\right)}{\left(10^{\frac{E_b/N_{o\,Ref}}{10}} \times BR_{Ref}\right)}\right] + 2$$

and

$$MaxDLCalculated(13.6\,\mathrm{kbps}) = PTxPrimaryCPICH - CPICHtoRefRABOffset + 10 \times \mathrm{LOG}\left[\frac{\left(10^{\frac{E_b/N_{o\,SRB}}{10}} \times 14.8\right)}{\left(10^{\frac{E_b/N_{o\,Ref}}{10}} \times BR_{Ref}\right)}\right]$$

The equation for the 3.4-kbps stand-alone SRB includes an additional 2 dB that is hard-coded within the RNC implementation. This factor has been included to help improve coverage. The stand-alone SRB bit rates used in these equations include the MAC and RLC layer overheads. The MAC layer uses a 4-bit header to indicate which SRB is being used at any point in time. The RLC layer uses an 8-bit header for unacknowledged mode SRB and a 16-bit header for acknowledged mode SRB. The primary content of the RLC header is the sequence number.

The minimum downlink transmit power equation is the same for both stand-alone SRB bit rates. The minimum downlink transmit power is computed using the equation:

$$StandaloneSRBMinDLPower = \text{Max}(StandaloneSRBMaxDLPower - PCrangeDL, PtxMax + PtxDPCH\text{min})$$

The initial downlink transmit power is computed using the equations:

$$InitialSRBDLPower(3.4\text{ kbps}) = 10 \times \text{LOG}\left\{\frac{\left(10^{\frac{E_b/N_o\,SRB}{10}} \times 3.7\right)}{ChipRate} \times \left[\frac{10^{\frac{PTxPrimaryCPICH}{10}}}{10^{\frac{CPICH\_E_c/I_o-6}{10}}} - \alpha \times 10^{\frac{PtxTotal}{10}}\right]\right\}$$

and

$$InitialSRBDLPower(13.6\text{ kbps}) = 10 \times \text{LOG}\left\{\frac{\left(10^{\frac{E_b/N_o\,SRB}{10}} \times 14.8\right)}{ChipRate} \times \left[\frac{10^{\frac{PTxPrimaryCPICH}{10}}}{10^{\frac{CPICH\_E_c/I_o}{10}}} - \alpha \times 10^{\frac{PtxTotal}{10}}\right]\right\}$$

These equations rely upon the UE providing a CPICH E_c/I_o measurement within the RRC: RRC Connection Request message and the NodeB providing a PtxTotal measurement within an NBAP: Radio Resource Indication message. If the UE does not provide a CPICH E_c/I_o measurement, these equations cannot be applied and the initial downlink transmit power is set equal to the maximum downlink transmit power. The equation for the 3.4-kbps stand-alone SRB includes an additional 6 dB, which is subtracted from the CPICH E_c/I_o measurement. This 6 dB, is hard-coded within the RNC implementation and has been included to help the performance of initial synchronization across the air interface. Similar to the equations used for the maximum downlink transmit power, the MAC and RLC overheads are included as part of the stand-alone SRB bit rates. A NodeB transmits at the initial downlink transmit power until both downlink and uplink air-interface synchronization has been achieved. Inner loop power control manages the downlink transmit power once synchronization has been achieved.

4.2.7.2 UE Moves from CELL_FACH to CELL_DCH

UE are instructed to move from CELL_FACH to CELL_DCH when either an uplink or downlink capacity request has been triggered by exceeding specific RLC buffer thresholds. These thresholds are defined by the RNC parameters NASsignVolThrUL, NASsignVolThrDL, TrafVolThresholdULLow, and TrafVolThresholdDLLow. The first two of these parameters are based upon only SRB 3 and 4. The third parameter is based upon all SRB plus any non-real time radio bearer traffic. The fourth parameter is based upon SRB 3 and 4 plus any non-real time radio bearer traffic. UE are also instructed to move from CELL_FACH to CELL_DCH when there is a request for a CS domain service.

When a UE moves from CELL_FACH to CELL_DCH, it establishes a new radio link. The new radio link may encapsulate only four stand-alone signaling radio bearers (stand-alone SRB) multiplexed into a single DCH transport channel and a single DPCH physical channel. This is the case when the transition is triggered for signaling purposes or for SMS (SMS make use of SRB 3 and 4). If the new radio link encapsulates only four stand-alone SRB, then the downlink transmit power calculations are the same as those used when a UE moves from RRC Idle to CELL_DCH.

Alternatively, the new radio link may encapsulate four SRB multiplexed into one DCH transport channel while a user plane bearer is fed into a second DCH transport channel. The two DCH transport channels are multiplexed into a single physical channel. It is also possible for there to be more than a single user plane transport channel. For example, if the transition from CELL_FACH to CELL_DCH has been triggered by a request for the speech service, there will be three user plane transport channels (for the class A, class B, and class C bits), plus the SRB transport channel, multiplexed into a single physical channel. Admission control calculates the maximum, minimum, and initial downlink transmit powers for the new radio link by accounting for both the SRB and the user plane transport channel(s). The calculation for the maximum downlink transmit power is dependent upon the traffic class of the user plane transport channel(s) (i.e., whether it is, or they are, being used to support conversational, streaming, interactive, or background services). The NodeB is informed of the results using an NBAP: Radio Link Setup Request message. The maximum downlink transmit power is computed using the equations:

$$MaxDLPower(Conversational/Streaming) = \mathrm{Min}(Ptxmax + PtxDPCHmax, MaxDLCalculated)$$

and

$$MaxDLPower(Interactive/Background) = \mathrm{Min}(Ptxmax + PtxDPCHmax, MaxDLCalculated, PtxDLabsMax)$$

where,

$Ptxmax = 43$ dBm for a 20-W cell and $Ptxmax = 46$ dBm for a 40-W cell

$$MaxDLCalculated = PTxPrimaryCPICH - CPICHtoRefRABOffset + 10 \times \mathrm{LOG}\left[\frac{\left(10^{\frac{E_b/N_{o\,SRB}}{10}} \times 3.7\right) + \left(10^{\frac{E_b/N_{o\,Serv}}{10}} \times BR_{Serv}\right)}{\left(10^{\frac{E_b/N_{o\,Ref}}{10}} \times BR_{Ref}\right)}\right]$$

The SRB bit rate includes the MAC and RLC layer overheads (i.e., the bit rate is 3.7 kbps rather than 3.4 kbps). Only the 3.4-kbps SRB bit rate is applicable when the SRB is multiplexed with a user plane transport channel. This is independent of whether or not the RNC database parameter StandAloneDCCHBitRate has been configured with a value of 13.6 kbps. The service bit rates used in this equation do not include any layer 2 overheads (i.e., the bit rates should be 12.2 kbps, 64 kbps, 128 kbps, or 384 kbps). In the case of the speech service, the E_b/N_o requirement for the class A bits is used. If dynamic link optimization (DyLO) is applicable to a specific interactive or background service, the maximum downlink transmit power is effectively 2 dB less than the calculated result. This is because DyLO is triggered once the RNC is informed that the downlink transmit power has exceeded a threshold that is 2 dB less than the maximum. Short term increases above this threshold can be experienced without triggering DyLO if the average downlink transmit power reported to the RNC remains below the threshold.

The minimum downlink transmit power equation is the same for all traffic classes. The minimum downlink transmit power is computed using the equation:

$$MinDLPower = \mathrm{Max}(MaxDLPower - PCrangeDL, Ptxmax + PtxDPCHmin)$$

The initial downlink transmit power is computed using the equation:

$$InitialDLPower = 10 \times \mathrm{LOG}\left\{\frac{\left(10^{\frac{E_b/N_{o\,SRB}}{10}} \times 3.7\right) + \left(10^{\frac{E_b/N_{o\,Serv}}{10}} \times BR_{Serv}\right)}{ChipRate} \times \left[\frac{10^{\frac{PTxPrimaryCPICH}{10}}}{10^{\frac{CPICH\_E_c/I_o}{10}}} - \alpha \times 10^{\frac{PtxTotal}{10}}\right]\right\}$$

This equation relies upon the UE providing a CPICH E_c/I_o measurement and the NodeB providing a PtxTotal measurement within an NBAP: Radio Resource

Indication message. If the UE does not provide a CPICH E_c/I_o measurement, then these equations cannot be applied and the initial downlink transmit power is set equal to the maximum downlink transmit power. Similar to the equation used for the maximum downlink transmit power, the MAC and RLC overheads are included as part of the SRB bit rate (i.e., the bit rate is 3.7 kbps rather than 3.4 kbps). The service bit rate does not include any layer 2 overheads (i.e., the bit rate should be 12.2 kbps, 64 kbps, 128 kbps, or 384 kbps). In the case of the speech service, the E_b/N_o requirement for the class A bits is used. A NodeB transmits at the initial downlink transmit power until both downlink and uplink air-interface synchronization has been achieved. Inner loop power control manages the downlink transmit power once synchronization has been achieved.

4.2.7.3 Configuration of an Existing Radio Link is Changed

If a UE has moved from RRC Idle to CELL_DCH for the transfer of user plane data (i.e., for the speech service, CS data service, or PS data service), then the initial radio link configured for the stand-alone SRB is reconfigured during RAB establishment. This reconfiguration allows the radio link to accommodate the user plane data (i.e., user plane transport channels are added). Alternatively, a radio link may be reconfigured to accommodate a change in an existing user plane data rate. For example, the use of Dynamic Link Optimization (DyLO) may reduce the bit rate during an interactive or background packet switched data session. Packet switched bit rates may also be reduced if pre-emption is triggered or if the cell enters the overload state. The user plane packet switched bit rate may be increased if RLC buffer thresholds are exceeded and an associated capacity request is generated. Reconfiguration of the radio link involves changes to the maximum, minimum, and initial downlink transmit powers.

The calculation for the maximum downlink transmit power is dependant upon the traffic class of the user plane transport channel(s) (i.e., whether it is, or they are, being used to support conversational, streaming, interactive, or background services). The NodeB is informed of the results using an NBAP: Radio Link Reconfiguration Prepare message. The maximum downlink transmit power is computed using the equations:

$$\begin{aligned} MaxDLPower(Conversational/Streaming) = {} & \mathrm{Min}(Ptxmax + PtxDPCHmax, \\ & MaxDLCalculated) \end{aligned}$$

and

$$\begin{aligned} MaxDLPower(Interactive/Background) = {} & \mathrm{Min}(Ptxmax + PtxDPCHmax, \\ & MaxDLCalculated, PtxDLabsMax) \end{aligned}$$

where,

$Ptxmax = 43$ dBm for a 20-W cell and $Ptxmax = 46$ dBm for a 40-W cell.

$$MaxDLCalculated = PTxPrimaryCPICH - CPICHtoRefRABOffset + 10 \times \mathrm{LOG}\left[\frac{\left(10^{\frac{E_b/N_o\,SRB}{10}} \times 3.7\right) + \left(10^{\frac{E_b/N_o\,NewServ}{10}} \times BR_{NewServ}\right)}{\left(10^{\frac{E_b/N_o\,Ref}{10}} \times BR_{Ref}\right)}\right]$$

The service bit rates used in this equation do not include any layer 2 overheads (i.e., the bit rates should be 12.2 kbps, 64 kbps, 128 kbps, or 384 kbps). This is in contrast to the SRB bit rate which does include the layer 2 overheads. If the reconfiguration is for the addition of user plane transport channels to a 13.6-kbps stand-alone SRB, then the SRB bit rate is reduced to 3.4 kbps. In the case of the speech service, the E_b/N_o requirement for the class A bits is used. If dynamic link optimization (DyLO) is applicable to a specific interactive or background service, the maximum downlink transmit power is effectively 2 dB less than the calculated result. This is because DyLO is triggered once the RNC is informed that the downlink transmit power has exceeded a threshold which is 2 dB less than the maximum. Short-term increases above this threshold can be experienced without triggering DyLO if the average downlink transmit power reported to the RNC remains below the threshold.

The minimum downlink transmit power is computed using the equation:

$$MinDLPower = \mathrm{Max}(MaxDLPower - PCrangeDL, PtxMax + PtxDPCH\mathrm{min})$$

Under normal circumstances, the initial transmit power of the reconfigured radio link is based upon the PtxAverage measurements provided to the RNC from the NodeB. Assuming these measurements are available, the initial transmit power is calculated using one of the following equations. The first equation is applied when the reconfiguration is from a stand-alone SRB and there are no existing user plane transport channels.

$$InitialDLPower = 10 \times \mathrm{LOG} \times \left[\frac{\left(10^{\frac{E_b/N_o\,SRB}{10}} \times 3.7\right) + \left(10^{\frac{E_b/N_o\,NewServ}{10}} \times BR_{NewServ}\right)}{\left(10^{\frac{E_b/N_o\,SRB}{10}} \times BR_{SRB}\right)}\right] + PtxAverage\_DPDCH$$

The following equation is applied when there are existing user plane transport channels (e.g., a change in bit rate resulting from DyLO).

$$InitialDLPower = 10 \times \mathrm{LOG} \times \left[\frac{\left(10^{\frac{E_b/N_o\,_{SRB}}{10}} \times 3.7\right) + \left(10^{\frac{E_b/N_o\,_{NewServ}}{10}} \times BR_{NewServ}\right)}{\left(10^{\frac{E_b/N_o\,_{SRB}}{10}} \times 3.7\right) + \left(10^{\frac{E_b/N_o\,_{OldServ}}{10}} \times BR_{OldServ}\right)} \right] + PtxAverage\_DPDCH$$

The service bit rates used in these equations do not include any layer 2 overheads (i.e., the bit rates should be 12.2 kbps, 64 kbps, 128 kbps, or 384 kbps). This is in contrast to the SRB bit rate which does include the layer 2 overheads. In the case of the speech service, the E_b/N_o requirement for the class A bits is used. The value of PtxAverage_DPDCH is an adjusted value of the measurement result provided by the NodeB. The measurement result is based upon the DPCCH pilot bits and needs to have the value of PO3 subtracted to obtain the DPDCH power. The value of PO3 is configured using the hidden RNC database parameter PowerOffsetDLdpcchPilot. This parameter is currently configured with a value of 0 dB. If the PtxAverage measurement is not available, then PtxAverage_DPDCH is replaced by the last calculated value for the initial downlink transmit power. In the case of radio link reconfiguration, the UE and NodeB are already synchronized across the air-interface. This means that inner loop power control is able to change the initial downlink transmit power as soon as it is applied.

4.2.7.4 Soft Handover Radio Link Additions

When a new radio link is added to a UE's active set, the RNC admission control calculates the maximum, minimum, and initial downlink transmit powers. The new radio link may be for either soft or softer handover. Radio links can be added while there is a stand-alone SRB or an SRB multiplexed with user plane data. Radio link addition can be triggered by either measurement reporting event 1a or 1c. The latter involves a radio link deletion as well as a radio link addition. These measurement reporting events are only applicable while a UE is in CELL_DCH. If a new radio link is added while there is only a stand-alone SRB configured, the maximum and minimum downlink transmit powers are calculated according to the equations used when a UE moves from RRC_Idle to CELL_DCH. If a new radio link is added while there are both SRB and user plane transport channels configured, then the maximum and minimum downlink transmit powers are calculated according to the equations used when a UE moves from CELL_FACH to CELL_DCH.

In the case of soft handover, if a new radio link is added while there is only a stand-alone SRB configured, then the initial downlink transmit power is computed according to the equations used when a UE moves from RRC_Idle to CELL_DCH. In the case of soft handover, if a new radio link is added while there are both SRB and user plane transport channels configured, then the initial downlink transmit power is computed according to the equations used when a UE moves from CELL_FACH to CELL_DCH. These equations rely upon the UE providing a CPICH E_c/I_o measurement within the RRC: RRC Connection Request message and the NodeB providing a PtxTotal measurement within an NBAP: Radio Resource Indication message. If the UE does not provide a CPICH E_c/I_o measurement, then these equations cannot be applied and the initial downlink transmit power is set equal to the highest measured downlink transmit power from the existing active set radio links. In the case of soft handover, when a new radio link is added to the active set the NodeB does not start transmitting at the initial transmit power. The NodeB software instructs the NodeB to transmit at the minimum transmit power until uplink synchronization has been achieved. Once uplink synchronization has been achieved, the NodeB transmits at the initial transmit power and inner loop power control is able to operate in its normal way. This approach has been adopted to help reduce differences in the downlink transmit powers during soft handover and to help avoid potential increases in uplink interference. If a NodeB was to start transmitting at its initial transmit power before achieving uplink synchronization, the UE would achieve synchronization and its downlink SIR would increase. The UE would then instruct the active set cells to power down. All of the active set cells except the new cell would power down. The new cell would not power down because it had not achieved uplink synchronization and would not be receiving power control commands. The original radio links would then have relatively poor quality and the UE would start to discard their TPC commands. The UE would obey the relatively high-quality TPC commands from the new radio link which, during the synchronization phase, instruct the UE to increase its transmit power. Transmitting at the minimum downlink transmit power during the period prior to uplink synchronization means that the UE is less likely to achieve downlink synchronization, and if it does then the increase in its SIR will be less.

In the case of softer handover, the initial downlink transmit power is set equal to the transmit power of the existing radio link belonging to the same NodeB. A NodeB maintains a single uplink synchronization state machine for all radio links associated with a specific UE (i.e., all radio links are combined within the same RAKE receiver). This means that a NodeB is already synchronized when a softer handover radio link is added to the active set. The issues associated with soft handover do not exist for softer handover and it is not necessary for the NodeB to start transmitting at the minimum downlink transmit power.

4.2.8 Uplink Open Loop Power Control

Uplink open loop power control is completed by the UE as part of the random access procedure required whenever the UE communicates with the network in RRC states

Table 4.15 Node B Commissioning Parameters Relevant to Uplink Open Loop Power Control

Name	*Range*	*Default (MHA Present)*	*Default (MHA not Present)*
MHA_in_use	Yes, No	Yes	No
L_Feeder	0 to 99, step 0.1 dB	3 dB or actual feeder loss	0 dB
G_mha	0 to 99, step 0.1 dB	12 dB	0 dB

IDLE mode, CELL_FACH, CELL_PCH, or URA_PCH. Uplink open loop power control is also required whenever the UE is instructed to establish a dedicated channel and enter RRC state CELL_DCH. If the uplink open loop power control calculations are inaccurate, then the UE may transmit with too little power and messages may be lost, or transmit with too much power and excessive interference may be generated at the NodeB receiver.

4.2.8.1 NodeB Commissioning Parameters

The NodeB commissioning parameters relevant to uplink open loop power control are presented in Table 4.15.

4.2.8.2 PRACH Concepts and Approaches

3GPP TS25.331 [11] specifies the open loop power control equation for the PRACH. This equation, presented next, determines the transmit power that the UE applies to the first PRACH preamble.

$$\begin{aligned} \mathit{PRACHInitialPower} &= \mathit{ReceiverPowerRequirement} + \mathit{LinkLoss} \\ &= (\mathit{PRACHRequiredReceivedCI} + \mathit{ULRSSI}) \\ &\quad + (\mathit{PtxPrimaryCPICH} - \mathit{CPICHRSCP}) \end{aligned}$$

The PRACHRequiredReceivedCI is an RNC database parameter with a default value of –25 dB. It is read by the UE from System Information Block 5 of the BCH. Its value can be compared with the C/I requirement for the speech service, which is typically –21 dB. The C/I requirement for the PRACH preamble can be lower because the NodeB receiver only needs to detect the PRACH scrambling code and PRACH signature. There are no data bits to decode. In addition, TS25.214 specifies that if the first PRACH preamble is not acknowledged by a downlink AICH, then further preambles can be transmitted. Each preamble is transmitted with a power greater than the preceding preamble. The increase in power between consecutive preambles is defined by the RNC database parameter PowerRampStepPRACH-preamble. This parameter has a default value of 1 dB. The maximum number of preambles that can be transmitted in a single power ramping cycle is defined by the RNC database parameter PRACH_preamble_retrans. This parameter has a default value of 8, which means that the final preamble belonging to a cycle would have a transmit power that is 7 dB greater than the first. A preamble cycle may be stopped before transmitting the final preamble if the commanded preamble power is 6 dB greater than the maximum allowed PRACH transmit power. If the UE does not receive an AICH during a preamble cycle, then it may start further preamble cycles. The maximum number of preamble cycles is defined by the RNC database parameter RACH_tx_Max. The power ramping that occurs during a PRACH preamble cycle means it is not critical if the open loop power control calculation generates an initial transmit power which is 2 or 3 dB too low. However, generating an initial power that is lower than this could reduce the reliability of the PRACH procedure when the UE is in a fading radio environment. The requirement for further PRACH preambles and preamble cycles has a relatively small impact upon the connection establishment delay.

The uplink RSSI figure that appears in the PRACH open loop power control equation is read from System Information Block 7 of the BCH. This figure is measured and provided by the NodeB. If this measurement is inaccurate, it will make the PRACH open loop power control calculation inaccurate. The Release 99 version of TS25.215 specifies that the measurement reference point for the uplink RSSI is at the output of the NodeB's pulse shaping filter. The Release 4 versions of TS25.215 and TS25.104 specify that if an MHA is not installed, then the measurement reference point for the uplink RSSI is at the NodeB cabinet antenna connector, and that if an MHA is installed, then the measurement reference point is at the air-interface side of the MHA. In the case of this measurement, the Release 4 versions of TS25.215 and TS25.104 have been implemented. This means that when an MHA is installed, the NodeB must complete a power measurement within its WTR and then calculate the corresponding power at the air-interface side of the MHA. The NodeB completes this calculation using the parameters configured in the NodeB commissioning file. If these parameters are configured inaccurately, then the resultant RSSI figure will also be inaccurate. The calculation completed by the NodeB involves compensating for the automatic gain control within the WTR, the gain provided by the WAF, cable and connector losses between the NodeB cabinet and the MHA, and the MHA gain. If it is assumed that the NodeB has accurate information regarding the automatic gain control within the WTR and the gain provided by the WAF, then the calculation becomes:

$$ULRSSI(MHA\_installed) = RSSI\_NodeB\_Cabinet + FeederLoss - MHA\_Gain$$

The feeder loss and MHA gain in this equation are the values that have been configured during the NodeB commissioning. If NodeB commissioning has been completed using a default feeder loss value rather than the actual feeder loss value, then inaccuracies will be introduced across the network. If the commissioned feeder loss value is less than the actual feeder loss value, then the uplink RSSI result will be too low. If the commissioned feeder loss value is greater than the actual feeder loss value, then the uplink RSSI result will be too high. The same argument can be applied to the MHA gain although an MHA has a fixed gain of 12 dB with a narrow tolerance. When an MHA is not installed, then the measurement reference point should be kept at the NodeB cabinet antenna connector. In this case, the calculation of uplink RSSI depends upon the version of the NodeB software being used. In the case of WN1 software, the NodeB completes the uplink RSSI calculation according to the following equation:

$$ULRSSI(MHA\_not\_installed\_WN1) = RSSI\_NodeB\_Cabinet + FeederLoss$$

The feeder loss value configured during NodeB commissioning should be assigned a value of 0 dB. If a non-zero value is assigned, then the uplink RSSI will appear too high and will cause the UE to transmit with an increased initial power. In the case of WN2 or later software, the NodeB completes the uplink RSSI calculation according to the following equation:

$$ULRSSI(MHA\_not\_installed\_WN2) = RSSI\_NodeB\_Cabinet$$

In this case, the feeder loss value configured during NodeB commissioning is not involved in the uplink RSSI measurement.

The CPICH transmit power figure that appears in the PRACH open loop power control equation is read from System Information Block 5 of the BCH. This figure reflects the value of the RNC database parameter PtxPrimaryCPICH. The measurement reference point for this parameter is at the NodeB cabinet antenna connector. This means that when an MHA is installed then the uplink RSSI and the downlink CPICH transmit power have different measurement reference points. This is a source of error in the PRACH open loop power control calculation. Future RAN releases will make use of the RNC database parameters CableLoss and MHA to move the CPICH transmit power measurement reference point out to the air-interface side of the MHA. This is not implemented for the PRACH open loop power control in RAN1.5.2.ED2. The use of different measurement reference points means that the UE overestimates the link loss, which leads to an increased initial PRACH transmit power. This can be viewed as introducing some margin into the calculation, or the PRACHRequiredReceivedCI parameter could be reduced by the value of the feeder loss to compensate. It should also be noted that a downlink link loss measurement is being used for an uplink transmit power calculation without accounting for propagation differences between the uplink and downlink frequencies. The propagation loss is greater in the downlink direction, meaning that further margin is introduced.

The CPICH RSCP figure that appears in the PRACH open loop power control equation is measured by the UE. 3GPP TS25.133 specifies that under normal conditions the CPICH RSCP measurement can be up to 6 dB greater than or 6 dB less than the actual value. This is a further potential source of error that indicates it is beneficial to include some margin in the PRACH open loop power control calculation.

4.2.8.3 DPCH Concepts and Approaches

The uplink DPCH open loop power control is not standardized in as much detail as the PRACH open loop power control. 3GPP TS25.331 [11] specifies the open loop power control equation for the uplink DPCH as:

$$UL\_DPCHInitialPower = DPCCHPowerOffset - CPICHRSCP$$

The definition of the DPCCHPowerOffset is vendor dependent. For example, Nokia's implementation is given by:

$$DPCCHPowerOffset = PtxPrimaryCPICH + RSSI + UL\_SIR - SF - Pcomp$$
$$MHA = 0 \Rightarrow Pcomp = 0$$
$$MHA = 1 \Rightarrow Pcomp = CableLoss$$

This implementation can be understood if the complete equation is presented in the same format as the PRACH equation, that is:

$$\begin{aligned} UL\_DPCHInitialPower &= ReceiverPowerRequirement + LinkLoss \\ &= (UL\_SIR - SF + RSSI) + (PtxPrimaryCPICH - Pcomp - CPICHRSCP) \end{aligned}$$

The combination of the uplink SIR and the spreading factor is equivalent to the PRACH C/I requirement. The spreading factor is subtracted from the SIR because the 3GPP definition for the uplink SIR is given in TS25.215 as:

$$UL\_SIR = RSCP/ISCP + SF$$

Including the spreading factor in this definition normalizes the SIR values for different bit rates and reduces the range of SIR values that the network is required to handle. The uplink SIR values are calculated by the RNC from the E_b/N_o tables, which are hidden parameters within the RNC database. In general, the E_b/N_o values in these tables are relatively high and introduce some margin in the DPCH open loop power control calculation. The E_b/N_o values in the RNC database tend to be applicable to harsh propagation conditions.

The uplink RSSI figure which appears in the DPCH open loop power control equation is read from System Information Block 7 of the BCH. This figure is the same as that used during the PRACH open loop power control and has the same issues associated with inaccuracies in the NodeB commissioning file. If an MHA is installed, then the NodeB commissioning parameter for feeder loss should be configured accurately. If an MHA is not installed, then the NodeB commissioning parameter for feeder loss should be configured with a value of 0 dB.

The CPICH transmit power figure which appears in the DPCH open loop power control equation is read from System Information Block 5 of the BCH. This figure is the same as that used during the PRACH open loop power control. However, the RAN1.5.2.ED2 implementation for the DPCH open loop power control includes compensation to move the CPICH transmit power reference point to the same point as the uplink RSSI measurement reference point. This is done using the variable Pcomp and the RNC database parameters MHA and CableLoss. If MHA is configured with a value of 1 to indicate that an MHA is installed, then the CPICH transmit power is reduced by a quantity equal to the value of the cable loss parameter. This increases the accuracy of the DPCH open loop power control relative to the accuracy of the PRACH power control.

The CPICH RSCP figure that appears in the DPCH open loop power control equation is measured by the UE in the same way it is measured for the PRACH open loop power control. 3GPP TS25.133 specifies that under normal conditions the CPICH RSCP measurement can be up to 6 dB greater than or 6 dB less than the actual value.

An important issue raised regarding the initial uplink DPCH transmit power is that the inner loop power control behavior can potentially lead to uplink power spikes at the NodeB receiver. 3GPP TS25.214 specifies that during radio link initialization the NodeB can transmit TPC commands that are either a continuous stream of power-up commands, or a pattern of alternating power-up and power-down commands but with an additional power-up command in every 10-ms radio frame. Usually the latter is implemented, which has the effect of increasing the UE transmit power by the power control step size once every 10 ms. If the NodeB does not synchronize rapidly, then a UE can significantly increase its transmit power. Figure 4.8 illustrates the signaling and timing associated with uplink and downlink air-interface synchronization.

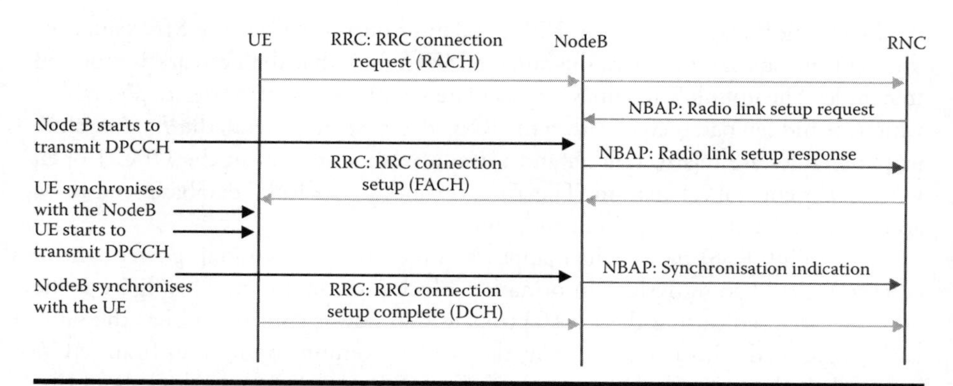

Figure 4.8 The signaling and timing associated with uplink and downlink air-interface synchronization.

The NodeB starts to transmit its DPCCH once it has received the NBAP radio link setup message from the RNC. This DPCCH includes the pattern of TPC commands which instruct an increase in the uplink power once every 10 ms. The UE receives information regarding the format of the downlink DPCCH in the RRC Connection Setup message. The UE is then able to synchronize with the downlink DPCCH. Once the UE has synchronized, it starts to transmit its uplink DPCCH. The initial power of the uplink DPCCH is determined by the DPCH open loop power control calculation, but the UE then obeys the TPC commands appearing in the downlink DPCCH. The UE thus starts to increase its transmit power by one step size every 10 ms. The NodeB attempts to synchronize with the uplink DPCCH. If it achieves synchronization, then inner loop power control starts to behave in the normal way. If synchronization is not achieved or if the NodeB takes a relatively long time to achieve synchronization, then the UE transmit power can be increased to the extent that interference is incurred by the NodeB.

Usually it is suggested that the CableLoss RNC parameter should be increased by 15 dB to help avoid generating spikes of uplink interference. Increasing the cable loss parameter by 15 dB meant that the UE will start to transmit its DPCCH with 15 dB less power. Initially this will make it more difficult for the NodeB to achieve uplink synchronization, but it will also give the NodeB more time to synchronize before incurring large quantities of uplink interference. It has been shown by lab tests that the connection establishment time was increased by approximately 70 ms when adopting this approach. However, product lines have now withdrawn Technical Note 47 and replaced it with Technical Note 62. Technical Note 62 explains that the additional 15 dB is unlikely to be required and that the CableLoss parameter should be configured with a value equal to the feeder loss if an MHA is being used.

4.2.9 Soft Handover

The soft handover procedure makes use of intra-frequency reporting events 1a, 1b, and 1c. These events are specified by 3GPP TS25.331 [11]. They can be based upon either path loss, CPICH RSCP, or CPICH E_c/I_o. In most cases, CPICH E_c/I_o is used because the associated measurements are more accurate than either CPICH RSCP or path loss. Reporting events 1a, 1b, and 1c are evaluated by the UE at layer 3 after layer 3 filtering has been completed. Layer 3 of the UE receives CPICH E_c/I_o measurements from layer 1 of the UE. Layer 1 of the UE records these measurements over a finite period of time. This introduces some filtering at layer 1 to help remove the effects of fast fading. In general, further filtering is required at layer 3. Layer 3 filtering means that soft handover events are less likely to be triggered by short-term changes in the radio conditions. It also means that soft handover is less responsive. Scenarios which have a requirement for very rapid soft handover (e.g., high-speed trains) can have layer 3 filtering disabled. Layer 3 filtering is configured by the RNC, which broadcasts the filter coefficient within SIB11 and SIB12. The RNC also informs the UE of the filter coefficient using a dedicated measurement control message when entering connected mode state CELL_DCH. Layer 3 filtering is specified by 3GPP TS25.331 using the equation:

$$Filter\_Result_n = (1 - a) \times Filter\_Result_{n-1} + a \times Meas_n$$

where,

$$a = 0.5^{Filter\_Coefficient/2}$$

The Release 99 version of TS25.331 does not specify the units of the measurements that form the input to the filter (i.e., they could be linear or logarithmic). The Release 5 version of TS25.331 specifies that the units should be logarithmic. This is intended to make the reporting events more responsive. If the filter coefficient is configured with a value of 0, then a = 1 and layer 3 filtering is disabled (i.e., the filter result equals the input measurement). As the value of the filter coefficient increases, so too does the memory of the filter, and previous input measurements have a greater influence upon the filter result. The measurement period for the filter is specified by 3GPP TS25.133 to be 200 ms. This means that the layer 3 filter expects an input measurement every 200 ms and generates an output result every 200 ms. However, the time-to-trigger associated with reporting events 1a, 1b, and 1c can be specified to have durations as small as 10 ms. If layer 3 filtering generates a result every 200 ms, then all time-to-trigger values between 10 and 200 ms would lead to equal performance. In practice, layer 1 of the UE provides layer 3 of the UE with measurements more frequently than every 200 ms. The precise rate is implementation dependant and will vary from one UE vendor to another. If a UE implementation generates filter results every 50 ms, then time-to-trigger values of 10, 20, and 40 ms would lead to equal performance, but time-to-trigger values of 60, 80, and 100 ms would lead to a different performance. The rate at which layer 1 of a UE generates measurements may depend upon the number of active set cells and neighbors being measured (i.e., if less cells are being measured, then those cells may be reported more frequently). When layer 3 filtering is completed at a rate that is more rapid than the 200 ms specified by 3GPP, then the equation used to complete the filtering must be modified to ensure the impulse response of the filter remains consistent. The impulse response of the filter represents the time history of the output when a single "1" is provided as an input. Figure 4.9 illustrates the impulse response of the filter when input measurements are provided at a rate of 200 ms and when input measurements are provided at a rate of 100 ms. The latter assumes that the filter equation has been modified appropriately.

If the filter equation was not modified when the rate of input measurements was increased to once every 100 ms, the memory of the filter would halve. In practice, it is understood that most UE use measurement input rates of between 50 and 100 ms. This means that time-to-trigger values should not be optimized with a resolution of less than 50 ms (i.e., 10, 20, and 40 ms would lead to the same performance).

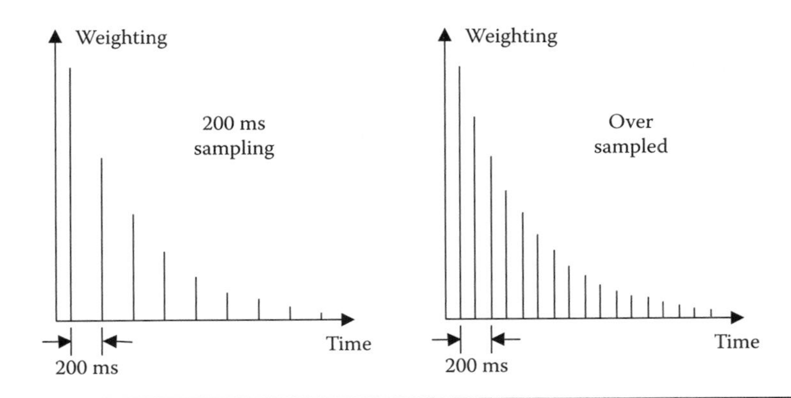

Figure 4.9 Impulse response of the layer 3 filter when inputs are provided every 200 and 100 ms.

Reporting event 1a is used to add a neighbor to the active set. Event 1a is only applicable to monitored set cells and when the active set size is less than its maximum. Usually the active set size is limited to 3. This helps avoid excessive soft handover overheads and places the emphasis upon planning areas of good dominance. 3GPP TS25.331 specifies the criteria for reporting event 1a as:

$$10\mathrm{LOG}[Meas_{neigh}] + Offset_{neigh} \geq W \times 10\mathrm{LOG}\left[\sum_{i=1}^{N} Meas_i\right] + (1 - W) \times 10\mathrm{LOG}\,[Meas_{best}] - Add\_Win$$

where,

$Meas_{neigh}$ is the candidate monitored set cell CPICH E_c/I_o measurement after layer 3 filtering.
$Offset_{neigh}$ is the RNC database parameter AdjsEcNoOffset for the candidate cell.
W is the RNC database parameter ActiveSetWeightingCoefficient.
N is the number of active set cells that are not forbidden to affect the reporting range.
$Meas_i$ is the active set cell CPICH E_c/I_o measurement after layer 3 filtering.
$Meas_{best}$ is the best active set cell CPICH E_c/I_o measurement after layer 3 filtering.
Add_Win is the RNC database parameter AdditionWindow.

The default value of the RNC database parameter ActiveSetWeightingCoefficient is 0 and so the equation for event 1a simplifies to:

$$10\mathrm{LOG}[Meas_{neigh}] + Offset_{neigh} \geq 10\mathrm{LOG}[Meas_{best}] - Add\_Win$$

Active set cells can be forbidden from affecting the reporting range by configuring the RNC database parameter AdjsDERR. The parameters for event 1a are broadcast within SIB11 and SIB12. They are also transmitted using a dedicated measurement control message when entering connected mode state CELL_DCH. Within these messages the addition window is signaled using a value twice as large as the actual value (i.e., the signaled value should be divided by 2 to obtain the actual value).

Reporting event 1b is used to remove a cell from the active set. Event 1b is only applicable to active set cells and when the active set size is greater than 1. 3GPP TS25.331 specifies the criteria for reporting event 1b as:

$$10\text{LOG}\,[Meas_{active}] + Offset_{active} \leq W \times 10\text{LOG}\left[\sum_{i=1}^{N} Meas_i\right]$$
$$+(1 - W) \times 10\text{LOG}\,[Meas_{best}] - Drop\_Win$$

where,

$Meas_{active}$ is the candidate active set cell CPICH E_c/I_o measurement after layer 3 filtering;
$Offset_{active}$ is the RNC database parameter AdjsEcNoOffset for the candidate cell;
W is the RNC database parameter ActiveSetWeightingCoefficient;
N is the number of active set cells that are not forbidden to affect the reporting range;
$Meas_i$ is the active set cell CPICH E_c/I_o measurement after layer 3 filtering;
$Meas_{best}$ is the best active set cell CPICH E_c/I_o measurement after layer 3 filtering;
$Drop\_Win$ is the RNC database parameter DropWindow.

The default value of the RNC database parameter ActiveSetWeightingCoefficient is 0 and so the equation for event 1b simplifies to:

$$10\text{LOG}[Meas_{active}] + Offset_{active} \leq 10\text{LOG}[Meas_{best}] - Drop\_Win$$

Active set cells can be forbidden from affecting the reporting range by configuring the RNC database parameter AdjsDERR. The parameters for event 1b are broadcast within SIB11 and SIB12. They are also transmitted using a dedicated measurement control message when entering connected mode state CELL_DCH. Within these messages, the drop window is signaled using a value twice as large as the actual value (i.e., the signaled value should be divided by 2 to obtain the actual value).

Reporting event 1c is used to replace an existing active set cell by a higher-quality neighbor. Event 1c involves both active set and monitored set cells. It is only applicable when the active set size is at its maximum. 3GPP TS25.331 specifies the criteria for reporting event 1c as:

$$10\mathrm{LOG}[Meas_{neigh}] + Offset_{neigh} \geq 10\mathrm{LOG}[Meas_{active\_lowest}]$$
$$+ Offset_{active\_lowest} + Replace\_Win/2$$

where,

$Meas_{neigh}$ is the candidate monitored set cell CPICH E_c/I_o measurement after layer 3 filtering;
$Offset_{active}$ is the RNC database parameter AdjsEcNoOffset for the candidate monitored set cell;
$Meas_{active\_lowest}$ is the candidate active set cell CPICH E_c/I_o measurement after layer 3 filtering;
$Offset_{active\_lowest}$ is the RNC database parameter AdjsEcNoOffset for the candidate active set cell;
$Replace\_Win$ is the RNC database parameter ReplacementWindow.

It is important to recognize that the replacement window is divided by a factor of 2. It is common to think of the replacement window as the difference between the candidate active set cell CPICH E_c/I_o and the candidate monitored set cell CPICH E_c/I_o. For example, if replacement is completed when a monitored set cell has a quality which is 1 dB greater than an active set cell, it is common to think of the replacement window as 1 dB. However, the actual value of the replacement window in this example is 2 dB. In addition, the signaled value of the replacement window is twice as large as the actual value. This means that the RNC database parameter ReplacementWindow could be configured with a value of 2 dB and a value of 4 is then signaled to the UE within SIB11, SIB12, or a dedicated measurement control message. The UE divides the signaled value by a factor of 2 to obtain the actual value of 2 dB and this corresponds to what is commonly interpreted as a 1 dB replacement window.

The performance of soft handover is optimal when the active set radio links are received with equal signal-to-noise ratios (SNR). If the radio links have unequal SNR, then the benefits of soft handover decrease. If the difference in SNR becomes too great, then weaker radio links are likely to fail. If two macrocells are configured with different CPICH transmit powers, then the soft handover area will be closer to the macrocell, which has the lower CPICH transmit power (assuming that offsets have not been applied to the UE CPICH measurements). This means that the dedicated channel radio link to the cell with the higher CPICH transmit power will be weaker. Alternatively, if neighboring cells are unequally loaded, then the cell with the higher load will experience weaker SNR during soft handover. If a cell configured with an MHA is neighbored to a cell not configured with an MHA, then the cell not configured with an MHA will experience weaker SNR during soft handover.

In addition, UE that are in soft handover between an indoor solution and a macrocell are likely to experience differences in their radio link SNR. Each of these scenarios should be studied on a case-by-case basis. The following paragraphs illustrate how an analysis can be completed.

The point at which a UE enters and leaves soft handover is defined by the CPICH E_c/I_o. CPICH E_c/I_o is computed by the UE using CPICH RSCP and RSSI measurements:

$$CPICH\_E_c/I_o = CPICH\_RSCP - RSSI$$

The RSSI measurement can be assumed to be equal for all CPICH RSCP measurements recorded at a specific time. This means that the point at which a UE enters and leaves soft handover is effectively defined by the CPICH RSCP. The CPICH RSCP received by a UE is given by:

$$CPICH\_RSCP = PtxCPICH - DL\_Link\_Loss$$

A UE enters soft handover when a neighboring cell CPICH RSCP approaches the best active set cell CPICH RSCP within a range defined by the addition window. This leads to the expression:

$$(PtxCPICH_{active1} - DL\_Link\_Loss_{active1}) - (PtxCPICH_{neigh} - DL\_Link\_Loss_{neigh}) = Addition\_Window$$

Consider the example of a UE that has one cell in the active set and one neighboring cell. Assume that the CPICH transmit power assigned to one of the cells is 3 dB less than the CPICH transmit power assigned to the other. Also assume that the addition window has been configured with its default value of 4 dB. It is then possible to specify two conditions for radio link addition:

If the active set cell has 3 dB less CPICH TX power,

$$DL\_Link\_Loss_{neigh} - DL\_Link\_Loss_{active1} = 7$$

If the active set cell has 3 dB greater CPICH TX power,

$$DL\_Link\_Loss_{neigh} - DL\_Link\_Loss_{active1} = 1$$

The first of these equations indicates that there is a 7 dB difference in downlink link loss when a UE adds a neighboring cell that has 3 dB greater CPICH transmit power. Assuming that both cells have equal loading, both are configured with MHA and both make use of uplink receive diversity, then the two cells are likely to have similar uplink link budgets. The UE transmit power will be controlled by the cell that has 7 dB less link loss. This means the other cell will, on average, receive

7 dB less power than necessary. This radio link will be more susceptible to loss of synchronization and subsequent failure.

A UE leaves soft handover when one of the active set cells has a CPICH RSCP that falls outside the range defined by the soft handover drop window. This leads to the expression:

$$(PtxCPICH_{active2} - DL\_Link\_Loss_{active2}) - (PtxCPICH_{active1} - DL\_Link\_Loss_{active1}) = Drop\_Window$$

Consider the example of a UE that has two macrocells in the active set. Assume that the CPICH transmit power assigned to one is 3 dB less than the CPICH transmit power assigned to the other. Also assume that the drop window has been configured with its default value of 6 dB. It is then possible to specify two conditions for radio link removal from the active set:

When moving away from a cell with 3 dB less CPICH power,

$$DL\_Link\_Loss_{active1} - DL\_Link\_Loss_{active2} = 3$$

When moving away from a cell with 3 dB greater CPICH power,

$$DL\_Link\_Loss_{active1} - DL\_Link\_Loss_{active2} = 9$$

The second of these equations indicates there is a 9 dB difference in downlink link loss when a UE is removing a cell from the active set that has a 3 dB greater CPICH transmit power. Assuming that both cells have equal loading, both are configured with MHA and both make use of uplink receive diversity, then the two cells will have similar uplink link budgets. The UE transmit power will be controlled by the cell that has 9 dB less link loss. This means the other cell will, on average, receive 9 dB less power than necessary. This radio link will be more susceptible to loss of synchronization and subsequent failure.

These imbalances in soft handover radio links form the justification for why it is recommended that neighboring cells should not be configured with CPICH transmit powers that have a difference of more than 3 dB. For scenarios where neighboring cells have different uplink link budgets, the difference in those link budgets should also be accounted for within the analysis. For example, if a cell configured with an MHA is neighbored to a cell not configured with an MHA, then differences in the uplink noise figure have an impact upon the soft handover radio link imbalance. A cell with an MHA typically has an uplink noise figure of 3 dB (referenced to the air-interface side of the MHA). A cell without an MHA typically has an uplink noise figure of 5 dB (referenced to the air-interface side of the feeder and assuming a feeder loss of 2 dB). This introduces a 2 dB difference in uplink sensitivity and could increase the 9 dB imbalance shown above to 11 dB.

The use of soft handover generates overheads in terms of Iub, NodeB hardware resources and NodeB transmit power. The use of softer handover generates overheads in terms of NodeB transmit power. These overheads mean that the quantity of UE in soft and softer handover should be limited. The quantity of soft and softer handover can be limited by careful radio network planning (i.e., site location, antenna type, height, azimuth and downtilt, or by adjusting the RNC database). The radio network plan should be used to manage the quantity of soft handover whenever possible. The RNC database should only be used when absolutely necessary. If the radio network plan has areas of poor dominance, then the quantities of soft handover will be relatively high. The radio network plan should minimize areas of poor dominance and thus minimize the soft handover overhead. The soft handover addition window could be reduced to limit the quantity of soft handover. However, doing so is likely to cause inefficient use of the air-interface (i.e., UE will not be able to add cells into the active set, which would normally provide a benefit). Instead, these cells will cause interference. It may be better to incur increased levels of soft handover overhead rather than to force a reduced overhead at the cost of degraded air-interface performance. If the RNC database is used to reduce the quantity of soft handover, then the soft handover parameter set for non-real-time services should be considered before the parameter set for real-time services. Non-real-time services typically have higher bit rates and thus create larger overheads although the quantity of non-real-time traffic may be less than the quantity of real-time traffic. Non-real-time services are also more tolerant to the relatively poor radio conditions that may be generated by reducing the soft handover addition window. The quantity of soft handover should be monitored during prelaunch optimization using drive test data. The quantity of soft handover should be monitored during post-launch optimization using RNC counters and their associated KPI. It is possible to monitor the quantity of soft handover using either an average active set size KPI or a soft handover overhead KPI. These KPI have different definitions but provide similar information. The target soft handover overhead is typically 40%. This may increase to 50% in urban areas where it is more difficult to plan the radio network with good dominance.

Soft handover gains and their associated definitions are discussed in Section 4.2.5. In general, there are three types of soft handover gains: reduction in E_b/N_o requirement as a result of greater diversity in the RAKE receiver; reduction in the fast fade margin as a result of not having to track all of the fast fades; and a reduction in the slow fade margin as a result of not having to track all of the slow fades. Differences exist between the soft handover gains assumed during link budget analysis and those assumed during 3G simulations. These differences result from 3G simulations explicitly modeling slow fading (i.e., the gain that results from slow fading should not be included within the soft handover gain figures, otherwise it would be double counted).

4.2.10 Scrambling Code Planning

3GPP TS25.213 [11] specifies that there are 512 downlink primary scrambling codes. Each primary scrambling code has 15 associated secondary scrambling codes. Additional scrambling codes are available for use during compressed mode. Each cell within the radio network plan must be assigned a primary scrambling code. There is no need for planners to assign either secondary scrambling codes or the compressed mode scrambling codes. The most important rule for scrambling code planning is that the isolation between cells which are assigned the same scrambling code should be sufficiently great to ensure that a UE never simultaneously receives the same scrambling code from more than a single cell.

3GPP TS25.213 specifies that the 512 downlink primary scrambling codes are organized into 64 groups of 8. Scrambling codes 0 to 7 belong to the same group, as do scrambling codes 8 to 15 and scrambling codes 16 to 23. The organization of scrambling codes into groups allows the UE to complete a three-step cell synchronization procedure using the primary and secondary synchronization channels (P-SCH and S-SCH) and the CPICH. This procedure is applied whenever a UE needs to access a cell or measure the quality of a cell (i.e., during cell selection, cell re-selection, and soft handover). The three-step synchronization procedure is:

- **Step 1**: UE uses the P-SCH to achieve slot synchronization.
- **Step 2**: UE uses the S-SCH to achieve frame synchronization and identify the scrambling code group.
- **Step 3**: UE uses the CPICH to identify the primary scrambling code.

Step 2 involves selecting 1 group out of 64, whereas Step 3 involves selecting 1 code out of 8. Step 3 is likely to be more reliable but also requires more UE processing (i.e., has a greater potential impact upon UE battery life).

It is possible to adopt a scrambling code planning strategy that places the emphasis upon either Step 2 or Step 3. Placing the emphasis upon Step 2 can be achieved by planning the scrambling codes such that neighbors belong to different scrambling code groups. Placing the emphasis upon Step 3 can be achieved by planning the scrambling codes such that neighbors belong to the same scrambling code group. Placing the emphasis upon Step 2 is intended to reduce UE power consumption, whereas placing the emphasis upon Step 3 is intended to improve the reliability of the cell synchronization procedure. The precise difference between the two strategies will depend upon the UE implementation. The difference has not been quantified in the field, and in practice is likely to be very small. It is recommended to plan the scrambling codes such that neighbors belong to the same scrambling code group (i.e., placing the emphasis upon Step 3 to increase reliability). However, other scrambling code strategies can be accepted.

One approach to scrambling code planning, which has been tested in the field and found to be problematic, is to plan the scrambling codes such that each cell belonging to the same NodeB belongs to the same scrambling code group, and also to configure the Tcell RNC database parameter with a value of 0 chips for all cells. If the cells belong to the same scrambling code group, then both their P-SCH and S-SCH will be the same. Configuring Tcell with a value of 0 for all cells means the P-SCH and S-SCH belonging to a specific NodeB will be transmitted simultaneously. This increases the signal strength of the P-SCH and S-SCH in the softer handover regions

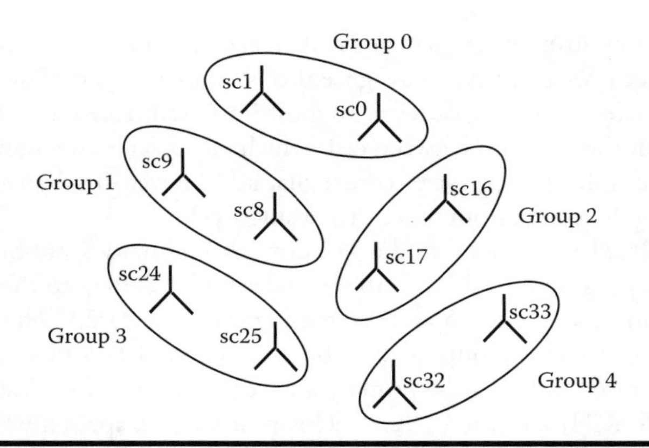

Figure 4.10 An example of assigning scrambling code groups.

and is intended to improve the reliability of the cell synchronization procedure. However, it has been discovered that some UE implementations do not function properly with this strategy. Transmitting the P-SCH simultaneously from three cells belonging to a NodeB means that a UE observes a single P-SCH rather than three P-SCH (assuming that differences in the propagation distance are small). Some UE interpret this as meaning there is only a single cell present (i.e., they detect the first cell and then stop searching). Other UE complete an exhaustive search and detect all three cells. It is recommended to avoid this scrambling code planning strategy and apply the default configuration for the Tcell RNC database parameter.

The scrambling code planning strategy should account for future network expansion. Future network expansion could mean the inclusion of additional NodeB, increased sectorization of existing NodeB, or the evolution of ROC NodeB to CEC NodeB. Scrambling codes should be excluded from the original plan so they can be assigned when additional cells are introduced. For example, a ROC NodeB scrambling code plan could be based upon using the first two scrambling codes in each group. Neighboring ROC NodeB could be paired and then assigned a pair of scrambling codes from the same group. This is illustrated in Figure 4.10.

Then when the ROC NodeB are upgraded to CEC NodeB, an additional four scrambling codes can be assigned from each group. This would leave two scrambling codes in each group for further network expansion. Making allowances for network expansion in this way means it is more difficult to maximize the number of neighbors which belong to the same scrambling code group. However, it is more important to have a practical process for coping with network expansion and being able to introduce additional scrambling codes without having to re-plan large sections of the network. An approach for a network of CEC NodeB could be to limit the initial scrambling code plan to code groups 0 to 33 and to reserve code groups 34 to 63 for future network expansion. NodeB could be paired in the same way shown in Figure 4.10, but in the case of CEC NodeB assigning six codes from groups 0 to 33.

The isolation between cells assigned the same scrambling code should be sufficiently great to ensure a UE never receives a scrambling code from one cell while it is expecting to receive the same scrambling code from a second cell (i.e., the second cell is neighbored while the first is not). This scenario can lead to failed active set updates and the network releasing the RRC connection with cause "unspecified."

Additional rules for scrambling code planning are required at locations close to international borders where there may be another 3G operator using the same RF carrier. Local regulatory bodies should be consulted for national regulations that may have been stipulated. In the case of European countries, the ERC has formulated recommendation "01-01 Border Coordination of UMTS/IMT-2000 Systems." This recommendation prioritizes the use of specific scrambling code groups on either side of an international border. It also recommends maximum allowed signal strengths for transmissions that cross international borders. In practice, operators are also coordinating the use of RF carriers at international borders.

Scrambling code planning can be completed independently for different RF carriers. If a radio network includes NodeB, which is configured with two or three RF carriers, then it is recommended that the same scrambling code plan be assigned to each carrier. This reduces system complexity and helps reduce the work associated with planning and optimizing the network.

Scrambling code planning should be completed in conjunction with neighbor list planning. Audits should be completed to determine whether or not the neighbor lists reflect the intended scrambling code planning strategy (e.g., neighbors belong to the same scrambling code group). Audits should also be completed to ensure there are no duplicate scrambling codes in any of the neighbor lists. A check should be made to ensure no cells are neighbored to two or more cells that have neighbor lists including the same scrambling code. Cells neighbored to two or more cells that have neighbor lists including the same scrambling code can lead to scrambling code conflicts when neighbor lists are combined for UE that are in soft handover (i.e., one of the neighbors has to be removed from the composite neighbor list). Audits can also be combined with path loss predictions to identify cells that have the same scrambling code and relatively low isolation.

4.3 Site Deployment and Configuration

This section examines the site deployment problem with typical configuration assumptions for WCDMA network. Optimal site locations can be found in an automatic manner by applying site rollout and site selection algorithms, in which a trade-off between coverage, capacity, and costs (CAPEX and OPEX) is achieved.

4.3.1 Site Selection

The site selection targets at optimizing mobile radio networks. It provides flexible methods to modify the configuration of a network in such a way that key performance measures are optimized and pivotal performance targets are met [12].

The working engine behind the site selection is a fast construction and solution of downlink and/or uplink cell equations for a given network configuration from which most performance measures can be deduced. Based on an analysis of the "permitted configurations," a preprocessing tree is constructed that contains all partial coefficients of the cell equations that might be relevant for the construction of the downlink/uplink/combined cell equations of an arbitrary permitted network configuration. This preprocessing tree then allows for an efficient accumulation of the coefficients of the equation system, which in turn yields a fast calculation of the cell transmit/interference powers/activities from which most performance measures can be deduced. One can observe how many configurations are evaluated by looking at the number of evaluations in the status messages (which are issued about once a second) when the algorithm is running after preprocessing.

Based on this fast evaluation, an optimization of the network is performed by evaluating all neighboring configurations of the current configuration and moving to the best of them if this improves the current network configuration. As long as an improvement can be obtained, this optimization step is iterated. When no improvement can be found, the last configuration is stored as the end design. On top of this basic procedure (construct preprocessing tree—perform descent to local minimum), there is the possibility for a repetition (since when many sites are switched off the coefficients in the preprocessing tree might become more and more irrelevant and yield too conservative cell equations) and for a division of large problems into smaller ones, where one part of the network is optimized after another in a circular way.

A key role in the simple local descent iteration is played by the comparison operator between configurations. Here the algorithm allows choosing the relevant performance measures/constraints by assigning priorities to them. For example, one often assigns the monetary cost the highest priority and the coverage constraint the second highest (non-zero) priority.

The site selection can be used, for optimizing cell parameters and selecting sites/cells. It can start from given configurations or empty/full networks. It can look through neighborhoods with selectable depths and iterate through these neighborhoods with a deterministic steepest or gradient descent, or a greedy random descent. However, this flexibility is at the price of a more complex user interaction. Through a large number of parameters, the user has to tell the program what kind of optimization is intended. In addition, in large optimization problems, the preprocessing tree requires a huge amount of computer memory and if a large number of partial interferer coefficients are stored, the evaluation can become slow. Partial coefficients that cannot be stored are accumulated in a so-called "remainder coefficient." This remainder coefficient can accumulate a large number of small interferers and then makes sure these interferers are not neglected in the cell equations. This saves computation time and computer memory, but might yield over-conservative cell transmit powers/channel activities, when the number of stored coefficients becomes too small or when the interferers that belong to the stored coefficients are not active (e.g., are switched off). In these cases, one might have to recalculate the preprocessing tree or allow for a larger number of coefficients within the nodes of the tree. Or one could divide a large problem into smaller ones, which also reduces the preprocessing tree.

4.3.1.1 Candidate Sites and Permitted Configurations

The site selection algorithm starts from a description of the permitted network configurations. Such descriptions consist of a set of potential cells (configured cells which might appear in a network design) and subsets of these cells which we call selection sets and choice sets.

The meaning of a selection set is that all potential cells in this set have to appear together in a permitted network configuration or none of them. In a site selection problem, one can think of a selection set as the set of cells that belong to a site: One has to select all of them or none of them. Since selection sets can consist of a single cell, one can also define "cell selection" problems, where individual cells are selected by putting each potential cell into an individual selection set.

The meaning of a choice set is that exactly one of the potential cells in this set has to appear in a permitted network configuration. In a tilt choice problem, such a set would consist of all potential cells that model the same cell but with different tilt settings. Defining such a choice set means to tell the optimization kit it should choose exactly one of these cells for a solution design. Through a choice set that contains only one potential cell, one can tell the program that this cell must appear in a solution design.

In fact, these sets are hidden from the user: The user just tells the algorithm what kind of problem should be solved by setting the "Permitted Configurations" parameter to either "site selection," "cell selection," "tilt choice," or "azimuth choice." The program then automatically generates the corresponding selection and choice sets. But the underlying concept is extensible: It is not difficult to use it for further network optimization problems in the future.

4.3.1.2 Optimization Algorithm

Given this definition of permitted configurations and the results of the preprocessing, the algorithm can walk through the permitted configurations and try to improve the performance measures of interest during this walk. The user can choose between several different manners of walking toward an optimized configuration by either "local steepest descent", "local gradient descent," "random greedy descent," or "local immediate descent."

The "local steepest descent" method evaluates all neighbor network configurations and then replaces the current with the best one found, where the comparison checks the selected performance measures exactly according to their priority. The walk stops if no neighbor configuration improves the current network or if the maximum number of configuration steps has been performed.

The "local gradient descent" method evaluates all neighbor network configurations and then replaces the current one with the best one, where the comparison is based on gradients instead of the absolute values of the performance measures. The walk stops if no neighbor configuration improves the current network, or if the maximum number of configuration steps has been performed.

The "random greedy descent" method chooses randomly one new configuration in the neighborhood of the current one and moves to it if it performs better than the old one. The walk stops if the maximum number of configuration steps has been performed or if one has evaluated the maximum number of random neighbor configurations without finding one better than the current one.

The "local immediate descent" is an exotic procedure where improvements are immediately accepted and the search continues from the improved configuration. This can speed up the algorithm, but at the cost of perhaps not giving as good results as the preceding more-thorough search methods.

In all cases, the walk moves from the current configuration to a configuration within a certain neighborhood of the current one. The neighborhood is defined as all configurations that can be reached by a given number of permitted elementary steps. An elementary step is either changing a given permitted configuration by switching one selection set on or off (adding its cells to the configuration or removing them) or modifying one cell in a choice set (replacing it by another one of the same choice set).

However, increasing the search depth to more than 1 usually dramatically increases the computation time because a large number of configurations might have to be considered before accepting a move. This increase can be controlled through algorithm parameters.

4.3.1.3 Performance Evaluation and Comparison

Two groups of performance measures exist: ordinary performance measures and performance constraints, which additionally have a target value that should be reached. The ordinary performance measures are:

- **ActiveCells**: The number of active cells in the evaluated network configuration. Small values are better than large ones.
- **ActiveSites**: The number of active sites. Small values are better than large ones.
- **MaxOtherCoefficient**: The maximum of the non-diagonal coefficients in the coupling matrix. Small values are better than large ones.
- **MaxOwnCoefficient**: The maximum diagonal coefficient in the coupling matrix. Small values are better than large ones.

- **MaxSiteCostPerAccess**: The maximum first-year cost of one basestation location (including hardware) divided by its sum of area and traffic access coverage in percent. Small values are better than large ones.
- **MaxUserLoadInPercent**: The maximum user load of a cell in the network. Small values are better than large ones. If this measure has high priority, the optimization seeks a configuration in which the maximum user load of a cell is as small as possible.
- **OverloadedCells**: The number of overloaded cells. Small values are better than large ones.
- **TotalFirstYearCost**: The most important (nontechnical) performance measure in site selection, namely the total first-year cost of a given network configuration. Small values are better than large ones.

The performance constraints are:

- **AreaCoverageInPercent**: The percentage of the area that can be served (= considered pixels that have a large enough receive signal strength from at least one of the active cells) by the network configuration. Its target value is the area coverage defined in the underlying optimization profile. Large values are better than small ones.
- **AccessCoverageInPercent**: The number of users on the served area of a configuration as a percentage of the number of users in the total considered area. The target value is the access coverage from the underlying optimization profile. Large values are better than small ones.
- **MinCoverageGapInPercent**: Here one looks at the differences of the area coverage in percent and its target value and the traffic coverage in percent and its target value. The smaller of these two differences is the MinCoverageGapInPercent. Large values are better than small ones. The target value of this constraint is 0.
- **TrafficCoverageInPercent**: After solving the cell equations, one can assess the number of customers that can be served by the network. The target value is the traffic coverage from the underlying optimization profile. Large values are better than small ones.

Only performance measures are calculated and displayed during the optimization for which the priority is strictly positive. The one with the largest positive value has highest priority; the one with the lowest positive value, the lowest priority. The positive priorities should be different.

When two evaluations of network configurations are compared, first the number of violations (i.e., number of performance constraints, where the target is not met) is compared. The configuration with fewer violations is better than the other one.

Next, if both configurations have the same nonzero number of violations, the violation with the highest priority is considered. The configuration where the corresponding performance measure has a better value is considered better than the other one. If both values agree, the performance measure with a violation and next highest priority is considered, and so on.

If there are no violations or the performance measures of all violated constraints agree, the performance measure with the highest priority is considered: The configuration that has the better value of this performance measure is considered to be better than the other one. If both values agree, one looks at the performance measure with the second highest priority, and so on.

Since the steepest descent algorithm accepts only "better" permitted configurations, this evaluation first tries to reach the performance targets and then to optimize the high-priority measures.

If the optimization method is set to "local gradient descent," the decision whether a neighbor of the current configuration is better than another neighbor of the current configuration is slightly more complex: Namely, instead of the absolute values, one considers gradients with respect to the current configuration.

4.4 Optimization of Cell Configuration

During the optimization phase after the first rollout, the network is still subject to be further optimized. The optimization problem still remains NP hard [13]. The parameters can be finely tuned and targets are analyzed and modeled for suitable heuristic search algorithms. A multi-dimensional optimization process is performed in order to configure the WCDMA system optimally.

4.4.1 Optimization Parameters and Targets

The targets of the radio network optimization are mainly twofold. First target is to minimize the interference caused by the individual cells, while a sufficient coverage over the planning area is maintained. This is in general a trade-off and needs to be balanced (e.g., tilting down the antenna causes lower coverage, but also lower interference in neighboring cells and thus a potentially higher network capacity). Second target is the traffic distribution between cells. It is desirable to maintain similar cell loading of neighboring cells in order to minimize blocking probabilities and maximize spare capacity for traffic fluctuations and a future traffic evolution.

The most effective parameter in network optimization is the antenna tilt. Antenna tilts need to be set such that the traffic within the "own" cell is served with maximum link gain, but at the same time the interference in neighboring cells is minimized. The possible tilt angles are typically restricted because of technical and civil engineering reasons. Especially in the case of collocated sites with multiband antennas, there might be strong restrictions on the possible tilt angles to be taken into account during optimization.

The transmitted pilot channel power and the other common channel powers, which are typically coupled by a fixed offset, are also vital parameters of network optimization. It needs to be assured that these channels are received with sufficient quality by all users in the serving cell. At the same time, a minimization of the common channel powers yields significant capacity gains: First, additional power becomes available for other (user traffic) channels, and second, the interference is reduced. The gains obtained from reducing the pilot power are often underestimated. It is important to note that in a capacity-limited WCDMA network (e.g., in urban areas) the reduction of pilot power levels by a certain factor also reduces the total transmit power of cells and as a consequence the cell loading by up to the same factor.

Optimization of azimuth angles of sectored sites is of great importance in particular in case of antennas with rather small horizontal beamwidth (e.g., 65 degree versus 90 degree in case of three-sectored sites). In this case, the difference between antenna gains in the direction of the main lobe and the half-angle between neighboring sectors is comparatively large, and cells of neighboring sites might need to be adjusted such that maximum coverage is achieved. It is observed that during optimization, azimuth changes are in particular introduced in order to reduce coverage problems. For possible azimuth angles, typically even stronger restrictions apply than for the tilt angles.

The antenna height is also often a degree of freedom for the optimization. Higher antennas can provide better coverage, but on the other hand also cause more interference in neighboring cells. Additional important parameters are the antenna type and the number of deployed sectors at a site. Both parameters are closely coupled, as a larger number of sectors also suggest the use of an antenna pattern with smaller horizontal beamwidth. The choice of sectorization is typically a trade-off between increased network capacity and higher monetary cost.

4.4.2 Advanced Search Algorithms

The optimization method described in the previous section is well suited for an initial planning and in cases where the level of completeness or accuracy of input data is limited. For a detailed optimization that takes the full set of input data into account, a search approach is proposed [14]. That is, the space of possible configurations denoted as search space is explored in order to find the point in the search space that is optimal with respect to a certain criterion. This point is denoted as the global optimum. An exhaustive search traverses the complete search space in a systematic manner. As all points in the search space are visited, with exhaustive search it is guaranteed to find the global optimum. The search space is very large for typical applications in network planning. For each site there can easily be several hundreds of possible configurations. Furthermore, configurations of different sites cannot be considered independently, so the amount of possible network configurations grows exponentially with the number of sites. Hence targeting this area, an exhaustive search is too time-consuming and local search algorithms are commonly used for network optimization purposes.

Local search algorithms start at some point in the search space denoted as initial solution and subsequently move from the present to neighboring solutions, if they fulfill some criterion (i.e., appear to be better or more promising). Local search algorithms cannot guarantee to find the global optimum. The objective of the search algorithm—developed or tailored for a particular problem—is to find a solution that is at least close to the global optimum. Local search algorithms are thus often classified as heuristics [15].

The basic procedure of the local search is independent of the actual search algorithm applied. Starting from the initial solution in each search step first a search neighborhood is generated. The search neighborhood is a subset of points from the search space that are close to the current solution (i.e., that have some attributes in common with the current solution). The point that is most appropriate with respect to some criterion is selected from the search neighborhood and accepted as a new initial solution for the next search step. If no appropriate new solution is found (or some other stop criterion is fulfilled), the search is terminated. The comparison of points from the search space is carried out by means of cost values associated with them. The cost values are generated from a cost function, in the literature often also referred to as objective function. The objective function maps a given point in the search space to a cost value. The cost value can be a scalar but could also be represented by a vector. In the latter case, an appropriate function to compare cost values needs to be defined.

Local search algorithms are very much application-specific. However, several search paradigms have been developed in the last three decades. The simplest search paradigm is the descent method. This method always selects the solution from the neighborhood that has lowest cost. If this value is lower than the lowest value in the last search step, the solution is accepted as a new solution, otherwise the algorithm is terminated. The algorithm hence explores the search space by always moving in the direction of the greatest improvement, so it typically gets trapped in a local minimum. A local minimum is a solution that is optimal with respect to the search neighborhood but which generally is worse than the global optimum. In order to escape from local minima, among several others, one widely applied approach is to carry out restarts—that is, the local search is restarted from a new solution selected from a different area of the search space. The new start solutions are often selected randomly. If restarts are applied, the algorithm, strictly speaking, is not a local search algorithm anymore.

Another option for escaping from local minima is to also accept cost-deteriorating neighbors under certain conditions. The most prominent local search paradigms that apply this strategy are simulated annealing and tabu search [15, 16].

Simulated annealing is based on an analogy with the physical annealing process. In simulated annealing, improving points from the neighborhood are always selected when exploring the search neighborhood, while non-improving points are accepted as new solutions with a certain probability. The probability of acceptance is a function of the level of deterioration, but also gradually decreases during the algorithm execution. The reduction of the probability of acceptance is determined by the cooling scheme [16].

In contrast to the simulated annealing which comprises randomness, classical tabu search is deterministic. The basic operation is equivalent to the descent method, with the difference that the best point from the neighborhood is also accepted if it is worse than the current solution. In this way, the search is directed away from local minima. In order to avoid a move back to already visited solutions, a tabu list is introduced. The tabu list typically contains sets of attributes of solutions that have already been visited. If a point from the neighborhood exhibits one of the sets of attributes stored in the tabu list, the point is only accepted as a new solution if its quality (i.e., cost) exceeds a certain aspiration level. The tabu list is updated, keeping the individual entries only for a number of iterations. The size of the tabu list is a very important design parameter of the tabu search. It, in particular, needs to be chosen large enough to prevent cycling, but a too large list might introduce too many restrictions. Several enhancements to the basic operation of the tabu search have been introduced, most of which modify the handling of the tabu list. These include intensification and diversification schemes [16].

4.4.3 Optimization Process

The basic structure of the local search algorithm that has been developed for an optimization of WCDMA networks is depicted in Figure 4.11.

The algorithm comprises the basic elements of a local search method that have been presented in the previous section. The local search starts from an initial solution, which can for example be the current configuration of the network to be optimized, a manually planned solution, or a solution suggested by the fast heuristics presented in the previous section.

At the beginning of each search step, the search neighborhood is generated. At first, a cluster of cells is selected for which parameter changes are considered. Based on the selected cells, the search neighborhood is generated and explored to yield a new solution.

The quality of a certain solution is assessed by a performance analysis. The choice of the method depends on the particular application and is a trade-off between accuracy and the speed of the optimization process. From the results of the performance analysis, a cost value is generated by means of a cost function. The cost function is basically a linear combination of the evaluated quantities. In addition, penalty components are added if certain thresholds are exceeded for any of the different cost function components (e.g., coverage probability below some design target). The search process can be guided by appropriately setting weights for the different cost function components.

As a search paradigm, either a descend method or a tabu search can be applied. The tabu list is maintained independent of the selected search paradigm. The search paradigm only influences the way in which the search is terminated. In case of the

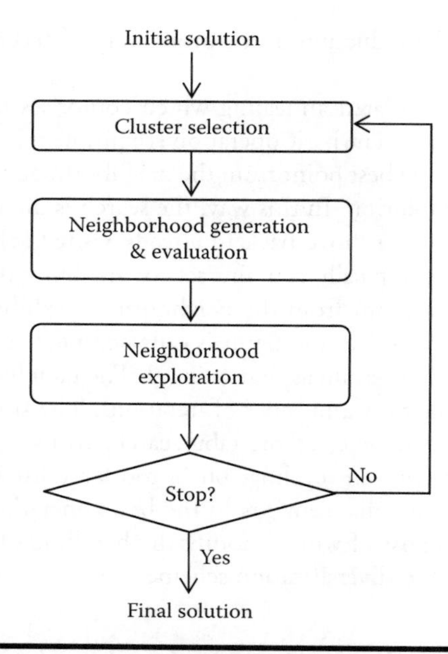

Figure 4.11 Local search algorithm for WCDMA network optimization.

descend method, the search process is terminated if no improvement can be found. For the tabu search, non-improving moves are accepted in order to escape from local minima. The tabu search is terminated once no improving moves are found for a certain number of search steps.

The performance of the local search method strongly depends on the applied performance evaluation. The choice between the methods is a trade-off between accuracy and running time. The basic static method is the fastest, but as was shown, it has some weaknesses in terms of the accuracy of results. The presented statistical methods significantly outperform the latter method in accuracy of results, but even if implemented efficiently are of higher computational complexity, especially if the experimental analysis is applied for evaluating quantities per pixel. The local search optimization presented in this section can be extended to yield a hybrid method that makes use of two methods for performance evaluation. The exploration of the neighborhood is split into two parts. In a first step, the neighborhood is explored by the use of a simple and fast basic performance evaluation method. As a result, a list of candidate solutions is generated. The list is sorted with respect to cost values. The next candidate solution is selected from this list using a more accurate but also more time-consuming advanced performance evaluation. Either the first improving solution from the list or the best solution from a subset of most promising solutions ("short list") is selected.

4.5 Summary

WCDMA radio network planning has to take a set of system parameters into account, and poses the problem to select antenna locations and configurations with respect to contradictory goals: low costs versus high performance. This chapter discusses the radio network planning objective, process, and challenges. Two fundamental approaches to WCDMA radio network planning path loss-based approach and simulation-based approach are described. The important factors that affect the coverage and capacity analysis are discussed as well, including site selection, propagation modeling, link budgets, power planning, soft handover, and scrambling code planning. It has been analyzed that WCDMA radio network planning is no more straightforward than the classic network planning method.

Furthermore, this chapter presents automated methods for selecting optimal antenna locations, configuring antenna parameters and optimizing transmit powers. Choosing a powerful optimization tool in order to search for the optimal network design in mass combinations of possibilities is always placed as the first task before the commercial launch of the mobile network. The network planning methods described in this chapter are basically applicable to other similar air interfaces, such as CDMA2000, TD-SCDMA, LTE, and WiMAX. A new trend exists that network operators utilize a common platform embedded with multistandard multiband technologies. In such cases, the proposed methods can be installed in the common Operation and Maintenance Center for the purpose of automatically optimizing the multi-standard system in a real-time manner.

References

[1] X. Huang, "Automatic Cell Planning for Mobile Network Design: Optimization Models and Algorithms," *Forschungsberichte aus dem Institut fuer Hoechstfrequenztechnik und Elektronik der Universitaet Karlsruhe* (TH), Band 30, Karlsruhe, Germany, 2001.

[2] H. Holma and A. Toskala, *WCDMA for UMTS: Radio Access for Third Generation Mobile Communications, Third Edition*, John Wiley & Sons, 2004.

[3] U. Tuerke, R. Perera, E. Lamers, T. Winter, and C. Goerg, "An Advanced Approach for QoS Analysis in UMTS Radio Network," in *Proc. ITC-18*, Berlin, Germany, 2003.

[4] M. J. Nawrocki, M. Dohler, and A. H. Aghvami, *Understanding UMTS Radio Network Modelling, Planning and Automated Optimisation: Theory and Practice*, John Wiley & Sons, 2006.

[5] X. Huang, U. Behr, and W. Wiesbeck, "A New Approach to Automatic Base Station Placement in Mobile Networks," in *Proc. of International Zurich Seminar on Broadband Communication*, Zurich, Switzerland, 2000.

[6] J. Laiho, A. Wacker, and T. Novosad, *Radio Network Planning and Optimisation for UMTS, Second Edition*, John Wiley & Sons, 2006.

[7] Nokia, *NetAct Planner—MultiRadio Planner 5.0—User Reference Guide*, 2008.

[8] T. Ojanperae and R. Prasad, *Wideband CDMA for Third Generation Mobile Communications*, Artech House, 2001.

[9] J. D. Parsons, *The Mobile Radio Propagation Channel*, Pentech Press, 1992.

[10] A. Davidson and C. Hill, “Measurement of Building Penetration into Medium Buildings at 900 and 1500 MHz,” *IEEE Transactions on Vehicular Technology*, vol. 46, pp. 161–168, February 1997.

[11] 3rd Generation Partnership Project (3GPP) website, http://www.3gpp.org.

[12] K. Majewski, “Network Optimization Kit,” Technical Report, Siemens AG, Munich, Germany, 2006.

[13] I. Siomina, and D. Yuan, “Minimum Pilot Power for Service Coverage in WCDMA Networks,” *ACM/Kluwer Wireless Networks Journal* (WINET), 2007.

[14] U. Tuerke, Efficient Methods for WCDMA Radio Network Planning and Optimization, Deutscher Universitaets-Verlag, Wiesbaden, Germany, 2007.

[15] C. R. Reeves, Modern Heuristic Techniques for Combinatorial Problems, McGraw-Hill, 1995.

[16] E. Aarts and J. K. Lenstra, *Local Search in Combinatorial Optimization*, John Wiley & Sons, 1997.

Chapter 5

TD-SCDMA Network Planning and Optimization

Jianhua Zhang and Guangyi Liu

Contents

5.1 Physical Layer of TD-SCDMA

5.1.1 Transport Channels

A typical TD-SCDMA wireless network can be divided into two parts: radio access network (RAN) and core network (CN) [1]. The physical layer offers data transport services to higher layers via the transport channel. A transport channel is defined by how the data is transferred over the air interface. In 3GPP specifications, transport channels are classified into two groups [2]: dedicated transport channel (DTrCH) and common transport channel (CTrCH).

DTrCH uses inherent addressing of user equipment (UE). The data on the DTrCH is only transmitted to a single user per time slot. The user is identified by the physical channel (i.e., code, time slot, and frequency) in TD-SCDMA systems. Only one type of DTrCH in TD-SCDMA systems is used to bear the user information or control information between the network and the specific user of either up- or downlinks.

CTrCH uses explicit addressing of UE if addressing is needed. The CTrCH is always shared among many users. And the user's ID is needed only while the CTrCH offers services to a specific user. Six types of CTrCH exist in TD-SCDMA systems: broadcast channel (BCH), paging channel (PCH), forward access channel (FACH), random access channel (RACH), uplink shared channel (USCH), and downlink shared channel (DSCH). The main characteristics of these channels are listed next.

- BCH, a downlink transport channel used to broadcast system or cell information to all UEs in a cell.
- PCH, a downlink transport channel used to transmit control information to a mobile station when the system does not know the location of the mobile station.
- FACH, a downlink transport channel used to transmit control information to a mobile station when the system knows the location cell of the mobile station; may also carry short user packets.
- RACH, an uplink transport channel used to carry control information from the mobile station; may also carry short user data packets.
- USCH, an uplink transport channel shared by several UEs carrying dedicated control or traffic data.
- DSCH, a downlink transport channel shared by several UEs carrying dedicated control or traffic data.

5.1.2 Physical Channels

The physical channel in TD-SCDMA systems takes a four-layer structure with respect to time slots/codes, subframe, radio frames, and system frame. The radio frame and time slot resources can be flexibly configured according to the service requirement by the upper layer radio resource management (RRM) entity.

A physical channel in TD-SCDMA systems is a burst, which also appears as a time slot in a preconfigured radio frame. The radio frame configuration can be continuous (i.e., the time slots in all frames are distributed to a certain physical channel) or discontinuous that is, only the time slots of part frames are distributed to a physical channel. The time slots are used in the sense of a TDMA component to separate different user signals in the time domain. A typical time slot is the combination of a data part, a midamble and a guard period. A transmitter could send several bursts at a time. However, in this case, the data part is distinguished by different orthogonal variable spreading factor (OVSF) channelization codes, but should be scrambled by the same scrambling code. The midamble parts of different data are obtained by the different shifts of the same basic midamble sequence. Therefore, the codes are used to separate different user signals in the code domain. Above all, a physical channel in TD-SCDMA systems is defined by frequency, time slot, channelization code, burst type, and radio frame number.

The specific radio frame structure in TD-SCDMA systems is depicted in Figure 5.1. A subframe contains seven 675 μs normal time slots and three special time slots. The three special time slots are downlink pilot time slot (DwPTS), guard period (GP), and uplink pilot time slot (UpPTS). The seven normal time slots are assigned to either up- or downlink for data transmission. In TD-SCDMA systems, TS0 is always allocated for downlink while TS1 is always allocated for uplink. The time slots for up- and downlink are separated by switching points. In each subframe of 5 ms, there are two switching points. One is the special time slot GP, the other is located in time slots TS1 to TS6. The flexible time slot resource distribution between up- and downlink depends on the second switch point setting among TS1 to TS6. In future service requirements, spectrum efficiency for asymmetric data transmission is a key challenge. Asymmetric data services are distinguished by very different traffic loads on up- and downlink. Consequently, WCDMA and CDMA2000 with symmetric paired frequency bands lead to low utilization of the spectrum. In contrast, by a flexible switching point configuration, the TD-SCDMA scheme allows full frequency utilization and application of both symmetric and asymmetric traffic loads.

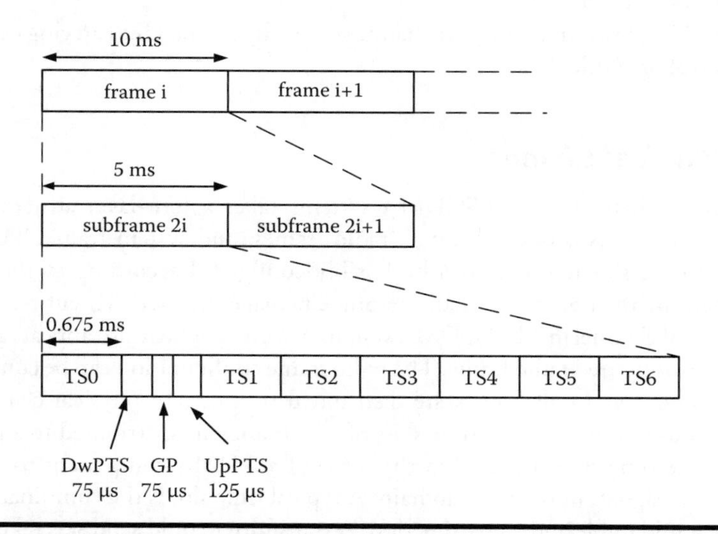

Figure 5.1 TD-SCDMA radio frame structure.

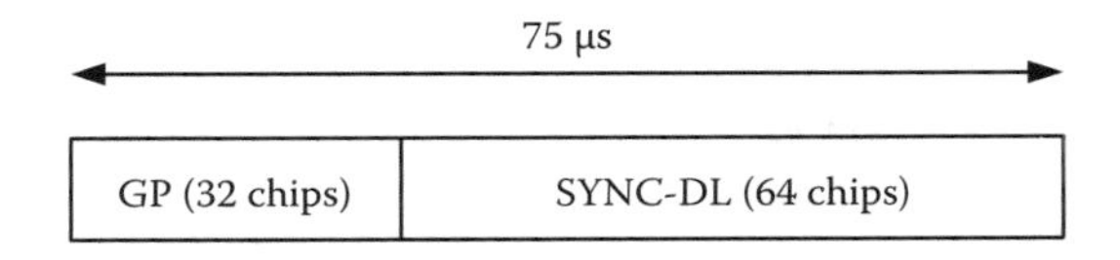

Figure 5.2 DwPTS structure.

From the preceding introduction, it is known that there are four kinds of time slots in TD-SCDMA systems: DwPTS, UpPTS, GP, and the normal time slot. Their characteristics and structures are listed as follows.

- **DwPTS:** DwPTS is used for downlink synchronization and channel estimation. The time slot contains a synchronization downlink sequence (SYNC-DL) of 64 chips and a guard period of 32 chips, whose structure is shown in Figure 5.2. The SYNC-DL in TD-SCDMA systems is a group of pseudo-noise (PN) codes, which is also used for cell identification. Two main advantages can be obtained via configuring a single time slot for DwPTS: (1) getting downlink synchronization quickly and (2) mitigating the interference to/from other downlink signals.
- **UpPTS:** UpPTS is used for uplink synchronization. The time slot contains a synchronization uplink sequence (SYNC-UL) of 128 chips and a guard period of 32 chips, whose structure is shown Figure 5.3. The SYNC-UL in TD-SCDMA systems is also a group of pseudo-noise (PN) codes, which is used for user identification in the random access procedure.
- **GP:** GP is the switching point from transmitting to receiving at base station (BS) node. The duration of GP is 96 chips, from which we can calculate the basic cell radius:

$$R = \frac{1}{2} \times \frac{96 \text{ chips}}{1.28 \times 10^6 \text{ chips/s}} \times 3 \times 10^5 \text{ km/s} = 11.25 \text{ km}$$

  where 1.28×10^6chips/s is the chip rate of TD-SCDMA. Basically, the GP should be long enough to avoid interference between up- and downlink.
- **Normal Time Slot:** A normal time slot consists of two data symbol fields, a midamble of 144 chips and a guard period. The data fields of the burst are

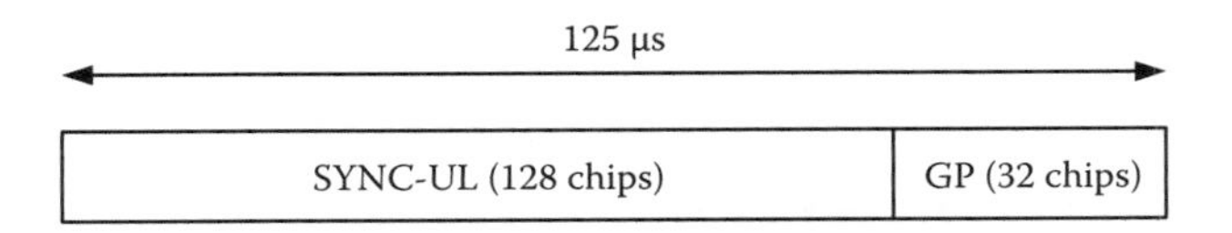

Figure 5.3 UpPTS structure.

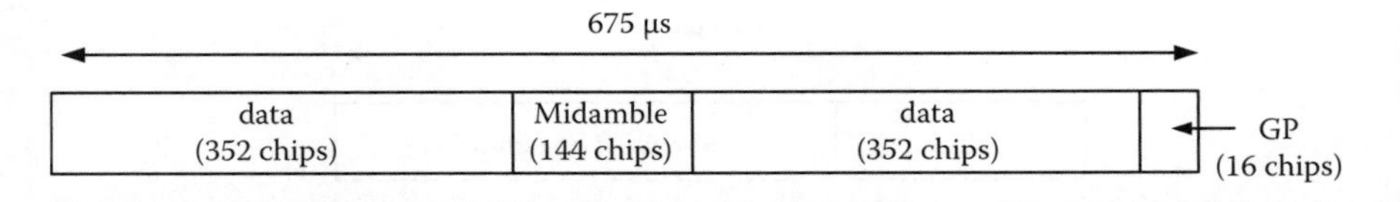

Figure 5.4 A normal time slot structure.

352 chips long. The length of the guard period is 16 chips. A normal time slot structure is presented in Figure 5.4. The midamble doesn't carry any data information. It is only a pilot for channel estimation, synchronization, and radio resource measurement, etc.

5.1.3 Mapping of Transport Channels to Physical Channels

The physical channels in TD-SCDMA systems can be divided into two types: the dedicated physical channel (DPCH) and common physical channel (CPCH). The details of physical channels are introduced in Figure 5.5.

The mapping of transport channels to physical channels in TD-SCDMA can be obtained from Figure 5.6.

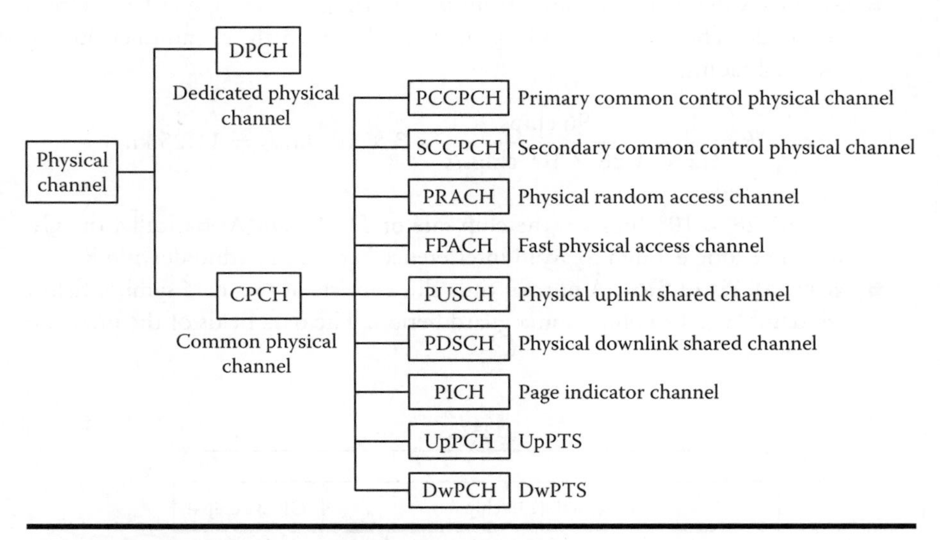

Figure 5.5 Physical channels in TD-SCDMA.

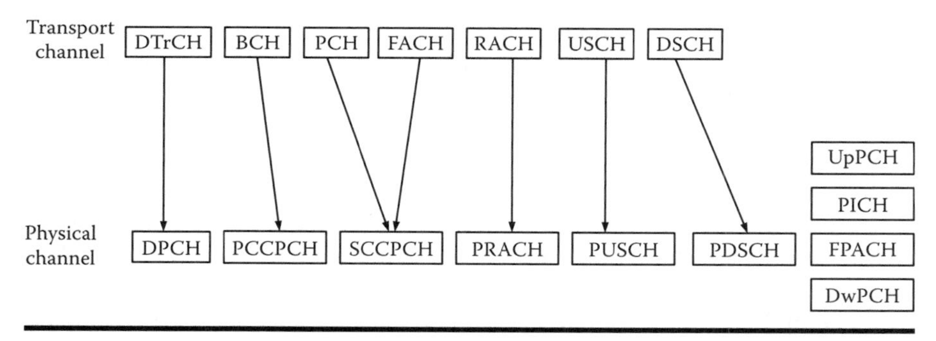

Figure 5.6 The mapping of transport channels to physical channels.

5.1.4 Physical Layer Procedures in TD-SCDMA Systems

5.1.4.1 Power Control

In TD-SCDMA systems, the system capacity is limited due to the near/far problem, code asynchronicity and the spreading code non-orthogonality. In order to solve these problems, the power control technology has been adopted. The basic purpose of power control is to limit the interference level within the system, thus mitigating the inter-cell interference and reducing the power consumption in the UE. The characteristics of power control in TD-SCDMA systems are summarized in Table 5.1.

5.1.4.2 Uplink Synchronization

In CDMA systems, the downlink signals are broadcast to all users. Thus, the downlink synchronization at all UE nodes is perfect. In contrast, the uplink synchronization requires that the signals transmitted from different users who are distributed at different geographic positions arrive at the BS node simultaneously. Therefore, to obtain uplink synchronization in TD-SCDMA systems is a key step that influences the whole system performance.

Table 5.1 Power Control Characteristics of TD-SCDMA

	Uplink	*Downlink*
Power control rate	Variable closed loop: 0–200 cycles/sec. Open loop: about 200 us–3575 us delay	Variable closed loop: 0–200 cycles/sec.
Step size	1, 2, 3 dB (closed loop)	1, 2, 3 dB (closed loop)

The uplink synchronization procedure contains preparation, establishment, and maintenance. When a UE is powered on, it first needs to establish the downlink synchronization with the targeted cell. Only after the UE has established the downlink synchronization, it will start the uplink synchronization procedure. The establishment of uplink synchronization is done during the random access procedure and involves the UpPCH and the PRACH. Although the UE can receive the downlink signal from the BS, the distance to the BS is still uncertain. This would lead to unsynchronized uplink transmission. Therefore, the first transmission in the uplink direction is performed in a special time slot UpPTS or other uplink access position indicated by RAN to reduce interference in the traffic time slots. After the detection of the synchronization uplink sequences in the searching window, the BS will evaluate the timing, and reply by sending the adjustment information to the UE to modify its timing for the next transmission. This is done with the FPACH within the following four subframes. After sending the PRACH, the uplink synchronization is established. The uplink synchronization procedure will also be implemented when uplink is out of synchronization. Uplink synchronization is maintained in TD-SCDMA systems by sending the uplink advanced in time with respect to the timing of the received downlink. For the maintenance of the uplink synchronization, the midamble field of each uplink burst can be used. In each uplink time slot, the midamble for each UE is different. The BS may estimate the timing by evaluating the channel impulse response of each UE in the same time slot. Then, in the next available downlink time slot, the BS will signal synchronization shift (SS) commands to enable the UE to adjust its transmission timing properly.

5.1.4.3 Cell Search

The UE in the cell search procedure is aimed at obtaining the DwPTS synchronization, scrambling code, and basic midamble code of the targeted cell, meanwhile controling the multiframe synchronization and reading the BCH. The entire procedure is carried out in four steps:

- **Search for DwPTS:** By the SYNC-DL (in DwPTS), UE requires downlink synchronization to the targeted cell. This step is typically realized by one or more filters (or any similar device) matched to the received SYNC-DL, which is chosen from the PN sequences set. During this procedure, the UE needs to identify which of the 32 possible synchronization downlink sequences is used.
- **Scrambling and Basic Midamble Codes Identification:**The UE receives the midamble of the P-CCPCH, which is followed by the DwPTS. In TD-SCDMA systems, each DwPTS code corresponds to a group of four different basic midamble codes. Therefore, there are a total of 128 midamble codes and these codes do not overlap. Since the 32 SYNC-DL and the group of 32 basic midamble codes of the P-CCPCH are related (i.e., once the SYNC-DL is detected, the four midamble codes can be determined), the UE knows which four basic midamble codes are used after detecting the SYNC-DL. Then the UE

can determine the used basic midamble code using a trial-and-error technique. The same basic midamble code will be used throughout the frame. Since each basic midamble code is associated with a scrambling code, the scrambling code is also known by that time. According to the result of the search for the right midamble code, UE may go to the next step or go back to the previous step.

- **Control Multiframe Synchronization:** The UE searches for the master indication block (MIB) of the multiframe of the BCH. The control multi-frame is positioned by a sequence of QPSK symbols modulated on the Dw-PTS. Consecutive DwPTSs are sufficient for detecting the current position in the control multiframe. According to the result of the control multiframe synchronization for the right midamble code, UE may go to the next step or go back to the previous step.
- **Read the BCH:** The broadcast information of the found cell in one or several BCHs is read. According to the result, the UE may move back to previous steps or the initial cell search will be finished.

5.1.4.4 Random Access Procedure

The physical random-access procedure is performed as follows:

UE side:

1. Set the signature re-transmission counter to M (allocate eight SYNC-UL sequences to UpPTS).
2. Set the signature transmission power to Signature_Initial_Power.
3. Based on the type of random access and the transport format indicated by the media access control (MAC) layer, an E-RUCCH or unique RACH used for the radio access is chosen. Then randomly select one UpPCH subchannel and one signature, respectively, from the available ones for the given access service class (ASC). The random function shall be such that each of the allowed selections is chosen with equal probability.
4. Transmit the signature at UpPCH or other uplink access positions indicated by higher layers using the selected UpPCH subchannel at the signature transmission power. Should the commanded signature transmission power exceed the maximum allowed value, set the signature transmission power to the maximum allowed power.
5. After sending a signature, listen to the relevant FPACH for the next N (the maximum signature subframe number ensured on the network side) subframes to get the network acknowledgement. The UE will read the $FPACH_i$ associated with the transmitted UpPCH only in the subframes fulfilling the following relation:

$$(SFN' \, mod \, L_i) = n_{RACH_i}; n_{RACH_i} = 0, \ldots, N_{RACH_i} - 1 \qquad (5.1)$$

Here, FPACH to which UE should listen is decided according to the following formula:

$$FPACH_i = N mod M \tag{5.2}$$

where $FPACH_i$ denotes the ith FPACH, L_i is the length of the RACH transport block, n_{RACH_i} is the PRACH that corresponds to the ith FPACH, N_{RACH_i} is the PRACH total number that the ith FPACH corresponds to, N denotes the signature number (0 . . . 7) and M denotes the maximum number of FPACHs defined in the cell.

6. In case no valid answer is detected in the due time, increase the signature transmission power by $\triangle P_0 = Power\ Ramp\ Step[dB]$, decrease the signature re-transmission counter by one and if it is still greater than 0, then repeat the steps starting from step 3; otherwise report a random access failure to the MAC sublayer.
7. In case a valid answer is detected in the due time:
   a. Set the timing and power level values according to the indication received by the network in the $FPACH_i$.
   b. Send at the subframe coming two subframes after the one carrying the signature acknowledgement, the RACH message on the relevant PRACH. In case L_i is bigger than one and the subframe number of the acknowledgement is odd, the UE will wait one more subframe. The relevant PRACH is the n_{RACH_i}th PRACH associated to the $FPACH_i$ if the following equation its fulfilled:

$$(SFN' modL_i) = n_{RACH_i} \tag{5.3}$$

Here SFN' is the subframe number of the arrival of the acknowledgement.

On both UpPCH and PRACH, the transmit power level should never exceed the indicated value signaled by the network.

Network side:

1. The Node B will transmit the $FPACH_i$ associated with the received UpPCH only in the subframes fulfilling the following relation:

$$(SFN' modL_i) = n_{RACH_i}; n_{RACH_i} = 0, \ldots, N_{RACH_i} - 1 \tag{5.4}$$

where FPACH number i is selected according to the following formula based on the acknowledged signature:

$$FPACH_i = NmodM \tag{5.5}$$

where *N* denotes the signature number (0 . . . 7) and *M* denotes the maximum number of FPACH defined in the cell.
2. The Node B will not acknowledge UpPCHs transmitted more than *N* subframes ago.
3. At the reception of a valid signature, measure the timing deviation with respect to the reference time, T_{ref} of the received first path in time from the UpPCH and acknowledge the detected signature sending the FPACH burst onto the relevant FPACH.

5.1.5 Physical Layer Techniques in TD-SCDMA Systems

Because of the special physical-layer structure of TD-SCDMA, many advanced wireless transmission techniques could be applied to improve the spectral efficiency. The techniques also have an impact on network planning and optimization. Some key techniques of TD-SCDMA are introduced in this section. And we also briefly point out how these radio transmission techniques influence the strategies in network planning and optimization.

- **TDD:** In the TDD mode [3], the paired frequency bands for up- and downlink is not necessary. The time slots in the TDD physical channel are divided into transmission and receiver types. Though compared with the FDD mode, the TDD brings new challenges in system design, such as small coverage, difficulty in supporting high-speed mobile users and the requirement of high-precision synchronization. The TDD mode also has many attractive advantages. For instances, channel state information (CSI) reciprocality on up- and downlink, higher spectral efficiency, lower power control requirements, high-quality receiver at UE, and a more convenient way to adopt smart antenna technology. Furthermore, the resources planning in TDD mode are more flexible than that in FDD mode. For instance, in TD-SCDMA systems, the time slots could be allocated to up- and downlink according to the service types.
- **Synchronization:** The main difference of the synchronization procedure between TD-SCDMA and other 3G standards is that a special time slot in the TD-SCDMA frame is used for quick and accurate synchronization. As a consequence, the synchronized signals are subject to lower interference in TD-SCDMA systems, and the UE and BS nodes can synchronize quickly. The synchronization procedure has been introduced before. It should be noted that the basic SYNC-UL and SYNC-DL codes distribution to different cells is always the first step in code planning, which is related to the planning of midamble and channelization codes. The specific details will be introduced in the section on resource planning.
- **Joint Detection:** Joint detection is one of the multiuser detection techniques. In CDMA systems, different users' signals overlap on both the

time- and frequency-domain. Thus an efficient signal processing technique is needed to distinguish the overlapped signals at the receiver. If the spreading codes of different users are non-orthogonal, the traditional single-user detection cannot eliminate the multiple access interference (MAI). Consequently, the recovered signal quality degrades severely. The joint detection technique takes into account all users' data and can separate all users' signals simultaneously at the BS node. Furthermore, the MAI can be eliminated totally and the near/far problem is solved perfectly. The specific details about joint detection can be found in [4–6]. As the joint detection technique is adopted in TD-SCDMA system, the interference among different users at the same time slot can be restrained effectively and the system capacity is improved, which should be considered fully when implementing network planning and optimization. For instance, because the MAI is eliminated in TD-SCDMA systems, the cell-breathing effect in TD-SCDMA is less severe than that in WCDMA. It also means the relation between coverage and capacity in TD-SCDMA is not as close as that in WCDMA. And different services always have the same coverage radius in TD-SCDMA, which means that in-network planning coverage planning and capacity planning can be separated.

- **Dynamic Channel Allocation:** The radio resources in TD-SCDMA systems contain frequency, time slot, code, power, and space. The physical channels are marked by their frequency, codes, and time slots. In the dynamic channel allocation (DCA) algorithm, the channels do not belong to a single cell inherently, but can be distributed to many cells. In realistic cases, the radio network controller (RNC) plays the role of RRM. The resources distribution should consider the whole network parameters, system load, and quality of service (QoS) parameters. The DCA technology is applied to TD-SCDMA conveniently due to the special frame structure. And higher spectral efficiency and system capacity can be obtained by applying the technique. Furthermore, the resources planning in TD-SCDMA systems is simplified because of the utilization of the DCA technique.
- **Baton Handoff:** It has been known that the soft handoff in CDMA systems is more reliable and efficient than the hard handoff, and the soft handoff technique has been used in WCDMA and CDMA2000 systems. However, a deficiency of soft handoff is that the procedure needs more channel resources because the UE still connects the primary BS when it switches to a new BS. Considering the trade-off between handoff quality and resources utilization, a handoff technique called baton handoff is used in TD-SCDMA systems. When a UE implements the baton handoff, it builds the signaling connection with a new BS at first. Then the UE cuts off the service connection with the primary BS and moves the service connection to the new BS. If the service handoff is successful, the UE cuts off the signaling connection with the primary BS. After these procedures, the baton handoff is finished and the UE only communicates with the new BS. As the UE could be positioned precisely due to smart antenna

and synchronization techniques in TD-SCDMA systems, the baton handoff can be implemented and the system loads will be reduced. Consequently, the system performance gets optimized.

- **Smart Antenna:** Smart antenna as a kind of space duplex multi-access (SDMA) technique can separate the signals occupying the same time-frequency-code resources. The smart antenna technique has been researched widely, and many specific technique details can be found in [7, 8]. Now in the three mainstream 3G standards, only TD-SCDMA has declared it would adopt the smart antenna technique. In contrast, the discussion is continued in WCDMA and CDMA2000. One major reason for the utilization of smart antenna technology in TD-SCDMA systems is that the TDD frame structure is suited to obtain CSI reciprocity between up- and downlink. Thus the implementation complexity is reduced in TDD systems. Because of the utilization of smart antenna, the system capacity and spectral efficiency are improved, and a great convenience is brought when we do network planning and optimization.
- **Software-Defined Radio:** In the software-defined radio technique [9, 10], all protocols and commands are implemented on a programmable hardware platform by software control. It means that we can provide different services via the same hardware platform, which is very attractive from the aspect of the operator. In TD-SCDMA systems asymmetric services are supported, which means the resources allocated to up- and downlink are also asymmetric. Because of the application of the software-defined radio technique, the resources allocation in different services is self-planning. Another important use of software-defined radio is that the smart antenna technology is dependant on software control according to the transmission environment.

5.2 Coverage Planning

A coverage planning is the initial step in network construction. The initial coverage quality always determines the increase in subscribers and the profit obtained from this network. Therefore, coverage planning is one of the decisive factors in whether the network operates in a healthy manner.

5.2.1 Coverage Region Partition

A coverage region partition is a complicated work. There needs to be a great deal of work done regarding data preparation before detailed implementation is carried out. Furthermore, influences from these kinds of factors should be taken into account so the optimal scheme is chosen. For instance, we should do some research to find out the potential user's number of the network, the population destiny in a region, the geographic environment information, the existing network coverage and their operation, the demand of each service, etc. After obtaining the information, we can estimate the minimum system capacity, the required BS node number, and the position and coverage radius of each BS in the region. Another important factor that should be considered in the initial coverage planning is that the network will be extended in the future. The initial planning should be convenient for the network extension.

Though many factors should be considered in coverage region partitions, two key characteristics of a region should have high priority: geographic environment and services distribution.

Information on the geographic environment should be obtained first in network planning, which contains terrain, distribution of buildings, building materials, and the traffic situation. And we should analyze the environment characteristics' impact on the transmission of electromagnetic waves. Then, according to the information, we can choose the proper wireless transmission channel model, and decide the BS position and the coverage radius. From the operators' aspect, the total environments can be classified into five categories: intensive urban, general urban, suburban, rural, and sparsely populated area. Each scenario has its own characteristics and should be treated differently.

The services distribution contains the network user's number in a region and their requirement. The services can be classified into different categories according to their service users and QoS requirement. For instance, the voice services have a relatively low transmission rate and can tolerate some distortion, but the total delay must be less than about 100 ms or else it becomes noticeable to the users. Therefore, these real-time services need to occupy the physical channel constantly. In contrast, the data services always require a high transmission quality with the low BERs (a BER of 10^{-6} or less) but do not need a small delay. The radio resources allocated for these non-real-time services can be noncontinuous and shared among users. For a specific service, the connection probability is always as the indicator of coverage quality. The connected probability is defined as the probability that the users communicate with the BS node successfully within the coverage region. Specifically, the edge coverage probability is an important indicator of coverage quality, for the cell-edge users always receive the worse QoS. In realistic cases, the coverage region partition according to the service requirements load can be classified as intensive service region, high-density service region, normal-density service region, and low-density service region.

We have mentioned before that low interference exists in a time slot due to the utilization of smart antenna and DCA techniques. Therefore, different services in TD-SCDMA systems almost have the same coverage radius. And the service planning can be separated from propagation environment for convenience. However, it doesn't mean that the service is independent with the propagation environment. Instead, the service and the geographic environment contact with each other closely and should be considered together for the optimal planning. For instance, in rural areas, the voice services are the main service requirement and the demand of data services is little, whereas both the data and voice services are the main demands in urban areas.

As the initial work in coverage planning, the coverage region partition determines the network size and the initial investment budget. So we should collect enough information and analyze this information comprehensively. Either way, it is aimed at obtaining as high a profit as possible with the most acceptable cost.

5.2.2 Link Budget

Link budget is an efficient method to analyze the wireless network coverage, which is also applied in the whole procedure of network planning and optimization. According to the pre-collected information and the initial coverage partition, the link budget can obtain the permitted maximum power loss so the maximum allowable path loss of each service can be calculated. The maximum allowable path loss is an important reference used for cell parameter configuration, such as the BS position and cell radius.

The specific procedures of up- and downlink budgets are almost the same. But their budget results are used for different purposes. The UE is constrained by transmitted power, so the uplink budget is always used to determine the cell coverage radius. At BS node, the total power will be shared among users. The downlink capacity is also constraint as the access users' numbers increase. Thus the downlink budget is always used to estimate the system capacity. The general link budget model is depicted in Figure 5.7. It is shown that the whole link budget procedure contains three parts: equivalent isotropically radiated power (EIRP) calculation, space propagation loss calculation, and the minimum input level sensitivity calculation.

5.2.2.1 EIRP

The EIRP is the effective power transmitted from BS, after variable gains and attenuations, which can be calculated as the following formula:

$$P_{\mathrm{EIRP}} = P_t + G_t - L_t, \tag{5.6}$$

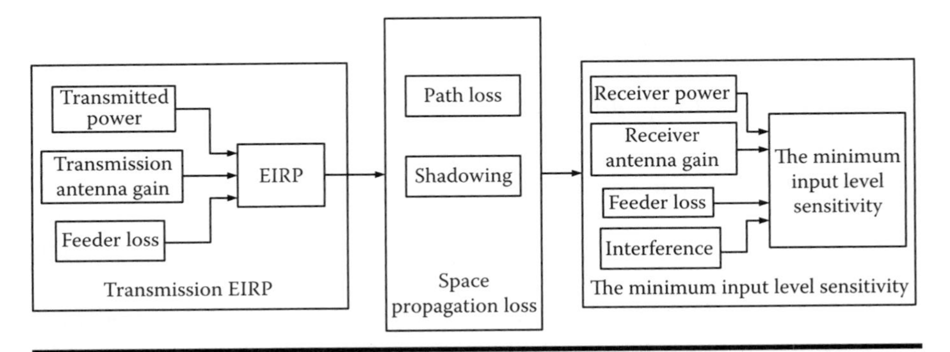

Figure 5.7 The link budget procedure.

Table 5.2 UE Power Classes

Power Class	*Nominal Maximum Output Power*	*Tolerance*
1	+30 dBm	+1 dB/−3 dB
2	+24 dBm	+1 dB/−3 dB
3	+21 dBm	+2 dB/−2 dB
4	+10 dBm	+4 dB/−4 dB

where P_t denotes the real transmitted power at BS node, G_t is the transmission antenna gain, and L_t is the feeder loss at transmission node. The downlink P_t denotes the transmitted power on physical channel at BS node. A time slot of TD-SCDMA contains 16 channels. Thus, a physical channel only transmits 1/16 of the total transmitted power. And the transmitted power of each service is determined by the channel number occupied by the service. For instance, 12.2 kbps voice service occupies two physical channels in a time slot and its transmitted power is 1/8 of the total transmitted power. In uplink, the P_t only denotes the DPCH transmitted power. According to the 3GPP specification [11], four types of power classes are defined for TD-SCDMA, which are given in Table 5.2.

5.2.2.2 Space Propagation Loss

The space propagation loss refers to the large scale fading caused by the wireless propagation environment and the transmission distance. The propagation loss contains path loss and shadowing and is modeled as a function of the transmission distance. The specific details about space propagation loss model can be found in [12, 13]. In coverage radius planning, the maximum permitted space propagation loss needs to be estimated first. Then, according to the transmission environment, a proper space propagation loss model should be chosen. Last, we obtain the permitted coverage radius by calculating the propagation loss function. In the following link budget procedure, L_s is used to denote the allowable space propagation loss.

5.2.2.3 Minimum Input Level Sensitivity

The minimum input level sensitivity contains two parts: the effective input power and the interference. The effective input power is given by

$$P_i = P_r + G_r - L_r \tag{5.7}$$

where P_r denotes the received power, G_r is the receiver antenna gain, and L_r is the feeder loss at receiver node. The interference can be divided into three parts: intra-cell interference, inter-cell interference, and noise.

$$I = I_{\text{intra}} + I_{\text{inter}} + N \tag{5.8}$$

where I_{intra}, I_{inter}, and N denotes intra-cell interference, inter-cell interference, and noise, respectively.

In realistic systems, each service has its required minimum input level sensitivity value γ. It should be guaranteed that the received signal-to-interference-plus-noise ratio (SINR) is larger than this sensitivity, i.e.,

$$\frac{P_i}{I} \geq \gamma \tag{5.9}$$

5.2.2.4 Maximum Coverage Radius Calculation

An important function of the link budget is to calculate the maximum coverage radius of a BS node. The calculation procedure follows the steps:

1. According to the transmitted power P_t, transmission antenna gain G_t, and transmission feeder loss L_t, calculate the EIRP P_{EIRP}.
2. According to the required minimum input level sensitivity γ and the estimated total interference I, calculate the effective input power $P_i = \gamma I$.
3. According to Equation 5.7, calculate the received power $P_r = P_i + L_r - G_r$.
4. Calculate the space propagation loss $L_s = P_{\text{EIRP}} - P_r$.
5. According to the propagation environment, choose the proper wireless propagation loss model $L_s = f(r)$, where variable r is the transmission distance. At last, calculate the coverage radius $r = f^{-1}(L_s)$.

5.2.3 Link Balance

The link budget only focuses on a single service coverage planning. In realistic TD-SCDMA coverage analysis, many factors should be considered simultaneously, which is also called link balance and contains three types: different service link balance, up- and downlink balance, control and service channel balance.

In a network, the QoS requirement and coverage region of each service is different. Consequently, the coverage radii of different services estimated by the link budget also are not the same. In practical coverage planning, all of the services should be available to all users in a cell. In order to achieve the requirement, the link budget of each service is calculated first. Then, we can choose the link budget result of the strictest requirement service as the network planning reference.

For a wireless network, the coverage radii of up- and downlink obtained by the link budget are always different. The UE performance is always limited by power, in contrast, the power is not a problem at BS node. As a result, there is a gap between the power transmitted from up- and downlink. If the uplink signals are too weak and the downlink signals are too strong, many problems will be caused in a handoff procedure. For example, the downlink pilot signal indicates the UE to implement the handoff, but the signals transmitted from UE are not strong enough to connect to another cell. As a result, the transmission link will be cut off. In order to avoid the unanticipated situation, the up- and downlink balance should be obtained after the link budget. As the uplink budget is always the reference of coverage, the uplink budget result is considered the coverage reference in system construction.

Another factor that should be taken into account in link coverage analysis is the balance between the control and service channel. In the TD-SCDMA system, the control information is broadcast to the whole cell. Thus, the control signals should be strong enough. Meanwhile, the power for the service channel also should be strong enough to guarantee the QoS will not be impacted. The balance of the control channel and the service channel not only denote the power allocation balance, but also the code and time slot allocation. Generally speaking, in the initial access period, the BS and UE need to transmit control signals to get cell search and synchronization, etc. Thus, the control channel should occupy more radio resources. In contrast, in the service providing period, the service channel should be allocated more radio resources.

5.3 Capacity Planning

5.3.1 Factors Influencing System Capacity

The system capacity of a wireless network is determined by the usable radio resources. A compound access scheme containing time duplex, frequency duplex, code duplex, and space duplex is adopted in TD-SCDMA systems. Therefore, the usable radio resources include time slot, frequency, code, and space. On the other hand, any technique that can improve the resource utilization efficiency can improve the system capacity. Except those general factors influencing system capacity in all 3G networks, some factors should be analyzed, especially in TD-SCDMA systems, which are caused by the special physical layer structure and techniques that we have introduced before. We list these main factors as follows:

- **Smart Antenna and Joint Detection:** Mitigate the MAI caused by the non-orthogonal codes and improve the QoS. Then, more users can be supported at the same time, thus increasing the system capacity.
- **Dynamic Channel Allocation:** Centralized management of the radio resources and improving the channel resources utilization efficiency.
- **Baton Handoff:** Improves the success probability of handoff; meanwhile, saves the resources consumed in the handoff process.
- **Uplink Synchronization:** Precise and quick synchronization makes it possible to use the smart antenna and joint detection techniques in TD-SCDMA systems.

- **Power Control:** Allocates the power resource among users reasonably so it improves the efficiency of power usage.

5.3.2 Service Types in TD-SCDMA Systems

According to the 3GPP specification [14], the provided services in 3G networks are categorized into four types in terms of the required QoS: conversational, streaming, interactive, and background. Many examples about how to apply these services in typical environments can be found in [15]. Another important step in capacity planning is to obtain the model of each service. In TD-SCDMA systems, according to the switch technologies, we always classify the services into two kinds: circuit service (CS) and packet service (PS). CS and PS denote that the services are transported by the circuit switch technique and the packet switch technique, respectively. More technique details about circuit switch and packet switch can be found in [16]. The services provided by TD-SCDMA systems according to the transmission rate include 12.2 kbps CS voice service, 64 kbps CS, 64 kbps PS, 128 kbps PS, and 384 kbps PS. In the enhanced TD-SCDMA standard, high-speed downlink packet access (HSDPA), 2 Mbps and even higher rate data services can be provided. In order to obtain the capacity planning information in a region, we should estimate the coverage and distribution of each service. The coverage analysis of each service has been introduced in section coverage planning. Thus, we mainly focus on the service distribution in a typical environment in this section. The distribution information of each service should be evaluated comprehensively, which includes application requirements, population density, economic levels, and so on. An example is given in Table 5.3, which shows the service distribution in some typical environments in TD-SCDMA systems, where the percent values denote the service demand probability that is required in a region. Taking this distribution information as a reference, the operator can pre-estimate the service types and demand for a certain region in network planning.

To obtain the service model accurately, many parameters should be taken into account. These parameters of CS and PS are introduced as follows.

5.3.2.1 CS

The CS in TD-SCDMA systems are related to the low rate voice services or circuit domain data services, such as audio service and video telephone, which includes 12.2 kbps and 64 kbps. When we model or evaluate a CS, the main indicators that should be considered are listed as follows:

- **Busy hour call attempts (BHCAs)**: The service conversation times per hour.
- **Average time length per call (T_s)**: The statistic of the average persistence time per service conversation; unit: second.

Table 5.3 An Example of Service Distribution in TD-SCDMA Network

Services	*Bearer (kbit/s) (Uplink/Downlink)*	*Intensive Urban*	*General Urban*	*Suburban*
Voice	CS (12.2/12.2)	100%	100%	100%
Video conference	CS (64/64)	30%	20%	5%
Internet services	PS (64/64)	70%	60%	30%
High-speed streaming media	PS (64/144)	30%	20%	0%
High-speed video	PS (64/384)	10%	0%	0%

- **Average voice volume per user (*V*: Erl/user)**: The value can be calculated by the formula $V = BHCA \times T_s/3600$; unit: Erl.
- **Activation factor (β)**: $\beta = \frac{R_s}{R_t}$, where R_s is the service data rate and R_t is the total data rate.
- **Blocking rate (η)**: $\eta = \frac{N_{\text{unsuccessful}}}{N_{\text{total}}}$, where $N_{\text{unsuccessful}}$ is the unsuccessful call times and N_{total} is the total call times.

5.3.2.2 PS

The PS in TD-SCDMA systems are related to data services, which includes 64 kbps, 144 kbps, and 384 kbps. The data services are asymmetrical between up- and downlink, and the downlink always affords more data transmission. Therefore, the up- and downlink should be taken into account respectively in capacity planning. The main indicators that should be considered in a PS model are listed as follows:

- **Packet size**: The transmission size per packet in PS service, unit: Byte.
- **Busy hour call attempts (BHCAs)**: The data packet transmission times per hour.
- **Average data streaming rate of the downlink R_d**: The downlink should take more data service than uplink, thus R_d is also an indicator in PS; unit: Byte.
- **Activation factor (β)**: $\beta = \frac{R_s}{R_t}$, where R_s is the service data rate and R_t is the total data rate.

5.3.3 Interference Analysis

In TD-SCDMA systems, there are three kinds of interferences that influence the QoS quality: noise, inter-system interference, and intra-system interference. The noise interference inherently exists in the comunication system. We cannot avoid it, only mitigate the noise affect via some advanced signal process techniques. The inter-system interference exists when other networks operate with TD-SCDMA systems in the same region. Thus in network planning, we also should take into account the frequency bands, coverage, and the operation state of the existing networks. The intra-system interference exists in the system itself, which is the main factor influencing the system performance and which should be paid more attention to in network planning. The intra-system interference is also the main content introduced in this section.

TD-SCDMA is a self-interference cellular system. The intra-system interference also can be categorized into inter-cell interference and intra-cell interference. The inter-cell interference mainly influences the cell-edge users' QoS, which degrades the whole system's performance. The intra-cell interference is a more severe problem compared to inter-cell interference because it impacts all users' QoS. Anyway, both the inter- and intra-cell interference decrease the whole system capacity and should be mitigated as much as possible. In the following, we will introduce some intra-system interferences that exist widely in TD-SCDMA systems, which may happen inter- or intra-cell.

5.3.3.1 Multiple Access Interference (MAI)

The MAI exists in all CDMA-based systems. Thus we only introduce it briefly and give some solutions adopted in TD-SCDMA systems. In CDMA-based systems, a UE is identified by its frequency, time slot, channelization code, and space beam director. The space duplex multiple access (SDMA) technique is only considered as an interference mitigation method now. Thus in TD-CDMA systems, the MAI generally refers to the interference among users who occupy the same time-frequency resources but their channelization codes are non-orthogonal or the channelization codes' orthogonality has been destroyed.

The MAI could be intra- or inter-cell interference. In the same cell, the orthogonal variable spreading factor code (OVSFC) is used as channelization codes to distinguish different users who share the same time-frequency resources in TD-SCDMA systems. The intra-cell MAI exists when the orthogonality of OVSFC is destroyed by synchronization errors. In contrast, the inter-cell MAI generally refers to the non-orthogonality among different users' scrambling codes. In TD-SCDMA systems, the solution for mitigating MAI is called the interference cancelation technique, which contains multiuser joint detection and smart antenna for mitigating the intra-cell MAI, the dynamic channel allocation technique for mitigating the inter-cell MAI.

5.3.3.2 Up- and Downlink Interference

In TD-SCDMA systems, the pair frequency band is not necessary. Thus, up- and downlink occupy the same frequency bandwidth. In some slots, some cell BSs or UEs are possible in the state "BSs transmitting and UEs receiving." Meanwhile, other cell BSs or UEs are in the state "BSs receiving and UEs transmitting." As a consequence, the up- and downlink interference exists (i.e., the up- and downlink time slots overlap), and the up- and downlink interference problem degrades the system QoS severely. The up- and downlink interference only happens in TDD mode systems. Therefore, compared with other FDD 3G standards, the up- and downlink interference is particular in TD-SCDMA systems.

Three things cause up- and downlink interference: (1) The up- and downlink switching points among neighborhood cells are conflicting. (2) The same up- and downlink switching point configuration is used, but the synchronization errors exist among neighborhood cells. (3) The transmission delay is larger than the guard period.

1. **The up- and downlink switching point among neighborhood cells is conflicting**: TD-SCDMA contains six service slots TS1–TS6 in each subframe. TS1 is always allocated to uplink, and TS2–TS6 can be allocated to either up- or downlink flexibly according to the corresponding services load. Two switching points exist in a TD-SCDMA subframe. The first switching point is located between DwPTS and UpPTS as GP. The second is flexibly located within TS2 and TS6. As shown in Figure 5.10, cell A adopts symmetrical time slots allocation that distributes three time slots to the up- and downlink, respectively. Then, the switching point is between TS3 and TS4. Cell B adopts an asymmetrical time slots allocation that distributes two time slots to uplink and four time slots to downlink. Then, the switching point is between TS2 and TS3. It is clearly shown in Figure 5.8 that the TS3

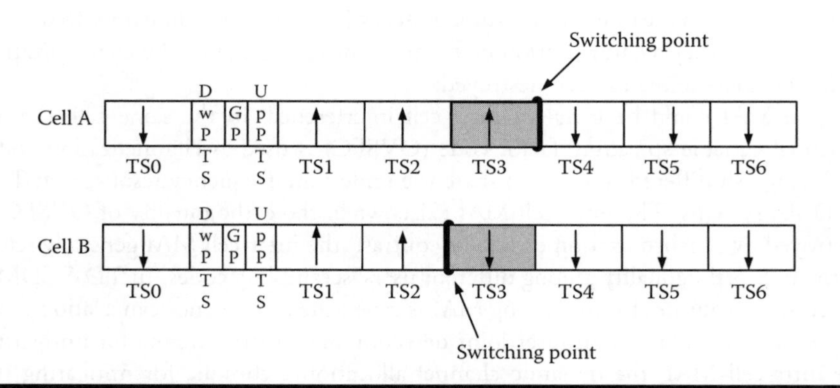

Figure 5.8 The switching point conflict among neighborhood cells.

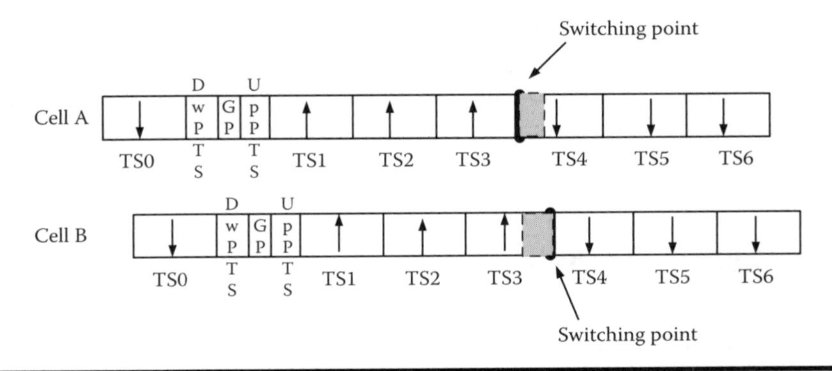

Figure 5.9 Synchronization error among neighborhood cells.

is "overlapped." When the BS in cell B is transmitting signals, the BS in cell A is receiving signals. Consequently, the up- and downlink interference exists in TS3. In order to avoid the conflicting switching point among cells situation, some protections should be used in network planning:

a. According to the up- and downlink services characteristics, all cells in a region have the same switching point configuration.
b. At the edge of neighborhood cells that possess different switching point configurations, if the testing shows the interference in the "overlapped" slot cannot be ignored, then discard these slots.

2. **With the same up- and downlink switching point configuration, the synchronization errors exist among neighborhood cells**: Although the neighborhood cells have the same switching point configuration, the up- and downlink interference may exist if these BSs cannot get transmitter synchronization. Figure 5.9 shows an example of the interference caused by synchronization errors between two neighborhood cells. Cell A and cell B have the same switching point configuration. However, the frame starting points of the two cells are different due to the synchronization errors, so the shadowed parts that denote the interference exist in each time slot. According to 3GPP specification [17], in realistic systems, the frame starting point error should be less than 3 μs (about 3.85 chips in TD-SCDMA) so that the synchronization can be ignored. In TD-SCDMA systems, BSs adopt the external reference clock (such as GPS or the Galileo satellite system) to get frame synchronization.
3. **The transmission delay is larger than the guard period**: We have introduced in the former section that the maximum cell coverage radius in TD-SCDMA systems is 11.25 km. If the coverage radius exceeds 11.25 km, the propagation delay between DwPTS and UpPTS will be longer than the guard period. This case will cause the unanticipated situation where a cell BS is transmitting DwPTS and another cell BS is receiving UpPTS. As a result, the

up- and downlink interference also exists among neighborhood cells. However, in sparsely populated areas, the cell coverage radius is always in excess of 11.25 km in order to obtain wide coverage. In these cases, because of fewer cell-edge users accessing the network, the interference has nearly no impact to the whole system capacity.

5.3.4 Capacity Analysis

System capacity generally has two kinds of definitions: (1) The maximum user number that a single cell can support simultaneously under required communication QoS conditions for all users. (2) The average data throughput of a single cell under required BER conditions.

Although many advanced interference mitigation techniques, such as smart antenna and joint detection, have been adopted in TD-SCDMA systems, the algorithm complexity and signal processing delay should be considered in realistic cases to achieve a compromise between performance and cost. Thus, the TD-SCDMA system capacity is also limited by interference. Besides interference, another important factor impacting system capacity in the TD-SCDMA system is the limited code resources. In the following sections, we will analyze the system capacity from two aspects: the interference-limited and the code-resource-limited situations.

The system upper bound capacity is a theoretical value and is only used as a reference in project implementation. In practical terms, the system capacity is smaller than the theoretical value due to the existing unpredicted interference. Therefore, we always use the ultimate capacity multiplied with a load factor η ($\eta < 1$) to estimate the realistic system capacity. The load factor η can be obtained from the testing or experience, which is omitted in the section. The reader can look up the content about load factor η calculation from some project manuals according to the specific system structure. Thus, the following capacity analysis are under ideal and theoretical conditions.

5.3.4.1 Interference Limited Capacity

Smart antenna and joint detection technologies can improve system capacity. Considering the performance gains brought by these advanced interference mitigation techniques, the single service SINR of a targeted user can be expressed as follows:

$$\frac{C}{I} = \frac{p}{(1-\beta+i)NpvA-(1-\beta)pvA+N_0} \tag{5.10}$$

where v is the activating factor; A is the performance gain brought by smart antenna; β is the joint detection efficient factor; N_0 is the thermal noise; N is the user number supported by the service simultaneously; p is the received power from a single user

Table 5.4 The Upper Bound of the TD-SCDMA Uplink Capacity in an Interference-Limited Situation

Services	*C/I (dB)*	β	*A*	*i*	*V*	$N_{ultimate}$	*Interference Margin*
AMR 12.2 kbps	−2.61	0.8	0.2	0.65	0.67	16.25	2.88
64 kbps CS	4.39	0.8	0.2	0.65	1	2.38	7.55
64 kbps PS	3.12	0.8	0.2	0.65	1	3.10	4.15
128 kbps PS	5.97	0.8	0.2	0.65	1	1.72	3.13
384 kbps PS	8.55	0.8	0.2	0.65	1	1.06	0.55

at BS node; and $i = P_{inter}/P_{intra}$, where P_{inter} and P_{intra} are the total received power from inter- and intra-cell, respectively.

From Equation 5.10, we can calculate the ultimate user number supported by a single service:

$$N_{upper} = \frac{(1 - \beta)vA + (C/I)^{-1}}{(1 - \beta + i)vA} \tag{5.11}$$

According to the Equations 5.10 and 5.11, the ultimate system capacities of different services in TD-SCDMA systems are listed in Tables 5.4 and 5.5, where "AMR" denotes the adaptive multirate voice coder.

The values in the two tables can be used as the reference in network planning. As the technology develops, the required C/I of each service will decrease and the system capacity will be further improved.

Table 5.5 The Upper Bound of the TD-SCDMA Downlink Capacity in an Interference-Limited Situation

Services	*C/I (dB)*	β	*A*	*i*	*V*	$N_{ultimate}$	*Interference Margin*
AMR 12.2 kbps	−2.61	0.8	0.2	0.65	0.67	16.25	2.88
64 kbps CS	4.39	0.8	0.2	0.65	1	2.38	7.55
64 kbps PS	4.32	0.8	0.2	0.65	1	2.41	7.24
128 kbps PS	6.77	0.8	0.2	0.65	1	1.47	4.18
384 kbps PS	10.05	0.8	0.2	0.65	1	0.53	0.80

5.3.4.2 Code Resources Limited Capacity

A channel is comprised of a carrier frequency, time slot, and spreading code, which is also called a base resource unit (BRU). In TD-SCDMA systems, the BRU is denoted as the radio resources occupied by a 16-length spreading code physical channel. Thus, a time slot can provide 16 BRUs. If the symmetrical time slot allocation scheme is adopted, there are three time slots in up- and downlink, respectively (i.e., both up- and downlink have $3 \times 16 = 48$ BRUs). Different services have different spreading codes, so they occupy different BRU numbers. For example, 12.2 kbps voice service occupies two 16-length spreading code channels, so a time slot only provides eight voice services and each service occupies 2 BRUs. The user number the downlink can support simultaneously is $48/2 = 24$. For uplink, the maximum voice service user number that can be supported is $8 \times 3 - 1 = 23$ (2 BRUs are allocated to the random access channel), which is also listed in Table 5.6. The supported user number of up- and downlink is very different in asymmetrical services. Take the 384 kbps data service, for example, one service occupies 40 BRUs in downlink and 8 BRUs in uplink, respectively. Thus, we can calculate the user number supported by up- and downlink as follows: uplink, $(48–2)/8 = 5.75$, the user number supported by uplink is 5; downlink, $48/40 = 1.2$; the user number supported by downlink is 1.

5.3.4.3 Capacity Estimation

In TD-SCDMA systems, there are only 16 channels per time slot. Because of the utilization of DCA, smart antenna, and joint detection techniques, the interference in a time slot is small. The self-interference of TD-SCDMA can be mitigated further due to the combination of time duplex and frequency duplex. Generally speaking, three factors influence the system capacity: network organization mode, the radio propagation environment, and the maturity of TD-SCDMA systems.

Table 5.6 The BRUs Occupied by Each Service

Services	*PS 12.2 kbps*	*CS 64 kbps*	*PS 64 kbps*	*PS 128 kbps*	*PS 384 kbps*
Downlink single service	2 BRUs	8 BRUs	8 BRUs	16 BRUs (one TS)	40 BRUs (three TSs)
Uplink single service	2 BRUs	8 BRUs	8 BRUs	8 BRUs	8 BRUs
Uplink PCCUCH	1–2 BRUs (RACH)	1–2 BRUs (RACH)	1–2 BRUs (RACH)	1–2 BRUs (RACH)	1–2 BRUs (RACH)
Downlink users number	24	6	6	3	1
Uplink users number	23	5	5	5	5

In network planning simulation and system testing, TD-SCDMA can operate in a full channel situation (i.e., every slot is allocated with 16 channels). That is because TD-SCDMA adopts TDD mode, and multiuser interference mitigation techniques are utilized. Therefore, the capacity of TD-SCDMA is mainly limited by the code resources instead of interference. However, because the network optimization and the maturity of TD-SCDMA techniques are still under development, it is impossible to allocate 16 channels per time slot in practical cases. We should also reserve some channels for the interference margin, even though the system capacity will decrease.

In order to make capacity planning clearly, we can estimate the system capacity roughly at first. In TD-SCDMA capacity estimation, the capacity calculation can follow some methods for simplification.

- The capacity estimation is mainly from the code resources planning aspect.
- The capacity estimation may be under the assumption that the TD-SCDMA operates in the ideal case (i.e., 16 channels in each time slot).
- In multiple carrier frequency cells, the interference among different carrier frequencies can be ignored.
- Set the symmetrical up- and downlink time slot allocation case as reference.

According to each service mode, the capacity in a typical cell sector is listed in Table 5.7, where "RAB" denotes radio access bearer.

Table 5.7 The Single Sector Capacity Analysis of Typical Regions

Region/Model	*Intensive Service Region*	*High-Density Service Region*	*Normal Density Service Region*	*Low-Density Service Region*
AMR 12.2 kbps RAB	92.00%	95.80%	98.40%	99.80%
64 kbps CS RAB	4.00%	2.52%	1.12%	0.16%
64 kbps PS RAB	2.40%	1.26%	0.48%	0.04%
128 kbps PS RAB	1.44%	0.42%	0.00%	0.00%
384 kbps PS RAB	0.16%	0.00%	0.00%	0.00%
Voice channel (Erl)	7.56	11.40	16.36	17.58
Video phone (Erl)	0.33	0.30	0.19	0.03
Downlink data throughput (kbps)	33.82	15.99	5.11	0.46

5.4 Radio Resource Planning

5.4.1 Radio Parameters in TD-SCDMA Systems

Before presenting more details about radio resources planning, many parameters in TD-SCDMA systems are introduced first. These parameters are related to the network identification, cell identification, and some important indicators that impact the network planning and operation.

5.4.1.1 Network Identity Parameters

These parameters are always used as the unique identity of a UE, so the core network can provide services to the UE effectively. The identity of a UE is a hierarchical structure. We briefly introduce these parameters from the upper layer structure to the lower layer structure.

- **Mobile Country Code:** The unique code identity of the country that a mobile user belongs to. The mobile country code is managed by International Telecommunications Union (ITU).
- **Mobile Network Code:** The unique code to identify which mobile network a user belongs to, which is managed by each country.
- **Location Area Code and Location Area Identity:** A location area is the smallest unit that the network pages a user. Thus, the coverage of a location area needs to be planned properly in the initial network coverage planning. A location area code is a unique identity of the corresponding location area. In mobile networks, the core network provides the CS by identifying a UE position according to the location area identity. The location area identity is composed of mobile country code, mobile network code, and location area code.
- **Route Area Code and Route Area Identity:** Similar to the location area code, a location area code is divided into many route areas. For packet services, the core network identifies a UE position according to the route area identity, which contains location area identity and route area code.
- **Radio Network Controller Identity:** The radio network controller (RNC) identity of each RNC is unique in a network. A RNC area can contain several location area identities.
- **Cell Identity:** The cell identity is used so a cell can be identified by a RNC. The cell identity is unique in a RNC, but cells in different RNCs could have the same cell identity.
- **Cell Identifier:** The cell identifier (CI) is the unique identity in a network, which is comprised of the RNC identity data (RNCID) and the cell identity data (CID):

$$CI = RNCID \times 65536 + CID \tag{5.12}$$

The CI is calculated automatically by the system.

- **Cell Global Identity:** As a worldwide cellular network, TD-SCDMA should allocate a unique global identity to each cell. The cell global identity is comprised of mobile country code, mobile network code, location area identity, and cell identifier.

5.4.1.2 Cell Basic Parameters

We have introduced the hierarchical structure that identifies a unique global cell in a TD-SCDMA network. In this section, more details related to the cell parameters are listed as follows.

- **Frequency:** The frequency parameters in a cell contains the center frequencies and bandwidths. Frequency planning among cells is an important step that determines the inter-cell interference level and impacts the QoS of the whole network, especially in TDD systems. The details of frequency planning will be introduced in Section 5.4.3.
- **Cell Parameter Identity Data:** The cell parameter identity data uniquely indicates the code resources that a cell uses, which includes the downlink synchronization sequence, the scrambling code, and midamble. The codes planning also impacts the interference among neighborhood cells.
- **Cell Carrier Frequency Priority:** In an N ($N > 1$) carrier frequencies network, a cell can utilize several carrier frequencies. The cell carrier frequency priority coefficient indicates the choice priority among carrier frequencies in a cell.
- **Cell Slot Priority:**The cell slot priority is the parameter imputed into the dynamic channel allocation algorithm. It indicates which time slot will be allocated for users with high priority by the dynamic channel allocation algorithm.

5.4.1.3 Power Coverage Parameters

The transmitted power from BS to UE should be strong enough so that the QoS can be maintained. On the other hand, we should control the transmitted power in order to avoid the inter-cell interference. In this section, the power coverage parameters that indicate the transmitted power level in cellular networks are introduced.

- **The Maximum Transmitted Power on the Up- and Downlink:** The downlink maximum transmitted power denotes the allowable total power on a carrier frequency transmitted from BS node, which determines the cell coverage as introduced in Section 5.2.2. The uplink maximum transmitted power is that allowable from the UE, which is related to the UE characteristics and service types.
- **PCCPCH Transmitted Power:** The PCCPCH transmitted power determines the coverage of the broadcasting message in a cell, which should be transmitted to all users in a cell.

- **DwPCH Transmitted Power:** The DwPCH transmitted power determines the coverage of the downlink pilot channel, which should also be configured properly.
- **SCCPCH Transmitted Power:** The SCCPCH transmitted power determines the coverage of the paging channel and the access channel.
- **PICH Transmitted Power:** The PICH transmitted power determines the power of the paging indicator channel, which impacts the coverage and performance of cell paging.
- **DPCH Transmitted Power:** The DPCH transmitted power contains the maximum and minimum transmitted power, which denotes the power control level of DPCH.
- **PRACH Transmitted Power:** The PRACH transmitted power determines the uplink access coverage and performance of a network, which is configured at the UE node.

5.4.1.4 Cell Access Parameters

The cell access parameters are related to the random access process. Two parameters should be noted.

- **The Required UpPCH Power:** The required UpPCH power should guarantee that the uplink synchronization sequence can be detected correctly by the BS node. If the required UpPCH power value is too high, the realistic coverage will be limited and the inter-cell interference will increase. But if the value is too low, the QoS performance will degrade.
- **The Maximum Transmitted FPACH Power on the Downlink:** After the BS node detects the effective UpPCH value, the BS node transmits acknowledgement information and measurement parameters in the FPACHs of the following four subframes. The downlink maximum transmitted FPACH power impacts the UE access success rate.

5.4.2 Radio Resources in TD-SCDMA Systems

It has been mentioned before that the multiple access schemes in TD-SCDMA systems includes frequency duplex, time duplex, and code duplex. In the future, the space duplex method will be adopted for higher spectral efficiency. Thus, the radio access resources in TD-SCDMA include frequency, time slot, code, and space resources. The space duplex multiple access (SDMA) is always considered as a future advanced technique and only as an interference mitigation technique now. Another kind of resource that should be considered is power resource, which also has an impact on the whole system capacity. But the power allocation is only implemented at the BS node and the power resource can be maintained and is not a problem in many cases. Therefore, the radio resource planning in TD-SCDMA systems mainly focuses on the frequency, time slot, and code resources planning now. The details will be introduced in the following sections.

5.4.3 Frequency Resource Planning

Sometimes only parts of frequency bands are suitable for transmitting wireless signals in the realistic propagation environment. Thus, the frequency is the most precious resource in wireless networks due to its scarcity. Furthermore, the users need large bandwidth for high data rate transmissions in the future. Thus, how to take full advantage of the existing frequency resource is a hot research issue in network planning.

The frequency bandwidth allocated to TD-SCDMA in China includes 1880–1920 MHz, 2010–2025 MHz, and 2300–2400 MHz. In the TD-SCDMA standard, one carrier frequency occupies a 1.6 MHz bandwidth. Take the frequency 2010–2025 MHz as an example. There are nine carrier frequencies, which are listed in Table 5.8.

In the following sections, we will introduce some frequency planning techniques in the TD-SCDMA system, according to the available frequency resources shown in Table 5.8.

5.4.3.1 Interference Brought by Frequency Reuse

In TD-SCDMA systems, the interference includes adjacent channel interference (ACI) or the co-channel interference (CCI). The ACI exists due to the non-ideal filter,

Table 5.8 TD-SCDMA Carrier Frequency Distribution in Band 2010—2025 MHz

Number	*Carrier Frequency (MHz)*	*Carrier Frequency Scope (MHz)*
f1	2010.8	2010–2011.6
f2	2012.4	2011.6–2013.2
f3	2014	2013.2–2014.8
f4	2015.8	2015–2016.6
f5	2017.4	2016.6–2018.2
f6	2019	2018.2–2019.8
f7	2020.8	2020–2021.6
f8	2022.4	2021.6–2023.2
f9	2024	2023.2–2024.8

which can be avoided by improving the equipment quality or using the frequency guard between adjacent channels. As the hardware technique continues developing, the ACI can be ignored in most cases. In contrast, the CCI is an important problem that should be paid more attention to in wireless networks. In order to obtain a high spectral efficiency and get more profit, the limited frequency will be reused in different cells. When the distance between two cells with the same carrier frequency is less than the minimum required value, the communication quality in both cells degrades severely. Therefore, planning the frequency resources in cells properly is the first step in resources planning. Of course, the frequency reuse factor may be 1 in the future with the interference cancelation scheme. In this case, the advanced scheduling algorithm should be adopted in order to mitigate interference among cells. However, if the user number is large and most of the frequency resources have been used, only a few resources can be scheduled. Consequently, the algorithm efficiency is too low. Thus, the multi-frequency network is considered at the initial network planning step. As the service requirement and user numbers increase, we can evolve the multi-frequency network into a single-frequency network, or the mixture of the two-network organization mode.

5.4.3.2 Single-Cell Carrier Frequency Configuration

Because a carrier frequency in a TD-SCDMA system only occupies a 1.6 MHz bandwidth, the number of carrier frequencies is abundant in TD-SCDMA compared with that in WCDMA or CDMA2000. The available carrier frequency number in a cell can also be flexible.

1. **Single carrier frequency configuration:** A single carrier frequency configuration denotes that each cell only is configured with a carrier frequency. Each carrier frequency has its own PCUCCH, broadcast channel, DwPTS, and UpPTS. If several carrier frequencies are allocated to the sectors of a cell, then each carrier frequency should be treated as a logical cell, which means these carrier frequencies are independent of each other. The single carrier frequency configuration is depicted in Figure 5.10.

   In realistic network organizations, many problems will appear in the single carrier frequency configuration as the network capacity increases, such as the difficulty of cell search and handoff, the high complexity of UE measurement, high interference, and small coverage radius.

2. **Multi-carrier frequency configuration:** Multi-carrier frequency configuration denotes that several carrier frequencies belong to the same cell. Choose one of these carrier frequencies as a primary carrier frequency; the others will be secondary carrier frequencies. All the carrier frequencies can carry different data information, but only the primary carrier frequency has the broadcast channel, DwPTS and UpPTS. All carrier frequencies have the same scrambling and midamble codes.

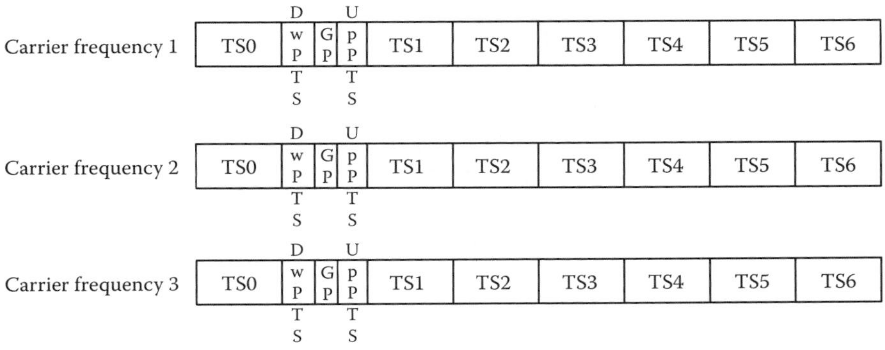

Carrier frequency 1	TS0	DwPTS	GP	UpPTS	TS1	TS2	TS3	TS4	TS5	TS6
Carrier frequency 2	TS0	DwPTS	GP	UpPTS	TS1	TS2	TS3	TS4	TS5	TS6
Carrier frequency 3	TS0	DwPTS	GP	UpPTS	TS1	TS2	TS3	TS4	TS5	TS6

Figure 5.10 The Single-carrier frequency configur ationcell. Frequency factor is 3.

The multi-carrier frequencies configuration is depicted in Figure 5.11. Adoption of the multi-carrier frequencies configuration can avoid problems that exist in single carrier frequency configuration with the sacrifice of spectral efficiency.

5.4.3.3 Network Organization Mode

How to reuse limited frequency resources among cells is the main subject of frequency resource planning. Generally speaking, two kinds of network organization modes also exist: the multi-frequency network (MFN) and the single-frequency network (SFN).

- **MFN:** The neighboring cells possess the different frequency bands. The MFN has small inter-cell interference, but the whole system capacity is limited because the spectrum efficiency is low. An example of MFN is shown in Figure 5.12, where the frequency reuse factor is 3.
- **SFN:** The neighboring cells possess the same frequency band. Thus, the inter-cell interference is a severe problem in SFN. In TD-SCDMA SFN, many

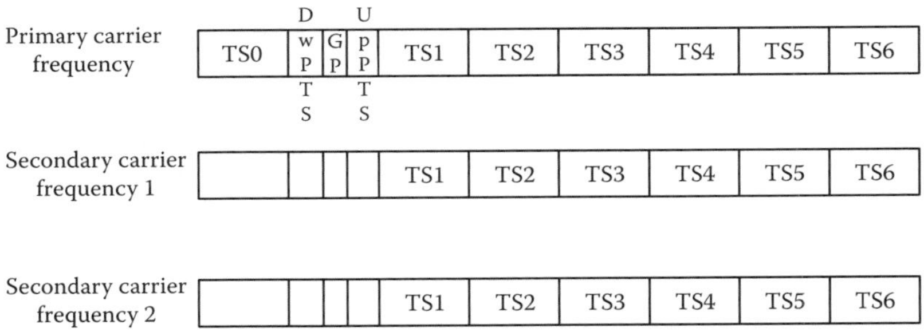

Primary carrier frequency	TS0	DwPTS	GP	UpPTS	TS1	TS2	TS3	TS4	TS5	TS6
Secondary carrier frequency 1					TS1	TS2	TS3	TS4	TS5	TS6
Secondary carrier frequency 2					TS1	TS2	TS3	TS4	TS5	TS6

Figure 5.11 The multi-carrier frequency configuration cell.

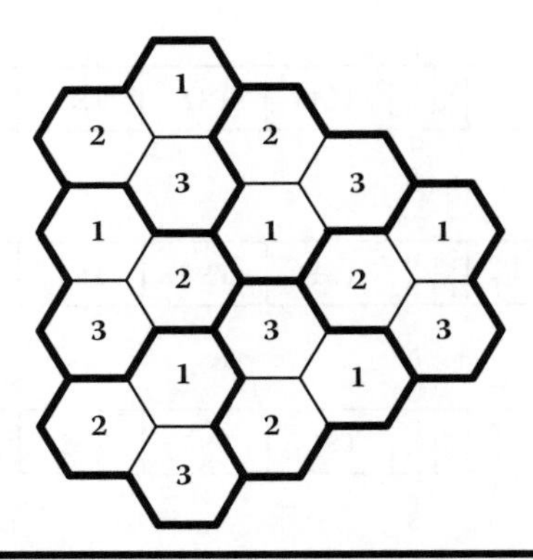

Figure 5.12 The MFN structure.

advanced interference mitigation and resource scheduling techniques are applied for inter-cell interference mitigation, such as DCA and smart antenna. The aim of SFN is to obtain a high system capacity and spectrum efficiency, but the network planning of SFN is also more complicated. Figure 5.13 shows an example of multi-carrier frequencies configuration SFN.

In network construction initial steps, because of the low service demand and only a fewer users accessing it, we can choose the MFN mode for simplification. As the network extends and users increase, the MFN can be evolved to SFN. In TD-SCDMA systems, the multi-carrier frequencies configuration in a cell and SFN organization is adopted, which can be seen in Figure 5.13.

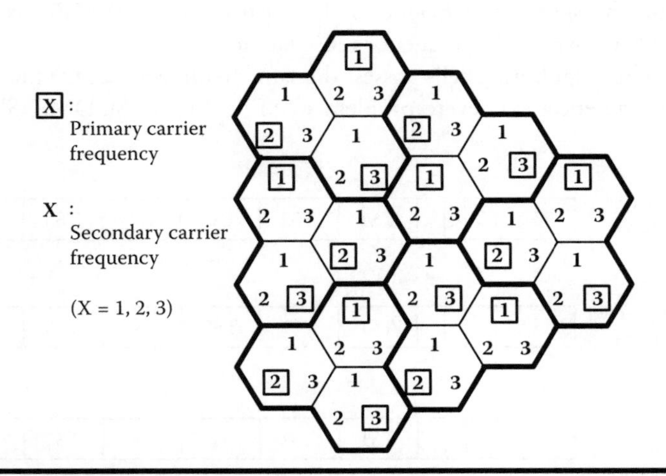

Figure 5.13 The SFN structure.

5.4.4 Time Slot Resource Planning

One of the characteristics of TD-SCDMA is that the time slot resources can be configured flexibly. This advantage makes it suitable for up- and downlink asymmetrical service transmission. We have introduced that there are two switching points in one TD-SCDMA subframe. One switching point is constantly located between DwPTS and UpPTS. The other switching point is located within TS1–TS6.

The time slot configuration only exists in TDD mode systems. Thus from the aspect of system capacity, TD-SCDMA systems can obtain a higher spectrum efficiency than that of WCDMA and CDMA2000 systems. However, if the time slot configuration in TD-SCDMA systems is not planned well, the extra interference, instead of performance gains, will occur.

In order to obtain performance gain and mitigate the negative influences brought by asymmetrical time slot allocation, some advice in time slot resources planning are given here.

- The up- to downlink slot ratio is constant in a cell, and there is only one switching point in TS1–TS6.
- For voice services, the time slot configuration in up- and downlink is always symmetrical. In contrast, the downlink should take more load in data services, thus the configuration can be asymmetrical for high spectrum efficiency.
- In cluster cells where the service requirement is similar, the configuration of these cells should be uniform to avoid the inter-cell interference between up- and downlink.
- In the future, data service requirements will always occur at an indoor environment. Furthermore, because the wall can weaken the signal interference from/to the outdoor environment, the configuration of indoor and outdoor can be different.

5.4.5 Code Resource Planning

The whole system capacity is severely restricted by the limited code resources in TD-SCDMA systems. Thus, how to utilize the code resources efficiently is one of the most important tasks in TD-SCDMA network planning. Because the correlation of 32 SYNC-DL sequences can satisfy the inter-cell interference requirement among neighborhood cells, we only ensure that different cells possess different SYNC-DL sequences. In contrast, the 128 scrambling codes need to be planned carefully. In the following sections, the code resources in the TD-SCDMA system are introduced first. Then, some code resource planning schemes are introduced.

5.4.5.1 TD-SCDMA Code Resources

Five kinds of code resources exist in TD-SCDMA systems: SYNC-DL, SYNC-UL, scrambling codes, midamble codes, and spreading codes. SYNC-DL, SYNC-UL, scrambling codes, and midamble codes are transmitted according to the common chip rate and do not need to be spread. These codes can be found in 3GPP specifications [18].

- **SYNC-DL:** The SYNC-DL sequence is the identification of a cell, which is transmitted in DwPTS. There are 32 different SYNC-DL sequences in TD-SCDMA systems. Each SYNC-DL sequence contains 64 chips. The SYNC-DL is used for cell search, downlink synchronization, codes identifying, and PCCPCH channel enabling. Thus, the SYNC-DL should be planned first.
- **SYNC-UL:** SYNC-UL is transmitted in UpPTS, which is used for uplink synchronization and random access. There are 256 usable SYNC-UL sequences in TD-SCDMA systems. Each SYNC-UL sequence contains 128 chips. The 256 SYNC-UL sequences are divided into 32 different groups, and each group contains eight SYNC-UL sequences. In TD-SCDMA network planning, each group SYNC-UL sequences correspond to a unique SYNC-DL sequence. A UE randomly chooses one of the eight known SYNC-UL sequences to transmit in the random access procedure.
- **Scrambling Codes and Midamble Codes:** There are 128 scrambling codes and 128 midamble codes in TD-SCDMA system. The 128 codes also are divided into 32 group. Each cell possesses four scrambling codes and four midamble codes.
- **Spreading Codes:** The OVSFC is used to spread transmitted signals in TD-SCDMA systems. The spreading factor scale is from 1 to 16. The spreading factor scale is determined by service type. The spreading codes are used to identify different users in a time slot.

Above all, one cell possesses one SYNC-DL sequence, eight SYNC-UL sequences, four scrambling codes, and four midamble codes.

5.4.5.2 Scrambling Code Configuration

The signal chips are multiplied by scrambling codes before being transmitted from the antenna. If the scrambling codes among neighborhood cells have a high correlation, the inter-cell interference will be a severe problem that degrades the QoS performance for cell-edge users. Thus, a regulation in the TD-SCDMA scrambling code configuration is that the high correlation codes cannot be allocated to neighborhood cells. For example, if a cell is allocated a scrambling code $\mathcal{R}$, then those codes that have a high correlation with code $\mathcal{R}$ cannot be allocated to its neighborhood cells.

In a specific planning process, we should calculate the correlation of the 128 scrambling codes in TD-SCDMA systems. Then, we can obtain a 128×128 matrix whose elements are the correlation coefficients of any two scrambling codes. According to the correlation matrix, the 128 scrambling codes are then divided into 32 groups. Each group containing four scrambling codes is allocated to a cell. Because four scrambling codes in a group is used in a cell, the four scrambling codes in a group could always have a high correlation, while the orthogonality of users in one cell is kept by channalization codes. Thus, the whole scrambling code planning procedure can be divided into two steps: (1) the correlation coefficient calculation and (2) the scrambling code groups allocation.

For a given 128 scrambling codes in a TD-SCDMA system, the correlation matrix is constant. The scrambling code groups configuration is the main issue in network planning. Two allocation strategies, the chain allocation algorithm and the parallel allocation algorithm, are introduced in [19].

5.5 Introduction of Network Optimization

TD-SCDMA is a terrestrial mobile communication network whose operation situation is changing all the time. That is because the network structure, UE distribution, and wireless propagation in mobile communication networks are unstable. In order to obtain long-term good performance, the network should be adjusted to satisfy the network's increasing requirements. Network optimization is targeted to improve the network performance during the development of the network.

Network optimization is explained as: According to the practical performance of the network, do the analysis of the network operation first. Then, based on this analysis, adjust the network resources distribution and system parameters to improve the system performance. In other words, under the existing network configuration condition, the network achieves optimal performance through network optimization.

5.5.1 Categories of Network Optimization

According to the different stages and targets, the network optimization is categorized into project and operation network optimization, respectively. After a new network is built, network optimization is started. The project network optimization always denotes a short-term period works and aims at solving the problem so the new network can operate normally. In contrast, the operation network optimization is implemented by the operator and is a long-term optimization to guarantee the network QoS.

According to the different optimization tasks, the network optimization is categorized into stage and persistence optimization, respectively. The stage optimization does a great job of changing the network in the short-term in order to change the network operation according to user requirements. The persistence optimization is aimed at improving the network performance from a long-term stand point and changes the network step-by-step.

5.5.2 Network Optimization of TD-SCDMA

Most of the network optimization procedures of TD-SCDMA are the same as that of other mobile communication networks. The uniqueness of network optimization of

TD-SCDMA is caused by the physical frame structure and the existing physical layer techniques. The unique network optimization in TD-SCDMA systems contains the link balance between the service and control channels, time slot configuration, the baton handoff, and the reasonable application of DCA and smart antenna or other key techniques. Consequently, the radio resource management (RRM) algorithm design in a TD-SCDMA system is more complicated and flexible.

Another important aspect is the parameters adjustment in TD-SCDMA systems. The reason is that TD-SCDMA systems have different physical techniques and procedures compared to WCDMA and CDMA2000. These parameters are as follows.

- **Pilot Power Parameter:** Contains the maximum FPACH power, PCCPCH power, and SCCPCH power.
- **Power Control Parameter:** Contains the initial SINR target value, SINR adjust step, service BLER target value, DPCH power allocation type, and the maximum transmitted power.
- **Synchronization Parameter:** Contains the continuous synchronization indicator and the discontinuous synchronization indicator.
- **Handoff Parameter:** Contains a handoff delay for intra-cell, the filter coefficient, report of measurement period, the self-cell added pilot threshold, and the neighboring cell added pilot threshold.
- **Channel Configuration Parameter:** Contains the time slot configuration and the slot dynamic channel allocation (SDCA) carrier frequency priority information.
- **Access Parameter:** Contains the UE required minimum receiver level, the UE maximum transmitted power permitted by RACH, the cell search threshold, and the cell selection threshold.
- **Frequency Parameter:** Contains network organization mode (SFN or MFN), the cell carrier frequency configuration, and the primary and secondary carrier frequency configuration.
- **DCA Parameter:** Contains the carrier frequency priority, the time slot priority and the channel scheduling algorithm.
- **Page Parameter:** Contains PICH, PCCH, and PCH.

Network optimization is huge and complex. Generally speaking, the normal network optimization procedures in all 3G standards are almost the same. Readers can refer to [20] about network optimization of WCDMA to find out more.

5.6 Conclusion

The chapter presents the basic methods about network planning and optimization in TD-SCDMA systems. It is shown that the difference between TD-SCDMA and other 3G standards is caused by the physical frame structure and techniques. Thus, the physical characteristics of TD-SCDMA specification are described briefly in part one. Then, based on these differences, the content of network planning is introduced from three aspects: coverage planning, capacity planning, and radio resource planning. Finally, some concepts of network optimization are also introduced.

Though the TD-SCDMA system can obtain higher spectral efficiency than WCDAM and CDMA2000 systems in theory, the technique and operation experience of TD-SCDMA are not mature enough. Many issues also need to be solved, such as sensitivity to synchronization error in TD-SCDMA systems, how to improve the cell coverage rate, how to utilize the limited code resources efficiently, the choice of network organization mode, etc. Of course, as the technique continues developing, the operation experience accumulates, and the hardware implementation accelerates, TD-SCDMA system performance will be improved and will receive greater attention.

Furthermore, as a response to high data rate services and the emergence of World interoperability for Microwave Access (WiMAX), 3GPP launched the 3G long-term evolution (3G LTE) project in November 2004 [21]. TD-SCDMA also evolved into TDD-LTE, and many new techniques, such as orthogonal frequency division multiplexing (OFDM) and multiple-input and multiple-output (MIMO), etc., have been adopted in TDD-LTE to support high data transmission rates and improve its spectral efficiency [22, 23]. So in the future, combining TDD specialized techniques, network planning, and optimization of TDD-LTE will also be enhanced and have its characteristics improved. However, one important tendency in 3GPP is that people always try to increase the commonality between TDD and FDD. For example, the frame structure of TDD-LTE systems is designed in harmony with FDD-LTE, which means they have a similar time slot and subframe length, etc. So network planning and optimization of TDD-LTE will have more commonality with FDD-LTE and it can benefit the future deployment of both systems.

References

[1] 3GPP TS 25.401 v8.1.0, UTRAN Overall Description (release 8), September 2008, http://www.3gpp.org/ftp/Specs/archive/25 series/25.401.

[2] 3GPP TS 25.221 v8.2.0, Physical Channels and Mapping of Transport Channels onto Physical Channels (TDD), September 2008, http://www.3gpp.org/ftp/Specs/archive/25 series/25.221.

[3] C. Gessner, R. Kohn, J. Schniedenharn, and A. Sitte, "UTRA TDD Protocol Operation," the 11th IEEE International Symposium on PIMRC, September 2000, vol. 2, pp. 1226–1230.

[4] J. Peter and B. Josef, "Joint Detection with Coherent Receiver Antenna Diversity in CDMA Mobile Radio Systems," *IEEE Transactions on Vehicular Technology*, vol. 44, no. 1, pp. 76–88, February 1995.

[5] J. Y. Yun and D. K. Sung, "Joint Detection for On/Off Uplink Traffic in the TD-SCDMA System," *IEEE VTC-spring*, no. 57, vol. 3, pp. 1677–1680, April 2003.

[6] V. Marius, H. Martin, and G. Jurgen, "Comparative Study of Joint-Detection Techniques for TD-CDMA-based Mobile Radio Systems," *IEEE Journal on Selected Areas in Communications*, vol. 19, no. 8, pp. 1464–1475, August 2001.

[7] A. Alexiou and M. Haardt, "Smart Antenna Technologies for Future Wireless Systems: Trends and Challenges," *IEEE Communications Magazine*, vol. 42, no. 9, pp. 90–97, September 2004.

[8] J. Razavilar, F. Rashid-Farrokhi, and K. J. R. Liu, "Software Radio Architecture with Smart Antennas: A Tutorial on Algorithms and Complexity," *IEEE Journal on Selected Areas in Communications*, vol. 17, no. 4, pp. 662–676, April 1999.

[9] C. Svensson, "Software Defined Radio-Vision or Reality," the 24th Norchip Conference, p. 149, November 2006.

[10] A. C. Tribble, "The Software-Defined Radio: Fact and Fiction," *2008 IEEE Radio and Wireless Symposium*, pp. 5–8, January 2008.

[11] 3GPP TS 25.102 v8.1.0, User Equipment (UE) Radio Transmission and Reception (TDD) (release 8), May 2008, http://www.3gpp.org/ftp/Specs/archive/25 series/25.102.

[12] T. S. Rappaport, "Wireless Communications Principles and Practice," *Second Edition*, Pearson Education Asia Limited, 2002.

[13] A. Goldsmith, "Wireless Communications," *First Edition*, Cambridge University Press, 2005.

[14] 3GPP TS 25.331 v8.6.0, Radio Resource Control (RRC): Protocol Specification (release 8), March 2009, http://www.3gpp.org/ftp/Specs/archive/25 series/25.331.

[15] 3GPP TS 22.105 v8.1.0, Service Aspects: Services and Service Capabilities (release 9), December 2008, http://www.3gpp.org/ftp/Specs/archive/22 series/22.105.

[16] R. Puzmanova, "Routing and Switching: Time of Convergence," *First Edition*, Addison Wesley/Pearson, 2004.

[17] 3GPP TS 25.123 v8.0.0, Requirements for Support of Radio Resource Management (TDD) (release 8), September 2008, http://www.3gpp.org/ftp/Specs/archive/25 series/25.123.

[18] 3GPP TS 25.223 v8.1.0, Spreading and Modulation (TDD) (release 8), May 2008, http://www.3gpp.org/ftp/Specs/archive/25 series/25.223.

[19] S. C. Yang and C. Samuel, 3G CDMA2000: "Wireless System Engineering," *First Edition*, Boston: Artech House, 2004.

[20] S. Kasera and N. Narang, "3G Mobile Networks: Architecture, Protocols, and Procedures," *First Edition*, New York: McGraw-Hill, 2005.

[21] 3GPP TS 25.814 v7.1.0, Physical Layer Aspects for Evolved Universal Terrestrial Radio Access (UTRA) (release 7), September 2006, http://www.3gpp.org/ftp/Specs/archive/25 series/25.814.

[22] P. Zhang, X. F. Tao, J. H. Zhang, Y. Wang, L. H. Li, and Y. Wang, "A Vision from the Future: beyond 3G TDD," *IEEE Communications Magazine*, vol. 43, no. 1, pp. 38–44, January 2005.

[23] G. Y. Liu, J. H. Zhang, Y. Wang, S. Li, and P. Zhang, "Evolution Map from TD-SCDMA to Future B3G TDD," *IEEE Communications Magazine*, vol. 44, no. 36, pp. 54–61, March 2006.

HSPA PLANNING AND OPTIMIZATION

Chapter 6

Capacity, Coverage Planning, and Dimensioning for HSPA

Anis Masmoudi and Tarek Bejaoui

Contents

6.1 Introduction

While in analog mobile networks and GSM (Global Service for Mobile communication), air interface planning consists in defining a frequency reuse pattern and a minimum spacing between frequencies allocated to the same cell, the RF problem is much simpler to solve in a UMTS network because the number of frequency carriers used by one operator is limited (two to three carriers). On the other hand, the frequency assignment difficulty in GSM is replaced by the planning of scrambling codes per cell for each carrier used.

TDMA (time-division multiple access)-based cellular networks such as GSM or GPRS (General Packet Radio Service) are based on breaking down the planning or dimensioning process into two tasks. Although the latter are jointly dependent, they allow an iterative dimensioning process that includes two phases. The first phase consists in predicting the needs in radio coverage (number of radio sites and coverage area per site) in the service area by taking into account terrain data and propagation laws. The traffic density is also taken into account at this stage in order to choose cell sizes adapted to the subscribers requirement needs. Resources assignment is included next in the second phase. More commonly, we talk about a frequency plan whose objective is to assign frequencies to base stations while minimizing interference, or maximizing C/I (carrier-to-interference) ratio. For TDMA-based systems, the dimensioning process should converge iteratively toward solutions offering acceptable qualities of service (C/I control per service) at the minimum cost (by minimizing the number of sites and saving the spectrum allocated to the operator using less carriers as possible). A first study of CDMA (code division multiple access) networks planning has been carried in [1] to apply the new radio technologies. A complementary study of frequency planning has also been achieved in [2].

In contrast to previous TDMA-based systems such as GSM where the dimensioning process can be split into several tasks, CDMA-based UMTS systems require a joint optimization: radio coverage and traffic capacity are mutually dependent. The problem is not yet to find an adequate frequency plan for a given coverage prediction, but an optimal power allocation for terminals. On the uplink of UMTS FDD (frequency division duplexing) mode, it is mainly a range problem. The link budgets for several services (voice at 16 kbps, data services at 128 kbps and 384 kbps) on the radio link indicate that the maximum range doesn't exceed 2.5 km. It is actually the downlink that limits site coverage due to limitations in the total power of NodeB. The cell size decreases versus traffic load to maintain the quality of service (QoS) or reduce the interference. A high traffic density can thus require a site's densification. The dimensioning consists essentially in predicting or finding the suitable cell sizes after solving the joint optimization problem.

The mentioned joint aspect of coverage and capacity in UMTS is due to the use of only one carrier in the CDMA multiple access technique (without frequency assignment), thus the capacity limitation of the system by the received interference level. The cell breathing mechanism in CDMA illustrates this interdependence.

Besides UMTS uses not only circuit-switched mode, such as in GSM, but also the packet-switched mode.

The initial dimensioning or preliminary planning of a WCDMA-based UMTS network is the first step of the global planning process, allowing a first evaluation of the density and configuration of required sites, the capacity offered to the planned air interface network, and the coverage to estimate jointly with the capacity. This step is before the detailed planning often based on sophisticated professional tools. It allows giving an initial general but important idea on the number of radio sites to deploy and the costs of the required infrastructure. The initial radio dimensioning includes:

- The link budget determining the maximum path loss the cell can support both in UL (uplink) and in DL (downlink). It is realized for each service and includes WCDMA-specific aspects such as interference margin due to the increase of the noise level caused by the present traffic, the fast fading margin due to fast power control, and SHO (soft handover) gain.
- Joint analysis of the coverage and the capacity.
- Estimation of the required number of radio sites and NodeBs.

The input parameters of dimensioning are:

- Services distribution among the total active users.
- Traffic density (number of users per unit area for each service area to plan).
- Estimation of the chronological increase of the number of subscribers for each service.
- Required QoS in terms of E_b/N_o ratio (signal energy per bit to noise power density per hertz) per service and of the required coverage probability.
- And the required GoS (grade of service) in terms of the tolerated blocking rate (for circuit-switched services) or of the minimum guaranteed bit rate at a required service time percentage (for packet-switched services).

6.2 Comparison between Basic Rel'99 UMTS and HSPA Planning Rules

Radio network planning of the classical WCDMA-based UMTS systems (Release 99) relies on PC, which is the main radio phenomenon acting on cell breathing and generating UMTS coverage and capacity—one depending on the other. The evolution of UMTS networks by integrating HSPA techniques [3–8] is based mainly on the AMC technique rather than power control in causing the cell breathing phenomenon [9].

Furthermore, the other HSPA specific mechanisms such as fast scheduling and HARQ (hybrid automatic repeat request) added to the HS (high speed) shared structure of data channels make the planning rules different for HSPA-based UMTS networks. For example, the choice of the scheduling technique has an impact on the cell size and resource share among services. Besides, the dedicated channels are not used in HSPA. Moreover, the SHO is not supported by the HSPA shared channel. In fact, the HHO (hard handover) between cells is done, owing to FCS (fast cell selection).

HSPA dimensioning allows determining the bit rate per user, cell capacity, and the covered cell range versus the scheduling technique. This latter is chosen according to the deployed services. For example, the FT (Fair Throughput) scheduling technique is adapted more to circuit-switched services requiring a given guaranteed bit rate for all the users, whereas the FR (Fair Resource) technique is preferred for interactive or background services, providing a compromise between the total capacity of the cell in terms of bit rate and users fairness.

6.3 Dimensioning Procedures for HSPA-based UMTS Networks

6.3.1 Coverage-Limited Dimensioning

An approximation of the relation between the CQI (channel quality indicator) reported on the UL of an HSPA cell and the SINR (signal-to-interference-and-noise ratio) received by the mobile has been elaborated in [10] for a required BLER equal to 10%. It is given according to the following linear function based on the standard [11] and on the AMC technique:

$$CQI = \begin{cases} 0 & \text{if } SINR \leq -16 \text{ dB} \\ \left[\dfrac{SINR}{1.02} + 16.62 \right] & \text{if } -16 \text{ dB} < SINR < 14 \text{ dB} \\ 30 & \text{if } 14 \text{ dB} \leq SINR \end{cases} \tag{6.1}$$

where $[\cdot]$ denotes the integer part (by lower value trunking). Taking this, the true expression should include the delay of the radio measures taken by the mobile. So it is written as follows (for a $SINR$ value between -16 dB and 14 dB) [12]:

$$CQI = \left[\frac{SINR(t - CQI_{\text{delay}})}{CQI_{\text{ratio}}} + Offset \right] \tag{6.2}$$

where CQI_{delay} is the delay between the time of CQI computation at the mobile [UE (user equipment)] and that needed to reflect effectively the channel at the NodeB (at the scheduler level), $CQI_{\text{ratio}} = 1.02$ and $Offset = 16.62$. The integer trunking is done to the lower value because the block error rate (BLER) shouldn't exceed the given value (10%).

For simplicity and just for initial dimensioning purposes, we have neglected the impact of measures delay by assuming that the received SINR doesn't change during CQI_{delay}. This hypothesis is justified by the deterministic propagation model, without considering fading, and by the fixed distance of mobile in movement versus NodeB. Thus the expression is simply restricted to the following relation:

$$CQI = \left\lfloor \frac{SINR}{CQI_{\text{ratio}}} + Offset \right\rfloor \tag{6.3}$$

Hence, for a given CQI_0 (from 1 to 30), there is a given bounded interval $[SINR_{\min}, SINR_{\max}]$ such that for each SINR belonging to it, we have the CQI_0 value. It is easy to determine the limits of this interval:

$$SINR_{\min} = CQI_{\text{ratio}} \cdot (CQI_0 - Offset) \tag{6.4}$$

and

$$SINR_{\max} = CQI_{\text{ratio}} \cdot (CQI_0 + 1 - Offset) \tag{6.5}$$

The SINR definition can be written in logarithmic scale (dB) as follows:

$$\begin{aligned} SINR &= P_{TX} - L_{\text{Total}} - 10\log_{10}\left(10^{\frac{I_{\text{intra}} - L_{\text{Total}}}{10}} + 10^{\frac{I_{inter}}{10}}\right) \\ &= P_{TX} - 10\log_{10}\left(10^{\frac{I_{\text{intra}}}{10}} + 10^{\frac{I_{\text{inter}} + L_{\text{Total}}}{10}}\right) \end{aligned} \tag{6.6}$$

where P_{TX} is the transmitted power per code in dBm (individual transmission power emitted by the NodeB on the HS shared channel), L_{Total} is the total path loss (in dB) due to distance and the shadowing effect, I_{inter} is the extracellular interference received by the mobile (in dBm), and I_{intra} is the emitted intracellular interference in dBm (total transmitted power by the NodeB multiplied by the non-orthogonality factor). It should be proportional to the number of active mobiles.

By taking into account the last expression, we can extract the tolerable total path losses L_{Total} referring to each value of $SINR_{\min}$ and $SINR_{\max}$, and thus the minimum and maximum distances to the NodeB $r_{0,\min}$ and $r_{0,\max}$ bounding the ring referring to CQI_0 (using an appropriate deterministic propagation model). Yet the 3GPP (Third Generation Partnership Project) standard [11] provides the correspondence tables between the different CQI values and the referred TBS (transport block size) with an indication of the corresponding number of codes (physical shared channels) and the used modulation type [QPSK (quadrature phase shift keying) or 16-QAM (16-quadrature amplitude keying)], and this for different terminal categories. Therefore we can determine, in the coverage-limited case, the number of HSPA codes $n(r)$ and the transport block size' $TBS(r)$ referring to this CQI_0, both depending on the distance r to the NodeB (between $r_{0,\min}$ and $r_{0,\max}$). In fact, a given distance r refers to a SINR level received by the mobile, which is associated to a *CQI* value referring to a determined number $n(r)$ of shared channel codes and a transport block size $TBS(r)$. Assuming a uniform traffic and a deterministic propagation model, the latter expressions will be defined as constant by pieces versus the distance r to the NodeB.

6.3.2 Capacity-Limited Dimensioning (Code Limitation)

Assuming a uniform traffic (within the cell), with area density ρ representing the number of simultaneous active users per area unit (density), the scheduling technique is "Fair Resource" (or "Round Robin"), and neglecting channel fluctuations, then the condition of limitation in the number of physical codes—not exceeding, for example, 15 allocated codes per cell in case of HSDPA (high-speed downlink packet access)—can be written along the cell area as follows:

$$\int_0^{2\pi} d\alpha \int_0^{R} n(r)\, \rho\, r\, dr \le 15 \tag{6.7}$$

where R is the cell radius, or:

$$\int_0^{R} n(r)\, r\, dr \le \frac{15}{2\pi\rho} \tag{6.8}$$

yet the function of the number $n(r)$ is defined as constant by pieces. Thus, we have:

$$\sum_i \int_{r_{i,\min}}^{r_{i,\max}} n_i\, r\, dr \le \frac{15}{2\pi\rho} \tag{6.9}$$

or finally:

$$\sum_i n_i\,(r_{i+1}^2 - r_i^2) \le \frac{15}{\pi\rho} \tag{6.10}$$

with $r_i = r_{i,\min}$ and $r_{i+1} = r_{i+1,\min} = r_{i,\max}$, where r_i and r_{i+1} denote, respectively, the lower and upper bounds of the range of the cell having the same CQI_i value (same modulation, coding rate, and number of shared channels n_i used by the mobile, thus the same block size TBS_i). The highest bound r_i (for $CQI = 1$) refers to the radius R of the coverage-limited cell (maximum size guaranteeing coverage).

The expression (Equation 6.10) can be generalized for a cell whose whole size is R as follows:

$$\sum_{i \le k_0} n_i\,(r_{i+1}^2 - r_i^2) + n_{k_0+1}(R^2 - r_{k_0+1}^2) \le \frac{15}{\pi\rho} \tag{6.11}$$

such that CQI_{k_0+1} is the *CQI* value referring to the ring, including the border of the cell with size *R*. The equality is verified for $R = R_{cap}$ (the radius limited by capacity in the number of codes).

6.4 Dimensioning Modeling of HSPA-based Networks

6.4.1 "Fair Resource"-based Dimensioning

In this paragraph, we give explanations on the iterative dimensioning process by applying the "Fair Resource" scheduling technique. In particular, we employ the analytical expressions used and their conditions to obtain the exact cell size, allowing a given bit rate. Figure 6.1 gives a general example of a dimensioned cell according to the Fair Resource method, including the subcells referring to the different services, as well as the different rings having the same *CQI*, and the capacity-limited cell.

In this technique, services analysis is made from the most limiting (in bit rate) toward the least limiting service, and the traffic density is decreasing when moving far away from the NodeB. The size of the capacity-limited cell is determined for each studied service (since traffic density is changed) until the total system capacity is reached (number of HSPA codes equal to 15 or total NodeB nominal power is reached) that is, until the capacity-limited cell radius becomes included in the current studied ring. The value of 15 is taken as an example here and until the end of this chapter.

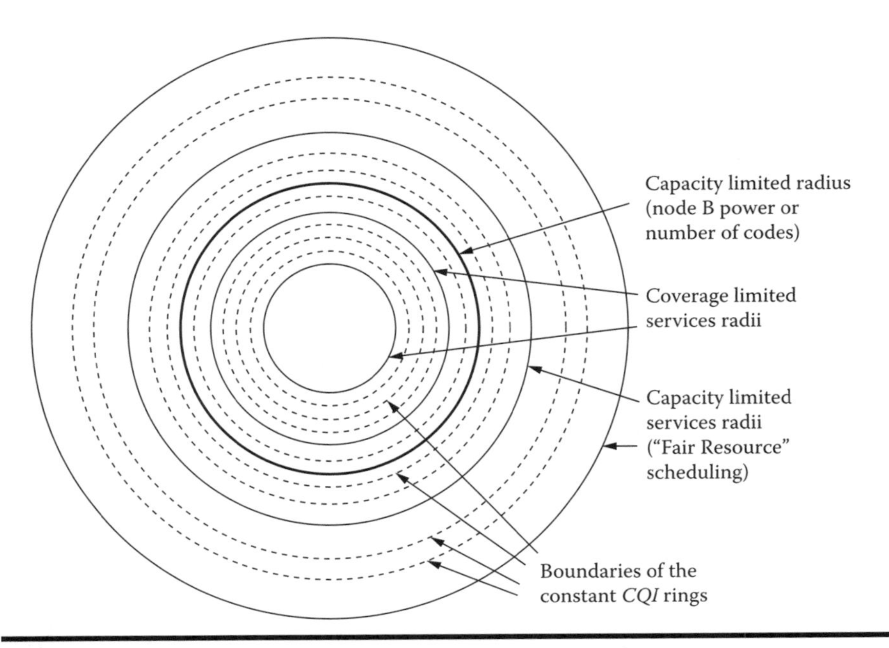

Figure 6.1 Dimensioned cell scheme according to the Fair Resource method.

In terms of the analytical formulation of the code-limited capacity, if we are determining the cell size referring to the ith service, we have the following procedure: While the number of HSPA codes N_{code_1} [allocated to users belonging to cells 1 to $(i-1)$ as they are ordered from the most coverage limiting service to the least limiting one] is smaller than 15, we calculate the cell size R_c limited by the number of codes versus the cell referring to service i (assuming uniform traffic) as follows:

$$R_c = \sqrt{r_m^2 + \frac{15 - N_{\text{code}_0}}{2.6 \cdot \sigma_i \cdot n_m} N_{\text{car}}} \tag{6.12}$$

(since the code-limited cell refers to that including exactly 15 codes)

where:

- N_{car} is the number of carriers used for dimensioning
- r_m is the lower size limit of the ring having the same $CQI = m$ ($\forall r$, the distance between the user and the NodeB), $r \in [r_m, r_{m-1}]$, $CQI = m$ ($r_{m-1} \geq r_m$) by assuming that $r_m = r_{m,\min} = r_{m+1,\max}$ (i.e., the minimal size r_m of the ring of $CQI = m$ is the maximal size of that with $CQI = m + 1$); and we denote n_m the number of HSPA codes assigned per user located in this ring; and m is the code-limited boundaries of the constant CQI rings that:

$$\begin{cases} N_{\text{code}_1} + 2.6\,\sigma_i\, n_{j_{i-1}+1}\left(r_{j_{i-1}}^2 - R_{i-1}^2\right) + \sum\limits_{k=m}^{j_{i-1}-1} 2.6 \cdot \sigma_i \cdot n_{k+1} \cdot \left(r_k^2 - r_{k+1}^2\right) \leq 15 \\ N_{\text{code}_1} + 2.6\,\sigma_i\, n_{j_{i-1}+1}\left(r_{j_{i-1}}^2 - R_{i-1}^2\right) + \sum\limits_{k=m-1}^{j_{i-1}-1} 2.6 \cdot \sigma_i \cdot n_{k+1} \cdot \left(r_k^2 - r_{k+1}^2\right) > 15 \end{cases}$$

$$15 - 2.6\,\sigma_i\, n_m (r_{m-1}^2 - r_m^2) < N_{\text{code}_1} + 2.6\,\sigma_i\, n_{j_{i-1}+1}\left(r_{j_{i-1}}^2 - R_{i-1}^2\right)$$
$$+ \sum_{k=m}^{j_{i-1}} 2.6\,\sigma_i\, n_{k+1}\left(r_k^2 - r_{k+1}^2\right) \leq 15$$

- σ_i is the users density of the subcell referring to the ith service, expressed as: $\sigma_i = \sum_{k=i}^{s} \rho_k$ where ρ_k is the area users density of service k assuming s is the number of services (and referring to the least limiting service). If activity factors are different from unity, we weigh area densities ρ_k by activity factors v_k of the respective services.
- N_{code_0} is the number of codes assigned to the users located within the CQI ring limit, followed immediately by the subcell referring to the $(i-1)$th service, given by:

$$N_{\text{code}_0} = N_{\text{code}_1} + \sum_{k=m}^{j_{i-1}-1} 2.6\,\sigma_i\, n_{k+1}(r_k^2 - r_{k+1}^2) + 2.6\,\sigma_i\, n_{j_{i-1}+1}\left(r_{j_{i-1}}^2 - R_{i-1}^2\right)$$

where j_{i-1} is the *CQI* that immediately follows the one that limits the cell referring to service $i-1$ (or the radius R_{i-1}, referring to cell $i-1$ which belongs to the ring with $CQI = j_{i-1}+1$ such that $r_{j_{i-1}+1} \leq R_{i-1} < r_{j_{i-1}}$).

- We used here "2.6" instead of π for the computation of the area because the cell shape is assumed to be hexagonal (and not circular).

N_{code_1} and N_{code_0} are updated for each service (from the most to the least limiting), and the capacity-limited cell radius (in code number) is finally determined when N_{code_1} is at least equal to 15.

In order to determine the cell radius limited by the NodeB nominal power, we can make the same procedure as we have done for the code number limited cell radius (by applying it successively to cells referring to the different services) except that we replace the number of codes by the individual transmit power per physical shared traffic channel, and the maximum HSPA code number (= 15) by the NodeB nominal power. Furthermore, we can determine for each service iteration, the maximum number of users N_p that the total NodeB nominal power can serve, then compare it to the number of active users N_c included inside the radius limited by the codes number (by taking into account the subcells referring to the different services) to find again, for each iteration of service i, the capacity-limited cell radius R_{cap} as follows:

$$\begin{cases} R_{\text{cap}} = R_p = \sqrt{R_{i-1}^2 + \frac{N_p - N_1}{2.6 \cdot \sigma_i}} & \text{if} \quad N_p < N_c \\ R_{\text{cap}} = R_c & \text{else} \end{cases} \tag{6.13}$$

where N_1 is the number of users included in the subcell referring to service $(i-1)$. In other words, the capacity-limited size R_{cap} is the minimum of the power-limited radius R_p and the code number limited R_c. The final value of the power-limited size (and thus R_{cap}) is found when the number N_1 reaches at least N_p.

In the iterative process, we compare the nominal bit rate of the given service i to the bit rate $(R_{\text{ens}})_j$ ensured at the current ring $CQI = j$ (coverage bit rate shared—in Fair Resource—by the total number of users from all services (i.e., inside the cell referring to the least limiting service if HSPA capacity is reached). The bit rate $(R_{\text{ens}})_j$ ensured at the ring $CQI = j$ is determined by the following expression if the radius of the least limiting service cell is above the capacity-limited one:

$$\begin{aligned} (R_{\text{ens}})_j &= \frac{TBS_j \cdot n_{\text{cap}}}{TTI_{\text{delay}} \cdot (n_{\text{Tot}})_j} \\ &= \frac{TBS_j \cdot \sum_{k=1}^{k_0-1} \left[\left(\sum_{l=k}^{s} \rho_l \right) \left(R_k^2 - R_{k-1}^2 \right) \right] + \left(\sum_{l=k_0}^{s} \rho_l \right) \left(R_{\text{cap}}^2 - R_{k_0-1}^2 \right)}{TTI_{\text{delay}} \cdot \sum_{k=1}^{s} \left[\left(\sum_{l=k}^{s} \rho_l \right) \left(R_k^2 - R_{k-1}^2 \right) \right]} \end{aligned} \tag{6.14}$$

where:

- n_{cap} is the number of users in the capacity-limited cell (determined by the minimum of the NodeB power-limited cell radius R_p and the codes number limited one R_c).
- $(n_{Tot})_j$ is the total number of users inside the border cell (external subcell referring to the least limiting service), and this by assuming the radius R_i of the cell referring to service $i-$ equal to the size r_j, lower bound of the ring at $CQI = j$.
- s is the total number of services and k_0 is the index of the service whose subcell radius contains the capacity-limited radius (i.e., the next index of the most limiting coverage-limited service).
- TTI_{delay} is the transmit time interval of HSPA (equal to 2 ms), and TBS_j denotes the transport block size referring to $CQI = j$.

The $n_{cap}/(n_{Tot})_j$ ratio updates and corrects the coverage bit rate TBS_j/TTI_{delay} with the Fair Resource policy, which consists in assigning the same fraction of time resources to all users. We assume $R_0 = 0$ and $R_1 \leq R_2 \leq \ldots \leq R_{s-1} \leq R_s$ (i.e., s is the least limiting service, thus having the highest dimensioned radius R_s). All services activity factors are assumed to be total (all equal to 1), otherwise (if some of them are partial), the area densities ρ_k are weighted by activity factors v_k of each of the services.

The computation of the range R_i of a subcell referring to service i and having exactly the nominal bit rate D_i is carried out owing to the following expression:

$$R_i = \sqrt{r_j^2 + \frac{n^*_{Tot} - (n_{Tot})_j}{2.6 \cdot \rho_i}} = \sqrt{r_j^2 + \frac{n^*_{Tot} - (n_{Tot})_j}{2.6 \cdot (\sigma_i - \sigma_{i+1})}} \tag{6.15}$$

where:

- r_j is the lower bound of the ring at $CQI = j$, and j is chosen such that:

$$\begin{cases} \dfrac{TBS_j \cdot n_{cap}}{TTI_{delay} \cdot (n_{Tot})_j} \geq D_i \\ \\ \dfrac{TBS_{j-1} \cdot n_{cap}}{TTI_{delay} \cdot (n_{Tot})_{j-1}} < D_i \end{cases} \tag{6.16}$$

  If, by applying expression (Equation 6.15) to the conditions (Equation 6.16), we find the calculated radius R_i is above r_{j-1}, then we maintain the final value of R_i equal to r_{j-1}.
- n^*_{Tot} is the total number of users in the external subcell (referring to the least limiting service) guaranteeing the nominal bit rate D_i. It is computed according

to the following expression:

$$n^*_{\text{Tot}} = \frac{TBS_j \cdot n_{\text{cap}}}{TTI_{\text{delay}} \cdot D_i} \tag{6.17}$$

- $(n_{\text{Tot}})_j$, n_{cap}, TTI_{delay}, TBS_j, ρ_i, and σ_i denote the same notations used previously.
  The update of the total number n_{Tot} of the users in the border subcell (after modifying the value of one or many radii) is made through the following equation:

$$n_{\text{Tot}} = 2.6 \cdot \sum_{k=1}^{s} \left[\sigma_k \cdot \left(R_k^2 - R_{k-1}^2\right)\right] = 2.6 \cdot \sum_{k=1}^{s} \left[\left(\sum_{l=k}^{s} \rho_l\right)\left(R_k^2 - R_{k-1}^2\right)\right] \tag{6.18}$$

If, at a given iteration, the size of a subcell referring to a given service is below one or several radii of subcells referring to more limiting services, then each of these radii is assigned the value of the radius referring to the previous service (subcell sizes shrinking), and the lowest subcell radius is assigned the value found for the studied service. Inversely, if the computed size is above one or several radii of subcells referring to less limiting services, then each of these radii is assigned the value of the radius referring to the following more limiting service (subcell sizes inflation), and the highest subcell radius is assigned the value found for the studied service.

The process will be applied for all services at each iteration until its convergence to obtain the sizes of the different subcells referring to each service, and thus the different radii values over which the Fair Resource dimensioning method is based. The least limiting radius (the maximum) gives the dimensioning according to the service's partial availability. The most limiting radius (the minimum) provides the dimensioning result according to full services availability.

Note that expressions in Equations 6.14 and 6.16 are not valid except in the case that the least limiting service is capacity-limited and without codes multiplexing (i.e., without allocation of many codes or multi-codes per user). In order to take into account the codes multiplexing if the least limiting service is coverage-limited (in particular, the number of codes assigned inside the coverage-limited border subcell is below 15), we can replace in both expressions, the ratio $n_{\text{cap}}/(n_{\text{Tot}})_j$ by $15/\sum_k n_k$ (ratio of the total number of HSPA codes available or allowable per UE terminal by the sum of the numbers of codes assigned to users inside the external subcell referring to the least limiting coverage-limited service) while including the summation over the *CQI* rings in expression (Equation 6.14), and by weighting the area densities of users (inside summation symbols) by the numbers of codes n_j referring to rings j at constant *CQI*. The total number of codes in expression (Equation 6.18) is therefore similarly updated instead of the total number of users. In order to compute the subcell radius referring to the nominal bit rate D_i, the expressions (Equations 6.15

and 6.17) are rewritten by replacing n_{cap} by the total number of codes (assumed equal to 15), $(n_{\text{Tot}})_j$ by the sum of the numbers of codes assigned to users inside the external subcell (without codes multiplexing) always by assuming the radius R_i referring to the service i equal to the lower bound of the ring at $CQI = j$, and by weighting the users density inside the ring j by the number n_j of codes referring to the *CQI* of this ring. In this case, the computation of n^*_{Tot} will be replaced by that of the codes number assigned inside the external subcell guaranteeing the nominal bit rate D_i.

6.4.2 "Fair Throughput"-based Dimensioning

This paragraph provides expressions of maximum bit rate per user ensured by "Fair Throughput" scheduling technique with and without consideration of codes multiplexing. Thus, we can determine Fair Throughput dimensioning procedure with its analytical support.

By applying the Fair Throughput technique, the resources allocated to the different users is not the same for those located at different distances from the NodeB such that the mobiles disadvantaged by the channel have the same throughput than those favored by the propagation channel. In fact, the further users (having lower TBS sizes) will have more time resources so as to have the same throughput for all the users. So the Fair Throughput technique tries to balance throughputs of the different users by providing more resources (thus more priority) to transport blocks (TBs) having a lower size such that the micro-flows from the different waiting queues have almost equal instantaneous throughputs per user at each time, or in terms of analytical expression:

$$\frac{TBS_1}{TTI_{\text{delay}}} \cdot p_1 = \frac{TBS_2}{TTI_{\text{delay}}} \cdot p_2 = \frac{TBS_3}{TTI_{\text{delay}}} \cdot p_3 = \ldots$$

$$= \frac{TBS_i}{TTI_{\text{delay}}} \cdot p_i = \ldots = R_{\text{ens}}; \forall i \qquad (6.19)$$

where $p_1, p_2, p_3, \ldots, p_i, \ldots$ are the time proportions (below 1) assigned to micro-flows with respective transport block sizes TBS_1, TBS_2, TBS_3, $\ldots$, TBS_i, $\ldots$ of each of the users in the cell according to their *CQIs*, thus $\sum_i p_i = 1$; besides there may be two or more equal TBS_is. The R_{ens} constant is the maximum ensured bit rate per user by Fair Throughput independently of the number of HSPA codes and their multiplexing (i.e., with the minimum number of codes as possible).

We have the following relation, valid at each time:

$$n_1 \cdot p_1 + n_2 \cdot p_2 + n_3 \cdot p_3 \leq 15 \qquad (6.20)$$

It expresses the fact that the number of codes weighted by their assigned resource amounts doesn't always exceed the maximum capacity provided by HSPA (15 codes

HS-PDSCH: high-speed—physical downlink shared channel). The condition in Equation 6.20 is always valid since $n_1 \cdot p_1 + n_2 \cdot p_2 + n_3 \cdot p_3 \leq \max_i \{n_i\} \leq 15$. By generalizing Equation 6.20 for whatever the number of users in the cell, we obtain:

$$\sum_i n_i \cdot p_i \leq 15 \tag{6.21}$$

where p_i is the time proportion allocated to the user i such that all the users have the same minimum bit rate. By solving the equations system (Equation 6.19) in p_i, we obtain:

$$p_i = \frac{R_{\text{ass}} \cdot TTI_{\text{delay}}}{TBS_i}; \forall i \tag{6.22}$$

Yet $\sum_i p_i = 1$, the maximum ensured bit rate by each of the users (without codes multiplexing) can be written as follows:

$$R_{ens} = \frac{1}{TTI_{\text{delay}} \cdot \sum_j \frac{1}{TBS_j}} \tag{6.23}$$

We can easily check that $R_{\text{ens}} \leq \frac{TBS_i}{TTI_{\text{delay}}}; \forall i$. In particular, R_{ens} is always below or equal to the most limiting coverage bit rate at the cell border $R_{\text{cov}} = \min_i \left(\frac{TBS_i}{TTI_{\text{delay}}} \right)$.

The bit rate in expression (Equation 6.23) is thus the minimum guaranteed bit rate independently of the number of codes and multi-codes available in HSPA. (It refers to the minimum required number of codes always below the number of available HSPA shared codes assumed to be equal to 15.) Otherwise, we can ensure a user bit rate (in Fair Throughput) above that given by Equation 6.23 by using more OVSF (orthogonal variable spreading factor) HSPA codes (with codes multiplexing).

Now we will determine the maximum bit rate per user while using the totality of the available HS-PDSCH channels (referring to the 15 allowed codes). For that purpose, we divide the minimum bit rate calculated in Equation 6.23 by the number of codes used to ensure it $\left[\text{i.e., } \left(\sum_i \frac{n_i}{TBS_i} \Big/ \sum_k \frac{1}{TBS_k}\right)\right]$ and we multiply by the available number of codes (assumed to be equal to 15); we establish the maximum ensured bit rate per user (by considering codes multiplexing) as follows:

$$(R_{\text{ass}})_{FT} = \frac{15}{TTI_{\text{delay}} \cdot \sum_j \frac{1}{TBS_j} \cdot \left(\sum_i \frac{n_i}{TBS_i} \Big/ \sum_k \frac{1}{TBS_k} \right)}$$

$$= \frac{15}{TTI_{\text{delay}} \cdot \sum_i \frac{n_i}{TBS_i}} \tag{6.24}$$

where n_i is the corresponding number of codes referring to the user i position within the cell (given by 3GPP standard [11]). Note that the last summation symbol applies to all the mobiles within the cell.

The maximum ensured bit rate in Equation 6.24 can be determined by assuming the equality in the condition (Equation 6.21) (using the maximum number of available channels), then combine it with Equation 6.22; thus, we obtain directly the final expression of Equation 6.24 (including the whole available HSPA physical channels).

6.4.3 "Enhanced Fair Throughput"-based Dimensioning

In this paragraph, we give explanations on the dimensioning procedure by applying an "Enhanced Fair Throughput" scheduling technique (Figure 6.2). This technique

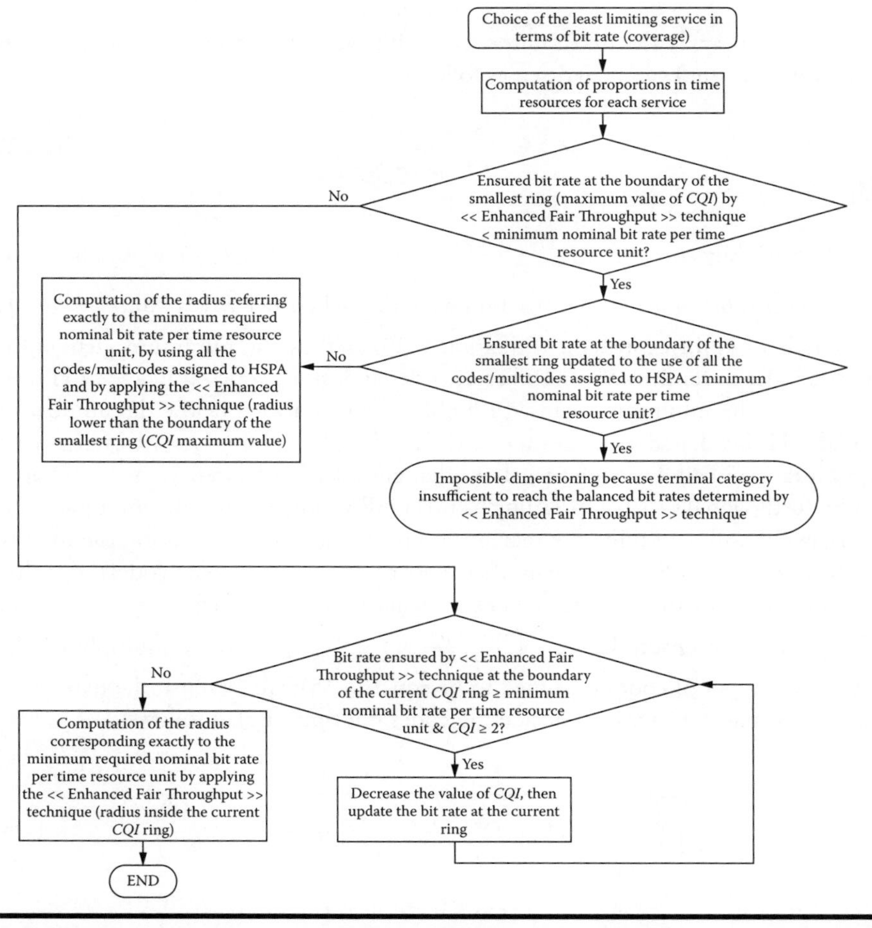

Figure 6.2 Flow chart of the "Enhanced Fair Throughput" dimensioning procedure.

is suggested by the authors for operators in order to improve planning performance and reduce costs. The expression used to obtain the exact size of the cell allowing a given bit rate is also provided.

Sites dimensioning according to the Enhanced Fair Throughput dimensioning method is based on the expression in Equation 6.24. The ensured bit rate $(D_{\text{ens}})_m$ at the higher bound of the ring at $CQI = m$ can be written by applying the basic Fair Throughput technique with a uniform traffic as follows:

$$(D_{\text{ens}})_m = \frac{15}{2.6 \cdot \rho \cdot TTI_{\text{delay}} \sum_{j=m}^{CQI_{\max}} \frac{n_j}{TBS_j}(r_{j-1}^2 - r_j^2)} \tag{6.25}$$

(we replace π by 2.6 to adapt the area to the hexagonal form of the cell) where $CQI_{\max}$ is the maximum CQI (belonging to the smaller ring CQI of the cell; it's the central subcell such that $r_{CQI_{\max}} = 0$), which has been established to be determined and limited by the intracellular interference level, ρ is the users area density.

The enhancement of the basic Fair Throughput technique is in the fact that instead of guaranteeing the same bit rate for all the mobiles, we guarantee the required bit rate of the service used by each user by applying an additional weighting to the resources that would be normally assigned in the ordinary Fair Throughput. By considering the basic bit rate D_{i_0} the one required for the least limiting service i_0, the additional weight α_i assigned to service i (with nominal bit rate D_i) is:

$$\alpha_i = \frac{D_i}{D_{i_0}} \tag{6.26}$$

In order to take into account this weight, the equipment provider should implement a device in the NodeB allowing the identification of the used service at the application layer and recognition of its required nominal bit rate. So the NodeB can determine the required weights for each user versus the used service in order to communicate them to the MAC-HS (medium access layer—high speed) layer to take them into account at the packets scheduler of each user.

In the case of the Fair Throughput technique that we have enhanced, the expression in Equation 6.25 remains valid by updating the expression of ρ to ρ_{upd} (updated ρ) as follows:

$$\rho_{\text{upd}} = \sum_i \alpha_i \cdot \rho_i = \sum_i \frac{D_i}{D_{i_0}} \cdot \rho_i \tag{6.27}$$

(with the same notations and by assuming all the activity factors equal to unity). The nominal bit rate D_{i_0} of the least limiting service is thus the nominal bit rate per time resources unit by applying the basic Fair Throughput with the updated area density ρ_{upd}. The users will have enough time resources to reach the nominal bit rates of their respective services (i.e., equal to the respective weights α_i).

The cell radius R (common to all services) dimensioned by this Enhanced Fair Throughput method is computed with the following expression (extracted from expression in Equation 6.25 such that the ensured bit rate is equal to D_{i_0}):

$$R = \sqrt{r_j^2 + \frac{TBS_j}{n_j}\left[\frac{15}{2.6 \cdot \rho_{\text{act.}} \cdot TTI_{\text{delay}} \cdot D_{i_0}} - \sum_{k=j+1}^{CQI_{\max}} \frac{n_k}{TBS_k}\left(r_{k-1}^2 - r_k^2\right)\right]} \tag{6.28}$$

where r_j are the lower bounds of the rings at constant CQI with the same notations as previously, and $r_{CQI_{\max}} = 0$. Likewise, for TBS_j, n_j, and TTI_{delay}.

j is chosen such that:

$$\begin{cases} (D_{\text{ens}})_{j+1} \geq D_{i_0} \\ (D_{\text{ens}})_j < D_{i_0} \end{cases} \tag{6.29}$$

where $(D_{\text{ens}})_j$ and $(D_{\text{ens}})_{j+1}$ are the ensured bit rates (by Enhanced Fair Throughput) at the higher bounds of the respective rings at $CQI = j$ and at $CQI = j+1$. These bit rates are computed by Equation 6.25 while taking into account the users density given by Equation 6.27.

If the bit rate at the boundary of the central (internal) subcell (having maximum CQI) is smaller than D_{i_0}, that is, $(D_{\text{ens}})_{CQI\max} < D_{i_0}$ then expression (Equation 6.28) remains valid (by taking $j = CQI_{\max}$ and $r_k = 0 \forall k \geq CQI_{\max}$), and thus the dimensioned radius becomes:

$$R_{\text{dim}} = \sqrt{\frac{15 \cdot TBS_{CQI_{\max}}}{2.6 \cdot \rho_{\text{act.}} \cdot TTI_{\text{delay}} \cdot D_{i_0} \cdot n_{CQI_{\max}}}} \tag{6.30}$$

The last expression is valid provided that the ensured bit rate at the boundary of the smallest ring ($CQI_{\max}$) updated to the use of all the codes and multi codes assigned to HSPA is above or equal to the bit rate D_{i_0}:

$$\frac{15 \cdot TBS_{CQI_{\max}}}{n_{CQI_{\max}} \cdot TTI_{\text{delay}}} \geq D_{i_0} \tag{6.31}$$

If this last condition is not realized, the dimensioning is impossible because the terminal category and the radio conditions (intracellular interference) don't allow the required bit rates achievement ($CQI_{\max}$ depends on the intracellular interference).

6.4.4 Multiple Aggregated Services Dimensioning Process

Let's now study the multiservice case. It is evident that the higher the service bit rate, the smaller the cell size due to the cell breathing effect caused by the AMC mechanism. Hence, we obtain concentric cells for each of the services, with the subcells referring to the higher bit rate, services are nearer to the NodeB, and vice versa (as for the basic UMTS).

Let's assume s services denoted by $1, 2, \ldots, i, \ldots, s$ in the increasing order of their required bit rates (the service 1 having the lowest bit rate, and the service s having the highest bit rate). Therefore, the mobiles out of the subcell referring to service i are not served by services $i, i+1, \ldots, s$ whose required bit rates are above or equal to that of service i. We call R_i the cell size referring to service i. Thus, we have $R_s \le R_{s-1} \le \ldots \le R_2 \le R_1$. Let's assume that $R_{s+1} = 0$. Dimensioning should therefore be accomplished according to the most constraining service s (having the highest bit rate) since its referring cell radius R_s is the most limiting (the smallest radius among those of the other services).

Assuming that the users distribution among the different services is such that $p_i^{\%}$ is the percentage of mobiles using the service i, and ρ_i is the area density of simultaneous users of the service i (with uniform distribution). Thus, we have: $\rho_i = p_i^{\%} \cdot \rho$ where ρ is the global area density of users of all the services $\left(\sum_{i=1}^{s} \rho_i = \rho\right)$. The dimensioning process is accomplished by determining cell size referring to services i one by one in the decreasing order of the required nominal bit rates by starting with the most constraining service s until the least limiting one 1. In order to guarantee a minimum bit rate $R_{\min,i}$ at the cell border referring to service i (where $R_{\min,i} \le R_{\min,i+1}$; $\forall i = 1, 2, \ldots, s$ and assuming the Fair Resource scheduling technique), the transport block size TBS_0 at the border of the subcell i without codes multiplexing should be given, with the same notations, by:

$$\begin{cases} TBS_j \ / \ TBS_j \in \text{tables [11] such that:} \\ TBS = \min_j \\ \left[TBS_j \ge R_{\min,i}\, TTI_{\text{delay}} \right. \\ \left. \times \max\left(1, \dfrac{\sum_{m=1}^{s} \left[\left(\sum_{l=1}^{m} \rho_l\right)\left(R_m^2 - R_{m+1}^2\right)\right]}{\sum_{m=k_0+1}^{s} \left[\left(\sum_{l=1}^{m} \rho_l\right)\left(R_m^2 - R_{m+1}^2\right)\right] + \left(\sum_{l=1}^{k_0} \rho_l\right)\left(R_{\text{cap}}^2 - R_{k_0+1}^2\right)}\right)\right] \end{cases} \tag{6.32}$$

where $(k_0 + 1)$ index is that of the most limiting service (coverage-limited), or in other words $R_{k_0+1} \le R_{\text{cap}} < R_{k_0}$, or $(k_0 + 1)$ is the service index whose dimensioned radius is that containing the capacity-limited radius R_{cap}. The number of users simultaneously served in the ring limited by radii R_{i+1} and R_i is $\pi\left(R_i^2 - R_{i+1}^2\right)\sum_{l=1}^{i} \rho_l$ since the bit rates of services $i + 1$ to s are not guaranteed for a size above R_{i+1}. The intracellular interference in this ring is proportional to this number; hence, the total intracellular interference of the cell derived from this NodeB is proportional to

$\pi \sum_{i=1}^{s} [(R_i^2 - R_{i+1}^2) \sum_{l=1}^{i} \rho_l]$. The servitude rate of this cell by this NodeB will thus be

$$\frac{\sum_{i=1}^{s} \left(R_i^2 - R_{i+1}^2\right) \sum_{l=1}^{i} \rho_l}{\left[\sum_{i=1}^{s} \left(R_i^2 - R_{i+1}^2\right)\right] \left(\sum_{i=1}^{s} \rho_i\right)}$$

by assuming the users of all the services uniformly distributed in the service area.

Expression (Equation 6.32) is valid for both coverage and capacity-limited dimensioning (through the "max" sign). The cell size can be concluded from TBS_0 as for the case of one service. Yet the intracellular interference depends on the number of active mobiles in the cell (thus on its size), and as the expression (Equation 6.32) of TBS_0 depends on the cell size (through the summation) since TBS_0 depends on the link quality at the border of the subcell referring to the concerned service, we should have recourse to either use an iterative process by dichotomy in order to converge to the exact cell radius (for each service i in the decreasing order) or use a mathematical deduction. For each iteration, once the radius referring to service i is determined, we can restart the same work to find the size of the subcell referring to service $i-1$ by using the previous results relative to services i to s.

6.4.5 Shadowing Impact on Dimensioning

Now, let's model the bit rates and the different HSPA parameters using the shadowing effect. Yet the coverage bit rate (or the transport block size) is directly related to *CQI* (cf. Section 6.3.1), thus the determination of the distribution model of CQI is enough to model that of the bit rate as well as the other related parameters. This modeling is the direct result of the AMC mechanism. It is at the origin of the general simplified HSPA dimensioning methodology that will be described and applied in the following paragraphs.

6.4.5.1 Adaptation in the Modulation and Coding (AMC) Model

Let's start with the expression (Equation 6.6) giving the *SINR* received by the mobile in dB at the logarithmic scale (with the same notations):

$$SINR = P_{TX} - 10 \cdot \log_{10}\left(10^{\frac{I_{\text{intra}}}{10}} + 10^{\frac{I_{\text{inter}} + L_{dB}}{10}}\right) \tag{6.33}$$

The linear path loss L follows a log-normal law, thus its distribution PDF (probability distribution function) is:

$$f_L(x) = \frac{\xi}{\sqrt{2\pi}\,\sigma\, x} e^{-\frac{[\xi\, Ln(x) - \mu]^2}{2\sigma^2}} \tag{6.34}$$

where

$$\xi = \frac{10}{Ln(10)}, \quad \mu = \xi \cdot Ln(\overline{L}) = 10 \cdot \log_{10}(\overline{L})$$

is the average path loss (logarithmic) in dB referring to the distance path loss, and σ is the standard deviation (logarithmic) of the shadowing effect in dB. Thus, the CDF (cumulative distribution function) of L can be written as:

$$F_L(x) = \int_1^x \frac{\xi}{\sqrt{2\pi}\,\sigma\, t} e^{-\frac{[\xi\, Ln(t)-\mu]^2}{2\sigma^2}} dt = \frac{1}{2} erf\left(\frac{\xi \cdot Ln(x) - \mu}{\sqrt{2}\sigma}\right) + \frac{1}{2} erf\left(\frac{\mu}{\sqrt{2}\sigma}\right) \tag{6.35}$$

where $erf(\cdot)$ is the error function defined by:

$$erf(t) = \frac{2}{\sqrt{\pi}} \int_0^t e^{-u^2} du \tag{6.36}$$

By taking into account Equation 6.33, we can rewrite Equation 6.3 into:

$$CQI = \left[\frac{P_{TX} - 10 \cdot \log_{10}\left(10^{\frac{I_{\text{intra}}}{10}} + 10^{\frac{I_{\text{inter}}+L_{dB}}{10}}\right)}{CQI_{\text{ratio}}} + Offset \right] \tag{6.37}$$

where $[\cdot]$ denotes the integer part (by lower value trunking). Assume that:

$$Y = \frac{P_{TX} - 10 \cdot \log_{10}\left(10^{\frac{I_{\text{intra}}}{10}} + 10^{\frac{I_{\text{inter}}+L_{dB}}{10}}\right)}{CQI_{\text{ratio}}} + Offset \tag{6.38}$$

We elaborate the CDF distribution law of Y as follows:

$$F_Y(y) = Prob(Y < y) = 1 - F_L\left[10^{-\frac{I_{\text{inter}}}{10}}\left(10^{\frac{P_{TX}-CQI_{\text{ratio}}(y-Offset)}{10}} - 10^{\frac{I_{\text{intra}}}{10}}\right)\right] \tag{6.39}$$

where F_L is given by Equation 6.35. The probability density of Y is written as follows:

$$f_Y(y) = \frac{\partial\, F_Y(y)}{\partial\, y} = \frac{CQI_{\text{ratio}}}{\xi} 10^{\frac{P_{TX}-I_{\text{inter}}-CQI_{\text{ratio}}(y-Offset)}{10}}$$

$$\times f_L\left[10^{-\frac{I_{\text{inter}}}{10}}\left(10^{\frac{P_{TX}-CQI_{\text{ratio}}(y-Offset)}{10}} - 10^{\frac{I_{\text{intra}}}{10}}\right)\right] \tag{6.40}$$

So the PDF law (discrete) of $CQI = E[Y]$ is concluded as follows:

$$p_k = Prob(CQI = k) = \int_k^{k+1} f_Y(y)\,dy$$

$$= F_L\left[10^{-\frac{I_{\text{inter}}}{10}}\left(10^{\frac{P_{TX}-CQI_{\text{ratio}}(k-Offset)}{10}} - 10^{\frac{I_{\text{intra}}}{10}}\right)\right]$$

$$- F_L\left[10^{-\frac{I_{\text{inter}}}{10}}\left(10^{\frac{P_{TX}-CQI_{\text{ratio}}(k+1-Offset)}{10}} - 10^{\frac{I_{\text{intra}}}{10}}\right)\right] \quad (6.41)$$

where F_L is given by Equation 6.35.

6.4.5.2 Use of Tables and Abacuses for HSPA Dimensioning

Starting with the *CQI* analytical model established in the previous Section 6.4.5.1, we can construct a simple and general dimensioning methodology for a UMTS radio access network based on the HSPA technique. For that, we use the correspondence model quality/bit rate resulting from the AMC mechanism (through the table of 3GPP standard [11] referring to the adequate terminal category). For example, Table 6.1 provides this correspondence for a category of terminals equal to 10. Then, we describe the method for coverage-limited dimensioning of HSPA-based UMTS cellular networks while taking into account the shadowing effect.

The calculations done by a dimensioning tool with HSPA functionalities can generate tables providing the maximum range of the cell (coverage-limited) versus the required area coverage probability, the minimum bit rate to guarantee by the user, and the shadowing standard deviation. The scheduling policy applied is the "Best Effort," providing the best bit rate (maximum) that the coverage-limited HSPA link can offer in certain conditions. We can include extra margins (fast fading, human body loss, etc.) to the dimensioning result by adding the margin in dB to the maximum allowed path loss (MAPL) obtained from the distance given by the tables and a simple appropriate propagation model (with one decay).

For example, with a shadowing standard deviation of 14 dB, in order to have a minimum bit rate of 128 kbps in the cell at a probability of 75%, the maximum (allowed) cell radius is equal to 836 m according to Table 6.2, referring to the indicated shadowing standard deviation. We can eventually include other margins (fast fading, etc.) by subtracting them from the MAPL equivalent to the distance 836 m (for a given propagation model), then by recalculating the corresponding maximum distance (below 836 m).

As another example, Figure 6.3 provides the maximum allowed cell size versus the maximum offered bit rate at the cell border for different coverage probability and shadowing standard-deviation values. We note that the smaller the coverage probability (case of 70%), the less the impact of shadowing standard deviation is

Table 6.1 Correspondence Table of *CQI* for a Category 10 UE Terminal

CQI	*Transport Block Size (TBS)*	*Number of Shared Codes*	*Modulation Type*
0	N/A	Out-of-Cell Coverage	
1	137	1	QPSK
2	173	1	QPSK
3	233	1	QPSK
4	317	1	QPSK
5	377	1	QPSK
6	461	1	QPSK
7	650	2	QPSK
8	792	2	QPSK
9	931	2	QPSK
10	1262	3	QPSK
11	1483	3	QPSK
12	1742	3	QPSK
13	2279	4	QPSK
14	2583	4	QPSK
15	3319	5	QPSK
16	3565	5	16-QAM
17	4189	5	16-QAM
18	4664	5	16-QAM
19	5287	5	16-QAM
20	5887	5	16-QAM
21	6554	5	16-QAM
22	7168	5	16-QAM
23	9719	7	16-QAM
24	11418	8	16-QAM
25	14411	10	16-QAM
26	17237	12	16-QAM
27	21754	15	16-QAM
28	23370	15	16-QAM
29	24222	15	16-QAM
30	25558	15	16-QAM

Table 6.2 Dimensioning Table (Cell Size Versus Maximum Offered Bit Rate and Area Coverage Probability with a Shadowing Std-Deviation of 14 dB)

Standard Deviation = 14 dB																						
Maximum Distance (m)		*Area Coverage Probability (%)*																				
		60	*61*	*62*	*63*	*64*	*65*	*66*	*67*	*68*	*69*	*70*	*71*	*72*	*73*	*74*	*75*	*76*	*77*	*78*	*79*	*80*
Maximum Offered Bit Rate at the Border of the Cell (kbps)	*68.5*	1557	1517	1477	1439	1401	1364	1327	1291	1256	1221	1187	1153	1120	1087	1055	1022	990	959	928	897	867
	86.5	1456	1419	1381	1345	1310	1275	1241	1208	1175	1142	1110	1078	1047	1016	986	956	926	897	868	839	811
	116.5	1361	1326	1292	1258	1225	1192	1161	1129	1098	1068	1037	1008	979	950	922	894	866	839	812	784	758
	158.5	1273	1240	1208	1176	1145	1115	1085	1056	1026	998	970	943	915	889	862	836	810	784	759	734	709
	188.5	1190	1159	1129	1100	1070	1042	1014	987	960	933	907	881	856	831	806	781	757	733	709	686	662
	230.5	1112	1084	1055	1028	1001	974	948	922	897	872	848	823	800	776	753	730	708	685	663	641	619
	325	1039	1012	986	961	935	911	886	862	839	815	792	770	748	726	704	683	661	640	620	599	578
	396	971	946	922	897	873	850	828	805	783	762	740	719	698	678	658	637	618	598	579	559	540
	465.5	907	883	861	838	816	794	773	752	731	711	691	672	652	633	614	595	577	559	540	523	505
	631	847	825	803	783	762	742	722	702	683	664	645	627	609	591	573	556	539	522	505	488	471

	741.5	790	770	750	730	711	692	673	655	637	619	602	585	568	551	535	519	503	487	471	455	440
	871	736	717	699	681	662	645	628	611	594	578	561	545	530	514	498	484	469	454	439	425	410
	1139.5	686	668	651	634	617	601	584	569	553	538	523	508	494	479	464	450	436	423	409	395	382
	1291.5	638	622	606	590	574	559	544	529	515	500	486	472	459	445	432	419	406	393	380	368	355
	1659.5	592	577	562	548	533	519	505	492	478	465	451	439	426	414	401	389	377	365	353	342	330
	1782.5	549	535	521	508	494	481	468	456	443	431	419	407	395	383	372	361	350	338	327	317	306
	2094.5	508	494	481	469	456	445	433	421	409	398	387	376	365	355	344	333	323	312	303	292	283
	2332	467	455	443	431	420	409	398	387	376	366	356	346	336	326	316	307	297	287	278	269	260
	2643.5	428	416	406	395	384	375	364	355	345	335	326	317	308	298	289	281	272	263	255	246	238
	2943.5	388	378	368	358	349	340	331	322	313	304	295	287	279	271	262	255	247	239	231	223	216
	3277	347	338	330	321	312	304	296	288	280	272	265	257	250	242	235	228	221	214	207	200	193
	3534	304	296	288	281	273	266	259	252	245	238	231	225	219	212	206	199	193	187	181	175	169
	4859.5	253	246	240	233	227	221	215	209	204	198	192	187	181	176	171	165	161	156	150	145	140
	5709	176	172	167	162	158	154	150	146	142	138	134	130	126	123	119	115	112	108	105	101	98

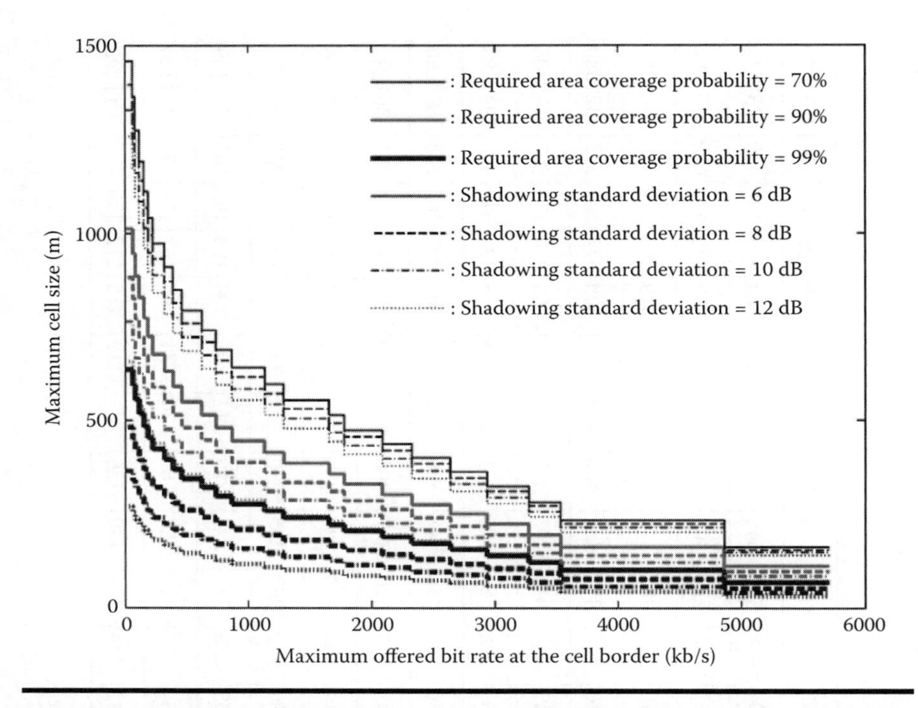

Figure 6.3 Abacuses of the cell size versus offered bit rate for different coverage probability and shadowing standard-deviation values.

important on the dimensioned cell size (due to the impact of coverage probability on the shadowing margin). Those abacuses can help for an initial HSPA network dimensioning with basic radio parameters.

6.5 HSPA RF Planning Reference Scenarios and Procedure

Dimensioning procedure uses two packet scheduling techniques: Fair Resource [13] and the new introduced technique Enhanced Fair Throughput improving upon the classical Fair Throughput scheduling technique.

By applying the Fair Resource scheduling technique to dimension NodeB sites, we have to choose between two alternatives:

- **D1 Dimensioning Method:** It consists in dimensioning sites according to the most limiting service (in terms of coverage and capacity) in order to have access simultaneously to all services and in each point of the concerned service area. However, this method requires an important number of sites to deploy, with a reduced efficiency especially for the less limiting services, which users may be available over a greater area. Moreover, this method has the drawback of generating a great deal of intracellular interference.

- **D2 Dimensioning Method:** It consists in dimensioning sites according to the least limiting service in such a way that access to different services is accomplished through concentric subcells where some services are available near the NodeB. If we move from one subcell to another away from the NodeB, we gradually lose access to services one by one (the guaranteed bit rate decreases by moving from one subcell to another away from the NodeB). In other words, in order to have access to some services demanding more bandwidth, users must not be far away from the NodeB. This adaptive technique is similar to the Wi-Fi (wireless fidelity) access points principle (partial availability in hot spots) and to EDGE (enhanced data rates for GSM evolution) (rate adaptation). Nevertheless, access to lower-rate services is maintained in wider areas. This method allows deploying a lower number of sites while guaranteeing an acceptable access to different services and optimizing bandwidth and radio resource usage in order to increase capacity and improve the range of the whole cell (limited by the external subcell) for a given traffic density.

D1 and D2 are complementary methods since the last one may be an intermediate step between different cellular densification phases after eventual subscriber traffic evolutions in order to provide all services even in limited areas (hot spots). We can also adopt an intermediate method between D1 and D2 by choosing a service with an intermediate nominal bit rate as a reference for sites dimensioning so as to set services to be dimensioned at 100% and the ones to be dimensioned partially as hot spots or coverage concentric subcells.

In order to enhance air interface dimensioning performance in terms of cell size and capacity, we introduce a third dimensioning method (D3) consisting in a modification of the Fair Throughput scheduling technique by incorporating weights to allocated resources proportionally to nominal bit rates of different services such that each mobile is satisfied by its required service bit rate in each point of the cell. This method is called Enhanced Fair Throughput, adapted both to services and propagation conditions. We obtain a common dimensioning for all services (the same subcell sizes), which is better than D1 and D2.

We have presented various dimensioning methods or scenarios in order to give HSPA mobile operators a reasonable margin to select the suitable alternative according to their priorities in terms of cost and/or QoS. Figure 6.4 shows the general flow chart of the HSPA dimensioning methodology.

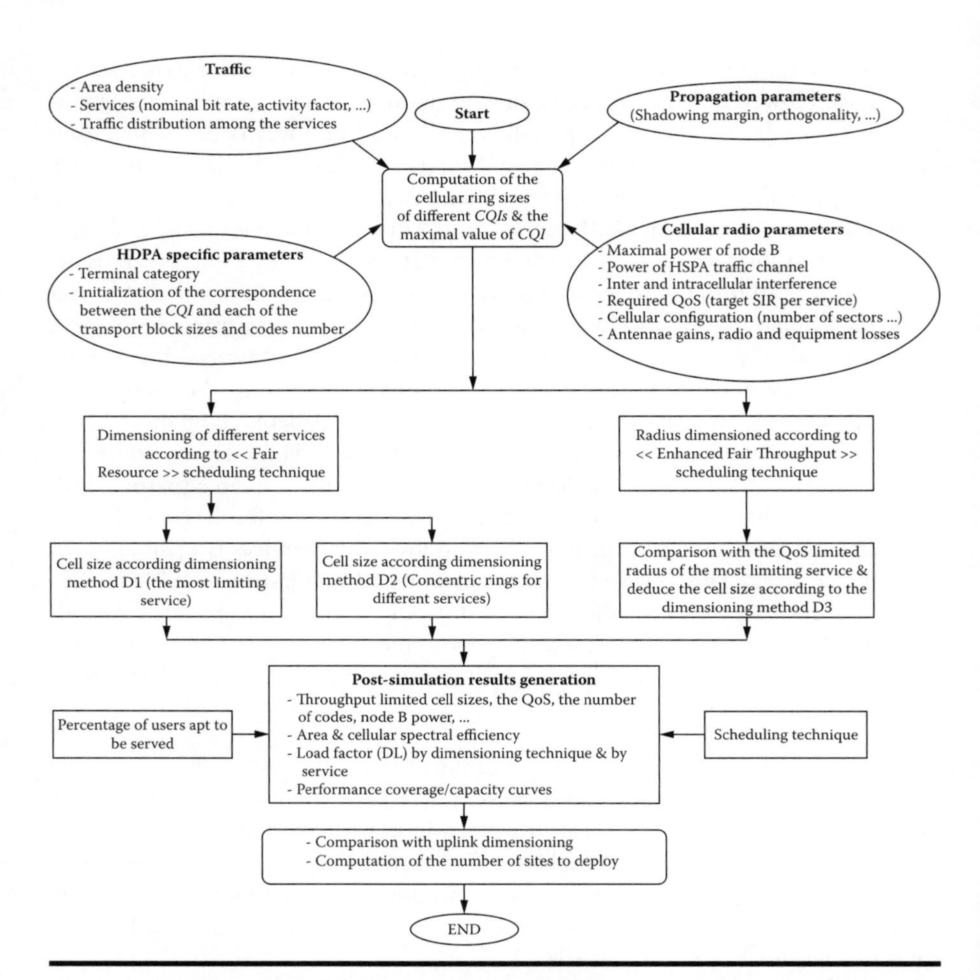

Figure 6.4 Flow chart of the HSPA dimensioning procedure according to the different methods D1, D2, and D3.

6.6 Dimensioning Performance Comparison between UMTS (Rel'99) and HSPA Evolution

The dimensioning performance is measured in terms of coverage (bit rate and cell range) and capacity (in number of HSPA shared codes or limited by the total power of the NodeB).

Our comparison is performed using two different configurations of service distribution: the first is configuration A in which most mobiles use the usual services

Table 6.3 Traffic Distribution According to Services in the Configurations A and B

Services	*Configuration A: (users %)*	*Configuration B: (users %)*
Voice (16 kbps): RAB 16	94.0%	35%
Packet (14 kbps): RAB 14	1.5%	25%
Packet (384 kbps): RAB 384	0.7%	15%
Packet (2 Mbps): RAB 2 Mbps	1.6%	10%
Circuit (128 kbps): RAB 128	2.2%	15%

with low bit rates, and the second is configuration B in which the traffic distribution is balanced between the services (the traffic load above the one in configuration A). See Table 6.3.

Figures 6.5 and 6.6 show the impact of the inclusion of HSPA with its different scheduling techniques versus basic UMTS (without HSPA). Note that HSPA

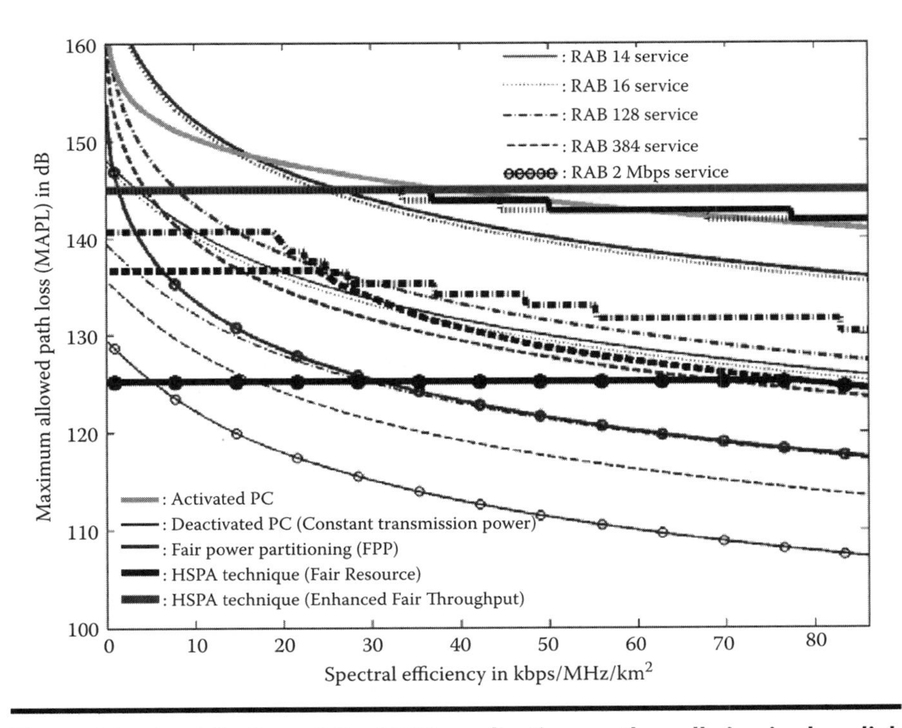

Figure 6.5 Contribution of the HSPA application on the cell size in downlink (low traffic case and services distribution according to configur ation A).

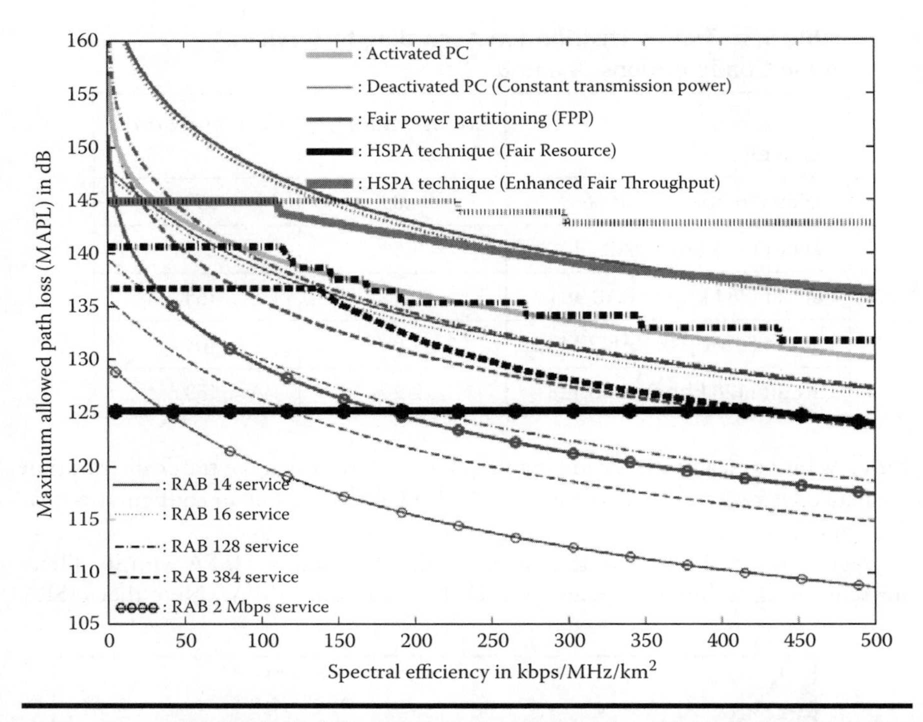

Figure 6.6 Contribution of the HSPA application on the cell size in downlink (low traffic case and services distribution according to configuration B).

improves sensibly the coverage and thus dimensioning versus basic UMTS for a high users area density (area spectral efficiency typically above about 30 kbps/MHz/km^2 for the configuration A and above about 150 kbps/MHz/km^2 for the configuration B) referring to a capacity-limited dimensioning. For the case of coverage-limited dimensioning (low users area density), the contribution of HSPA versus basic (Rel'99) UMTS is not realized, especially versus the case of activated power control (PC) and versus the FPP technique.

This contribution and performance enhancement appear especially versus the case of basic UMTS without PC [i.e., at constant transmitted power (with a gain from 7 to 18 dB according to the service type), which remains an unrecommended solution in terms of optimization of the number of sites]. The HSPA gain versus the FPP dimensioning technique is lower but remains important (from 3 to 7 dB according to the service). The performance of the Enhanced Fair Throughput technique

(D3 method) is eventually above that of basic UMTS especially for an important traffic of users. The D3 dimensioning method based on Enhanced Fair Throughput in HSPA has a higher performance than the FPP method in UMTS Rel'99, even versus the lower bit rate services—for a high area spectral efficiency above 400 kbps/MHz/km^2.

The dimensioning with activated PC is not recommended except if HSPA is not available since the Enhanced Fair Throughput technique has more performance while having a common size of the subcells referring to all the services.

In the case of a high traffic of the users (see Figures 6.7 and 6.8), the activated PC and FPP methods (versus some services with lower bit rates) present more economical dimensioning results than the method D1 with HSPA, but the Enhanced Fair Throughput technique (method D3 in HSPA) is effectively

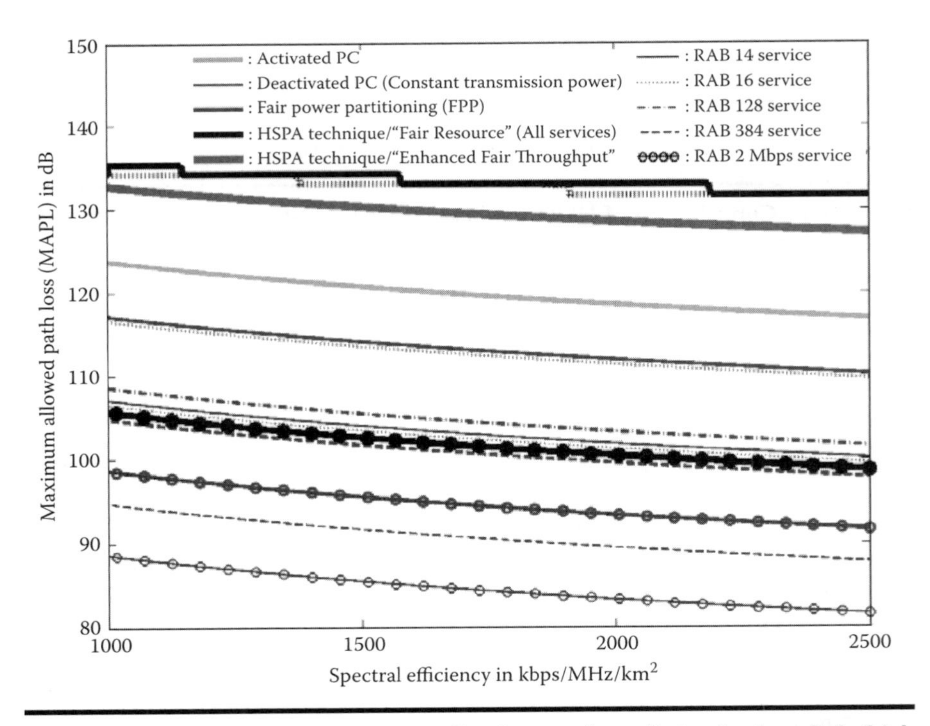

Figure 6.7 Contribution of HSPA application on the cell size in downlink (high traffic case and services distribution according to configuration A).

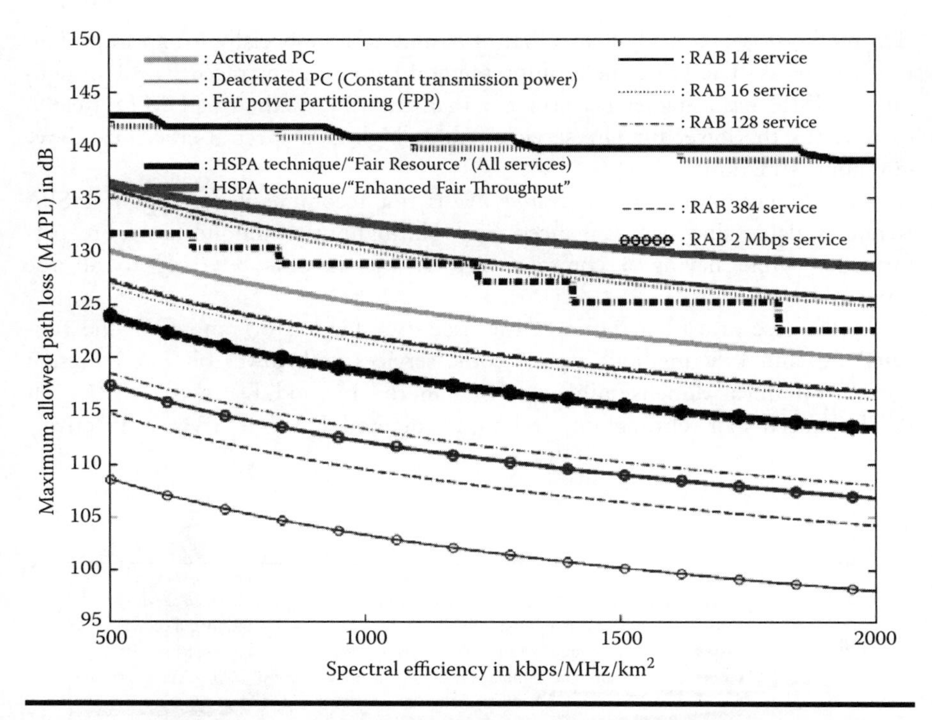

Figure 6.8 Contribution of HSPA application on the cell size in downlink (high traffic case and services distribution according to configuration B).

more economical and has more performance than all the methods or options of dimensioning that have been adopted for basic UMTS (Rel'99). Those latter are not over-performed in terms of minimization of the number of radio sites except by the dimensioning technique D2 (with HSPA) according to the least limiting service (and except for the two lower bit rate services), and that realizes, in addition, a common dimensioning for all the services (same MAPL or cell size for the different services). However, this last method (D2) has the drawback that all services are satisfied except partially in some limited areas (hot spots).

According to Figures 6.9 and 6.10, the techniques used in methods D2 and D3 (in HSPA) provide a maximum capacity per cell above that for the other methods.

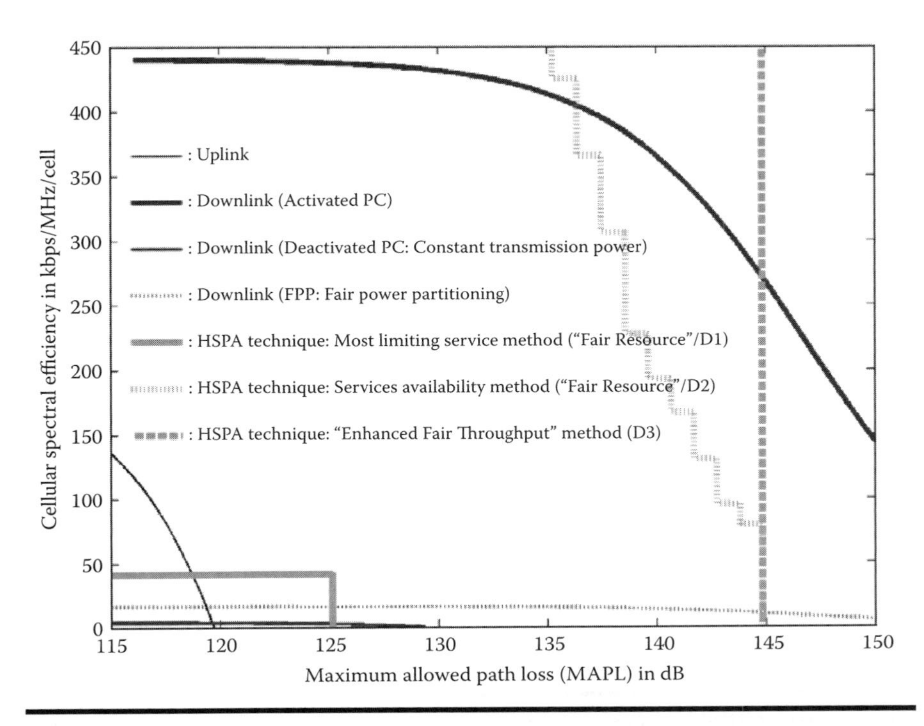

Figure 6.9 Cellular spectral efficiency according to the most limiting service (services distribution according to configuration A).

Nevertheless, the method D1 presents a cell capacity below that with the activated PC method (at constant transmission power). Although the FPP technique provides a maximum capacity per cell that is relatively low (about 20 kbps/MHz/cell in configuration A and 100 kbps/MHz/cell in configuration B), the cell size is high (acceptable range according to Figures 6.5 to 6.8 and their respective comments).

Figures 6.9 and 6.10 show also the cell capacity improvements achieved through methods D2 and D3 (in HSPA) and the activated PC (in UMTS Rel'99). Those enhancements are performed only for the uplink (UL), which limits capacity per cell and thus limits the dimensioning of radio sites.

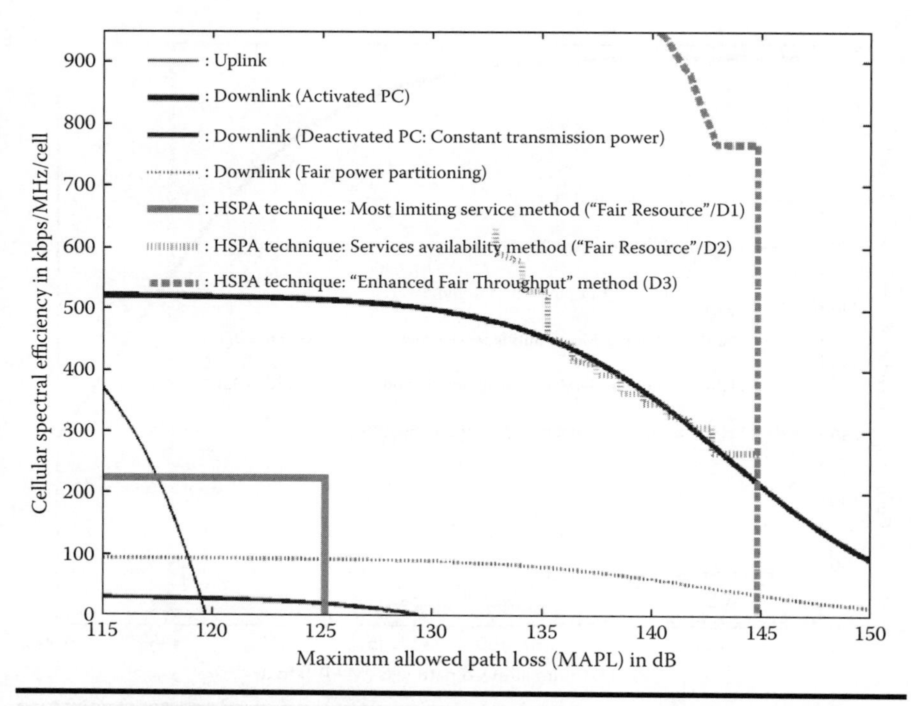

Figure 6.10 Cellular spectral efficiency according to the most limiting service (services distribution according to configuration B).

6.7 Conclusion

This chapter presented dimensioning procedures in HSPA-based UMTS networks, including the two limitation cases: coverage-limited and capacity-limited. Three planning reference scenarios have been described and compared with the necessary analytical support and modeling. As shown, the Enhanced Fair Throughput dimensioning alternative offers a common cell size for all services. Dimensioning process has also been described in the case of multiple aggregated services as well as by the consideration of the shadowing effect. At the end of the chapter, we made a performance comparison between dimensioning techniques of an HSPA-based network and some planning techniques for a basic UMTS Rel'99.

The conclusions show under which conditions HSPA dimensioning techniques are better than those of Rel'99 UMTS, and in which case Rel'99 UMTS deployment is more efficient and better than implementing HSPA-based UMTS infrastructure.

References

[1] A. Masmoudi, "Outil de Planification d'un Réseau Cellulaire de Type CDMA," Engineering end studies project report, Sup'Com, Tunis, July 1999.
[2] A. Masmoudi, "Affectation de Fréquences dans les Réseaux Cellulaires de Troisième Génération," Master Report, ENIT, Tunis, July 2001.
[3] 3GPP TS 25.321 V6.0.0, "Technical Specification Group Radio Access Network; "MAC Protocol Specification," 3GPP Release 6, January 2004.
[4] R. Caldwell and A. Anpalagan, "HSDPA: An Overview," *IEEE Canadian Review*, Spring 2004.
[5] T. E. Kolding, K. I. Pederson, J. Wigard, F. Frederiksen, and P. E. Mogensen, "High Speed Downlink Packet Access: WCDMA Evolution," *IEEE Vehicular Technology Society News*, IEEE, February 2003.
[6] Holma Harri and Antti Toskala, *WCDMA for UMTS, Third Edition* John Wiley & Sons, Ltd, 2004.
[7] T. E. Kolding, F. Frederiksen, and P. E. Mogensen, "Performance Aspects of WCDMA Systems with High Speed Downlink Packet Access (HSDPA)," Proceedings, VTC, vol. 1, pp. 477–481, September 2002.
[8] 3GPP TS 25.308, "UTRA High-Speed Downlink Packet Access (HSDPA); Overall Description," 3GPP Release 5, March 2002.
[9] A. Masmoudi, D. Zeghlache, and S. Tabbane, "Dimensionnement des réseaux UMTS basés sur la technique HSDPA," Sciences of Electronic, Technologies of Information and Telecommunications (SETIT 2007), Hammamet, Tunisia, pp. 25–29 March 2007.
[10] F. Brouwer, et al. "Usage of Link-Level Performance Indicators for HSDPA Network-Level Simulations in E–UMTS," ISSSTA 2004, Sydney, 2004.
[11] 3GPP TS 25.214, "Physical Layer Procedures (FDD)," 3GPP Release 6, December 2003.
[12] EURANE (Enhanced UMTS Radio Access NEtwork) User Guide, "User Manual for Eurane v 1.6."
[13] Thomas Bonald, Philippe Olivier, and James Roberts, "Dimensionnement de Liaisons d'accès IP Transportant du Trafic de Données," *Annales des Télécommunications*, vol. 59, no. 11–12, November–December, 2004.

References

Chapter 7

Radio Resource Optimization and Scheduling Techniques for HSPA and LTE Advanced Technologies

Tarek Bejaoui, Anis Masmoudi, and Nidal Nasser

Contents

7.1 Introduction

In the near future, a broad range of multimedia applications with guaranteed quality of service (QoS) is expected to be provided by new evolved UMTS networks. The 3rd Generation Partnership Project (3GPP) has standardized in Release 6 and 7 new cellular-based systems denoted respectively as high-speed packet access (HSPA) and evolved high-speed packet access (eHSPA or HSPA+). Another area of focus defined in 3GPP Release 8 is the introduction of a new OFDM-based technology through the long-term evolution (LTE) work item, often referred to as the evolved UMTS terrestrial radio access network (EUTRAN). It is the next generation cellular wireless standard that is considered the prominent path to the 4G cellular wireless system.

7.1.1 Objective and Context

The next-generation cellular systems rely on new technologies. They make it possible to bring improved support and performance for constrained services, thanks to important new additions such as enhanced receivers, multiple-input multiple-output, continuous packet connectivity, higher order modulations, fast cell selection, and fast packet scheduling.

The long-term evolution network is introduced through the definition of the new flatter-IP core network, and will improve performance by providing higher data rates, reduced latency, and improved spectral efficiency. The focus was on enhancement of the radio-access technology (UTRA) and the optimization and simplification of the radio access network (UTRAN) as well.

The EUTRAN uses a simplified single node architecture consisting of the EUTRAN NodeB denoted eNB, which communicates with the evolved packet core (EPC), the mobility management entity (MME), and the user plane entity (UPE). In the EUTRAN, this eNB supports all the functions in a typical radio network such as radio bearer control, connection mobility management, admission control and scheduling, dynamic resource allocation, inter-cell interference coordination, load balancing, and inter-radio access technology functions. The access stratum resides then completely at this node.

Radio resource management (RRM) is therefore one of the key design features of HSPA, HSPA+, and LTE. The objective of the RRM techniques is to optimize the use of radio resources while fulfilling the quality requirements of the largest possible number of users. The most representative of these techniques is the packet scheduling. A packet scheduler controls the allocation of channels to users within the system coverage area by deciding which user should transmit during a given time interval.

7.1.2 Radio Resource Management for Advanced Wireless Systems

The next-generation cellular networks are expected to support a broad spectrum of multimedia services with guaranteed QoS. The resource access protocol that defines how the wireless medium is shared among contending users, is therefore a pioneer element on which depends the overall performance of these networks. In this context, an efficient resource allocation scheme should handle a wide range of information bit rates as well as various types of real-time and non-real-time services with different traffic characteristics and QoS guarantees. In addition, the protocol must operate under different constraints of mobility, dynamic traffic load variations, and highly sensitive wireless links. Under these constraints, QoS provisioning becomes a challenging task and difficult to ensure. Effective management of the limited radio resources is therefore important to enhance the network performance. Some cross-layered radio resource management algorithms are then designed and proposed to optimally adapt to channel conditions and specific applications requirements. Their purpose is to solve the issue of the lack of built-in mechanisms for protocol layers that makes it very difficult to provide guaranteed QoS for multimedia applications.

The packet scheduling constitutes one of the fundamental RRM techniques for QoS provisioning to the evolved UMTS networks. It controls the allocation of channels to users who have data to transfer within the coverage area of the system and, to a large extent, it determines the overall behavior of the system.

In this chapter, we will investigate some of the radio resource management features and discuss the various QoS requirements in evolved UMTS networks, and some of the solutions proposed for effective management of the limited radio resources to enhance their performance. In this context, the chapter concentrates on the packets scheduling schemes proposed for QoS provisioning in such networks. It gives an accurate analytical modeling of some of these protocols like Fair Resource, Fair Throughput, Proportional Fair algorithms, and "the maximum CIR scheme," which were proposed for HSPA systems. These scheduling protocols that are defined today as "conventional" are not suitable for aggregated multiple services with different profiles and required QoS parameters (data, voice, video, etc.).

Indeed, many of these proposed protocols focus on different layers separately. Each layer communicates with its peer using a set of rules and conventions collectively known as layered protocol, and should perform its own defined functions, without knowledge of details on the services' implementation in the other layers. In implementing protocols in these layers, control is passed from one layer to the next. The interactions between layers are controlled, each layer has the property that it only uses the functions of the layer below, and only exports functionality to the layer above. In wireless networks, wireless channels and networks are dynamic in behavior, such as temporal and spatial changes quality and user distribution. Furthermore, meeting the end-to-end performance requirements of demanding applications is extremely challenging without interaction between protocol layers. The conventional layered protocol architecture is inflexible and unable to adapt to such dynamically changing network behaviors, since the various protocol layers can only communicate with each other in a strict and primitive manner. In such a case, the layers are most often designed to operate under worst conditions, rather than adapting to changing conditions. This eventually leads to inefficient use of spectrum and energy.

Thus, new optimized packet scheduling techniques adapted for multiple services (multi-class of real-time and non-real-time services) in the next-generation evolved UMTS networks are presented. Some novel approaches based on cross-layered radio resource management protocols that attempt to focus the radio channel conditions are then explored.

7.2 Radio Resource Management for Evolved UMTS Networks

Several RRM functions are defined for the evolved UMTS networks HSPA/HSPA+ and LTE. In LTE, these functions are assigned to eNB(s) and mapped over the layers 1, 2, and 3. They include the radio bearer control (RBC), the radio admission control (RAC), the connection mobility management (CMM), the dynamic resource allocation (DRA) or packet scheduling, the inter-cell interference coordination (ICIC), and load balancing (LB).

Compared to HSPA, LTE introduces new functionalities in base stations like the radio link control layer (RLC), radio resource control (RRC), and the functions defined for the packet data convergence protocol (PDCP) as ciphering and header compression. The medium access control (MAC) layer functionality is similar to HSPA operation and remains in the eNB as shown in Figure 7.1 [1].

It depicts the architecture for downlink (as for uplink) of the layer 2 of the radio access protocol in the eNodeB, which is constituted by the PDCP/RLC/MAC sublayers supporting the radio resource management.

An overview on services and functions provided by each sublayer are presented in the following sections [1].

7.2.1 MAC Sublayer

The MAC sublayer is a protocol layer that arbitrates and controls access to the shared transmission medium. It runs in both the UE and the eNB and has different behaviors when running in each, generally giving commands in the eNB and responding to them in the UE.

Thus, the main functions of the MAC sublayer includes mapping between logical channels and transport channels, multiplexing/demultiplexing of RLC packet data units (PDUs) belonging to one or different radio bearers into/from transport blocks (TB) delivered to/from the physical layer on transport channels, traffic volume measurement reporting, error correction through HARQ, priority handling between logical channels of one UE, priority handling between UEs by means of dynamic scheduling, MBMS service identification, and transport format selection and padding.

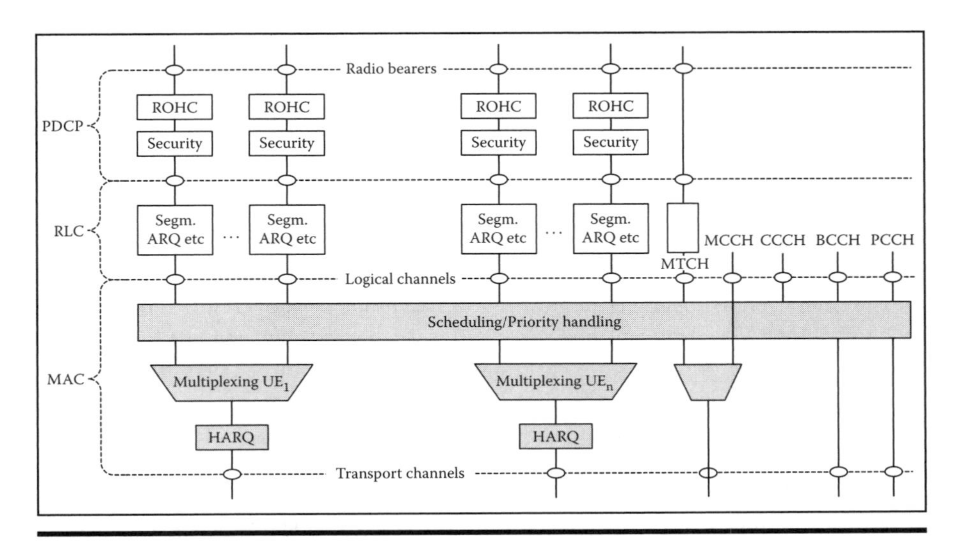

Figure 7.1 Layer 2 structure for DL.

Those marked with circles at the interface between sublayers in Figure 7.1, are the service access points (SAPs) for peer-to-peer communication. The SAP between the physical layer and the MAC sublayer provides the transport channels. The SAPs between the MAC sublayer and the RLC sublayer provide the logical channels.

7.2.2 RLC Sublayer

The main services and functions of the RLC sublayer includes the transfer of upper layer PDUs supporting acknowledged mode (AM) or unacknowledged mode (UM), transparent mode (TM) data transfer, error correction through automatic repeat request (ARQ), segmentation according to the size of the TB, re-segmentation of PDUs that need to be retransmitted, concatenation of SDUs for the same radio bearer, in-sequence delivery of upper layer PDUs except at handover, duplicate detection, protocol error detection and recovery, and SDU discard and reset.

7.2.3 PDCP Sublayer

The PDCP sublayer performs both user plane and control plane main functions. The PDCP sublayer functions in the user plane include header compression and decompression (ROHC only), transfer of user data, in-sequence delivery of upper layer PDUs at handover for RLC AM, duplicate detection of lower layer SDUs at handover for RLC AM, retransmission of PDCP SDUs at handover for RLC AM, ciphering and timer-based SDU discard in uplink. The PDCP sublayer functions in the control plane include ciphering and integrity protection and transfer of control plane data.

7.2.4 RRC Sublayer

The radio resource control (RRC) protocol is being used to configure and control the radio resource between the eNB and the user equipment. The RRC sublayer performs the following control plane main functions: broadcast of system information related to access stratum (AS) and non-access stratum (NAS), paging, establishment, maintenance and release of an RRC connection between the UE and EUTRAN, signaling radio bearer management, security handling, mobility management, including UE measurement reporting and configuration, active mode handover, idle mode mobility control, MBMS notification services and radio bearer management for MBMS, QoS management, and NAS direct message transfer to/from NAS from/to UE.

The RRC specifications defined for LTE are slightly different from those defined for legacy 3G-RNC systems like HSPA. The following describes a few:

- The number of RRC states: two states in LTE and five in 3G-RNC system.
- The number of signaling radio bearers: The LTE has three signaling radio bearers and 3G-RNC system has four.

- MAC entity: Only one MAC entity is defined for LTE, whereas in 3G-RNC systems there are four different MAC entities based on different types of transport channels, and then less signaling is involved.
- As there is no common transport channel defined in LTE, the radio bearer mapping is much simpler.
- No RRC connection mobility is defined in LTE, like cell update and ura update.
- Instead of having two domain identities (CS and PS domains) as in 3G-RNC systems, only the PS domain identity is specified, with less complexity and signaling overhead.
- As there is only the PS domain in LTE, there is no signaling connection release procedure.
- In LTE, a limited number of most frequently transmitted parameters is included in the MIB, and the scheduling information that mainly indicates when the SI messages are transmitted is contained in the SIB type I, whereas in 3G-RNC systems, MIB includes both the frequently transmitted parameters and the scheduling information.
- Only one type of paging is required for LTE and two types are required in 3G-RNC systems.
- In case of reconfiguration, only one reconfiguration message is used in LTE to reconfigure all logical, transport, and physical channels, and then fewer signaling messages are exchanged.
- In LTE, the latency of the RRC connection establishment is reduced since no NBAP protocol is used.
- In LTE, there is no need to specify the RRC state in an RRC message.
- In LTE, there is no need to define activation time. This leads to a significant reduction in the latency during establishment and reconfiguration of radio bearers.
- In LTE, only a shared channel is defined, and there is no need to define the downlink transport channel configuration in the RRC reconfiguration message. This will reduce the signaling message size effectively. All DL-SCH transport channel information is broadcast in system information.

7.3 Overview of Packet Scheduling in HSPA and Beyond

One of the most important features of HSPA is packet scheduling. The main goal of packet scheduling is to maximize the system throughput while satisfying the QoS requirements of the users. The packet scheduler determines which user the shared channel transmission should be assigned to at a given time. In HSDPA, the packet scheduler can exploit the short-term variations in the radio conditions of different users by selecting those with favorable instantaneous channel conditions for transmission, which is illustrated in Figure 7.2. This idea is based on the fact that good channel conditions allow for higher data rates (R) by using a higher order modulation and coding schemes [2], which results in increasing the system throughput.

In order to quickly obtain up-to-date information on the channel conditions of different users, the functionality of the packet scheduler has been moved from the radio network controller (RNC) in UMTS to the medium access control high-speed (MAC-hs) sublayer at the NodeB [2], as shown in Figure 7.3. The MAC-hs is a new sublayer that is added to the MAC layer at the NodeB in HSDPA in order to execute the packet scheduling algorithm. In addition, the time transmission interval (TTI) (i.e., the time between two convective transmissions) has been reduced from 10 ms in UMTS Release 99 to 2 ms in Release 5 that includes HSDPA. This is because it allows the packet scheduler to better exploit the varying channel conditions of different users in its scheduling decisions and to increase the granularity in the scheduling process. It should be noted that favoring users with good channel conditions may prevent those with bad channel conditions from being served and may, therefore, result in starvation. A good design of a scheduling algorithm should take into account not only maximization of the system throughput through service differentiation, but also being fair to users who use the same service and pay the same amount of money. That is, scheduling algorithms should balance the trade-off between maximizing throughput and fairness.

The packet scheduler for HSDPA implemented at the MAC-hs layer of NodeB works as follows (see Figure 7.3). Every TTI, each user regularly informs the NodeB of his channel quality condition by sending a report known as a channel quality indicator (CQI) in the uplink to the NodeB. The CQI contains information about the instantaneous channel quality of the user. This information includes the size

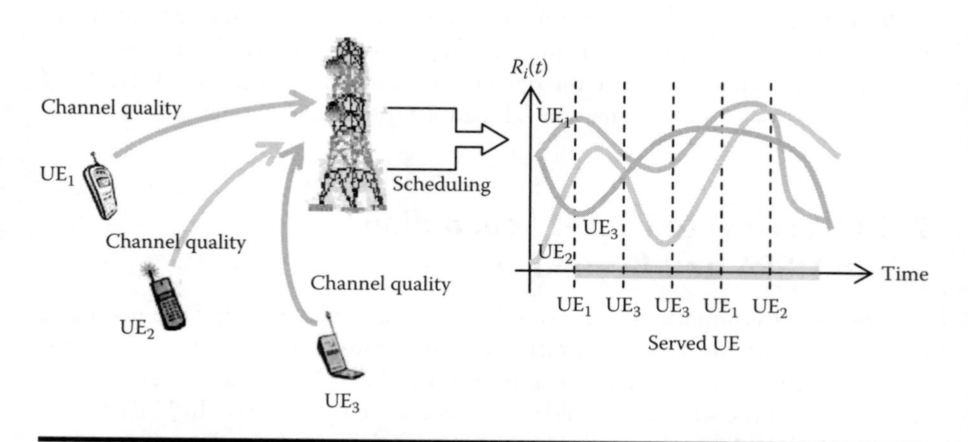

Figure 7.2 Exploiting the user channel quality for scheduling decisions.

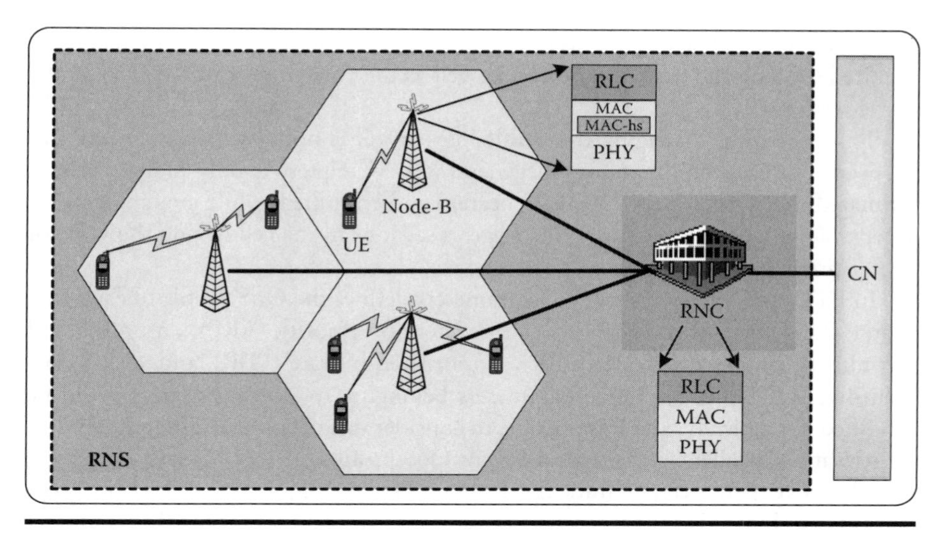

Figure 7.3 The MAC-hs at the NodeB in HSDPA.

of the transport block that the NodeB should send to the user, the number of simultaneous channel codes, and the type of modulation and coding schemes that the user can support. NodeB then would select the appropriate mobile user according to the adopted scheduling discipline and send data to the selected user at the specified rates. The user is able to measure his current channel conditions by measuring the power of the received signal from the NodeB and then using a set of models described in [3], determine his current supportable data rates (i.e., the rates that he can receive data from the NodeB given his current channel condition). Therefore, users with good channel conditions will enjoy potentially higher supportable data rates by using higher modulation and coding rates, whereas users with bad channel conditions will experience lower data rates instead of adjusting their transmission power.

7.4 Overview on Packet Scheduling in LTE

The packet scheduling constitutes one of the RRM functions defined for LTE. In this chapter, the focus will only be on the scheduling and admission control functions. The radio admission control, the QoS management, and the persistant scheduling functions are defined for layer 3.

7.4.1 Radio Admission Control

In order to decide about the acceptance of the requests for new evolved packet system (EPS) bearers in the cell, the admission control algorithm takes into consideration several parameters like the resource availability in the cell, the priority level, and the required QoS by the new EPS bearer, as well as the currently provided QoS to the active sessions.

This algorithm managing the admission control is not specified by 3GPP but is specific to each eNB vendor. In general, a new request is only granted if it is estimated that QoS for the new EPS bearer can be fulfilled, while guaranteeing an acceptable service to the existing in-progress sessions in the cell having the same or higher priority.

In LTE, a set of associated QoS parameters defines the QoS profile of each EPS bearer. This set consists of an allocation retention priority (ARP), an uplink and downlink guaranteed bit rate (GBR), a prioritized bit rate (PBR), and a QoS class identifier (QCI) [4, 5]. These parameters belonging to existing bearers could be modified dynamically, and it is possible to consider simultaneously different services by activating parallel bearers with different QoS profiles.

The ARP parameter defines the level of priority required for the admission control decision. It is an integer that ranges between 1 and 16. The GBR parameter is specified only for EPS GBR bearers. For non-GBR bearers, an aggregate MBR (AMBR) is specified. PBR is a QoS parameter specified for the uplink per bearer introduced to avoid the uplink scheduling starvation problem that may occur for UE with multiple bearers.

As for QCI, 3GPP specifications define a mapping table for nine values and their corresponding typical services (cf. Table 7.1). This parameter includes other parameters like the layer 2 packet delay budget, packet loss rate, and scheduling priority.

To ensure high spectral efficiency in the LTE cell while providing the required QoS, much more focus should be on dynamic packet scheduling and link adaptation.

The link adaptation is performed to adapt the selection of modulation and channel coding schemes to current channel conditions on the basis of CQI feedback from the users in the cell. This leads to the definition of the data rate and the error probabilities of each link.

In this chapter, only packet scheduling techniques for LTE will be discussed.

7.4.2 Uplink Packet Scheduling

On the uplink, LTE uses an approach called SC-FDMA, which is somewhat similar to OFDMA but has a 2-to-6-dB peak-to-average ratio of the signal advantage over the OFDMA.

An uplink packet scheduler has to share the available radio resources between users while taking into account requirements and limitations imposed by other RRM functionalities.

The LTE uplink is a constrained link due to single-carrier FDMA transmission. It limits both frequency and multi-user diversity. Therefore, the packet scheduler has to fulfill the hard QoS requirements by users having data to transmit over an

Table 7.1 LTE QCI (QoS Class Identifier), as Defined by 3GPP TS 23.207

QCI	*Resource Type*	*Priority*	*Packet Delay Budget*	*Packet Error Loss Rate*	*Example Services*
1	GBR	2	100 ms	10^{-2}	Conversational voice
2	GBR	4	150 ms	10^{-3}	Conversational video (live streaming)
3	GBR	3	50 ms	10^{-3}	Real-time gaming
4	GBR	5	300 ms	10^{-6}	Non-Convers. video (buffered streaming)
5	GBR	1	100 ms	10^{-6}	IMS signaling
6	GBR	6	300 ms	10^{-6}	Video (buffered streaming)
7	Non-GBR	7	100 ms	10^{-3}	Voice, live streaming, interactive gaming
8	Non-GBR	8	300 ms	10^{-6}	"Premium bearer" for video (buffered)
9	Non-GBR	9	300 ms	10^{-6}	"Default bearer" for video

interface typically characterized by high interference variability. On the basis of information conveyed by the buffer status reports (BSRs), the scheduler can handle the prioritization between these users. Since synchronous HARQ is used for LTE uplink, this packet scheduler interacts closely with the HARQ manager, and then user equipment (UE) must be scheduled if an earlier transmission has failed. Its power capabilities must be considered when, for example, packet scheduler allocates the uplink transmission bandwidth to a specific UE. As multi-antenna transmission techniques are used in LTE, enhancing system performance and service capabilities, the uplink packet scheduler can simultaneously allocate resources to several users. However, in LTE uplink, users cannot be scheduled for transmission on a physical uplink shared channel (PUSCH) unless they are listening to the L1/L2 control channel.

Compared to eNB, these users are low-power devices and consequently they cannot be allocated a high transmission bandwidth to compensate the effects of radio environment conditions, especially in macro-cellular network configuration.

7.4.3 Downlink Packet Scheduling

LTE implements OFDM in the downlink. Its basic principle is to split a high-rate data stream into a number of parallel low-rate data streams, each narrowband signal carried by a subcarrier. They are generated in the frequency domain and combined to form the broadband stream while using the inverse fast Fourier transform (IFFT) algorithm. To avoid any performance degradation in high-speed conditions, the subcarriers have a 15-kHz spacing from each other, maintained regardless of the overall channel bandwidth. In LTE, the number of subcarriers ranges from 75 in a 1.25-MHz channel to 1200 in a 20-MHz channel. Thanks to OFDMA, different users could be assigned different subcarriers over time. Over both time and frequency, a minimum resource block that the system can assign to a user transmission consists of 12 subcarriers over 14 symbols, as shown in Figure 7.4.

In the downlink, the dynamic packet scheduler performs scheduling decisions every TTI and allocates to the users both physical resource blocks (PRBs) and selected modulation and coding schemes. They are signaled to the scheduled users on the PDCCH. In LTE, an active user with an EPS bearer has several data flows. It has a control plane data flow for the RRC protocol and one or multiple user plane data flows for EPS bearers; each of them are uniquely identified with a 5-bit logical channel identification (LCID) field. On the basis of the scheduled transport block size (TBS) for a particular user, the medium access control protocol decides the amount of data sent from each LCID.

Even though a user has several data flows, the scheduling decisions are carried out on a per-user basis. As in uplink, the packet scheduler interacts closely with the HARQ manager since it is responsible for scheduling retransmissions. Asynchronous adaptive HARQ is supported, and then the scheduler dynamically schedules pending HARQ retransmissions in time and frequency domain. However, it is not allowed to send at the same time a new and pending HARQ transmission to each scheduler user.

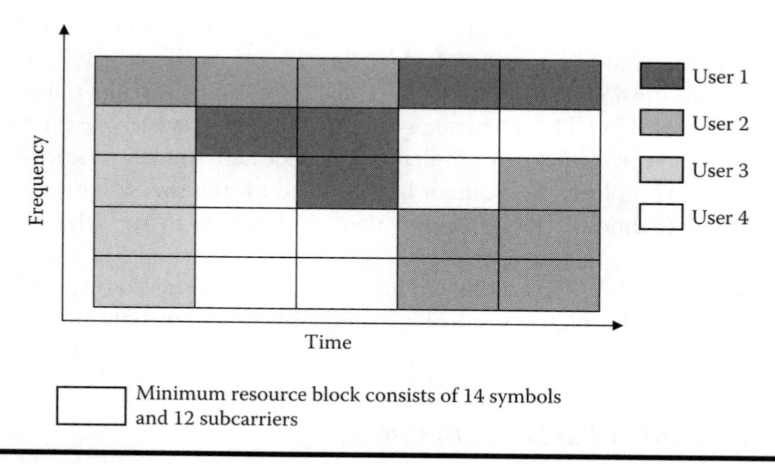

Figure 7.4 LTE OFDMA downlink resource assignment in frequency and time.

On the basis of CQI feedback from terminals operating within the LTE cell, the downlink packet scheduler is informed through the link adaptation about the supported modulation and coding scheme for a user depending on the selected set of PRB. As in HSDPA, an outer loop link adaptation algorithm could be applied to control the block error rate of the first transmissions.

7.4.4 Time and Frequency Domains Packet Scheduling

To improve the LTE system capacity, 3GPP has proposed efficient techniques called time domain packet scheduling (TDPS) and frequency domain packet scheduling (FDPS). In case of frequency fast fading, the TDPS can provide multiuser diversity gains that depend on the amount and speed of the fading. In LTE, TDPS gains are relatively low due to the typically used large bandwidths and both the mobile and base stations' antenna diversity capability. These gains can also be affected by high mobile speed and multipath propagation. Indeed, on the basis of the CQI feedback, the packet scheduler selects only the pool of PRB having the highest channel quality on which the buffered data in the eNB will be transmitted.

With the FDPS technique that exploits frequency selective power variations on either the useful signal or the interference, users are scheduled on the PRB with high channel quality. The PRB where users are experiencing deep fades are avoided. A high FDPS gain is achieved when the effective coherence bandwidth of the radio channel is less than a system bandwidth equal to or larger than 5 MHz. The main drawbacks of this technique are high scheduler complexity and increased signaling overhead in both the uplink and downlink.

7.4.5 Scheduling and Persistent Scheduling

In LTE, transmissions can be carried out with higher data rates while the scheduling of the resources is performed on the basis of the channel conditions knowledge. In addition to dynamic scheduling applied both to the uplink and downlink, LTE supports persistent scheduling for which radio resources are allocated to a user for a given set of subframes. This is due to the fact that for services with small payloads and regular packet arrivals, the control signaling required for dynamic scheduling might be disproportionately large relative to the amount of user data transmitted.

7.5 HSPA and LTE Scheduling Techniques Models

In this section, some HSPA scheduling techniques models proposed in literature are presented. Some of them could be extended to LTE.

7.5.1 Fair Resource Scheduling Technique Protocol

The maximum bit rate $R_{\max,i}$ ensured by user i is as follows:

$$R_{\max,i} = \begin{cases} \min\left(\dfrac{TBS_i}{TTI_{\text{delay}}}, \dfrac{TBS_i}{TTI_{\text{delay}}}\dfrac{n_{\text{cap}}}{n_{\text{Tot}}}\right) & \text{(without codes multiplexing)} \\ \min\left(\dfrac{TBS_i}{TTI_{\text{delay}}}, \dfrac{TBS_i}{TTI_{\text{delay}}}\dfrac{n_{\text{cap}}}{n_{\text{Tot}}}\right) \cdot \max\left(\dfrac{15}{\sum_{i=1}^{k} n_i}, 1\right) & \text{(with codes multiplexing)} \end{cases} \quad (7.1)$$

where n_i denotes the number of codes referring to the appropriate transport block size TBS_i (by AMC) for a given user i within the cell according to the quality of its link to the NodeB (tables correspondence [3] according to the terminal category), n_{Tot} is the total number of users within the served cell, and n_{cap} is the number of users included in the capacity-limited cell (for which the number of assigned codes is exactly equal to 15). The summation in Equation 7.1 is applied to the numbers of codes referring to the users of the served cell, and k is the number of users in the cell. The $n_{\text{cap}}/n_{\text{Tot}}$ ratio is equal to ${R_{\text{cap}}}^2/r^2$ in the case of a uniform traffic (in terms of user density), where R_{cap} is the size of the capacity-limited cell, and r is the size of the served cell.

Therefore, the bit rate $(R_{\text{ass}})_{FR}$ guaranteed by the "Fair Resource" technique protocol at the cell border is established by replacing in Equation 7.1 the coverage bit rate $R_{\text{cov},i} = \frac{TBS_i}{TTI_{\text{delay}}}$ with that at the cell border $\min\limits_i\left(\frac{TBS_i}{TTI_{\text{delay}}}\right)$ as follows:

$$(R_{\text{ens}})_{\text{FR}} = \begin{cases} \min\limits_i\left(\dfrac{TBS_i}{TTI_{\text{delay}}}\right) \cdot \min\left(\dfrac{n_{\text{cap}}}{n_{\text{Tot}}}, 1\right) & \text{(without codes multiplexing)} \\ \min\limits_i\left(\dfrac{TBS_i}{TTI_{\text{delay}}}\right) \cdot \min\left(\dfrac{n_{\text{cap}}}{n_{\text{Tot}}}, 1\right) \cdot \max\left(\dfrac{15}{\sum_{i=1}^{k} n_i}, 1\right) & \text{(with codes multiplexing)} \end{cases} \quad (7.2)$$

In this context, the expression of $(R_{\text{ens}})_{FR}$ includes both the coverage bit rate of the HSPA cell and a term of capacity limitation (number of HSPA codes of the cell) that is, the minimum between the coverage-limited bit rate and that limited by capacity without codes multiplexing. If we try to increase the number of users while keeping the same guaranteed bit rate $(R_{\text{ens}})_{FR}$, we should reduce the cell size in order to be at a higher border TBS_i: It's the cell breathing phenomenon, such as for a basic UMTS (WCDMA) network, but ensured in HSPA through adaptation in modulation and coding (AMC) instead of power control in basic UMTS (Rel 99). The fact of specifying a minimum (guaranteed) bit rate and the cell size (and thus the minimum transport block size or the coverage-limited bit rate) limits the maximum number of users served (subscriber density) with their respective codes n_i. Inversely, if the minimum guaranteed bit rate and subscriber density are given, then the cell size should be well determined (dimensioning). Moreover, at a given cell size, the ensured bit rate $(R_{\text{ass}})_{FR}$ is provided by Equation 7.2.

Since Fair Resource protocol tries to share the bandwidth and the available resources equally among the users while maximizing the cell bit rate versus Fair Throughput protocol, we can adopt this protocol for non-real-time (NRT) services. In fact, this method provides a good compromise between the fairness of users of different services (Web browsing, FTP, etc.) and the maximization of the global bit rate within the cell. Besides, NRT services don't require a minimum bit rate (the Fair Resource method doesn't ensure any guaranteed bit rate to the different users).

7.5.2 Fair Throughput Scheduling Technique

The maximum ensured bit rate by each of the users of a cell by applying the "Fair Throughput" scheduling technique protocol (without codes multiplexing) is given by the expression (Equation 7.19). We realized in Chapter 6 that the Fair Throughput protocol isn't, in any case, limited by coverage.

We also provided in Chapter 6 the maximum ensured bit rate per user with codes multiplexing (given by Equation 7.20). We conclude that the maximum ensured bit rate by applying Fair Throughput $(R_{\text{ens}})_{FT}$ corresponds exactly to the maximum number of available codes in HSPA (equal to 15 for the category 10 of mobile terminals), so by applying this scheduling protocol, the balanced bit rate is always capacity-limited (limited by the number of codes or physical channels).

Since Fair Throughput protocol tries to offer, if possible, the same bit rate for all the users, we can set the number of users and the cell size so as to guarantee a given bit rate. Therefore, this protocol is adapted for real-time (RT) services with guaranteed bit rates [such as for constant bit rate (CBR) users] more than the Fair Resource protocol, which doesn't guarantee a given bit rate, especially for users far away from the NodeB.

7.5.3 Maximum CIR Scheme (Max C/I)

In Max C/I scheduling, the channel is allocated in each TTI (transmit time interval) to the user having the best SINR (signal-to-interference-and-noise ratio), in other words the best channel quality. This scheduler maximizes the cell capacity but does not guarantee any QoS to the user. Users at the border of the cell always have poor channel conditions (due to the attenuation, interference, and absence of fast power control) and experience low bit rate [6].

The user bit rate achieved by this scheduler depends upon the wireless channel model. In this paragraph, we estimate the cell throughput and user bit rate in the case of uncorrelated and correlated Rayleigh fading channels [6].

In order to estimate the cell capacity and user bit rate, the probability that the shared channel is allocated to a given user (e.g., user i), denoted by $Pr(i)$, should be evaluated. $Pr(i)$ can be written as:

$$\begin{aligned} Pr(i) &= Prob(SINR_i > SINR_j \text{ for } j = 1\,..\,N_u \text{ and } j \neq i) \\ &= \prod_{j \neq i}^{N_u} Prob(SINR_i > SINR_j) \end{aligned} \tag{7.3}$$

where N_u is the total number of the users within the cell. The expression $SINR_i > SINR_j$ can be expressed as:

$$\left(\sum_{l_i=1}^{N_T} |\alpha_{l_i,i}|^2 \right) X_i > \left(\sum_{l_j=1}^{N_T} |\alpha_{l_j,j}|^2 \right) X_j \tag{7.4}$$

where X_i is given by Equation 7.5 for user i:

$$X_i = \frac{10^{bs_i/10}}{\sum_{l \neq i} \left(P_l \left(\frac{d_l}{d_i} \right)^{-\mu} 10^{bs_l/10} \right)} \tag{7.5}$$

where s_i corresponds to log-normal shadowing with zero mean and standard deviation σ (σ^2 between 8 and 12 dB). The shadowing loss s is correlated between the BSs [7]. This effect is usually modeled by considering the shadowing as a sum (in dB) of a component common to all base stations s_c and a component s_{si} specific to base station i noted BS_i. The shadowing loss expression is given as:

$$s_i = as_c + bs_{si} \tag{7.6}$$

where $a^2 + b^2 = 1$. The mean and variances of the log-normal variables are:

$$\begin{cases} E(s_i) &= E(s_c) = E(s_{si}) = 0 \\ Var(s_i) &= Var(s_c) = Var(s_{si}) = \sigma^2 \end{cases}$$

$$E(s_{si}s_{sk}) = 0 \text{ if } i \neq k \tag{7.7}$$

$\alpha_{l_i,i}$ is the complex path gain between the user i and its serving NodeB, N_T is the number of resolvable multipath components, and d_i is the distance between the user i and its serving NodeB.

The distribution function of the expression $v = \left(\sum_{l=1}^{N_T} |\alpha_l|^2\right) X_i$ can be approximated by:

$$pdf(v) = \sum_{l=1}^{N_T} \frac{(\Omega_l)^{N_T-2}}{\prod_{r\neq l}^{N_T}(\Omega_l - \Omega_r)} \frac{\varepsilon}{\sqrt{2\pi}\sigma_f v} e^{-\frac{(10\log v - 10\log\Omega_l - \mu_f)^2}{2\sigma_f^2}} \tag{7.8}$$

where $\Omega_l = E(|\alpha_l|^2)$, $\mu_f = -\varepsilon C + \mu_X$, and $\sigma_f^2 = \varepsilon^2\zeta(2.2) + \sigma_X^2$. $C = 0.5772$ is the Euler constant and $\zeta(2.2) = \pi^2/6$ is the Riemann-Zeta function.

The probability $Prob(SIR_i > SIR_j)$ is then given by:

$$Prob(SIR_i > SIR_j) = \sum_{l_i=1}^{N_T}\sum_{l_j=1}^{N_T} \frac{(\Omega_{l_i,i})^{N_T-2}}{\prod_{r_i\neq l_i}^{N_T}(\Omega_{l_i,i} - \Omega_{r_i,i})} \frac{(\Omega_{l_j,j})^{N_T-2}}{\prod_{r_j\neq l_j}^{N_T}(\Omega_{l_j,j} - \Omega_{r_j,j})} \times \left[Q\left(-\frac{(10\log\Omega_{l_i,i} + \mu_{f,i} - 10\log\Omega_{l_j,j} - \mu_{f,j})}{\sqrt{\sigma_{f,i}^2 + \sigma_{f,j}^2}}\right)\right] \tag{7.9}$$

Consequently, the bit rate of user i is given by:

$$R_i = \Pr(i)\sum_{CQI} \frac{R_{CQI}\, p_{CQI,i}}{N_s} = \Pr(i)\sum_{CQI} \frac{TBS_{CQI}}{TTI_{\text{delay}}} \frac{p_{CQI,i}}{N_s} \tag{7.10}$$

where R_{CQI} is the instantaneous bit rate referring to CQI (channel quality indicator), and $p_{CQI,i}$ is the (discrete) probability given by Equation 7.37 referring to CQI in the position of user i, N_s is the number of corresponding HARQ transmissions, TBS_{CQI} is the transport block size referring to CQI according to tables correspondence [3] versus terminal category, and TTI_{delay} is the transmit time interval [equal to 2 ms in the case of HSDPA (high-speed downlink packet access)].

In the case of correlated Rayleigh fading, the user bit rate can be estimated using the same method described earlier (i.e., in the case of uncorrelated Rayleigh fading). Consequently, the probability that the channel is allocated to user i is given by:

$$
\begin{aligned}
Pr(i) &= Prob(SINR_i > SINR_j \text{ for } j = 1 .. N_u \text{ and } j \neq i) \\
&= \prod_{j \neq i}^{N_u} Prob(SINR_i > SINR_j) \\
&= \prod_{j \neq i}^{N_u} \frac{1}{\prod_{l_i=1}^{N_T} \lambda_{l_i,i}} \frac{1}{\prod_{l_j=1}^{N_T} \lambda_{l_j,j}} \sum_{l_i=1}^{N_T} \sum_{l_j=1}^{N_T} \\
&\quad \frac{1}{\prod_{r_i \neq l_i}^{N_T} \left(\frac{1}{\lambda_{r_i,i}} - \frac{1}{\lambda_{l_i,i}}\right)} \frac{1}{\prod_{r_j \neq l_j}^{N_T} \left(\frac{1}{\lambda_{r_j,j}} - \frac{1}{\lambda_{l_j,j}}\right)} \\
&\quad \times \left[Q\left(-\frac{(10 \log \lambda_{l_i,i} + \mu_{f,i} - 10 \log \lambda_{l_j,j} - \mu_{f,j})}{\sqrt{\sigma_{f,i}^2 + \sigma_{f,j}^2}} \right) \right]
\end{aligned}
\tag{7.11}
$$

where λ_l, $l = 1, \ldots, N_T$ are the eigenvalues of the matrix DC. D and C are the $N_T \times N_T$ path power and covariance matrices given by:

$$
D = \begin{pmatrix} \Omega_1 & 0 & \cdots & 0 \\ 0 & \Omega_2 & \cdots & 0 \\ \cdot & \cdot & \cdot & \cdot \\ \cdot & \cdot & \cdot & \cdot \\ 0 & 0 & \cdots & \Omega_{N_T} \end{pmatrix} \tag{7.12}
$$

$$
C = \begin{pmatrix} 1 & \sqrt{\rho_{12}} & \cdots & \sqrt{\rho_{1N_T}} \\ \sqrt{\rho_{21}} & 1 & \cdots & \sqrt{\rho_{2N_T}} \\ \cdot & \cdot & \cdot & \cdot \\ \cdot & \cdot & \cdot & \cdot \\ \sqrt{\rho_{N_T1}} & \sqrt{\rho_{N_T2}} & \cdots & 1 \end{pmatrix} \tag{7.13}
$$

where $\rho_{l,l'}$ is the correlation parameter between the paths l and l'.

Consequently, the bit rate of user i is given by Equation 7.10 where $Pr(i)$ is given by Equation 7.11.

7.5.4 Fair Channel-Dependent Scheduling (FCDS) Protocol

The fair channel-dependent scheduling (FCDS) protocol introduced in [8–10] forms a trade-off between the two other extremes: low power use (system capacity) and fairness. In practice, the signal is fluctuating around a mean value that displays slow trends as well. This underlying slow fluctuation accounts for the distance from the base station.

The time scale of the so-called fading variations in the signal itself, due to multipath reception and/or shadow fading, is much smaller than that of the variations of local mean. The scheduling is done on the basis of the relative power (i.e., the instantaneous power relative to its own recent history). So, the transmission level of all mobile terminals is first translated with respect to their local means, and subsequently normalized with their local standard deviations. A transmission is scheduled to the UE that has the lowest value for the relative power.

The idea of a relative power, introduced earlier, needs the definition of a local mean, with *local* referring to the recent time history. Exponential smoothing weighs past observations with exponentially decreasing weights in order to update the value for the local mean. It takes the local mean of the previous period and adjusts it up or down based on what actually occurred in that period. By the choice of a weighting factor, this procedure can be made sensitive to a small or gradual drift in the process. This method is simple and therefore has low data storage requirements and data processing since only the actual (instantaneous) value and the old local mean value are needed to update the new local mean value. Comparing with, for example, moving averaging, the low storage and higher weights on more recent samples are two properties in favor of the FCDS method. The performance of these algorithms is considered with the parameters presented in Table 7.2.

Note that t either refers to the physical time unit or the corresponding integer index.

The local mean, introduced earlier, as well as the variance, are updated with each time unit according to the following algorithm [11]:

$$\begin{cases} \mu_t &= \alpha_1 \cdot P_t + (1 - \alpha_1) \cdot \mu_{t-1} \\ v_t &= \alpha_2 \cdot (P_t - \mu_t)^2 + (1 - \alpha_2) \cdot v_{t-1} \end{cases} \tag{7.14}$$

In other words, the new local mean (or variance) is a weighted average of the instantaneous contribution and the old mean (or variance). In the rest of this study, the parameter α refers to both α_1 and α_2 when not specified any further. As we

Table 7.2 Variables Used in FCDS

P_t	(Instantaneous) transmission power at time t
μ_t	Local mean of P_t based on the time interval $[t_0,t]$
v_t	Local variance of P_t based on the time interval $[t_0,t]$
σ_t	Local standard deviation, defined by $\sigma_t^2 = v_t$
α_1	Smoothing coefficient w.r.t. the local mean
α_2	Smoothing coefficient w.r.t. the local variance

will see later, extreme values of α (0, respectively 1) lead to the extreme variants in scheduling: C/I-based scheduling, respectively "Round Robin" scheduling.

The criterion that determines the optimal mobile node for the next downlink transmission at time t is formulated as follows, where the superscript i is used to denote the situation at node i:

$$\min_i \left\{ \left(P_t^i - \mu_t^i\right) / \sqrt{v_t^i} \right\} \tag{7.15}$$

Here, the data point, P_t, is translated with respect to the local mean, μ_t, and is next normalized with the corresponding local standard deviation, σ_t. This subsequently translated and normalized transmission power is referred to as the scaled power, P_s. At each time t, the value for P_s is compared for all nodes i and the most optimal node is selected for the downlink transmission.

7.5.5 Score-based Scheduling

The score-based (SB) scheduler, proposed by Bonald in [11], consists of allocating the channel to the user having the maximum transmission rate relative to its past rate statistics [6]. This algorithm can be explained as follows: Let us consider a HSPA system with two active users. Let $r_{1,v}$ and $r_{2,v}$ where $v = 1, \ldots, n$ be the past transmission rates for each user (even when the TTI is not attributed to this user) observed over a window size n. The idea is to classify the past rates of each user in decreasing order and to give a rank for each rate (e.g., rank 1 for the highest rate). During the TTI $n + 1$, if the possible rate of user 1 $r_{1,n+1}$ is classified in rank 1 relative to his own rate statistics and the rate of user 2 $r_{2,n+1}$ is classified in rank 3 relative to his own rate statistics, in this case the channel is allocated to user 1 even if $r_{2,n+1} > r_{1,n+1}$. This algorithm has the advantage of not suffering from asymmetric fading and data rate constraints, which is not the case with the proportional fair algorithm.

7.6 New Optimized Scheduling Techniques for Multiple Services Case

By using multiple services having different characteristics (some requiring a guaranteed bit rate such as streaming, voice, and so on at the opposite of interactive services), new scheduling protocols should be used. The following scheduling methods adapted for the multiple services case were first introduced by A. Masmoudi et al. [12] and Nasser and Bejaoui [13], and implemented in NS-2 *Eurane* simulators and tested with multiple services. For some of them, the cross-layer design is applied to optimize the use of bandwidth and improve their performance.

7.6.1 Concept of Cross-Layer Design

In next-generation wireless networks, the mechanisms and protocols at the different layers in the protocol stack will interact dynamically with each other to provide guaranteed services. This is central to cross-layer design. The cross-layer design concept (cf. Figure 7.5) does away with the rigid structure of the layered protocol architecture that has served extremely well in the development and implementation of both past and current communication systems. Such architectures, for which each layer is responsible to serve only the higher layer, had the advantage to exhibit a high degree of modularity, which allows an easy replacement and theoretically an arbitrary combination of protocols.

In cross-layer architecture, parameters have to be exchanged interactively to overcome the robust characteristics and constraints of multimedia traffic and wireless

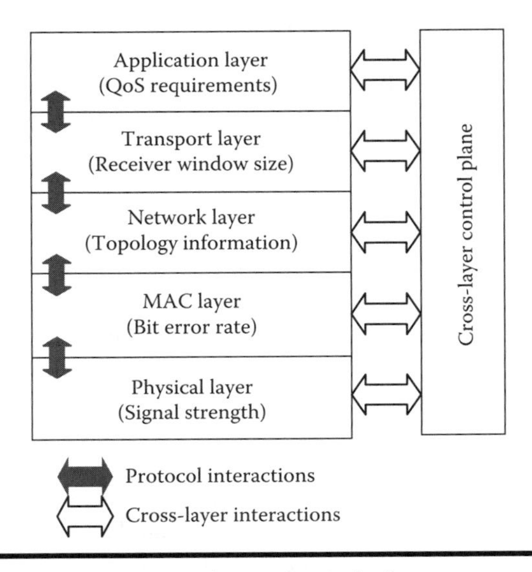

Figure 7.5 The general concept of cross-layer design.

networks. Thus, active exploration of the various synergies of exchanging information between lower layers and the upper layer is emphasized. Their main purpose is to help improve the end-to-end performance given network resources by meeting high data rates, higher performance gains, and QoS requirements for various real-time and non-real-time applications.

However, to preserve the layered structure as much as possible, the interlayer dependencies introduced by the cross-layer design should be kept to a minimum. Thus, cross-layering should be viewed as an enhancement or a complement, not an alternative, to layered design. Its ultimate goal is then to preserve the key characteristics of a layered architecture and to allow for performance improvements.

In practice, the purpose of using the cross-layer design is either to optimize the use of bandwidth through RRM algorithms in cellular networks, as in the wireless local area networks (WLANs), or to enhance the path selection process and achieve minimum energy consumption in ad-hoc and wireless sensor networks.

For HSPA, the cross-layer design is applied to improve the performance of scheduling techniques and to achieve bit-level, packet-level, and call-level QoS. To achieve a superior performance, the scheduling and control mechanisms can benefit from information coming from different layers as from the PHY one about the state of the wireless channel.

7.6.2 Description of the Introduced Techniques (Protocols)

7.6.2.1 Prioritized Differentiated Services Scheduling

This scheduling method differentiates between services requiring guaranteed bit rate (such as CBR services, video streaming, voice, etc.) and non-guaranteed bit rate ones (such as interactive Web browsing, FTP, email, etc.). It introduces a parameter specifying the maximum priority degree for which service is considered as a guaranteed bit rate and are scheduled at a Fair Throughput manner, while the remaining others (non-guaranteed bit rate flows) are scheduled at Fair Time fashion. Priority degrees among all services are taken into account.

7.6.2.2 Prioritized Rayleigh Peak Scheduling

This method attempts to schedule guaranteed bit rate services first only in Rayleigh peak instances [i.e., if their quality is good enough or, in terms of implementation, provided that their CQI is greater than or equal to a specified CQI threshold value). In this case, Fair Throughput scheduling is applied to guaranteed bit rate flows having an acceptable CQI value. Otherwise, if CQI is less than this CQI threshold, flows are treated at Fair Resource scheduling.

The basic Rayleigh peak scheduling protocol as simulated has the same concept as the Proportional Fair scheduling (Proportional Fair Throughput or Proportional Fair Resource) [5] except by applying another method consisting of a specified CQI threshold to schedule and serve mobiles using the guaranteed bit rate services so as to disadvantage the packets whose channel is suffering from deep Rayleigh peak fading. In prioritized Rayleigh peak scheduling, we apply the same principle as the Rayleigh peak scheduling while complying with the different services differentiation priority degrees.

7.6.2.3 Weighted Differentiated Services Scheduling

It is similar to the prioritized differentiated services scheduling except that instead of taking an absolute priority degree for the guaranteed bit rate services, and in order to increase the resources reserved to non-guaranteed bit rate services, this method assigns weights both to guaranteed and non-guaranteed bit rate services to balance the bandwidth dedicated for each of them being scheduled respectively in Fair Throughput and Fair Resource manners.

This method thus consists in differentiating guaranteed bit rate services to non-guaranteed bit rate ones through two weight coefficients for each one in order to keep a fixed proportion of radio resources guaranteed for NRT services often disadvantaged in terms of priority degrees in both previous protocols.

7.6.2.4 Weighted Rayleigh Peak Scheduling

It has the same principle as the previous method but adds the condition for CQI to be above a pre-determined threshold (at Rayleigh peaks) for guaranteed bit rate flows before deciding to serve them according to Fair Throughput scheduling. The aim of this method is to avoid wasting resources and dedicating them to bad TTI links (especially at Fair Throughput). If CQI is less than the threshold, non-guaranteed bit rate flows are scheduled at Fair Resource fashion.

Hence, it is a hybrid method (between Fair Throughput and Fair Resource) according to the service type, while complying with a cyclic weighted allocation between RT and NRT services. It has the same basis as the weighted differentiated services scheduling while applying the CQI threshold rule as defined in the Rayleigh peak scheduling.

7.6.2.5 ThreshOld-based Priority (TOP) Scheduling Algorithm

Only a few cross-layered resource allocation protocols have been proposed for the HSPA system. The most pioneering of such cross-layered protocols was developed by Nasser and Bejaoui [13]. This scheduling protocol that aims at enhancing the average cell throughput is called threshold-based priority (TOP) and serves users to access packets in the downlink. It is proposed to provide priority scheduling between non-real-time services of different QoS classes and fairness between users within the same class. Within each queue allowed respectively to the non-real-time interactive (class-1) and background (class-2) traffic classes, packet data units (PDUs) are supposed to be waiting for service in a first-in-first-out manner, and the service discipline is considered to be preemptive resume. According to the TOP scheme, the service is made by considering two steps. The first one is dedicated to the selection of a class to be served. Thus, after a specific user is served, the next user to be selected for service is a class-2 user if the number of PDUs of class-2 in buffer C denoted m_2 is greater than the threshold T_2 and m_1 of those of class-1, less than the threshold T_1. Otherwise, service is allocated to a class-1 user if present in the system. The second step is for the selection of which user of the selected class to be served. In the selected class, the user with the highest priority is selected for transmission where the priority for user i at time t is calculated as follows:

$$P_i(t) = \begin{cases} CQI & \text{if } S_i(t) \geq R \\ CQI * W & \text{otherwise} \end{cases} \tag{7.16}$$

where CQI is the channel quality indicator for user i that represents the current channel condition for this user, $S_i(t)$ average throughput for user i up to time t, R is the predefined minimum throughput (e.g., 64 kbps) and $W = R/S_i(t)$. TOP prioritizes users based on their radio channel quality. However, it increases the priority of those with average throughput below a certain threshold by W and hence increases their chance of getting served.

The CQI—used to determine the rate at which the user can support from the NodeB—is mapped using the signal-to-noise ratio (SNR)* according to the following equation [3]:

$$CQI = \begin{cases} 0 & SNR \leq 16 \\ \left\lfloor \dfrac{SNR}{1,02} + 16, 62 \right\rfloor & -16 < SNR < 14 \\ 30 & 14 \leq SNR \end{cases} \tag{7.17}$$

The choice of the channel model to be considered is important in evaluating the effectiveness of the protocol. It describes how much the radio signal attenuates on its way from the NodeB to the user and therefore describes how the channel condition of the user changes with time depending on the environment of the user and his speed. The propagation model used in the performance evaluation of TOP consists of five parts: path loss, shadowing, multipath fading, intra-cell interference, and inter-cell interference. The path loss is calculated as follows:

$$L(d) = 137, 4 + 10 \cdot \beta \log_{10}(d) \tag{7.18}$$

where d is the distance from the UE to the NodeB in kilometers, β is the path loss exponent and is equal to 3.52. Shadowing is modeled through a log-normal distribution and a correlation distance with a mean value of 0 dB. The multipath fading

* Signal-to-noise ratio (SNR) is the signal strength relative to background noise.

corresponds to 3GPP channel models for pedestrian and vehicle A environments. The intra-cell and inter-cell interferences are assumed to be constants and are set equal to 30 dBm and −70 dBm, respectively. Then at the user side, the SNR is extracted from the received signal from the NodeB to determine how strong the signal is according to the following formula:

$$\begin{aligned} SNR &= P_{Tx} - L_{\text{Total}} - 10 \cdot \log_{10}\left(10^{\frac{I_{\text{intra}} - L_{\text{Total}}}{10}} + 10^{\frac{I_{\text{inter}}}{10}}\right) \\ &= P_{Tx} - 10 \cdot \log_{10}\left(10^{\frac{I_{\text{intra}}}{10}} + 10^{\frac{I_{\text{inter}} + L_{\text{Total}}}{10}}\right) \end{aligned} \tag{7.19}$$

where P_{Tx} is the transmitted code power in dBm, L_{Total} is the sum of the path loss, shadowing, and multipath fading in dB, I_{intra} and I_{inter} are the inter-cell and intra-cell interferences, respectively, in dBm.

7.6.3 Scheduling Techniques Optimization According Services Profiles and Requirements

7.6.3.1 Prioritized Differentiated Services Scheduling Technique

Consider two services 1 and 2 with respective nominal guaranteed bit rates $R_{\min_1}$ and $R_{\min_2}$ and with respective served number of users N_1 (referring to transport block sizes TBS_i and to number of codes n_i) and N_2 (referring to transport block sizes TBS_i' and to number of codes n_i') since the cell size is already fixed. Some of the TBS_is and the TBS_i's can be equal if the traffic is not uniform. Service 1 is assumed to have the priority over service 2, and each of the services thus uses the Fair Throughput. Assuming that both nominal services bit rate $R_{\min_1}$ and $R_{\min_2}$ are CBR type, having the highest priority degrees versus the other services types, using the Fair Throughput protocol, and assuming N_1 and N_2 as their respective number of users, then the number of available codes n' remaining for the CBR service with priority degree 2 is:

$$n' = \max\left\{0;\ 15 - R_{\min_1} TTI_{\text{delay}} \cdot \sum_{j=1}^{N_1} \frac{n_j}{TBS_j}\right\} \tag{7.20}$$

where n_j is the number of codes referring—according to Table 6.1 from 3GPP standard [3]—to the CQI of user j with service 1 (in the cell), and whose appropriate transport block size is TBS_j. The expression (Equation 7.20) is valid with or without codes multiplexing.

The maximum bit rate ensured by multiplexing all the available HSPA codes (15 codes with spreading factor 16) is that given by Equation 7.20 by applying the Fair Throughput protocol only to the users of the CBR service with priority degree 1.

The bit rate $R_{\min_1}$ isn't reached by the users except if it doesn't exceed the maximum bit rate. However, if

$$R_{\min_1} \geq \frac{15}{TTI_{\text{delay}} \sum_{j=1}^{N_1} \frac{n_j}{TBS_j}},$$

then the number n' of the remaining codes for the service 2 is null, and the guaranteed bit rate for the CBR service with priority degree 1 is lower than the required bit rate $R_{\min_1}$ (while using all the 15 available codes). So we should reduce the cell size until having a minimum bit rate equal to $R_{\min_1}$.

For the service with priority degree 2, the maximum ensured bit rate per user (by using the n' remaining codes out of the radio resources) can be determined like in Equation 6.20 according to:

$$R'_{\text{ens}} = \frac{n'}{TTI_{\text{delay}} \cdot \sum_{j=1}^{N_2} \frac{n'_j}{TBS'_j}} \tag{7.21}$$

where n'_j and TBS'_j are respectively the number of codes and the transport block size referring, according to the terminal category (from 3GPP standard [3], cf. Table 7.1 for terminal category 10) to the CQI of user j with service 2.

The number of available codes remaining for NRT services can be determined with the same manner as in Equation 7.20 as follows:

$$n'' = \max\left\{0;\ n' - R_{\min_2} TTI_{\text{delay}} \cdot \sum_{j=1}^{N_2} \frac{n'_j}{TBS'_j}\right\} \tag{7.22}$$

with the same notations for n'_j and TBS'_j as in Equation 7.20 while considering the users of service with priority degree 2.

By generalizing, the bit rate per user i of the NRT services (the remaining resources and bandwidth being equally shared among the users) is provided, by applying the Fair Resource scheduling protocol, as in Equation 7.1, by:

$$R''_i = \begin{cases} \min\left(\frac{TBS''_i}{TTI_{\text{delay}}}, \frac{TBS''_i}{TTI_{\text{delay}}} \frac{n''_{\text{cap}}}{n''_{\text{Tot}}}\right) & \text{(without codes multiplexing)} \\ \min\left(\frac{TBS''_i}{TTI_{\text{delay}}}, \frac{TBS''_i}{TTI_{\text{delay}}} \frac{n''_{\text{cap}}}{n''_{\text{Tot}}}\right) \cdot \max\left(\frac{n''}{\sum_i n''_i}, 1\right) & \text{(with codes multiplexing)} \end{cases} \tag{7.23}$$

where n''_i and TBS''_i are respectively the number of codes and the transport block size referring to the CQI of the NRT user i according to the adequate terminal category from the standard [3] (cf. Table 7.1 for category 10), n''_{Tot} is the total number of NRT users in the served cell, and n''_{cap} is the number of users included in the cell whose size is capacity-limited (the number of codes n'' allocated to it is exactly equal to that remaining for NRT services).

So the guaranteed bit rate R''_{ens} for the users of interactive services is that present at the cell border (the smallest TBS_{i} size). It is given similarly to Equation 7.2 by:

$$R''_{\text{ens}} = \begin{cases} \min_i \left(\dfrac{TBS''_i}{TTI_{\text{delay}}}\right) \cdot \min\left(\dfrac{n''_{\text{cap}}}{n''_{\text{Tot}}}, 1\right) & \text{(without codes multiplexing)} \\[2ex] \min_i \left(\dfrac{TBS''_i}{TTI_{\text{delay}}}\right) \cdot \min\left(\dfrac{n''_{\text{cap}}}{n''_{\text{Tot}}}, 1\right) \cdot \max\left(\dfrac{n''}{\sum_i n''_i}, 1\right) & \text{(with codes multiplexing)} \end{cases} \tag{7.24}$$

R''_{ens} refers to the guaranteed bit rate at the cell border. The summation in each of the expressions (Equations 7.23 and 7.24) applies to users of all the NRT services in the cell.

Therefore, the fact of specifying a minimum bit rate for all the users limits the maximum cell size (or the allowed maximum path loss), thus the coverage of NRT services in HSPA translates the fact that the bit rate in each point of the cell is above a threshold, whereas their capacity is according to the "Best Effort" policy (with a fairness among the users according to the quality of their link and the number of HSPA codes available for NRT services).

7.6.3.2 Weighted Differentiated Services Scheduling Technique: Optimization According to Services Profiles and Requirements

The objective of this paragraph is to optimize the weighting coefficients of the "Weighted Differentiated Services Scheduling" protocol (called also "Weighted Round Robin" or WRR) in order to adapt it according to the different profiles of multiple services (NRT or CBR, etc.).

Some NRT services consume much radio resources often at the expense of other services in the multiservice case. This is mainly due to their high edge bit rate occupying the channel if only for a short time (such as the FTP service which hardly leaves enough bandwidth for the other services with lower edge bit rate such as the "Web Browsing" service). The solution is to balance the packets scheduling versus the edge bit rate and the load of each service such that the users have an equal probability to serve the minimum bit rate of each service. This solution consists in assigning to the user of a given service i a weighting coefficient $w_1^{(i)}$ taking into account the effective target load (the minimum) versus the edge bit rate [14]. The expression of this weight coefficient is given by the following equation:

$$w_1^{(i)} = \frac{\rho^{(i)} \frac{R_{\min}^{(i)}}{R_{max}^{(i)}}}{\sum_{k=1}^{s} \rho^{(k)} \frac{R_{\min}^{(k)}}{R_{\max}^{(k)}}} \tag{7.25}$$

where s is the number of NRT services with respective edge bit rates $R_{\max}^{(2.1)}$, $R_{\max}^{(2.2)}, \ldots, R_{\max}^{(s)}$ (maximum source bit rates), $R_{\min}^{(2.1)}, R_{\min}^{(2.2)}, \ldots, R_{\min}^{(s)}$ are the respective required minimum bit rates, and $\rho^{(i)}$ is the planned or effective number of users of service i (per area unit). Hence the service weight takes into account the real need of this service while avoiding the bad impact of the high edge bit rate services considered as great consumers of radio resources. This weight coefficient selected according to the "Best Effort" strategy to balance the load of the different services—is suitable if all the services are NRT or interactive. Weighting the services differently is the basis of this scheduling protocol.

On the other hand, if services are all RT (or having a required guaranteed bit rate for their QoS), we have seen that the Fair Throughput protocol is the most adapted in this case since it tries to guarantee a constant bit rate for all the users. This protocol is equivalent to assigning high weights to the users far from the NodeB or having the worst CQI values and weights lower than the most advantaged ones in terms of quality and position. More formally, and according to Section 7.5.2 earlier about Fair Throughput protocol, it is as if we use the weighted differentiated services scheduling protocol with a weight associated to each of the users inversely proportional to its referring transport block size TBS_i (i.e., the weight $w_{21}^{(i)}$ associated to the user with block size TBS_i is:

$$w_{21}^{(i)} = \frac{\frac{1}{TBS_i}}{\sum_{k=1}^{N} \frac{1}{TBS_k}} \tag{7.26}$$

where N is the number of mobiles using HSPA within the cell, and TBS_k is the block size referring to user k. The expression (Equation 7.26) is valid if the bit rate required by the RT service is above the minimum bit rate guaranteed by the Fair Throughput for each user independently of the available number of codes (given by Equation 7.19). In contrast, if the required bit rate is above it, then all the users will be satisfied without time multiplexing according to the Fair Throughput protocol or another technique.

Yet, the different types of RT services shouldn't have the same priority degrees. In fact, the voice service for example, should have more priority than streaming services due to the importance of this service and in order to maximize the number of served users. Moreover, if we choose to optimize the weights assigned to RT services so as to maximize the number of served users, we fall on exhaustive (hard) priority degrees avoiding the services requiring a more guaranteed bit rate. Otherwise, the resources won't be assigned to these last services (like the streaming) except if all the more priority services (such as the voice) are satisfied and if HSPA resources are available. This exhaustive priority may not satisfy low priority services or not provide them the required bit rates. For that, we give non-exhaustive weights for the priority degrees in order to obtain a compromise between the priorities required by different services and the satisfaction of users from different services profiles. As an example, we can choose uniform weights as follows: Given s RT services with decreasing priority degrees 1, 2, ..., s, then the weight $w_{22}^{(i)}$ associated to service with priority degree i is:

$$w_{22}^{(i)} = \frac{s+1-i}{\left(\frac{s(s+1)}{2}\right)} = \frac{2(s+1-i)}{s(s+1)} \tag{7.27}$$

So the global weight $w_2^{(i)}$ (RT service) will be:

$$w_2^{(i)} = \frac{w_{21}^{(i)} w_{22}^{(i)}}{\sum_{k=1}^{s} w_{21}^{(k)} w_{22}^{(k)}} \tag{7.28}$$

Moreover, let's now examine the case of aggregated services with different natures (at guaranteed and non-guaranteed bit rates: NRT, CBR, etc.). In fact, if there are together s_1 RT services and s_2 NRT services, the more natural is to assign the RT services (at guaranteed required bit rates) priorities higher than NRT ones by giving favor to those whose required bit rate is the lowest (such as voice). For NRT services, interactive services are assigned priorities higher than those in background, then these priority degrees are converted into weights $w_{22}^{(i)}$ according to a form function (uniform as in Equation 7.27 or other law). For weighting coefficients $w_1^{(i)}$, they are calculated as for RT services in Equation 7.25 except by including also in the summation of the denominator the terms of RT services. For these latter (e.g., CBR services), the traffic source transmit packets at the same cadence, and the nominal required bit rate is equal to the maximum edge bit rate from the source. Hence the coefficients $w_1^{(i)}$ are obtained by Equation 7.25 by assuming the equal bit rates $R_{\min}^{(i)}$ and $R_{\max}^{(i)}$ of RT services. Consequently, in the multiservice case (RT and NRT), the global weight coefficient $w^{(i)}$ associated to a NRT service i will be:

$$w^{(i)} = \frac{w_1^{(i)} w_{21}^{(i)} w_{22}^{(i)}}{\sum_{k=1}^{s_1+s_2} w_1^{(k)} w_{21}^{(k)} w_{22}^{(k)}} \tag{7.29}$$

where $w_{22}^{(i)} = \frac{1}{s_2}$ for all NRT services since we decided to apply the Fair Resource protocol for this type of service. In contrast, for RT services, we will also take into account coefficients $w_{21}^{(i)}$ from Equation 7.26 applied only to RT users in order to obtain—as possible—equal bit rates for all these users not exceeding the source bit rate (partial Fair Throughput), so the global coefficient associated to a RT service i will be given by Equation 7.29 except by assuming coefficients $w_{21}^{(i)}$ according to Equation 7.26.

For the global weighting coefficient of RT services users (CBR at guaranteed bit rates), we can replace the coefficient $w_{22}^{(i)}$ referring to priority degrees (given by Equation 7.27) by a coefficient proportional to the required bit rate of the concerned CBR service, as follows:

$$w_{22}^{(i)} = \frac{R_{\min}^{(i)}}{\sum_{k=1}^{s_1} R_{\min}^{(k)}} \tag{7.30}$$

where $R_{\min}^{(i)}$ is the required bit rate of service i, and s_1 is the number of CBR services at guaranteed bit rates. We can also define $w_{22}^{(i)}$ as inversely proportional to the required bit rate of the concerned CBR service i.

The coefficient $w_1^{(i)}$ can be ignored (taken equal to 1) in the case of guaranteed bit rate services since the load or the area density of users is already included in the coefficient expression $w_{21}^{(i)}$ (according to Equation 7.26) referring to the bit rate balancing according to Fair Throughput protocol. For the users of NRT services, we can either take into account the priority degrees of different services in the coefficient $w_{22}^{(i)}$ or introduce some fairness among the users (elimination of priority levels): the global coefficient $w^{(i)}$ becomes thus, in this case, equal to $w_1^{(i)}$. The expression of the global coefficient $w^{(i)}$ is written therefore according to Equation 7.29, except with the sub-coefficients in Table 7.3. In the notations of Table 7.3, two different users of the same service i have the same weight coefficients.

The combination of the sub-coefficients according to Equation 7.29 has the disadvantage of giving the same importance to the different factors (materialized by

Table 7.3 Values of Weighting Sub-coefficients for Multiple Services (Different Profiles)

	Service of User i Is at Guaranteed Bit Rate	*Service of User i Is at Non-guaranteed Bit Rate*	
$w_1^{(i)}$	Constant equal to 1	Equation 7.25	
$w_{21}^{(i)}$	Equation 7.26	Constant equal to $1/s_2$	
$w_{22}^{(i)}$	Equation 7.30	Constant equal to $1/s_2$	if fairness among users
		Equation 7.27	else (taking priority degrees)

the sub-coefficients) so as to result in the global weight coefficient. In order to take into account the weight dedicated to each factor (load, priority, etc.), we can rewrite the global weight coefficient linearly as follows (instead of Equation 7.29):

$$w^{(i)} = \alpha_1 w_1^{(i)} + \alpha_{21} w_{21}^{(i)} + \alpha_{22} w_{22}^{(i)} \tag{7.31}$$

where α_1, α_{21}, and α_{22} are the respective weights referring to sub-coefficients $w_1^{(i)}$, $w_{21}^{(i)}$, and $w_{22}^{(i)}$. Thus, we can give more importance to any of the factors influencing services differentiation by decreasing the weights referring to negligible criteria, for example, in order not to take into account the priorities within the coefficient $w_{22}^{(i)}$ referring to the users of non-guaranteed bit rate services, we can take a null coefficient α_{22}, etc.

In the case of the WRR protocol, to determine the maximum bit rate R_{ass} ensured for each of the WRR users independently of the number of available HSPA physical channels, we proceed similar to the Fair Throughput protocol. In order to determine the maximum bit rate ensured per user in WRR (by using the totality of the available HSPA codes for this scheduling protocol, WRR is studied in this paragraph), we find similarly as in the Fair Throughput protocol that:

$$(R_{\text{ens}})_{WRR} = \frac{15 \cdot TBS_i \cdot w^{(i)}}{TTI_{\text{delay}} \cdot \sum_j n_j \cdot w^{(j)}} \tag{7.32}$$

Note that the index j in the summation of Equation 7.32 refers to the user and not to the service as for the other notations of weight coefficient indexes in this paragraph. Thus, two different users i and j of the same service should have the same weights ($w^{(i)} = w^{(j)}$).

7.7 Summary and Open Problems

HSPA and LTE have been introduced in the new UMTS standard to provide high data rate services in wireless cellular networks. In this chapter, some scheduling protocols of literature used in HSPA- and LTE-based networks are presented. Since aggregation of multiple services is so important, this chapter also provides some optimized scheduling techniques adapted for the multiservice case as well as an overview of the optimization of the cross-layer architecture design.

As future problems are studied, comparison between scheduling protocols should be made on the basis of which one(s) are better for each service profile or requirements. Performance can be evaluated in terms of throughput, delay, and retransmissions. The impact of intra-cellular interference and "multiuser" diversity on the total HSPA capacity can also be studied and examined as open issues.

References

[1] 3GPP TS 36300 900, "Evolved Universal Terrestrial radio Access (E-UTRA) and Evolved Universal Terrestrial Radio Access Network (E-UTRAN); Overall description, stage 2, Release 9 (V9.0.0), June 2009.

[2] 3GPP TS 25.308, "High Speed Downlink Packet Access (HSDPA); Overall Description," Release 5, March 2003.

[3] T. E. Kolding, K. Pedersen, J. Wigard, F. Frederiksen, and P. E. Mogensen, "Performance Aspects of WCDMA Systems with High Speed Downlink Packet Access (HSDPA)," VTC, vol. 1, pp. 477–481, September 2002, Online: http://nds2.ir.nokia.com/downloads/.

[4] 3GPP TS 25.214, "Physical Layer Procedures (FDD)," 3GPP Release 6, December 2003.

[5] M. Assaad, "Cross Layer Study in HSDPA System," Ph.D. thesis, Ecole Nationale des Télécommunications, Paris, France, March 2006.

[6] K. S. Gilhousen, I. M. Jacobs, R. Padovani, A. J. Viterbi, L. A. Weaver, and C. E. Wheatley, "On the Capacity of a Cellular CDMA System," *Vehicular Technology, IEEE Transactions on*, vol. 40 Issue: 2, May 1991.

[7] I.C.C. de Bruin, G. Heijenk, M. El Zarki, and J. Lei Zan, "Fair Channel-Dependent Scheduling in CDMA Systems," 12th IST Summit on Mobile and Wireless Communications Summit 2003, Aveiro, Portugal, June 15–18, 2003, pp. 737–741.

[8] L. Zan, G. Heijenk, and M. El Zarki, "Fair and Power-Efficient Channel-Dependent Scheduling for CDMA Packet Networks," *Proceedings of International Conference on Wireless Networks (ICWN)*, June 2003, URL: http://www.home.cs.utwente.nl/heijenk/.

[9] L. Zan, G. Heijenk, and M. El Zarki, "A Real-Time Traffic Scheduling Algorithm in CDMA Packet Networks," *Proceedings of 14th IEEE International Symposium on Personal, Indoor, and Mobile Radio Communications PIMRC 2003*, September 2003, Beijing, China.

[10] S. W. Roberts, "Control Chart Test Based on Geometric Moving Averages," *Technometrics*, 1, 1959, pp. 239–250.

[11] Thomas bonald, "A Score-based Opportunistic Scheduler for Fading Radio Channels," *Proceedings of European Wireless*, 2004.

[12] A. Masmoudi, D. Zeghlache, and S. Tabbane, "Resource and Scheduling Optimization in HSDPA-based UMTS Networks," *Proceedings of World Wireless Congress (WWC 2005)*, San Francisco, California, May, 24–27, 2005.

[13] N. Nasser and T. Bejaoui, "User Satisfaction-based Scheduling Algorithm for High-Speed Wireless Networks," ACM International Wireless Communications and Mobile Computing Conference (IWCMC), Honolulu, Hawaii, August 2007, pp. 164–169.

[14] T. Bonald and A. Proutière, "Wireless Downlink Data Channels: User Performance and Cell Dimensioning," MobiCom'03, San Diego, California, September 14–19, 20.

Chapter 8

Teletraffic Engineering for HSDPA and HSUPA Cells

Maciej Stasiak, Piotr Zwierzykowski, and Mariusz Głąbowski

Contents

8.1 Introduction

The increasing popularity of data transfer services in mobile networks of the second and the third generations has been followed by an increasing interest in methods for the dimensioning and optimization of networks servicing multirate traffic. In traffic theory, these issues are in full swing. The problems concern primarily the special conditions of constructing the mobile networks, and the infrastucture of the radio access network—as its development, or extension, needs a precise definition and assessment of clients' needs and is relatively time-consuming. Cellular network operators define, on the basis of a service level agreement (SLA), a set of key performance indicator (KPI) parameters that serve as determinants in the process of network dimensioning and optimization. Dimensioning can be presented as an unending and ongoing process of analyzing and designing the network. To make this work effective, it is thus necessary to work out algorithms that would, in a reliable way, model the parameters of a designed network.

The dimensioning process for the third-generation Universal Mobile Telecommunications System (UMTS) should make it possible to determine such a capacity of individual elements of the system that will secure, with the assumed load of the system, a pre-defined level of grade of service (GoS). With the dimensioning of the UMTS system servicing R99 and HSPA traffic, the most characteristic constraints are the radio interface and the Iub interface.

Analytical modeling of radio and Iub interfaces is based on the assumption that a full-availability group carrying multirate traffic can be used as the fundamental traffic engineering model for those interfaces (i.e., [1–7]). The model of the full-availability group (FAG) is a well-known multirate model, which is characterized by the occupancy distribution [8]. The form of the occupancy distribution in the FAG depends mainly on the type* of multirate traffic serviced by the group (cf. [9, 10] or [11]).†

Two important properties of the interface can be distinguished in the modeling of radio interface: the level of interference and the limited number of users serviced by the interface.‡ Several papers have been devoted to traffic modeling in cellular systems with the WCDMA radio interface (i.e., [1–5,7]), but only in [12] and in [1] both properties are taken into consideration. The most general analytical model of the WCDMA interface is proposed in [7], where the authors model the WCDMA radio interface by the full-availability group servicing a mixture of multirate Erlang (infinite source population) and Engset (finite source population) traffic streams [7]. In the model, the dependence of mutual interference between cells on the decrease in the theoretical flow capacity of the radio interface is taken into account on the basis of a fixed-point methodology [13]. The characteristic feature of the developed method is the possibility to model, unlike in the previous models, a group of cells servicing different classes of users [7] as well as to take into consideration the interdependence of service processes in the uplink and downlink directions in the case of bi-directional services, both symmetrical and asymmetrical. In the models hitherto discussed in literature, it was assumed that the WCDMA interface carries only R99 traffic classes. In this chapter, we propose the application of this method to model the radio interface carrying both R99 and HSPA traffic streams.

The relevant literature proposes only one analytical model of the Iub interface. In [6], the authors discuss the influence of the Iub organization scheme on the efficiency of the interface. The paper describes two analytical models corresponding to the static and the dynamic organization scheme of the Iub interface. In the static scheme, it was assumed that Iub was divided into two separated links and one of them carried a mixture of R99, whereas the other an HSDPA traffic stream. In this scheme, each of the links was modeled by the full-availability group with multirate traffic. The second organization scheme assumed a dynamic constraint of the Iub interface resources for R99 traffic, accompanied by unconstrained access to the resources for HSDPA traffic. The dynamic organization scheme of Iub is analyzed in a new model of the full-availability group with constraint, proposed by the authors. The analysis presented by the authors in [6] is limited only to the downlink direction because in the uplink direction HSPA traffic is serviced based on R99 resources [12]. In all models, the average throughput per an HSPA user was

* Multirate traffic carried by the radio and Iub interfaces can be divided into the so-called Erlang multirate traffic, generated by an infinite, population of traffic sources, and the so-called Engset multirate traffic, generated by a finite population of traffic sources.

† In the chapter, we have presented models of the radio and Iub interfaces based on the most general and effective occupancy distribution [13].

‡ In [7], the authors show that the application of Erlang multirate traffic instead of Engset multirate traffic leads to a higher value of blocking probabilities for all traffic streams carried by the interface.

not discussed. The relevant literature discusses some analytical models for multirate traffic with compression (i.e., [14–16]), which can be applied for modeling HSDPA traffic. Models presented in [14, 15] are quite simple under the assumption that all classes of the carried traffic are characterized by the compression property. In any other case, when the system services classes which undergo and do not undergo compression simultaneously, the methods are characterized by a high complexity, which limits their practical application. In [16], an effective analytical model of the Iub interface carrying a mixture of Release 99 and HSPA traffic classes with adopted compression functionality was proposed. In this chapter, we will treat this model as the basis for modeling the HSPA traffic carried by the WCDMA and Iub interfaces.

The chapter is divided into seven sections. Section 8.2 presents the basic architecture of the system. In Section 8.3, we discuss the basic model (i.e., the full-availability group servicing a mixture of different multirate traffic streams), which will be used further on as a model of the WCDMA and the Iub interfaces. Section 8.4 presents a model of the full-availability group servicing multirate traffic with compression property, which is used in the following sections for modeling HSPA traffic behavior. The application of the described analytical methods for modeling the WCDMA and Iub interface, carrying R99 and HSPA traffic, is shown in Sections 8.5 and 8.6. Section 8.7 sums up the chapter.

8.2 System Architecture

Let us consider the structure of the UMTS network presented in Figure 8.1. The presented network consists of three functional blocks, designated respectively: user equipment (UE), UMTS terrestrial radio access network (UTRAN), and core network (CN). The following notation has been adopted in Figure 8.1: RNC is the radio

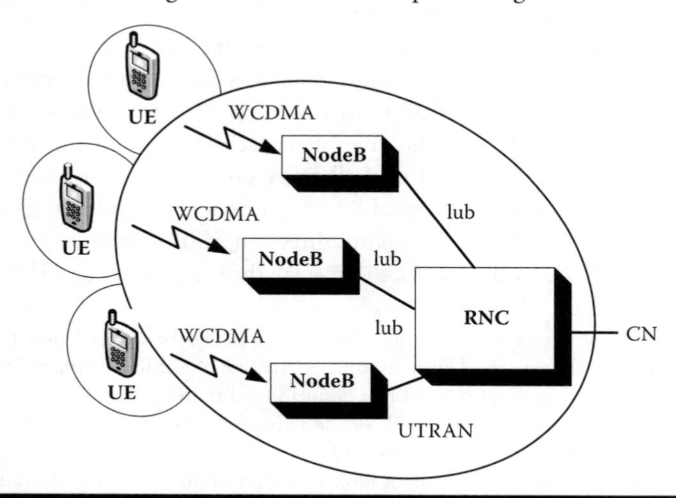

Figure 8.1 Elements of the UMTS network structure.

network controller, WCDMA is the radio interface, and Iub is the interface connecting NodeB and RNC.

High-speed downlink packet access (HSDPA) has been included by 3GPP into the system specification in version 5. The aim of its introduction is to increase transmission speed in the downlink and shorten delays in the network. An equivalent of HSDPA for the uplink is the HSUPA fast packet data transmission in the uplink, which became part of the UMTS system, in version 6 [17].

In successive versions of HSDPA, it is assumed that the users will be able to transmit data at the speed of 1.8 Mbps, 3.6 Mbps, 7.2 Mbps, and 14.4 Mbps. Therefore, new solutions have been worked out concerning the organization and management of transport and physical channels. The following channels have been defined in the system [17]:

- **High-Speed Downlink Shared Channel (HS-DSCH)**: A channel shared by many mobile stations, used for transmitting user's data from higher layers of the network and controlling information. The channel is an extension of the DCH channel for high-speed data transmission.
- **Physical Channels**:
  - High-speed physical downlink shared channel (HS-PDSCH): Used for data transmission with the constant spreading factor equal to 16.
  - Shared control channel (HS-SCCH): Used to inform the mobile station about a planned transmission in the HS-DSCH channel.
  - High-speed dedicated physical control channel (HS-DPCCH): Used in the uplink to confirm transmitted data and send the channel quality indicator.

Besides the definitions of new channels, the HSDPA technology introduces the following new mechanisms:

- **Adaptive Modulation and AMC Coding**: Apart from the QPSK modulation, HSDPA allows for the application, with a low level of interference and 16 quadrature amplitude modulation (16 QAM). Modulation and coding schemes can be changed depending on the quality of the signal and the load of the radio link.
- **High-Speed Packet Transmission from the Level of NodeB**: The HS-DSCH channel is shared by different users of the system to fully make use of the available resources of the radio link, depending on propagation conditions and the level of interference. On the basis of the signal level indicator CQI in the downlink is sent by mobile stations, the base station decides which user will be sent appropriate data.
- **High-Speed Retransmission from the Level of NodeB HARQ (hybrid automatic repeat request)**: HSDPA technology includes the function of retransmission in the physical layer. The function is located in the base station of NodeB, and therefore the process of retransmission that

does not get RNC involved is carried out much faster. In addition, HARQ introduces the concept of incremental redundancy. When the mobile station receives wrong data, the data is stored and reused by the decoder to restructure the received signal after the retransmission of redundant data to the mobile station. The base station sends incremental redundant data if in the previous transmission it was impossible to decode the received information.

- **Multicode Transmission**: HSDPA allows for multicode transmission. The base station can transmit a signal to a mobile station using simultaneously up to 15 channel codes with the spreading factor of 16.

High-speed uplink packet access (HSUPA) is a counterpart to HSDPA for the uplink. It enables data transmission from the subscriber to the base station with the speed of 5.76 Mbps. The HSUPA technology uses high-speed retransmission from the HARQ level of a mobile station with incremental redundancy, allows TTI (transmission time interval) between subsequent transmissions and introduces a new type of E-DCH (enhanced dedicated channel). The E-DCH, unlike the HS-DSCH used in the HSDPA technology, is not a shared channel but a dedicated one. This means each mobile station sets up, with the servicing NodeB, its own E-DCH. Additionally, HSUPA does not use the adaptive modulation (Table 8.1). Like in the R99 version of the UMTS system, modulation BPSK is used. High-speed HARQ retransmission for HSUPA operates in a similar way for HSDPA. The base station informs the mobile station if it has received data packets or not. When the base station receives packets erroneously, they are immediately retransmitted by the mobile station. Having received them, NodeB, also using the previously received signal, tries to re-create the data sent by the mobile station. The retransmission is then repeated until the packets sent by the mobile station have been received properly, or the number of admissible retransmissions has run out. The procedure for high-speed packet access in HSUPA is different than in HSDPA. In HSDPA, the

Table 8.1 A Comparison of the Properties of DCH Channels (R99), HS-DSCH (HSDPA), and E-DSH (HSUPA)

Feature	*DCH*	*HSDPA (HS-DSCH)*	*HSUPA (E-DCH)*
Variable spreading factor	Yes	No	Yes
Fast power control	Yes	No	Yes
Adaptive modulation	No	Yes	No
BTS based scheduling	No	Yes	Yes
Fast L1 HARQ	No	Yes	Yes
Soft handover	Yes	No	Yes
TTI length (ms)	80, 40, 20, 10	2	10, 2

HS-DSCH channel is shared by all participants serviced by a given cell. Due to this reason, the base station can allocate, though for a short time, all resources to exactly one mobile station when other mobile stations do not receive demanded data. In HSUPA, the E-DCH channel is a dedicated channel, which results in a situation when co-sharing is not possible. Therefore, the procedure of high-speed transmission in HSUPA operates in a similar way as packet scheduler for the R99 traffic. RNC informs all mobile stations about the maximum power they can use for transmission. If the interference level approaches the value that can cause instability in the system, the level of admissible transmission power allocated to all mobile stations is reduced.

In the dimensioning process for the UMTS network, an appropriate dimensioning of the connections in the access part (UTRAN) (that is, the radio interface between the user and NodeB, and the Iub connections between NodeB and the radio network controller (RNC), has a particular significance. Successive sections of the chapter describe the analytical models for the WCDMA and Iub interfaces in the uplink and downlink directions, carrying a mixture R99 and HSPA traffic streams.

8.3 Model of the Full-Availability Group with Multirate BPP Traffic

The full-availability group carrying a mixture of different multirate traffic streams is the analytical model of radio and Iub interfaces. In this section, we introduce an analytical model that is fundamental for the considerations presented in subsequent sections of the chapter.

8.3.1 *Basic Assumptions*

Consider the model of a full-availability group with the capacity of V basic bandwidth units (BBUs) presented in Figure 8.2 [11]. The group is offered two types of traffic streams: m_I Erlang streams from the set $I = \{1, \ldots, i, \ldots, m_I\}$, and m_J Engset

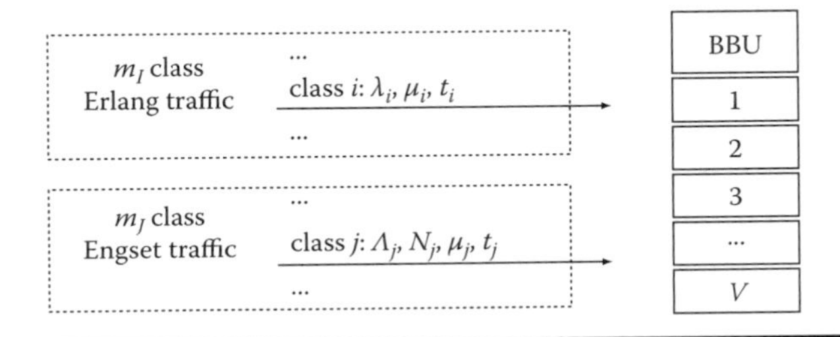

Figure 8.2 Full-availability group with the Erlang and Engset traffic stream. [With kind permission from Springer Science+Business Media: Głąbowski, M., Modelling of state-dependent multi-rate systems carrying BPP traffic. *Annals of Telecommunications,* 63(7–8): 393–407, August 2008.]

streams from the set $J = \{1, \ldots, j, \ldots, m_J\}$. It has been assumed that the letter "i" denotes any class of Erlang traffic, the letter "j" any class of Engset traffic, and the letter "c" any traffic class. The number of BBUs demanded by calls of class c is denoted by the symbol t_c.

The call intensity for Erlang traffic (Poisson distribution) of class i is λ_i. The parameter $\lambda_j(n_j)$ determines the call intensity for the Engset traffic stream of class j (binomial distribution). The intensity $\lambda_j(n_j)$ depends on the number of n_j of currently serviced calls of class j and decreases with the increasing number of serviced traffic sources:

$$\lambda_j(n_j) = (N_j - n_j)\Lambda_j \tag{8.1}$$

where N_j is the number of traffic sources of class j, while Λ_j is the call intensity of calls generated by a single free source of class j.

The total intensity of the Erlang traffic of class i offered to the group is:

$$A_i = \lambda_i/\mu_i \tag{8.2}$$

whereas the intensity of Engset traffic α_j of class j, offered by one free source, is equal to:

$$\alpha_j = \Lambda_j/\mu_j \tag{8.3}$$

In Formulae 8.2 and 8.3, the parameter μ is the average service intensity with the exponential distribution.

8.3.2 Multidimensional Erlang–Engset Model at the Microstate Level

Let us now consider the multidimensional Markov process in the full-availability group with the capacity of V BBUs, presented in Figure 8.3. The group is offered two types of call streams: Poisson and Engset call streams. Each microstate of the process $\{x_1, \ldots, x_i, \ldots, x_{m_I}, y_1, \ldots, y_j, \ldots, y_{m_J}\}$ is defined by the number of serviced calls of each class of offered traffic, where x_i denotes the number of serviced calls of the Poisson stream of class i (Erlang traffic), y_j denotes the number of serviced calls of the Engset stream of class j (Engset traffic). To simplify the description, the microstate probability will be denoted by the symbol $[p(\ldots, x_i, y_j, \ldots)]_V$.

The multidimensional service process in the Erlang–Engset model is a reversible process [11]. In accordance with Kolmogorov criterion, considering any cycle for the microstates is shown in Figure 8.3, we always obtain equality in the intensity of passing (streams) in both directions. The property of reversibility implies the local equilibrium equations between any of two neighboring states of the process. Such

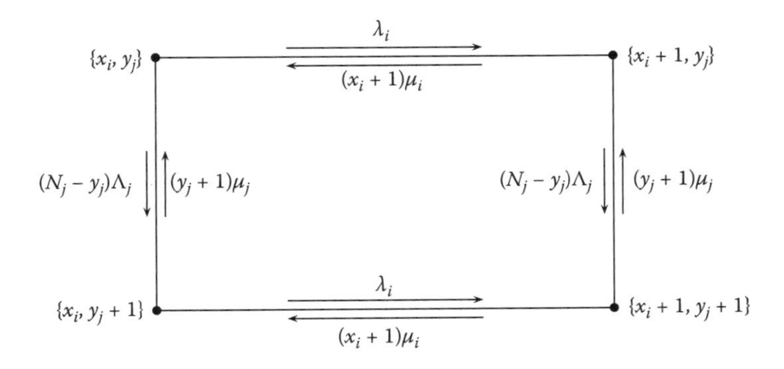

Figure 8.3 Fragment of a diagram of the Markov process in the full-availability group. [With kind permission from Springer Science+Business Media: Głąbowski, M., Modelling of state-dependent multi-rate systems carrying BPP traffic. *Annals of Telecommunications*, 63(7–8): 393–407, August 2008.]

an equation for an Erlang stream of class i and the Engset stream of class j can be written in the following way (Figure 8.3):

$$x_i \mu_i p(\ldots, x_i, y_j, \ldots) = \lambda_i p(\ldots, x_i - 1, y_j, \ldots) \tag{8.4}$$

$$y_j \mu_j p(\ldots, x_i, y_j, \ldots) = [N_j - (y_j - 1)]\Lambda_j p(\ldots, x_i, y_j - 1, \ldots) \tag{8.5}$$

Since the call streams offered to the group are independent, we can add up, for the microstate $\{\ldots, x_i, y_j, \ldots\}$, all m_I equations of the type (Equation 8.4) for the Erlang streams and m_J equations of the type (Equation 8.5) for the Engset streams. Additionally, taking into consideration traffic intensity (Figures 8.2 and 8.3), we get:

$$p(\ldots, x_i, y_j, \ldots)\left[\sum_{i=1}^{m_I} x_i t_i + \sum_{j=1}^{m_J} y_j t_j\right]$$
$$= \sum_{i=1}^{m_I} A_i t_i p(\ldots, x_i - 1, y_j, \ldots) + \sum_{j=1}^{m_J} [N_j - (y_j - 1)]\alpha_j t_j p(\ldots, x_i, y_j - 1 \ldots) \tag{8.6}$$

8.3.3 Full-Availability Group with BPP Traffic at the Macrostate Level

It is convenient to consider the multidimensional process occurring in the considered system at the level of the so-called macrostates. Each macrostate contains information about the number of busy BBUs in the considered group, regardless of the number of serviced calls of particular classes.

The macrostate probability $[P_n]_V$ is then the occupancy probability n BBU of the group and can be expressed as the aggregation of the probabilities of appropriate microstates:

$$[P_n]_V = \sum_{\Omega(n)} p(\ldots, x_i, y_j, \ldots) \tag{8.7}$$

where $\Omega(n)$ is a set of all such subsets $\{\ldots, x_i, y_j, \ldots\}$ that fulfill the equation:

$$n = \sum_{i=1}^{m_I} x_i t_i + \sum_{j=1}^{m_J} y_j t_j \tag{8.8}$$

The definition of the macrostate Equation 8.8 makes it possible to convert Formula 8.6 into the following form:

$$n\, p(\ldots, x_i, y_j, \ldots) = \sum_{i=1}^{m_I} A_i t_i\, p(\ldots, x_i - 1, y_j, \ldots) + \sum_{j=1}^{m_J} [N_j - (y_j - 1)]\alpha_j t_j\, p(\ldots, x_i, y_j - 1, \ldots)$$

Summing on both sides all microstates that belong to the set $\Omega(n)$, we get:

$$n \sum_{\Omega(n)} p(\ldots, x_i, y_j, \ldots) = \sum_{i=1}^{m_I} A_i t_i \sum_{\Omega(n)} p(\ldots, x_i - 1, y_j, \ldots) + \sum_{j=1}^{m_J} [N_j - (y_j - 1)]\alpha_j t_j \sum_{\Omega(n)} p(\ldots, x_i, y_j - 1, \ldots) \tag{8.9}$$

Following the definition of macrostate probability, expressed by Formula 8.7, we are in a position to convert Formula 8.9 as follows:

$$\begin{aligned} n[P_n]_V &= \sum_{i=1}^{m_I} A_i t_i [P_{n-t_i}] + \sum_{j=1}^{m_J} [N_j - (y_j - 1)]\alpha_j t_j \sum_{\Omega(n)} p(\ldots, x_i, y_j - 1, \ldots) \\ &= \sum_{i=1}^{m_I} A_i t_i [P_{n-t_i}] + \sum_{j=1}^{m_J} \alpha_j t_j \sum_{\Omega(n)} [N_j - (y_j - 1)] \\ &\quad \times \frac{p(\ldots, x_i, y_j - 1, \ldots)}{\sum_{\Omega(n)} p(\ldots, x_i, y_j - 1 \ldots)} \sum_{\Omega(n)} p(\ldots, x_i, y_j - 1, \ldots) \end{aligned} \tag{8.10}$$

In Formula 8.10 the sum:

$$\sum_{\Omega(n)}[y_j - 1]\frac{p(\ldots, x_i, y_j - 1, \ldots)}{\sum_{\Omega(n)} p(\ldots, x_i, y_j - 1, \ldots)} = y_j(n - t_j) \tag{8.11}$$

determines the value of the average number of calls of class j in the occupancy state $n - t_j$. When taking into consideration Equation 8.11, Formula 8.10 can be rewritten in the following way:

$$n[P_n]_V = \sum_{i=1}^{m_I} A_i t_i [P_{n-t_i}] + \sum_{j=1}^{m_J} \alpha_j t_j [N_j - y_j(n - t_j)][P_{n-t_j}]_V \tag{8.12}$$

where $[P_{n-t_c}]_V = 0$, if $n < t_c$, and the value $[P_0]_V$ results from the normative condition $\sum_{n=0}^{V} [P_n]_V = 1$.

Let us introduce the following notation for the offered traffic intensity in appropriate occupancy states of the group:

$$A_i(n) = A_i, \quad A_j(n) = \alpha_j [N_j - (y_j(n))] \tag{8.13}$$

Formula 8.12 can be now finally rewritten in the following form:

$$\begin{aligned} n[P_n]_V &= \sum_{i=1}^{m_I} A_i(n - t_i) t_i [P_{n-t_i}] + \sum_{j=1}^{m_J} A_j(n - t_j) t_j [P_{n-t_j}]_V \\ &= \sum_{c=1}^{m} A_c(n - t_c) t_c [P_{n-t_c}]_V \end{aligned} \tag{8.14}$$

The average number of calls of class c in the group in state $n + t_c$ can be written as follows:

$$y_c(n + t_c) = \begin{cases} A_c(n)[P_n]_V / [P_{n+t_c}]_V & \text{for } n + t_c \leq V \\ 0 & \text{for } n + t_c > V \end{cases} \tag{8.15}$$

Let us remark that if the system services the Erlang streams only, then Equation 8.12 can be simplified to Kaufman-Roberts recursion [9, 10]:

$$n[P_n]_V = \sum_{i=1}^{m_I} A_i t_i [P_{n-t_i}] \tag{8.16}$$

8.3.4 MIM-BPP Method

Let us now consider a full-availability group with Erlang and Engset multirate traffic Equation 8.12. Notice that in order to determine the parameter $y_c(n)$, it is necessary to determine first the occupancy distribution $[P]_V$. Simultaneously, in order to determine the occupancy distribution $[P]_V$, it is necessary to determine the value $y_c(n)$. This means Equations 8.14 and 8.15 form a set of confounding equations that can be solved with the help of iterative methods [11]. Let $[P^{(l)}]_V$ denote the occupancy distribution determined in step l, and let $y_c^{(l)}(n)$ denote the average number of serviced calls of class c, determined in step l. Then:

$$y_c^{(l+1)}(n) = \begin{cases} A_c^{(l)}(n-t_c)\left[P_{n-t_c}^{(l)}\right]_V / \left[P_n^{(l)}\right]_V & \text{for} \quad 0 \le n \le V \\ 0 & \text{in remaining instances} \end{cases} \tag{8.17}$$

where $A_c^{(l)} = \alpha_c[N_c - y_c^{(l)}(n)]$.

In order to determine the initial distribution $[P_{n-t_c}^{(0)}]_V$, it was assumed that:

$$A_c^{(0)}(n) = A_c = N_c\alpha_c \tag{8.18}$$

On the basis of the reasoning presented here, in [11] the MIM-BPP method for determining the occupancy distribution and the loss probability in the full-availability group with BPP traffic was proposed. The method can be presented in the following way:

8.3.4.1 Method MIM-BPP

1. Setting the starting point of the iteration at $l = 0$
2. Determination of initial values $y_j^{(l)}(n)$, $y_k^{(l)}(n)$:

$$\forall_{1\le j\le m_J}\forall_{0\le n\le V}\; y_j^{(l)}(n) = 0, \qquad \forall_{1\le k\le m_K}\forall_{0\le n\le V}\; y_k^{(l)}(n) = 0$$

3. Increase in each iteration step: $l = l + 1$
4. Determination of the value of Engset traffic of class j on the basis of Formula 8.13
5. Determination of the state probabilities $[P_n^{(l)}]_V$ (Formula 8.14)
6. Determination of the average number of serviced calls $y_j^{(l)}(n)$ i $y_k^{(l)}(n)$ on the basis of Formula 8.17
7. Repetition of steps 3–6 until a pre-defined occuracy ϵ of the iterative process is achieved:

$$\forall_{0\le n\le V}\left|\frac{y_j^{(l-1)}(n) - y_j^{(l)}(n)}{y_j^{(l)}(n)}\right| \le \epsilon \quad \forall_{0\le n\le V}\left|\frac{y_k^{(l-1)}(n) - y_k^{(l)}(n)}{y_k^{(l)}(n)}\right| \le \epsilon \tag{8.19}$$

8. Defining the blocking probability E_c for calls of class c and the loss probability B_i for Erlang calls of class i, B_j for Engset calls of class j

$$E_c = \sum_{n=V-t_c+1}^{V} [P_n]_V \quad B_i = E_i \tag{8.20}$$

$$B_j = \frac{\sum_{n=V-t_j+1}^{V} [P_n]_V[N_j - y_j(n)]\alpha_j}{\sum_{n=0}^{V}[P_n]_V[N_j - y_j(n)]\alpha_j} \tag{8.21}$$

8.4 Model of the Full-Availability Group with Traffic Compression

This section presents a model of the full-availability group, carrying a mixture of different R99 and HSPA traffic classes, which is also known as the model of the full-availability group with traffic compression. This model is applied in the chapter for modeling the radio and Iub interfaces, carrying both R99 and HSPA traffic streams.

Let us assume now that a full-availability group services a mixture of different multirate Erlang traffic streams with the compression property. This means the traffic mixture contains such calls for which a change in demands (requirements) is followed uniformly by overload of the system.

In this group, it is assumed that the system services simultaneously a mixture of different multirate Erlang traffic classes, while these classes are divided into two sets: classes with calls that can change requirements while being serviced, and classes that do not change their demands in the service time.

This section discusses two models of the systems with traffic compression. The presented models differ in the compression method. In the first model (Section 8.4.1), we assume that all traffic classes undergoing compression are compressed to the same degree (evenly). Whereas in the second model (Section 8.4.2), it is assumed that traffic classes with the compression property can be compressed to a different degree (unevenly).

In all the models considered, the following notation is used:

- $\mathbb{M}_k$ denotes a set of classes capable of compression, while $M_k = |\mathbb{M}_k|$ is the number of compressed traffic classes.
- $\mathbb{M}_{nk}$ is a set of classes without compression, and $M_{nk} = |\mathbb{M}_{nk}|$ denotes the number of classes without compression.*

---

* Further in the section, for simplicity of the description, we limited the considerations to PCT1 traffic classes.

8.4.1 Basic Model of the Full-Availability Group with Compression

It was assumed in the model that all classes undergoing compression were compressed to the same degree (evenly). The measure of a possible change in requirements is the maximum compression coefficient, that determines the ratio of the maximum demands to minimum demands for a given traffic class. The coefficient $K_{\max}$ can be determined on the basis of the dependence [16]:

$$\forall_{j\in\mathbb{M}_k} \ K_{\max} = \frac{t_{j,\max}}{t_{j,\min}}, \tag{8.22}$$

where $t_{j,\max}$ and $t_{j,\min}$ denote, respectively, the maximum and minimum number of basic bandwidth units (BBUs) demanded by a call of class j. We assume the system will be treated as a full-availability group with multirate Erlang traffic.*

Let us consider a system with maximum compression (i.e., under the assumption that the amount of resources required by calls of classes with the compression property is minimum. In the case of a system carrying a mixture of traffic streams that undergo and do not undergo compression, the occupancy distribution (Equation 8.16) will be more conveniently expressed after dividing the two types of traffic:

$$n[P_n]_V = \sum_{i=1}^{M_{nk}} A_i t_i [P_{n-t_i}]_V + \sum_{j=1}^{M_k} A_j t_{j,\min} [P_{n-t_{j,\min}}]_V \tag{8.23}$$

where $t_{j,\min}$ is the minimum number of BBUs demanded in a given occupation state of the system by a call of class j that belongs to the set $\mathbb{M}_k$.

The blocking and loss coefficient in the full-availability group will be determined on the basis of Equation 8.16:

$$E_i = B_i = \begin{cases} \displaystyle\sum_{n=V-t_i+1}^{V} [P_n]_V & \text{for} \quad i \in \mathbb{M}_{nk} \\ \displaystyle\sum_{n=V-t_{i,\min}+1}^{V} [P_n]_V & \text{for} \quad i \in \mathbb{M}_k \end{cases} \tag{8.24}$$

* This assumption simplifies the description of the system to Kaufman-Roberts recursion Equation 8.16. In the case of the service of Erlang as well as Erlang and Engset streams, it is necessary to apply the MIP-BPP method described in Section 8.3.4.

For Erlang and Engset traffic streams, after the application of the MIM-BPP method (Section 8.3.4), the blocking (loss) probability is determined on the basis of Equation 8.20.

In Equations 8.23 and 8.24, the model is characterized by the parameter $t_{i,\min}$, which is the minimum number of BBUs demanded by a call of class i in the conditions of maximum compression. Such an approach is indispensable in determining the blocking probabilities in the system with compression, since the blocking states will occur in the conditions of maximum compression. The maximum compression determines such occupancy states of the system in which a further decrease in the demands of class i calls is not possible.

In order to determine a possibility of the system compression, it is necessary to evaluate the number and kind of calls serviced in a given occupancy state of the system. For this purpose, we can use Formula 8.15, which makes it possible to determine the average number of calls of class i serviced in the occupancy state n BBUs. This dependence, under the assumption of the maximum compression, can be written in the following way:

$$y_i(n) = \begin{cases} \dfrac{A_i\left[P_{n-t_i}\right]_V}{\left[P_n\right]_V} & \text{for} \quad i \in \mathbb{M}_{nk} \\[2ex] \dfrac{A_i\left[P_{n-t_{i,\min}}\right]_V}{\left[P_n\right]_V} & \text{for} \quad i \in \mathbb{M}_k \end{cases} \tag{8.25}$$

On the basis of Formula 8.25, knowing the demands of individual calls, we can thus determine the total average carried traffic in state n, under the assumption of the maximum compression:

$$Y_{\max}(n) = Y^{nk}(n) + Y^{k}_{\max}(n) = \sum_{i=1}^{M_{nk}} y_i(n)t_i + \sum_{j=1}^{M_k} y_j(n)t_{j,\min} \tag{8.26}$$

where $Y^{k}_{\max}(n)$ is the average number of busy BBUs in state n, occupied by calls that undergo compression, whereas $Y^{nk}(n)$ is the average number of busy BBUs in state n, occupied by calls without compression.

Let us assume that the value of the parameter $Y^{nk}(n)$ refers to non-compressed traffic and is independent of the compression of remaining calls. The real values of carried traffic, corresponding to state n (determined in the conditions of maximum compression), will depend on the number of free BBUs in the system. We assume the real system operates in such a way as to guarantee the maximum use of the resources (i.e., a call of a compressed class always tends to occupy free resources and decreases its maximum demands to the least extent possible.) Thus, the real traffic value $Y(n)$ carried in the system in a given state, corresponding to state n (determined in maximum compression), can be expressed in the following way:*

$$Y(n) = Y^{nk}(n) + Y^{k}(n) = \sum_{i=1}^{M_{nk}} y_i(n)t_i + \sum_{j=1}^{M_k} y_j(n)t_j(n) \tag{8.27}$$

The parameter $t_j(n)$ in Formula 8.27 determines the real value of a demand of class j in state n:

$$\forall_{j \in \mathbb{M}_k} \quad t_{j,\min} < t_j(n) \leq t_{j,\max} \tag{8.28}$$

The measure of the compression degree in state n is the compression coefficient $\xi_k(n)$, which can be expressed in the following way:

$$t_j(n) = t_{j,\min}\xi_k(n) \tag{8.29}$$

When taking into consideration Equation 8.29, the average number of busy BBUs occupied by calls with compression can be written thus:

$$Y^{k}(n) = \sum_{j=1}^{M_k} y_j(n)t_j(n) = \xi_k(n)\sum_{j=1}^{M_k} y_j(n)t_{j,\min} \tag{8.30}$$

We assume that in the considered model the system operates in such a way that it guarantees the maximum use of available resources. This means that calls that undergo compression will always tend to occupy free resources, decreasing their demands to the least possible. Another parameter of the considered system, besides the blocking (loss) probability, is the average number of busy BBUs in the system, occupied by calls with compression (Formula 8.30). The knowledge of the compression coefficient $\xi_k(n)$ is indispensable to determine this parameter. This coefficient can also be defined as the ratio of resources potentially available for the service of calls with compression to the resources occupied by these calls in the state of maximum compression. Thus, we can write (Figure 8.4):

$$\xi_k(n) = \frac{V - Y^{nk}(n)}{Y^{k}_{\max}(n)} = \frac{V - Y^{nk}(n)}{n - Y^{nk}(n)} \tag{8.31}$$

The numerator in the Formula 8.31 expresses the total amount of system resources that can be occupied by calls of the class with compression. Whereas the denominator can be interpreted as the amount of resources that can be occupied by calls of the

* Further on in the description, to simplify the description, we will use the term "in state n" instead of "a given state n in maximum compression."

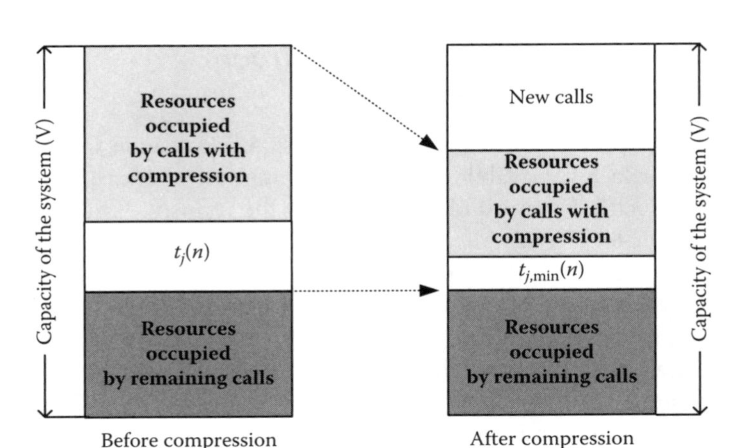

Figure 8.4 Exemplary system with compression, in which class *i* calls are maximally compressed.

class with compression, under the assumption that the system (FAG) is in the state n of busy BBUs. A constraint to the value of the coefficient 8.31 is the maximum compression coefficient, determined on the basis of the dependence 8.22. This constraint can be taken into account by formally defining the compression coefficient in the following way:

$$\xi_k(n) = \begin{cases} K_{\max} & \text{for} \quad \xi_k(n) \geq K_{\max} \\ \xi_k(n) & \text{for} \quad 1 \leq \xi_k(n) < K_{\max} \end{cases} \tag{8.32}$$

The compression coefficient determined by Formula 8.32 is not dependent on the traffic class. This results from the assumption adopted in the model of the same compression degree for all traffic classes that undergo the mechanism of compression.

Knowing the value of the compression coefficient in every state n, we can determine the average resources occupied by calls of class j with compression:

$$Y_j^k = \sum_{n=0}^{V} y_j(n)[\xi_k(n) t_{j,\min}][P_n]_V \tag{8.33}$$

On the basis of the average resources occupied by calls of class j, we can determine the average resources occupied by calls of all traffic classes with compression:

$$Y^k = \sum_{j=0}^{M_k} Y_j^k \tag{8.34}$$

Note that the value Y^k in Formula 8.34 is the average traffic carried in the system by calls that undergo compression.

8.4.2 Model of the Full-Availability Group with Uneven Compression

In the model of the FAG with uneven compression, we assume that the system will be treated as a full-availability group with multirate traffic. The occupancy distribution in such a system can be expressed by the recursive Kaufman–Roberts formula (Equation 8.16), under the assumption that the amount of resources required by calls of the classes with the compression property is minimum. The blocking coefficient in such a system will be determined by the dependence in Equation 8.24.

The basic assumption in this model is that classes undergoing compression can be compressed to a different degree. The measure of a possible change in requirements is the maximum compression coefficient $K_{j,\max}$, which can determine the ratio of maximum demands to minimum demands for a given traffic class [18]:

$$\forall_{j\in\mathbb{M}_k} \; K_{j,\max} = \frac{t_{j,\max}}{t_{j,\min}} \tag{8.35}$$

where $t_{j,\max}$ and $t_{j,\min}$ denote, respectively, the maximum and minimum number of basic bandwidth units (BBUs), demanded by a call of class j (cf. Equation 8.22).

The introduction of different values of the maximum compression coefficient also results in changes in the definition of the average compression coefficient, determined by Formula (8.32):

$$\xi_{j,k}(n) = \begin{cases} K_{j,\max} & \text{for} \quad \xi_k(n) \geq K_{j,\max} \\ \xi_k(n) & \text{for} \quad 1 \leq \xi_k(n) < K_{j,\max} \end{cases} \tag{8.36}$$

where the coefficient $\xi_k(n)$ is determined on the basis of Equation 8.31.

Knowing the value of the compression coefficient in every state n, we can determine the average resources occupied by calls of all traffic classes with compression, with the application of Equations 8.33 and 8.34:

$$Y^k = \sum_{j=0}^{M_k} \sum_{n=0}^{V} y_j(n)[\xi_{j,k}(n) t_{j,\min}][P_n]_V \tag{8.37}$$

where $y_j(n)$ is determined on the basis of Equation 8.25.

8.5 Modeling and Dimensioning of the Radio Interface

In this section, we will present traffic issues that refer to the UMTS mobile system, which can be analyzed with the application of the models with multirate traffic, presented in Section 8.3.

A single cell of the mobile system can be treated as a full-availability group with hard or soft capacity, depending on a possible influence of the environment upon the load of the radio interface. The GSM system is a system with a hard capacity of cells. In this system, the maximum number of subscribers serviced by one cell is determined unequivocally and depends exclusively on the number of used frequency channels. The UMTS system is a system with soft capacity. Soft capacity indicates a possibility of changing the capacity of a cell, depending on external influence, in which the element of essential importance is the degree of load in neighboring cells.

8.5.1 Resource Allocation in Mobile Systems with Soft Capacity

The wideband code division multiple access (WCDMA) radio interface applied in the UMTS system has a large theoretical flow capacity (throughput) of the separated interface. At the same time, the available throughput is limited by the admissible level of the interference volume in the frequency channel. In every cellular system with spread signal spectrum, the capacity of the radio interface is constrained as the result of a few types of interference [19]: co-channel interference within a cell—from concurrent users of a frequency channel within the area of a given cell; external co-channel interference within a cell—from the concurrent users of the frequency channel, working within the area of adjacent cells; adjacent channels interference—from the adjacent frequency channels of the same operator or other cellular telecommunication operators; and all possible noise and interference from other systems and sources, both broadband and narrowband.

Summing up, in the WCDMA radio interface, a growth in load is accompanied by a simultaneous growth in interference, generated by other users serviced by the same cell or other cells. To secure an appropriate level of service, it is necessary to limit the number of allocated resources by active traffic sources. It is estimated that the maximum usage of the radio interface resources without lowering the quality of service will be equal to about 50% to 80% [19]. For the same reason, the soft capacity of the WCDMA radio interface is defined as the noise limited capacity (noise limited).

Multirate traffic in the UMTS system is composed of a few classes, and each of them demands a certain bit rate to service its own call. In the probabilistic analysis of radio systems that are offered multirate traffic streams, it is necessary to take into consideration the class of call and the bit rate demanded by a call of this class. The UMTS system—in respect to the flow capacities of services carried out—can be then considered a discrete multiservice switching network. In the following analysis of the radio interface, we will use the universally accepted notion of BBU, which will be defined in Section 8.5.2.

Accurate signal reception in the receiver of the UMTS system is possible only when the ratio of energy per bit E_b to noise spectral density N_0 is appropriate. A too low value of E_b/N_0 will cause the receiver to be unable to decode the received signal, while a too high value of the energy per bit in relation to noise spectral density will be perceived by other users of the same radio channel as interference.

The ratio E_b/N_0 for a given traffic source of class i can be written as the following dependence [17]:

$$\left(\frac{E_b}{N_0}\right)_i = \frac{W}{\nu_i R_i} \frac{P_i}{I_{\text{total}} - P_i} \tag{8.38}$$

In Formula 8.38, the following notation is adopted: P_i, average signal power received from the traffic source of class i; I_{total}, total power of the received signal in the base station, with thermal noise taken into consideration; W, flow capacity of the spread signal (the so-called chip rate) (in the UMTS system it is conventionally 3.84 Mchip/s i.e., the speed at the input signal is spread (data signal or speech signal); R_i, throughput of the data signal from the traffic source of class i; ν_i, activity coefficient of the traffic source of class i, which denotes the percentage of occupancy time of the transmission channel in which the source is active (i.e., transmits a signal with the flow capacity R_i).

Formula 8.38 can be converted in such a way as to get the average power of the received signal from the traffic source of class i:

$$P_i = \frac{I_{\text{total}}}{1 + \frac{W}{\left(\frac{E_b}{N_0}\right)_i R_i \nu_i}} = L_i I_{\text{total}} \tag{8.39}$$

where L_i is the load factor, imposed by a class i call:

$$L_i = \frac{1}{1 + \frac{W}{\left(\frac{E_b}{N_0}\right)_i R_i \nu_i}} \tag{8.40}$$

Sample loads of the WCDMA radio interface by calls of different classes are shown in Table 8.2 [20].

The method for dimensioning the WCDMA interface proposed in [7] can be extended for the HSPA traffic. It should be noticed, however, that in the HSUPA technology, changes ensue at the required E_b/N_0 level in relation to R99, which is linked to the applied solutions. In HSUPA, the following factors will be conducive to E_b/N_0:

- Outer loop power control target block terror (BLER).
- **Transmit time interval (TTI)**: Transmit time of each block of data in HSUPA.
- **Transport block size (TBS)**: The number of bits transmitted in each "transport block."
- The number of HARQ transmissions.

In the modeling proccess, we assumed that the load factor for the HSPA traffic—based on [21]—can be determined by a simulation procedure. Sample values of load factors for an exemplary HSUPA traffic stream (service) are shown in Table 8.3 [21].

Table 8.2 Sample WCDMA Radio Interface Loads by Calls of Different Classes

	Service			
Parameters	*Speech (Voice)*	*Video*	*Data*	*Data*
W (Mchip/s)	3,84			
R_i (kbps)	12,2	64	144	384
ν_i	0,67	1	1	1
E_b/N_0 (dB)	4	2	1,5	1
L_i	0,005	0,026	0,050	0,112

Source: With kind permission from Springer Science+Business Media: Stasiak, M., Wiśniewski, A., Zwierzykowski, P., Blocking probability calculation in the uplink direction for cellular systems with WCDMA radio interface, In 3rd Polish-German Teletraffic Symposium, pp. 65–74, Dresden, 2004. © 2004 IEEE.

8.5.1.1 Uplink

Let us remark that the load coefficient is non-dimensional and defines the fraction of a possible interface load. The coefficient also shows the nonlinear dependence between the percentage load of the interface and the throughput of the traffic source of a given class. On the basis of the known load coefficients of single traffic sources,

Table 8.3 Sample HSPA Radio Interface Loads by Calls of Different Classes

	Service		
Parameters	*Service 1*	*Service 2*	*Service 3*
W (Mchip/s)	3,84		
R_i (kbps)	54,72	800,12	82,1
ν_i	1	1	1
E_b/N_0 (dB)	4,84	4,55	3,74
L_i	0,041624641	0,372667591	0,0481371632

Source: From Engineering Services Group, Aspects of HSUPA Network Planning, Qualcomm Incorporated, Technical Report, No. 80-W1159-1, Revision B, San Diego, 2007.

it is possible to determine the total load η_{UL} for the uplink:

$$\eta_{UL} = \sum_{i=1}^{M} N_i L_i \tag{8.41}$$

where N_i is the number of serviced traffic sources of class i in the uplink under consideration.

Dependence 8.41 determines the ideal maximum interface load in a system of one isolated cell. In real circumstances, however, the traffic generated in other cells, which also influences the capacity of the radio interface of a given cell, has to be taken into consideration. Hence, Formula 8.41 is complemented with a coefficient that takes into account interference from other cells. To achieve that, a parameter δ, defined as the ratio of the interference from other cells to the interference of the measured cell, is introduced. This coefficient, in the case of the uplink, is determined in the receiver of the base station [19]. The total load for the uplink can thus take on the following form [17]:

$$\eta_{UL} = (1 + \delta) \sum_{i=1}^{M} N_i L_i \tag{8.42}$$

It is generally assumed that the maximum usage of the resources of the radio interface, without lowering the quality of service, amounts to 50% to 80% of its theoretical capacity [17].

It should be emphasized that the influence of inter-cellular interference can also be taken into consideration by applying the so-called fixed-point methodology [1,7].

8.5.1.2 Downlink

The total load for the downlink can be written in the following way [17]:

$$\eta_{DL} = \sum_{i=1}^{M} N_i L_i (1 - \xi_i + \delta_i) \tag{8.43}$$

where ξ_i is the orthogonality factor for the class i traffic. It indicates the degree of interference reduction between the users of the same cell through the application of channel codes based on the OVSF (orthogonal variable spreading factor). This means they can have different dispersion coefficients and their mutual correlation is (theoretically) equal to zero [22]. Usually, coefficient values δ_i and ξ_i are similar [17], so the influence of the interference upon the decrease in loadability of the downlink can be omitted.

8.5.2 Allocation Units in the WCDMA Radio Interface

In systems with soft capacity, the available capacity of a system can vary and can be different from the theoretical maximum capacity—where the capacity of the ideal isolated cell, not exposed to external influences, can be regarded as the measure unit—to a certain minimum capacity, when the influence of the load of the neighboring cells is at its maximum. In the system under consideration, the use of bit rates as the measure for allocation is not very convenient. It is much more convenient to measure the state of allocated resources more appropriately in other units, reflecting the physical nature of a given system. Formulae 8.42 and 8.43 clearly indicate that the measure of resource allocation in the WCDMA radio interface can be the percentage of noise load of the interface. Therefore, in the radio interface: allocation is not based on adding bit rates but on adding noise loads.

A single interface load imposed by a traffic source can be applied as the allocation unit. The way of changing resource allocation, expressed in kbps, into the resource allocation, expressed in the percentage of the load of the radio interface, is shown in Figure 8.5 [18].

In the UMTS system, servicing many traffic classes with different flow capacities and treated as a multirate system, it is assumed that the value of a BBU should be lower or equal to the greatest common divisor of the resources demanded by individual call streams [23, 24]. For the WCDMA radio interface, we can write:

$$L_{BBU} = \mathrm{GCD}(L_1, L_2, \ldots, L_M) \tag{8.44}$$

Then, the interface capacity can be expressed by the number of the defined BBUs in Equation 8.44:

$$V = \lfloor \eta_{UL/DL} / L_{BBU} \rfloor \tag{8.45}$$

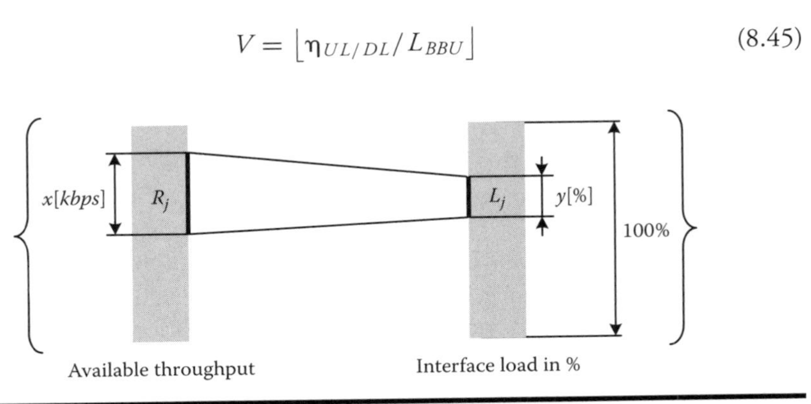

Figure 8.5 Resource allocation in the WCDMA radio interface. (From Stasiak, M. and Zwierzykowski, P., Modelling full availability groups with adaptive-rate. Internal report 9/2008, Poznan University of Technology, September 2008.)

where $\eta_{UL/DL}$ is the radio interface capacity for the uplink or the downlink. In a similar way we can express the number of BBUs required by a call of a given class:

$$t_i = \lceil L_i / L_{BBU} \rceil \tag{8.46}$$

In the considerations presented in this section, we have assumed, for simplicity, that the influence of interference on the flow capacity of the WCDMA radio interface can be determined by the parameters $\bar{\delta}$ and ξ_i [19].

8.5.3 Analytical Model of the WCDMA Interface

In this section, we will analyze four GoS parameters, important for the dimensioning and optimization process of the WCDMA interface, carrying R99 and HSPA traffic: blocking probability, loss probability, average throughput, and available throughput.

The WCDMA interface in a UMTS network can be treated as the full-availability group (FAG) with multirate traffic. In the model, we assume that the radio interface carries both R9 and HSPA traffic streams. We also assume that there are traffic classes belonging to the HSPA traffic that calls that can change occupied resources in the service time. Therefore, it is assumed that the system services simultaneously a mixture of different multirate traffic classes, while these classes are divided into two sets: $\mathbb{M}_k$ classes with calls that can change requirements while being serviced, and $\mathbb{M}_{nk}$ classes that do not change their demands in the service time. Let us assume that the total capacity of the group is equal to V basic bandwidth units (BBUs). The group is offered M* independent classes of Poisson traffic streams,† having the intensities: $\lambda_1, \lambda_2, \ldots, \lambda_M$. The class i call requires t_i BBUs to set up a connection. The holding time for calls of particular classes has an exponential distribution with the parameters: $\mu_1, \mu_2, \ldots, \mu_M$. Thus, the mean traffic offered to the system by the class i traffic stream is equal to:

$$A_i = \frac{\lambda_i}{\mu_i} \tag{8.47}$$

The resources demanded in the group for servicing particular classes can be treated as a call demanding an integer number of BBUs. The value of BBU (i.e., t_{BBU}) is calculated as the greatest common divisor (GCD) of all resources demanded by traffic classes offered to the system (Equation 8.44):

$$L_{BBU} = GCD(L_1, \ldots, L_M) \tag{8.48}$$

* $M = M_k + M_{nk}$, where $M_k = |\mathbb{M}_k|$ and $M_{nk} = |\mathbb{M}_{nk}|$

† In the analytical model, for simplicity, we assume that the system carries only Erlang traffic streams. In the case of Erlang and Engset traffic streams, we can use the MIM-BPP algorithm.

where L_i is the load factor for a user of the class i call (Table 8.2), defined in Equation 8.40.

The multidimensional Markov process in the FAG can be approximated by the one-dimensional Markov chain, which can be described by Kaufman-Roberts recursion (Equation 8.15):

$$n[P_n]_V = \sum_{i=1}^{M_{nk}} A_i t_i [P_{n-t_i}]_V + \sum_{j=1}^{M_k} A_j t_{j,\min} [P_{n-t_{j,\min}}]_V \tag{8.49}$$

where $[P_n]_V$ is the probability state of n BBUs being busy, and t_i and $t_{j,\min}$ are the numbers of BBUs required by classes not undergoing and undergoing compression, respectively (Equation 8.46):

$$t_i = \left\lfloor \frac{L_i}{L_{BBU}} \right\rfloor \qquad t_{j,\min} = \left\lfloor \frac{L_{j,\min}}{L_{BBU}} \right\rfloor \tag{8.50}$$

The interface capacity V is defined as follows [25]:

$$V = \begin{cases} \dfrac{\eta_{DL}}{1+\bar{\delta}-\xi_i} & \text{for downlink direction} \\ \dfrac{\eta_{UL}}{1+\bar{\delta}} & \text{for uplink direction} \end{cases} \tag{8.51}$$

where η_{DL} and η_{UL} are the physical capacities of the WCDMA interface in the downlink and in the uplink direction, respectively [7].

8.5.3.1 Blocking (and Loss) Probability

The blocking probability B_i for the class i of Erlang traffic streams can be expressed in the following form (Equation 8.24):

$$E_i = B_i = \begin{cases} \displaystyle\sum_{n=V-t_i+1}^{V} [P_n]_V & \text{for } i \in \mathbb{M}_{nk} \\ \displaystyle\sum_{n=V-t_{i,\min}+1}^{V} [P_n]_V & \text{for } i \in \mathbb{M}_k \end{cases} \tag{8.52}$$

The loss and blocking probabilities for Erlang traffic streams are determined by identical formulas 8.20.

8.5.3.2 Average Throughput

The radio interface carries both Release 99 and HSPA traffic streams. The classes belonging to R99 do not undergo compression. Therefore, the determination of the average throughput is important only for those traffic classes of the HSPA traffic that can undergo compression. Moreover, the application of a given analytical model depends on the mechanisms applied in the solutions used by the equipment manufacturers and providers of UMTS networks. Therefore, in this chapter we will discuss potential applications of models with compression to determine the average throughput separately for the uplink and for the downlink.

8.5.3.3 Downlink Direction

Let us consider a scenario in which the average bandwidth is allocated to all subscribers equally. Let us further assume that the subscribers have different classes of terminals at their disposal. This means the average throughput offered to a given subscriber depends mainly on the network load, while, with a small network load, the class of users' terminals is also a constraint. Assume that the subscribers with newer mobile user terminals can achieve higher maximum throughput. Such a scenario can be considered for use to describe the system, which can be modeled with even compression, presented in Section 8.4.1.

The first step to determine the average throughput is to determine the compression coefficient $\xi_k(n)$. The coefficient, on the basis of the dependence in Equations 8.31 and 8.32, takes on the following form:

$$\xi_k(n) = \begin{cases} K_{\max} & \text{for} \quad \dfrac{V - Y^{nk}(n)}{n - Y^{nk}(n)} \geq K_{\max} \\ \dfrac{V - Y^{nk}(n)}{n - Y^{nk}(n)} & \text{for} \quad 1 \leq \dfrac{V - Y^{nk}(n)}{n - Y^{nk}(n)} < K_{\max} \end{cases} \tag{8.53}$$

where the $Y^{nk}(n)$ parameter is expressed by Equation 8.27 and $Y^k(n)$ can be determined based on Equation 8.30.

In the next step, we can obtain the average resources occupied by calls of class j (average throughput) on the basis of the following Equation 8.33:

$$Y_j^k = \sum_{n=0}^{V} y_j(n)[\xi_k(n)t_{j,\min}][P_n]_V \tag{8.54}$$

8.5.3.4 Uplink Direction

Let us consider now a scenario in which the average bandwidth is allocated unevenly and a decrease in the throughput offered to a given subscriber depends on the current network load and on the kind of subscriptions assigned to them. Assume that the throughput will be decreased first to the group of users that generate the least profit for the operator. Therefore, the order in which the throughput will be decreased is directly dependable on the amount of the subscription fee. Additionally, the upper limit will also be the class of terminal operated by the user. This scenario is matched by the model of the system with uneven compression, described in Section 8.4.2.

The determination of the average throughput will be initiated, as earlier, by determining the compression coefficient $\xi_{k,j}(n)$. Thus, based on Equations 8.31 and 8.36, we obtain:

$$\xi_{k,j}(n) = \begin{cases} K_{j,\max} & \text{for} \quad \dfrac{V - Y^{nk}(n)}{n - Y^{nk}(n)} \geq K_{j,\max} \\ \dfrac{V - Y^{nk}(n)}{n - Y^{nk}(n)} & \text{for} \quad 1 \leq \dfrac{V - Y^{nk}(n)}{n - Y^{nk}(n)} < K_{j,\max} \end{cases} \tag{8.55}$$

where the $Y^{nk}(n)$ parameter is expressed by Equation 8.27 and $Y^{k}(n)$ can be determined based on Equation 8.30.

Finally, the average number of BBUs occupied by compressed traffic can be expressed with the following dependence (in Equation 8.33):

$$Y_j^k = \sum_{n=0}^{V} y_j(n)[\xi_{k,j}(n)t_{j,\min}][P_n]_V \tag{8.56}$$

8.5.3.5 Average Throughput Available for HSPA Users

To determine the average capacity of the interface available to the HSPA traffic, it is necessary to first determine the occupancy distribution:

$$n[P_n]_V = \sum_{i=1}^{M_{nk}} A_i t_i [P_{n-t_i}]_V + \sum_{j=1}^{M_k} A_j t_{j,\min}[P_{n-t_{j,\min}}]_V \tag{8.57}$$

For each occupancy state n BBUs, the average number of service calls of particular traffic classes is determined on the basis of Equation 8.15:

$$y_i(n) = \begin{cases} \dfrac{A_i[P_{n-t_i}]_V}{[P_n]_V} & \text{for} \quad i \in \mathbb{M}_{nk} \\ \dfrac{A_i[P_{n-t_{i,\min}}]_V}{[P_n]_V} & \text{for} \quad i \in \mathbb{M}_{k} \end{cases} \tag{8.58}$$

Knowing the average number of calls $y_i(n)$ of each of the traffic classes, we can, for state n, determine the bandwidth (the number of available BBUs) that can be used by the HSPA traffic as the difference between the total capacity of the cell and the number of BBUs occupied by the UMTS calls. The average throughput offered to

HSDPA calls is equal to:

$$T_x = \sum_{n=0}^{V} \left[V - \sum_{i=1}^{M_{nk}} y_i(n) t_i \right] [P_n]_V \tag{8.59}$$

8.5.3.6 Summary

The models presented in this section can be used for the analysis and dimensioning of the WCDMA interface that services a mixture of different R99 *i* HSPA traffic classes, both in the uplink and the downlink directions. The proposed models enable us to determine four different GoS parameters, to which different priorities can be assigned, depending on the preferred optimization and development policy of the UMTS network operator. Therefore, the interface dimensioning process calculations of the quality parameters are to be repeated iteratively, each time with an increase in the interface capacity and checking if the GoS parameters, significant for the operator, are correct. The dimensioning process is terminated when these requirements are met.

Trying to maximize the simplicity of the described analytical models, we assume that the WCDMA radio interface services traffic generated by an infinite number of users (Erlang traffic). When the radio interface services a number of users of a given class that is lower or only slightly higher than the interface capacity, the proposed models should also include Engset traffic. The method for determining the characteristics of the system with Erlang and Engset traffic is presented in Section 8.3.4.

The proposed analytical methods are based on the well-known and verified Kaufman-Roberts distribution. The calculations made with the formulas presented in the method are not complicated or complex; this is, undoubtedly, an advantage from the network designer's point of view.

8.6 Dimensioning of the Iub Interface with HSPA Traffic

8.6.1 Exemplary Architecture of the Iub Interface

Having in mind the duration time of network expansion and the huge costs involved, as well as possible savings in expenditures, the operators of cellular networks are inclined to implement technological solutions that optimize investments but still retain the complex quality of service. One such solution, frequently used in real networks, is the separation of links on the Iub interface. The operator is in a position to configure two virtual paths (VPs) of ATM (asynchronus transfer mode) system on the Iub interface and assign them respectively to real-time traffic and best-effort traffic. Assuming that the best effort VC (virtual channel) will not allocate the maximum demanded bandwidth in the same time, the total bandwidth can be co-shared among the VCs, which results in its better utilization. This method should thus be recommended

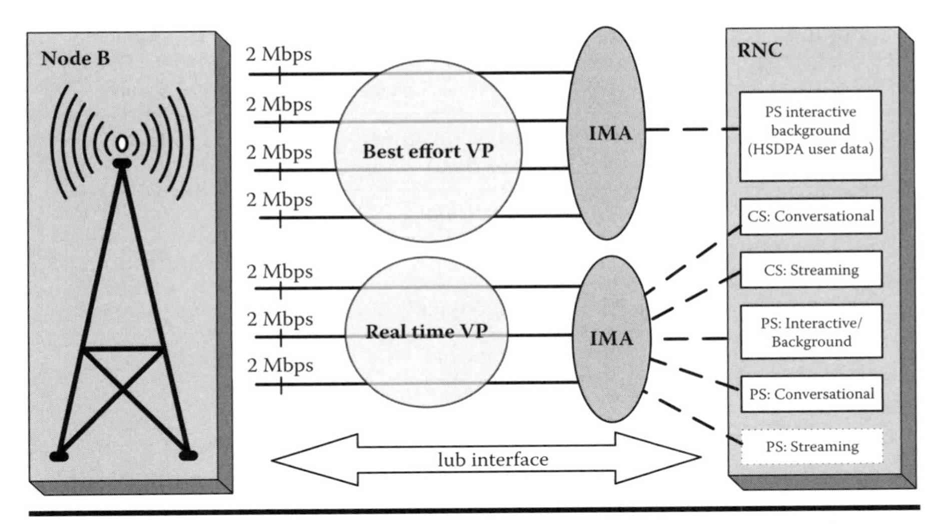

Figure 8.6 One of the most common ways of carrying out a connection between the UMTS base station and radio network controller with the application of IMA technology. (With kind permission from Springer Science+Business Media: Stasiak, M., Zwierzykowski, P., Wiewióra, J., and Parniewicz, D., European Performance Engineering Workshop, volume 5652 of LNCS, Analytical Model of Traffic Compression in the UMTS network, pp. 79–93. Springer, London, July 2009.)

even for distinguishing parameters needed for the designing/dimensioning of networks with different QoS requirements for different clients. Obviously, in the case of bandwidth overload, part of the ATM cells will be lost. An example of physical realization of a solution of this type on the Iub interface, with the application of IMA (inverse multiplexing for ATM) [26], is shown in Figure 8.6 [16]. The application of IMA makes it possible to create two logical ATM paths on the basis of separate physical links.* Table 8.4 shows an example of UMTS packet switched (PS) and circuit switched (CS) services, carried out by logical ATM paths dedicated to servicing best-effort traffic and real-time traffic, respectively, and corresponding to Figure 8.6.

Additionally, it should be mentioned that this solution paves the way for further optimization of capacity since with the application of traffic concentration devices between NodeB and RNC, the paths of the "real-time" type will be carried by the concentration device in the capacity ratio 1:1, while the paths of the "best-effort" type can be carried, for example, in the ratio 2:1 (a two-fold higher capacity at the input of the concentration device than at the output). Using the properties of offered traffic (e.g., different busy hours), we can get further savings, at least by means of developing or expanding RNC that has a limited number of input ports. A very good technology that ensures successful realization of the task, simultaneously facilitating

* Figure 8.6 assumes that the links constituting IMA have throughput of 2 Mbps

Table 8.4 An Example of Service Class Mapping into ATM Classes

ATM Class of Service	*UMTS Class of Service*	*Exemplary Service*
Best-effort VP	Interactive background (HSDPA user data)	Web browsing
Real-time VP	CS: Conversational	Voice
Real-time VP	CS: Streaming	Modem connection
Real-time VP	PS: Interactive/background	FTP, realtime gaming
Real-time VP	PS: Conversational	Speech (VoIP)
Real-time VP	PS: Streaming	Mobile TV

the construction of the Iub interface, is LMDS (local multipoint distribution service) [27].

Regrettably, this rapid pace in the development of relevant technologies is not appropriately matched by mathematical models that could enable us to plan and dimension networks in accordance with required service predictions.

8.6.2 Analytical Model of the Iub Interface

The Iub interface in a UMTS network can be treated as the full-availability group (FAG) with multirate traffic. In the model, we assume, similar to the WCDMA interface, that the Iub interface carries both R9 and HSPA traffic streams. We also assume there are traffic classes belonging to the HSPA traffic with calls that can change occupied resources in the service time. Therefore, it is assumed that the system services simultaneously a mixture of different multirate traffic classes, while these classes are divided into two sets: $\mathbb{M}_k$ classes whose calls can change requirements while being serviced, and $\mathbb{M}_{nk}$ classes that do not change their demands in the service time. Let us assume that the total capacity of the group is equal to V basic bandwidth units (BBUs). The group is offered M^* independent classes of Poisson traffic streams,[†] having the intensities: $\lambda_1, \lambda_2, \ldots, \lambda_M$. The class i call requires t_i BBUs to set up a connection. The holding time for calls of particular classes has an exponential distribution with the parameters: $\mu_1, \mu_2, \ldots, \mu_M$. Thus, the mean traffic offered to the system by the class i traffic stream is equal to:

$$A_i = \frac{\lambda_i}{\mu_i} \tag{8.60}$$

* $M = M_k + M_{nk}$, where $M_k = |\mathbb{M}_k|$ and $M_{nk} = |\mathbb{M}_{nk}|$

† In the analytical model, for simplicity, we assume the system carries only Erlang traffic streams.

The resources demanded in the group for servicing particular classes can be treated as a call demanding an integer number of BBUs. The value of BBU (i.e., R_{BBU}, is calculated as the greatest common divisor (GCD) of all resources demanded by the traffic classes offered to the system (Equation 8.44):

$$R_{BBU} = GCD\,(R_1, \ldots, R_M) \tag{8.61}$$

where R_i is the amount of the resources demanded by the class i call in $kbps$.

The multidimensional Markov process in the FAG can be approximated by the one-dimensional Markov chain, which can be described by Kaufman-Roberts recursion (Equation 8.15):

$$n[P_n]_V = \sum_{i=1}^{M_{nk}} A_i t_i [P_{n-t_i}]_V + \sum_{j=1}^{M_k} A_j t_{j,\min} [P_{n-t_{j,\min}}]_V \tag{8.62}$$

where $[P_n]_V$ is the probability state of n BBUs being busy, and t_i and $t_{j,\min}$ are the number of BBUs required by a class that is not undergoing, and a class that is undergoing, compression, respectively (Equation 8.46):

$$t_i = \left\lfloor \frac{R_i}{R_{BBU}} \right\rfloor \qquad t_{j,\min} = \left\lfloor \frac{R_{j,\min}}{R_{BBU}} \right\rfloor \tag{8.63}$$

where $R_{j,\min}$ is the minimum amount of resources demanded by class j traffic undergoing compression, in kbps. In Equation 8.62, the interface capacity V is defined as follows:

$$V = \lfloor V_{\text{phy}} / R_{BBU} \rfloor \tag{8.64}$$

where V_{phy} is the physical capacity of the group in kbps.

In this section, we will also analyze four GoS parameters: blocking probability, loss probability, average throughput, and available throughput.

8.6.2.1 Blocking (and Loss) Probability

On the basis of Formula 8.62, the blocking probability B_i for the class i Erlang traffic stream can be expressed in the following form [Equation 8.24]:

$$E_i = B_i = \begin{cases} \displaystyle\sum_{n=V-t_i+1}^{V} [P_n]_V & \text{for} \quad i \in \mathbb{M}_{nk} \\[2ex] \displaystyle\sum_{n=V-t_{i,\min}+1}^{V} [P_n]_V & \text{for} \quad i \in \mathbb{M}_k \end{cases} \tag{8.65}$$

8.6.2.2 Average Throughput

Determination of the average throughput is only important for those traffic classes of the HSDPA traffic* that can undergo compression. The application of a given analytical model depends on particular mechanisms used in the solutions provided by manufacturers of equipment for the UMTS network. Let us consider a scenario in which the average bandwidth is assigned to all users unevenly. Let us further assume that the subscribers in this network have terminals of different classes, while those subscribers that have newer terminals are capable of achieving higher maximum throughput. This scenario can be further considered with the application of the model with uneven compression, presented in Section 8.4.2.

In the first stage of the determination of average throughput we determine the compression coefficient $\xi_k(n)$. The coefficient, following the dependencies Equations 8.31 and 8.36, takes on the following form:

$$\xi_{k,j}(n) = \begin{cases} K_{j,\max} & \text{for} \quad \dfrac{V - Y^{nk}(n)}{n - Y^{nk}(n)} \geq K_{j,\max} \\ \dfrac{V - Y^{nk}(n)}{n - Y^{nk}(n)} & \text{for} \quad 1 \leq \dfrac{V - Y^{nk}(n)}{n - Y^{nk}(n)} < K_{j,\max} \end{cases} \tag{8.66}$$

where the $Y^{nk}(n)$ parameter is expressed in the following way (Equation 8.27):

$$Y^{nk}(n) = \sum_{i=1}^{M_{nk}} y_i(n) t_i \tag{8.67}$$

and $Y^k(n)$ can be determined based on Equation 8.30:

$$Y^k(n) = \xi_k(n) \sum_{j=1}^{M_k} y_j(n) t_{j,\min} \tag{8.68}$$

In Equations 8.67 and 8.68, the average number of calls of class i, serviced in the occupancy state n BBUs $[y_j(n)]$, can be determined as follows (Equation 8.25):

$$y_i(n) = \begin{cases} \dfrac{A_i \left[P_{n-t_i}\right]_V}{[P_n]_V} & \text{for} \quad i \in \mathbb{M}_{nk} \\ \dfrac{A_i \left[P_{n-t_{i,\min}}\right]_V}{[P_n]_V} & \text{for} \quad i \in \mathbb{M}_k \end{cases} \tag{8.69}$$

* HSPA traffic is limited only to the downlink direction, because in the uplink direction HSPA traffic is services-based on R99 resources [12].

In the next step, we can obtain the average resources occupied by calls of class j (average throughput) on the basis of the following formula (Equation 8.33):

$$Y_j^k = \sum_{n=0}^{V} y_j(n)[\xi_{k,j}(n)t_{j,\min}][P_n]_V \tag{8.70}$$

8.6.2.3 Average Throughput Available for HSDPA Users

The average capacity of the Iub interface available to the HSDPA traffic can be determined in a similar way as the available throughput of the WCDMA interface presented in Section 8.5.3.5.

8.7 Conclusion

This chapter presents analytical methods that allow us to determine such a capacity of individual elements of the UMTS system that will guarantee—with the assumed load of the system—a pre-defined level of GoS. The most characteristic constraints in the dimensioning of the UMTS system are the radio interface and the Iub interface. The chapter describes the application of the analytical models to these interfaces. In the models, it was assumed that the system carried a mixture of different R99 and HSPA traffic classes.

References

[1] M. Glabowski, M. Stasiak, A. Wiśniewski, and P. Zwierzykowski. *Performance Modelling and Analysis of Heterogeneous Networks*, chapter "Uplink Blocking Probability Calculation for Cellular Systems with WCDMA Radio Interface and Finite Source Population," pp. 301–318. Information Science and Technology. River Publishers, 2009.

[2] Y. Ishikawa, S. Onoe, K. Fukawa, and H. Suzuki. "Blocking Probability Calculation Using Traffic Equivalent Distributions in sir-based Power Controlled w-cdma Cellular Systems." *IEICE Transactions on Communications*, E88-B(1):312–324, 2005.

[3] V. B. Iversen and E. Epifania. "Teletraffic Engineering of Multi-band W-CDMA Systems." In *Network Control and Engineering for QoS, Security and Mobility II*, pp. 90–103, Norwell, MA, 2003. Kluwer Academic Publishers.

[4] I. Koo and K. Kim. "Erlang Capacity of Multi-service Multi-access Systems with a Limited Number of Channel Elements According to Separate and Common Operations." *IEICE Transactions on Communications*, E89-B(11):3065–3074, 2006.

[5] D. Staehle and A. Mäder. An Analytic Approximation of the Uplink Capacity in a UMTS Network with Heterogeneous Traffic. *18th International Teletraffic Congress*, pp. 81–91, Berlin, 2003.

[6] M. Stasiak, J. WiewiÛra, and P. Zwierzykowski. "Analytical Modelling of the Iub Interface in the UMTS Network. *Proceedings of the 6th Symposium on Communication Systems, Networks, and Digital Signal Processing*, Graz, Austria, July 2008.

[7] M. Stasiak, A. Wiśniewski, P. Zwierzykowski, and M. Glabowski. "Blocking Probability Calculation for Cellular Systems with WCDMA Radio Interface Servicing PCT1 and PCT2 Multirate Traffic." *IEICE Transactions on Communications*, E92-B(4):1156–1165, April 2009.

[8] H. Akimuru and K. Kawashima. *Teletraffic: Theory and Application.* Berlin-Heidelberg-New York, 1999.

[9] J. S. Kaufman. "Blocking in a Shared Resource Environment." *IEEE Transactions on Communications*, 29(10):1474–1481, 1981.

[10] J. W. Roberts. "A Service System with Heterogeneous User Requirements — Application to Multi-service Telecommunications Systems." In G. Pujolle, editor, *Proceedings of Performance of Data Communications Systems and Their Applications*, pp. 423–431, Amsterdam, 1981.

[11] M. Glabowski. "Modelling of State-Dependent Multirate Systems Carrying BPP Traffic." *Annals of Telecommunications*, 63(7-8):393–407, August 2008.

[12] H. Holma and A. Toskala. *HSDPA/HSUPA for UMTS: High Speed Radio Access for Mobile Communications.* John Wiley and Sons, 2006.

[13] F.P. Kelly. "Loss Networks." *The Annals of Applied Probability*, 1(3):319–378, 1991.

[14] I. D. Moscholios, M. D. Logothetis, and G. K. Kokkinakis. "Connection-dependent Threshold Model: A Generalization of the Erlang Multiple Rate Loss Model. *Performance Evaluation* 48:177–200, May 2002.

[15] S. Rácz, B. P. Gerö, and G. Fodor. "Flow Level Performance Analysis of a Multi-service System Supporting Elastic and Adaptive Services. *Performance Evaluation*, 49(1-4):451–469, 2002.

[16] M. Stasiak, P. Zwierzykowski, J. Wiewióra, and D. Parniewicz. *European Performance Engineering Workshop*, vol. 5652 of LNCS, chapter "Analytical Model of Traffic Compression in the UMTS Network," pp. 79–93. Springer, London, July 2009.

[17] J. Laiho, A. Wacker, and T. Novosad. *Radio Network Planning and Optimization for UMTS.* John Wiley and Sons, Ltd., 2006.

[18] M. Stasiak and P. Zwierzykowski. "Modelling Full Availability Groups with Adaptive-Rate." Internal report 9/2008, Poznan University of Technology, September 2008.

[19] H. Holma and A. Toskala. *WCDMA for UMTS. Radio Access for Third Generation Mobile Communications.* John Wiley and Sons, 2000.

[20] M. Stasiak, A. Wiśniewski, and P. Zwierzykowski. "Blocking Probability Calculation in the Uplink Direction for Cellular Systems with WCDMA Radio Interface," In *3rd Polish-German Teletraffic Symposium*, pp.65–74, Dresden, 2004.

[21] Engineering Services Group, *Aspects of HSUPA Network Planning*, Qualcomm Incorporated, Technical Report, No. 80-W1159-1, Revision B, San Diego, 2007.

[22] S. Faruque. *Cellular Mobile Systems Engineering.* Artech House, London, 1997.

[23] J. W. Roberts, ed. *Performance Evaluation and Design of Multiservice Networks, Final Report COST 224.* Commission of the European Communities, Brussels, 1992.

[24] J. W. Roberts, V. Mocci, and I. Virtamo, ed. *Broadband Network Teletraffic, Final Report of Action COST 242*, Springer, Berlin, 1996.

[25] M. Stasiak, P. Zwierzykowski, J. Wiewióra, and D. Parniewicz. *European Performance Engineering Workshop*, vol. 5261 of LNCS, chapter "An Approximate Model of the WCDMA Interface Servicing a Mixture of Multirate Traffic Streams with Priorities," pp. 168–180. Springer, Palma de Mallorca, September 2008.

[26] J. Bannister, P. Mather, and S. Coope. *Convergence Technologies for 3G Networks: IP, UMTS, EGPRS and ATM.* Wiley, March 2004.

[27] C. Smith. *LMDS: Local Mutipoint Distribution Service.* McGraw-Hill, August 2000.

Chapter 9

Radio Resource Management for E-MBMS Transmissions toward LTE

Antonios Alexiou, Christos Bouras, and Vasileios Kokkinos

Contents

9.1 Introduction

Indisputably, tomorrow's mobile marketplace will be characterized by bandwidth-hungry multimedia services that are already experienced in wired networks. Long-term evolution (LTE), the evolutionary successor of universal mobile telecommunication system (UMTS) and high-speed packet access (HSPA) networks, addresses this emerging trend, by shaping the future mobile broadband landscape. LTE promises a richer, more immersive environment that significantly increases peak data rates and spectral efficiency. However, the plethora of mobile multimedia services that are expected to face high penetration, poses the need for the deployment of a resource economic scheme. Multimedia broadcast/multicast service (MBMS), also called evolved MBMS (e-MBMS) in LTE terminology, constitutes an efficient way to compensate for this necessity since it allows resources' sharing during data transmission [1, 2].

The main requirement during the provision of MBMS multicast services is to make an efficient overall usage of radio and network resources. The system should conceive and adapt to continuous changes that occur in such dynamic wireless environments and optimally allocate resources. To this direction, a critical aspect of MBMS performance is the selection of the most efficient radio bearer, in terms of power consumption, for the transmission of multimedia traffic. The selection of the most efficient radio bearer is an open issue in today's MBMS infrastructure and several mechanisms have been proposed to this direction. Nevertheless, the selection of the most appropriate mechanism is plagued with uncertainty, since each mechanism may provide specific advantages. In this chapter, the prevailing radio bearer selection mechanisms are presented and compared in terms of power consumption so as to highlight the advantages each mechanism may provide.

Additionally, this chapter examines the operation and performance of several techniques, such as dynamic power setting (DPS), macro diversity combining (MDC), and rate splitting (RS) that could be utilized in order to further minimize the base station's total MBMS transmission power. This chapter examines the operation and performance of these techniques and demonstrates the amount of power that could be saved through their employment.

Furthermore, in this chapter the performance enhancements emerged from multiple-input multiple-output (MIMO) antennas used in next-generation mobile networks is highlighted. MIMO systems are a prerequisite for next-generation mobile networks and have the potential to address the unprecedented demand for wireless multimedia services and particularly for MBMS. In particular, the intention is to examine how the introduction of MIMO antenna systems affect the MBMS power planning strategy of next-generation cellular networks.

9.2 Multimedia Broadcast/Multicast Service

In MBMS, rich wireless multimedia data is transmitted simultaneously to multiple recipients by allowing resources to be shared in an economical way. MBMS efficiency is derived from the single transmission of identical data over a common channel without clogging up the air interface with multiple replications of the same data.

The major factor for integrating MBMS into UMTS networks was the rapid growth of mobile communications technology and the massive spread of wireless data and wireless applications. The increasing demand for communication between one sender and many receivers led to the need for point-to-multipoint (PTM) transmission. PTM transmission is opposed to the point-to-point (PTP) transmission, using the unicast technology, which is exclusively used in conventional UMTS networks (without the MBMS extension). Broadcast and multicast technologies constitute an efficient way to implement this type of communication and enable the delivery of a plethora of high-bandwidth multimedia services to a large number of users.

From the service and operators' point of view, the employment of MBMS framework involves both an improved network performance and a rational usage of radio resources, which in turn leads to extended coverage and service provision. In parallel, users are able to realize novel, high bit-rate services, experienced until today only by wired users. Such services include mobile TV, weather, or sports news as well as fast and reliable data downloading [3].

9.2.1 Operation

As the term MBMS indicates, there are two types of service modes: the broadcast mode and the multicast mode. Each mode has different characteristics in terms of complexity and packet delivery.

The broadcast service mode is a unidirectional PTM transmission type. Actually, with broadcast, the network simply floods data packets to all nodes within the network. In this service mode, content is delivered using PTM transmission, to a specified area without knowing the receivers and no matter whether there is any receiver in the area. As a consequence, the broadcast mode requires no subscription or activation from the users' point of view.

In the multicast operation mode, data are transmitted solely to users that explicitly request such a service. More specifically, the receivers have to signal their interest for the data reception to the network and then the network decides whether the user may receive the multicast data or not. Thus, in the multicast mode there is the possibility for the network to selectively transmit to cells, which contain members of a multicast group. Either PTP or PTM transmission can be configured in each cell for the multicast operation mode [2].

Unlike the broadcast mode, the multicast mode generally requires a subscription to the multicast subscription group and then the user joining the corresponding multicast group. Moreover, due to the selective data transmission to the multicast group, it is expected that charging data for the end user will be generated for this mode, unlike the broadcast mode.

9.2.2 Architecture

The MBMS framework requires minimal modifications in the current UMTS architecture. As a consequence, this fact enables the fast and smooth upgrade from pure UMTS networks to MBMS-enhanced UMTS networks. Actually, MBMS consists of a MBMS bearer service and a MBMS user service. The latter represents applications, which offer, for example, multimedia content to the users, while the MBMS bearer service provides methods for user authorization, charging, and quality of service (QoS) improvement to prevent unauthorized reception [2].

The UMTS network is split into two main domains: the user equipment (UE) domain and the public land mobile network (PLMN) domain. The UE domain consists of the equipment employed by the user to access the UMTS services. The PLMN domain consists of two land-based infrastructures: the core network (CN) and the UMTS terrestrial radio access network (UTRAN) (Figure 9.1). The CN is responsible for switching/routing voice and data connections, while the UTRAN handles all radio-related functionalities. The CN is logically divided into two service domains: the circuit-switched (CS) service domain and the packet-switched (PS) service domain [3]. The CS domain handles the voice-related traffic, while the PS domain handles the packet transfer. The remainder of this chapter will focus on the UMTS packet-switching mechanism.

The PS portion of the CN in UMTS consists of two kinds of general packet radio service (GPRS) support nodes (GSNs), namely gateway GSN (GGSN) and serving GSN (SGSN) (Figure 9.1). SGSN is the centerpiece of the PS domain. It provides routing functionality, interacts with databases [like home location register (HLR)] and manages many radio network controllers (RNCs). SGSN is connected to GGSN via the Gn interface and to RNCs via the Iu interface. GGSN provides the interconnection of UMTS network (through the broadcast multicast service center) with other packet data networks (PDNs), like the Internet.

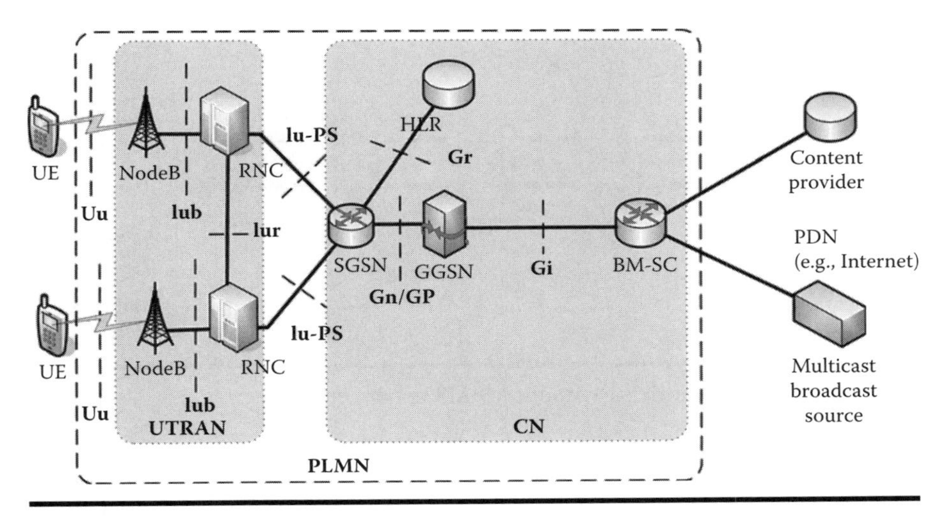

Figure 9.1 UMTS and MBMS architecture.

UTRAN consists of two kinds of nodes: the first is the RNC and the second is the NodeB. NodeB constitutes the base station and provides radio coverage to one or more cells (Figure 9.1). NodeB is connected to the UE via the Uu interface [based on the wideband code division multiple access (WCDMA) technology] and to the RNC via the Iub interface. One RNC with all the NodeBs connected to it is called radio network subsystem (RNS) [3].

The major modification in the existing UMTS platform for the provision of the MBMS framework is the addition of a new entity called broadcast multicast service center (BM-SC). Actually, BM-SC acts as an entry point for data delivery between the content providers and the UMTS network and is located in the PS domain of the CN. The BM-SC entity communicates with existing UMTS networks and external PDNs [1, 2].

The BM-SC is responsible for both control and user planes of a MBMS service. More specifically, the function of the BM-SC can be separated into five categories: membership, session and transmission, proxy and transport, service announcement, and security function. The BM-SC membership function provides authorization to the UEs requesting to activate a MBMS service. According to the session and transmission function, the BM-SC can schedule MBMS session transmissions and shall be able to provide the GGSN with transport associated parameters, such as QoS and MBMS service area. As far as the proxy and transport function is concerned, the BM-SC is a proxy agent for signaling over a Gmb reference point between GGSNs and other BM-SC functions. Moreover, the BM-SC service announcement function must be able to provide service announcements for multicast and broadcast MBMS user services and provide the UE with media descriptions specifying the media to be delivered as part of a MBMS user service. Finally, MBMS user services may use

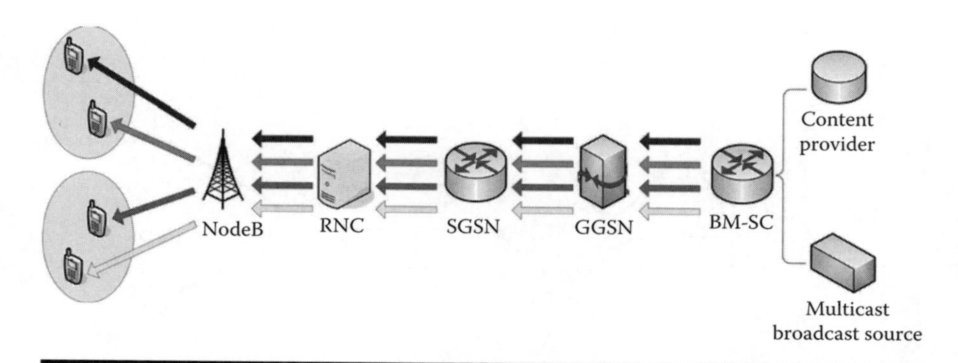

Figure 9.2 UMTS multicast without MBMS enhancement.

the security functions for integrity or confidentiality protection of the MBMS data, while the specific function is used for distributing MBMS keys (key distribution function) to authorized UEs.

9.2.3 Multicast Mode of MBMS

MBMS multicast efficiency improvement in UMTS networks can be derived from Figures 9.2 and 9.3. More specifically, these figures present the UMTS multicast functionality without and with MBMS enhancement, respectively.

Without the MBMS enhancement, multicast data is replicated as many times as the total number of multicast users in all interfaces. Obviously, a bottleneck is created when the number of users increases significantly. All interfaces are heavily overloaded due to the multiple transmissions of the same data. On the other hand, MBMS multicast benefits UMTS networks through the radio and network resources' sharing. Only a single stream per MBMS service of identical data is essential for the delivery of the multicast content, thus saving expensive resources. Conclusively,

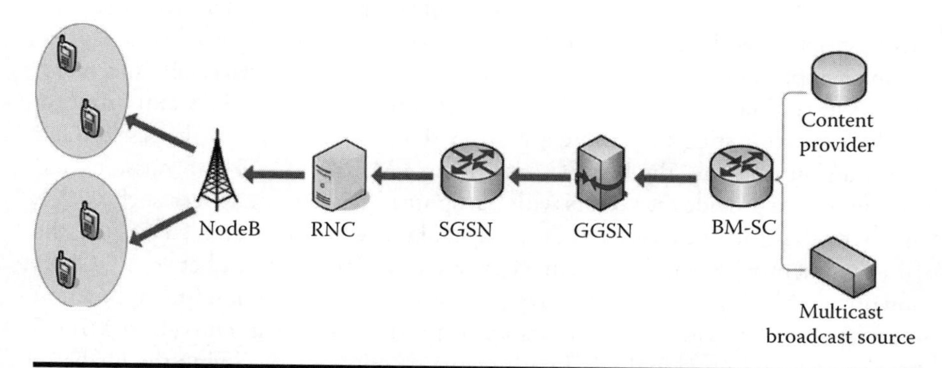

Figure 9.3 UMTS multicast with MBMS enhancement.

MBMS multicast data distribution is optimally configured throughout the UMTS network.

9.2.3.1 Packet Delivery Process

An overview of the multicast data flow procedure during a MBMS service provision is presented in this paragraph. Figure 9.4 depicts a subset of a UMTS-MBMS network. In this architecture, there are two SGSNs connected to a GGSN, four RNCs, and twelve NodeBs. Furthermore, eleven members of a multicast group are located in six cells. The BM-SC acts as the interface toward external sources of traffic. The presented analysis assumes that a data stream that comes from an external PDN, through BM-SC, must be delivered to the eleven UEs as illustrated in Figure 9.4.

The analysis presented in this paragraph, covers the forwarding mechanism of the data packets between the BM-SC and the UEs. With multicast, the packets are forwarded only to those NodeBs that have multicast users. Therefore, in Figure 9.4, the NodeB2, B3, B5, B7, B8, and B9 receive the multicast packets issued by the BM-SC. We briefly summarize the five steps needed for the delivery of the multicast packets.

Initially, the BM-SC receives a multicast packet and forwards it to the GGSN that has registered to receive the multicast traffic. Then, the GGSN receives the multicast packet and by querying its multicast routing lists, it determines which SGSNs have multicast users residing in their respective service areas. In Figure 9.4, the GGSN

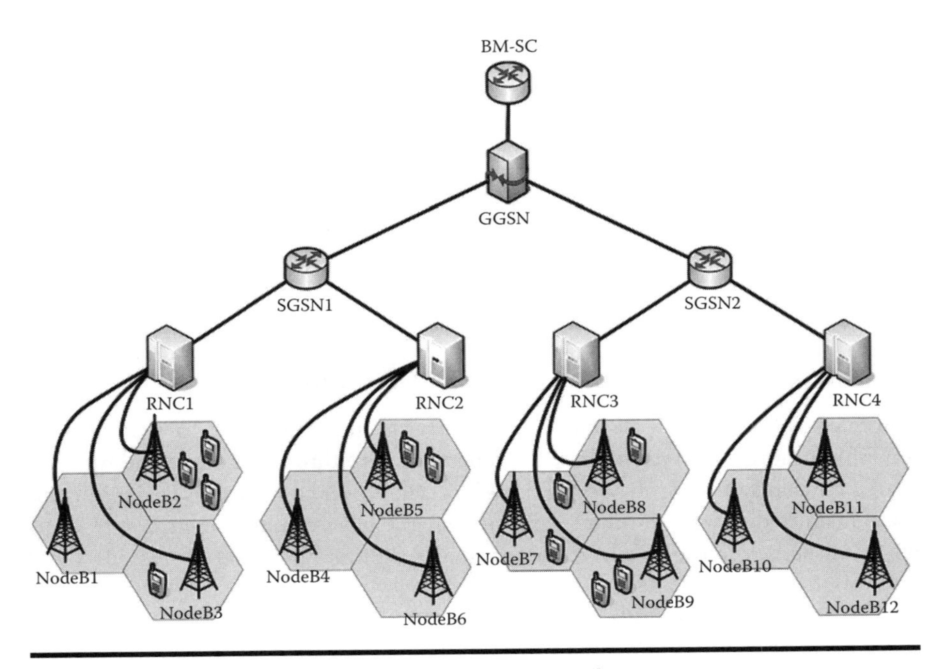

Figure 9.4 Packet delivery in MBMS multicast mode.

duplicates the multicast packet and forwards it to the SGSN1 and the SGSN2 [4]. Then, both destination SGSNs receive the multicast packets and, having queried their multicast routing lists, determine which RNCs are to receive the multicast packets. The destination RNCs receive the multicast packet and send it to the NodeBs that have established the appropriate radio bearers for the multicast application. In Figure 9.4, these are NodeB2, B3, B5, B7, B8, and B9. The multicast users receive the multicast packets on the appropriate radio bearers, by dedicated channels transmitted to individual users separately or by common channels transmitted to all members in the cell [4].

9.2.3.2 MBMS Multicast Mode Radio Bearers

According to current MBMS specifications, the transmission of the MBMS multicast packets over the Iub and Uu interfaces may be performed on common (forward access channel—FACH), on dedicated (dedicated channel—DCH) channels, or on the shared channel named high-speed downlink shared channel (HS-DSCH), introduced in Release 5. The main requirement is to make an efficient overall utilization of the radio resources: this makes a common channel the favorite choice, since many users can access the same resource at the same time.

More specifically, the transport channel that the 3rd Generation Partnership Project (3GPP) decided to use as the main transport channel for PTM MBMS data transmission is the FACH with turbo coding and quadrature phase shift keying (QPSK) modulation at a constant transmission power [1]. DCH is a PTP channel and hence, it suffers from the inefficiencies of requiring multiple DCHs to carry the data to a group of users. However, DCH can employ fast closed-loop power control and soft handover mechanisms, and generally is a highly reliable channel [3, 5]. The allocation of HS-DSCH as a transport channel affects the obtained data rates and the remaining capacity to serve Release'99 users (users served by DCH). High-speed downlink packet access (HSDPA) cell throughput increases when more HSDPA power is allocated, while DCH throughput simultaneously decreases [6].

9.3 Power Control in MBMS Multicast Mode

Power control is one of the most critical aspects in MBMS due to the fact that downlink transmission power in UMTS networks is a limited resource and must be shared efficiently among all MBMS users in a cell. Power control aims at minimizing the transmitted power, eliminating in this way the inter-cell interference. However, when misused, the use of power control may lead to a high level of wasted power and worse performance results.

On the PTP downlink transmissions, fast power control is used to maintain the quality of the link and thus provide a reliable connection for the receiver to obtain the data with acceptable error rates. Transmitting with just enough power to maintain the required quality for the link also ensures there is minimum interference affecting the neighboring cells. However, when a user consumes a high portion of power, more than actually required, the remaining power, allocated for the rest of the users, is dramatically decreased, thus leading to a significant capacity loss in the system.

During PTM downlink transmissions, NodeB transmits at a power level that is high enough to support the connection to the receiver with the highest power requirement among all receivers in the multicast group. This would still be efficient because the receiver with the highest power requirement would still need the same amount of power in a unicast link, and by satisfying that particular receiver's requirement the transmission power will be enough for all the other receivers in the multicast group. Consequently, the transmitted power is kept at a relatively high level most of the time, which in turn increases the signal quality at each receiver in the multicast group. On the other hand, a significant amount of power is wasted and moreover inter-cell interference is increased.

As a consequence, downlink transmission power plays a key role in MBMS planning and optimization. This section provides an analytical description of the HS-DSCH, DCH, and FACH power profiles that are employed during PTP and PTM transmission. The following analysis refers to a macrocell environment with parameters described in Table 9.1 [3, 7].

Table 9.1 Macrocell Simulation Assumptions

Parameter	*Value*
Cellular layout	Hexagonal grid
Number of cells	18
Sectorization	3 sectors/cell
Site-to-site distance	1 km
Cell radius	0.577 km
Maximum BS Tx power	20 watt (43 dBm)
Other BS Tx power	5 watt (37 dBm)
Common channel power	1 watt (30 dBm)
Propagation model	Okumura Hata
Multipath channel	Vehicular A (3 km/h)
Orthogonality factor	0.5
E_b/N_o target	5 dB

9.3.1 HS-DSCH Power Profile

HS-DSCH is a rate-controlled rather than a power-controlled transport channel. Although there are two basic modes for allocating HS-DSCH transmission power [6], this chapter will focus on a dynamic method in order to provide only the required marginal amount of power needed to satisfy all the served multicast users and, in parallel, eliminate interference. Two major measures for HSDPA power planning are the HS-DSCH signal-to-interference-plus-noise ratio (SINR) metric and the geometry factor (G). SINR for a single-antenna Rake receiver is calculated as in Equation 9.1 [6]:

$$\mathrm{SINR} = SF_{16} \cdot \frac{P_{HS-DSCH}}{p \cdot P_{\mathrm{own}} + P_{\mathrm{other}} + P_{\mathrm{noise}}} \tag{9.1}$$

where $P_{HS-DSCH}$ is the HS-DSCH transmission power, P_{own} is the own cell interference experienced by the mobile user, P_{other} is the interference from neighboring cells, and P_{noise} is the additive white gaussian noise. Parameter p is the orthogonality factor ($p = 0$: perfect orthogonality), while SF_{16} is the spreading factor of 16.

The geometry factor is another major measure that indicates the users' position throughout a cell. A lower G is expected when a user is located at the cell edge. G is calculated as in Equation 9.2 [3]:

$$G = \frac{P_{\mathrm{own}}}{P_{\mathrm{other}} + P_{\mathrm{noise}}} \tag{9.2}$$

There is a strong relationship between the HS-DSCH allocated power and the obtained MBMS cell throughput. This relationship can be disclosed in the three following steps. Initially, we have to define the target MBMS cell throughput. Once the target cell throughput is set, the next step is to define the way this throughput relates to the SINR [6]. Finally, we can describe how the required HS-DSCH transmission power ($P_{HS-DSCH}$) can be expressed as a function of the SINR value and the user location (in terms of G) as in Equation 9.3 [6]:

$$P_{HS-DSCH} \geq \mathrm{SINR} \cdot [p - G^{-1}] \cdot \frac{P_{\mathrm{own}}}{SF_{16}} \tag{9.3}$$

When MIMO is supported in HS-DSCH, multiple transmit antennas and receive antennas are used (different data streams are transmitted simultaneously over each antenna) and SINR is further improved [8]. Early requirements consider two transmit and receive antennas (MIMO 2x2) and approximately, double data rates are obtained with the same base station transmission power. Therefore, without loss of generality, half power is required, compared to conventional HS-DSCH single antenna systems, for the delivery of the same MBMS session. In other words,

MIMO further contributes in saving significant power resources and, in parallel, maximizing system capacity.

9.3.2 DCH Power Profile

The total downlink transmission power allocated for all MBMS users in a cell served by multiple DCHs is variable. It mainly depends on the number of served users, their location in the cell, the bit rate of the MBMS session and the experienced signal quality, E_b/N_o, for each user. Equation 9.4 calculates the NodeB's total DCH transmission power required for the transmission of the data to n users in a specific cell [9].

$$P_{DCH} = \frac{P_P + \sum_{i=1}^{n} \frac{P_N + x_i}{\frac{W}{(E_b/N_o)_i \cdot R_{b,i}} + p} \cdot L_{p,i}}{1 - \sum_{i=1}^{n} \frac{p}{\frac{W}{(E_b/N_o)_i \cdot R_{b,i}} + p}} \tag{9.4}$$

where P_{DCH} is the base station's total transmitted power, P_P is the power devoted to common control channels, $L_{p,i}$ is the path loss, $R_{b,i}$ is the ith user transmission rate, W is the bandwidth, P_N is the background noise, p is the orthogonality factor ($p = 0$ for perfect orthogonality), and x_i is the inter-cell interference observed by the ith user given as a function of the transmitted power by the neighboring cells P_{Tj}, $j = 1, \ldots, K$, and the path loss from this user to the jth cell L_{ij}. More specifically [9]:

$$x_i = \sum_{j=1}^{K} \frac{P_{Tj}}{L_{ij}} \tag{9.5}$$

DCH may be used for the delivery of PTP MBMS services, but can not be used to serve large multicast populations since high downlink transmission power would be required. Figure 9.5 depicts the downlink transmission power when MBMS multicast data are delivered over multiple DCHs (one separate DCH per user). Obviously, higher power is required to deliver higher MBMS data rates. In addition, an increased cell coverage area and larger user groups lead to higher power consumption.

9.3.3 FACH Power Profile

A FACH essentially transmits at a fixed power level since fast power control is not supported. FACH is a PTM channel and must be received by all users throughout the cell (or the part of the cell that the users reside in), thus, the fixed power should be high enough to ensure the requested QoS in the desired coverage area of the cell, irrespective of users' locations. FACH power efficiency strongly depends

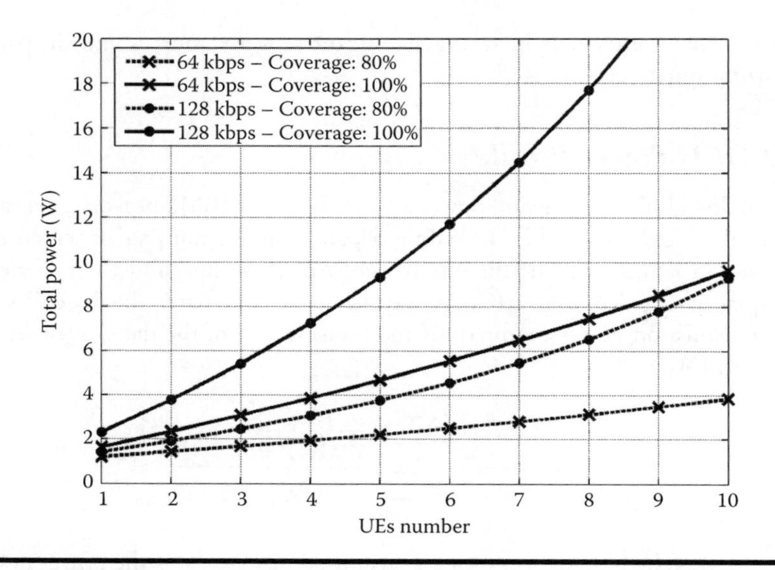

Figure 9.5 DCH transmission (Tx) power.

on maximizing diversity as power resources are limited. Diversity can be obtained by the use of a longer transmission time interval (TTI) in order to provide time diversity against fast fading (fortunately, MBMS services are not delay-sensitive) and the use of combining transmissions from multiple cells to obtain macro diversity [10, 11].

Table 9.2 presents some indicative FACH downlink transmission power levels obtained for various cell coverage areas and MBMS bit rates, without assuming diversity techniques [10]. A basic constraint is that the delivery of high data rate MBMS services over FACH is not feasible, since excessive downlink transmission power would be required (overcoming the maximum available power of 20 W). High bit rates can only be offered to users located very close to NodeB.

Table 9.2 FACH Tx Power Levels

Cell Coverage (%)	*Service Bit Rate (kbps)*	*Required Tx Power (W)*
50	32	1.8
	64	2.5
95	32	4.0
	64	7.6

9.4 Power Saving Techniques

The main problem during a MBMS session, in terms of power consumption, is the exceedingly high fixed power levels when allocating FACH as a transport channel. As an example, we mention that in order to provide a 128-kbps MBMS service with a FACH coverage set to 95% of the cell, 16 W of power is required. If we contemplate that the maximum transmission power of the NodeB is 20 W (which should be shared among all the users of the cell and among all the possible services), it becomes comprehensible that this level of power makes impossible the provision of services with such bit rates. The techniques stated in the remaining of this section partly overcome this problem, since they reduce the FACH transmission power levels.

9.4.1 Dynamic Power Setting

DPS is the technique where the transmission power of the FACH can be determined based on the worst user's path loss. This way, the FACH transmission power is allocated dynamically, and the FACH transmission power will need to cover the whole cell only if one or more users are at the cell boundary. To perform DPS, the MBMS users need to turn on the measurement report mechanism while they are in the Cell_FACH state. Based on such measurement reports, the NodeB can adjust the transmission power of the FACH [12].

This is presented in Figure 9.6, where the NodeB sets its transmission power based on the worst user's path loss (i.e., distance). The information about the path loss is sent to the NodeB via uplink channels. The examination of Figure 9.6 reveals that 4.0 W are required in order to provide a 32-kbps service to the 95% of the cell. However, supposing that all the MBMS users are found near the Node B (10% coverage) only 0.9 W is required. In that case, 3.1 W (4.0 W minus 0.9 W) can be saved while delivering a 32-kbps service, since with DPS the NodeB will set its transmission power so as to cover only the 10% of the cell. The corresponding power gain increases to 6.2 W for a 64-kbps service and to 13.4 W for a 128-kbps service. These high sums of power underline the need for using this technique.

9.4.2 Macro Diversity Combining

Diversity is a technique to combine several copies of the same message received over different channels. Macro diversity is normally applied as diversity switching where two or more base stations serve the same area, and control over the mobile is switched among them. Basically, the diversity combining concept consists of receiving redundantly the same information bearing signal over two or more fading channels, and combine these multiple replicas at the receiver in order to increase the overall received signal-to-noise ratio (SNR).

Figure 9.7 presents how the FACH transmission power level changes with cell coverage when MDC is applied. For the needs of the simulation, we considered

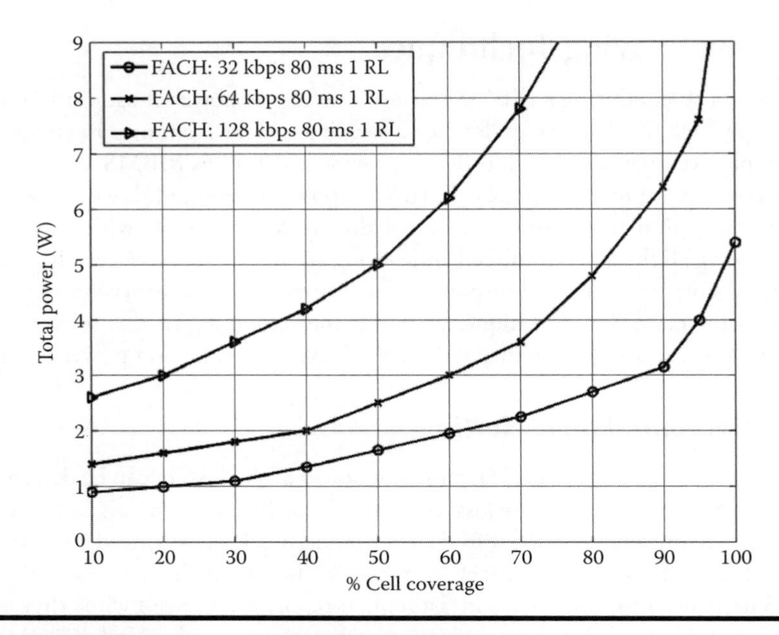

Figure 9.6 FACH Tx power with DPS (RL: Radio Link).

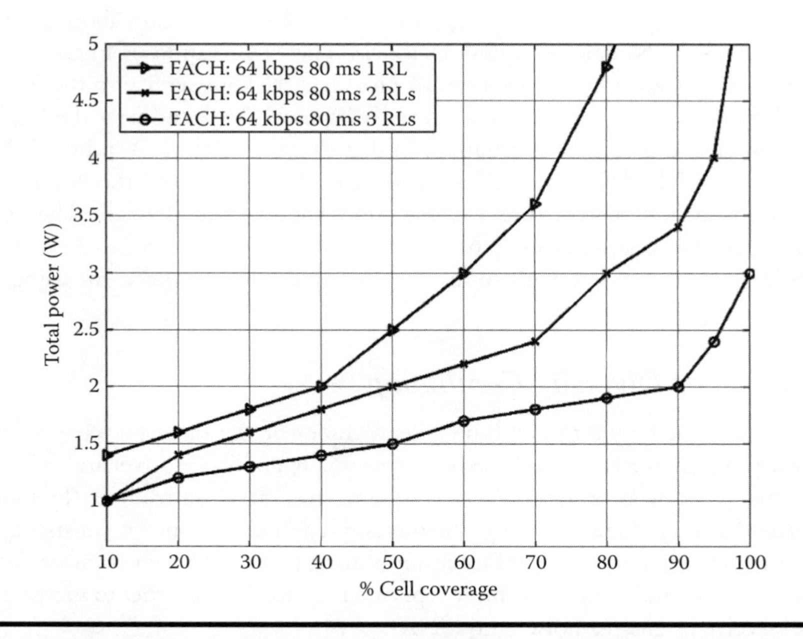

Figure 9.7 FACH Tx power with MDC (1 Radio Link [RL], 2 RLs, and 3 RLs).

that a 64-kbps service should be delivered, using one, two or three NodeBs (or radio links). TTI is assumed to be 80 ms. The main idea with regard to MDC is to decrease the power level from a NodeB when it serves users near the cell edge. However, as we assume three sectors per cell (see Table 9.1), this technique can also be used for distances near the NodeB, where each sector is considered as one radio link (RL). Succinctly, in Table 9.3 we mention some cases that reveal the power gains with this technique.

As the user receives data from two (or three) NodeBs simultaneously, the required power of each NodeB is decreased; however, the total required power remains the same and sometimes is higher. Nevertheless, this technique is particularly useful when the power level of a specific NodeB is high, while respectively the power level of its neighboring NodeB is low.

9.4.3 Rate Splitting

The RS technique assumes that the MBMS data stream is scalable, thus it can be split into several streams with different QoS. Only the most important stream is sent to all the users in the cell to provide the basic service. The less important streams are sent with less power or coding protection and only the users who have better channel conditions (i.e., the users close to NodeB) can receive those to enhance the quality on top of the basic MBMS. This way, transmission power for the most important MBMS stream can be reduced because the data rate is reduced, and the transmission power for the less important streams can also be reduced because the coverage requirement is relaxed [13].

In the following scenario, we consider that a 64-kbps service can be split into two streams of 32 kbps. The first 32-kbps stream (basic stream) is provided throughout the whole cell, because it is supposed to carry the important information of the MBMS service. On the contrary, the second 32-kbps stream is sent only to the users

Table 9.3 Indicative FACH Tx Power Levels with MDC

Cell Coverage (%)	*Radio Links (RLs)*	*Required Tx Power (watts)*
50	1	2.5
	2	2.0
	3	1.5
95	1	7.6
	2	4.0
	3	2.4

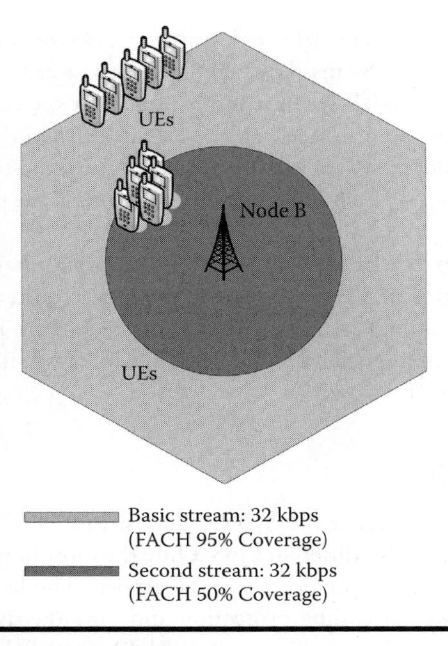

Figure 9.8 MBMS provision with RS.

who are close to the NodeB (50% of the cell area) providing the users in the particular region the full 64-kbps service. Figure 9.8 depicts the operation of the RS technique, in terms of channel selection and cell coverage.

From Table 9.2, it can be seen that this technique requires 5.8 W (4.0 W for the basic stream and 1.8 W for the second). On the other hand, in order to deliver a 64-kbps service using a FACH with 95% coverage, the required power would be 7.6 W. Thus, 1.8 W can be saved through the RS technique. However, it is worth mentioning that this power gain involves certain negative results. Some of the users will not be fully satisfied, as they will only receive 32 kbps of the 64-kbps service, even if these 32 kbps carry the important information. As the observed difference will be small, the NodeB should weigh between the transmission power and the users' requirements.

9.5 Existing Radio Bearer Selection Mechanisms

During the provision of MBMS multicast services, the system should conceive and adapt to continuous changes that occur in dynamic wireless environments and optimally allocate resources. Under this prism, a critical aspect of MBMS performance is the selection of the most efficient radio bearer for the transmission of MBMS multicast data. It is worth mentioning that this is still an open issue in today's MBMS infrastructure, mainly due to its catalytic role in radio resource management (RRM).

There exist two main research directions during the radio bearer selection procedure. According to the first approach, a single transport channel (PTP or PTM) can be deployed in a cell at any given time. In this case, a switching threshold is actually set that defines when each channel should be deployed. On the other hand, the second approach performs a simultaneous deployment of PTP and PTM modes. A combination of these modes is scheduled, and both dedicated and common bearers are established in parallel in a cell. In the following paragraphs, we present the main representative approaches of each of the two research directions.

The figures presented in the following paragraphs refer to the same scenario where a 64-kbps MBMS service is delivered to a constantly increasing number of MBMS users. The group initially consists of four users, and two users join the MBMS session every 5 s. Each user appears in a random position and moves randomly throughout the cell area with a speed of 3 km/h. The main target is to demonstrate the operation and power consumption of each mechanism.

9.5.1 MBMS Counting Mechanism (TS 25.346)

The 3GPP MBMS counting mechanism (or TS 25.346) constitutes the prevailing approach of switching between PTP (multiple DCHs) and PTM (FACH) radio bearers, mainly due to its simplicity of implementation and function [14]. According to this mechanism, a single transport channel (PTP or PTM) can be deployed in a cell at any given time. The decision on the threshold between PTP and PTM bearers is operator-dependent, although it is proposed that it should be based on the number of served MBMS users. In other words, a switch from PTP to PTM resources should occur, when the number of users in a cell exceeds a predefined threshold. Assuming that the threshold is 8 UEs (a mean value for the threshold proposed in the majority of research works), TS 25.346 will command NodeB to switch from DCH to FACH when the number of users exceeds this predefined threshold (at simulation time 10 s), since HS-DSCH is not supported (Figure 9.9).

Figure 9.9 also reveals the inefficiencies of TS 25.346. This mechanism provides a non-realistic approach because the mobility and current location of the mobile users are not taken into account. Moreover, this mechanism does not support FACH dynamic power setting. Therefore, when employed, FACH has to cover the whole cell area, leading to power wasting. Finally, TS 25.346 does not support the HS-DSCH, a transport channel that could enrich MBMS with broadband characteristics.

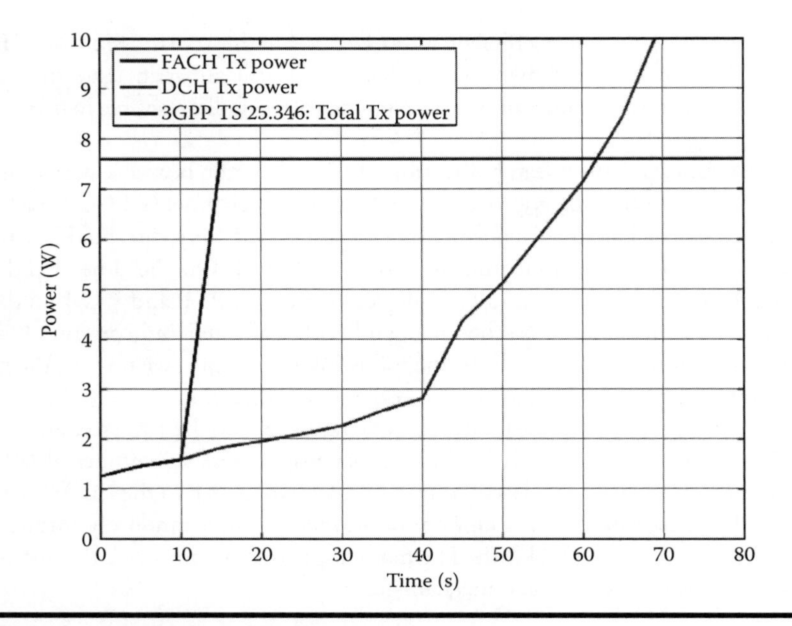

Figure 9.9 3GPP TS 25.346 Tx power levels.

9.5.2 MBMS PTP/PTM Switching Algorithm (TR 25.922)

The 3GPP MBMS PTP/PTM switching algorithm, or TR 25.922 [15], assumes that a single transport channel can be deployed in a cell at any given time. However, contrary to TS 25.346, it follows a power-based approach when selecting the appropriate radio bearer, aiming at minimizing the NodeB's power requirements during MBMS transmissions.

In TR 25.922, instead of using solely DCHs, HS-DSCH can also be transmitted. However, the restricted usage of either DCH or HS-DSCH (Figure 9.10) in PTP mode may result in significant power losses. In both cases, the PTP (DCH or HS-DSCH, since the switching between HS-DSCH and DCH is not supported in this mechanism) and the PTM power levels are compared, and the case with the lowest power requirements is selected. In general, for a small number of multicast users, PTP bearers are favored. As the number of users increases, the usage of a PTM bearer is imperative.

Even though TR 25.922 overcomes several inefficiencies of the TS 25.346 mechanism, it still does not support FACH dynamic power setting, leading in turn to increased power consumption in PTM transmissions.

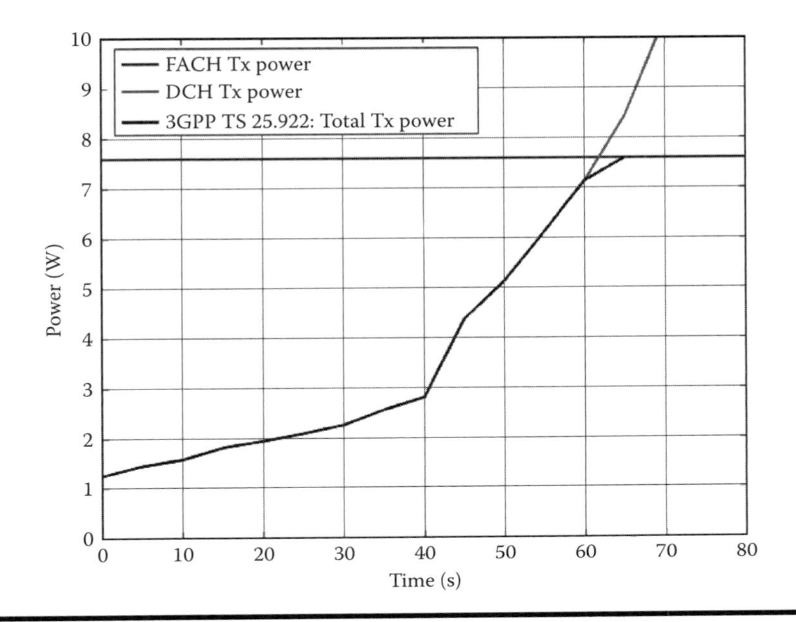

Figure 9.10 3GPP TR 25.922 (with DCH) Tx power levels.

9.5.3 Mechanism Proposed in 3GPP TSG RAN1 R1-02-1240

The preceding mechanisms allow a single PTP or PTM transport channel deployment at any given time. In [16], an alternative idea is presented, which is based on the simultaneous/combined usage of PTP and PTM bearers for MBMS transmissions. In particular, this approach considers the mixed usage of DCHs and FACH for the transmission of the MBMS data over the UTRAN interfaces. According to this approach, the FACH channel only covers an inner area of a cell/sector and provides the MBMS service to the users that are found in this part. The rest of the users are served using DCHs to cover the remaining outer cell area. The power for serving the outer part users is calculated as in Equation 9.4. The total downlink power consumption, including FACH and dedicated channels, is the sum of these two power levels (Figure 9.11).

However, as clearly concluded in [16], this approach is only beneficial when the number of outer part users that use the DCHs is extremely small (less than 5). This suggests that the use of DCH in association with FACH for MBMS services is rather limited for real-world traffic scenarios.

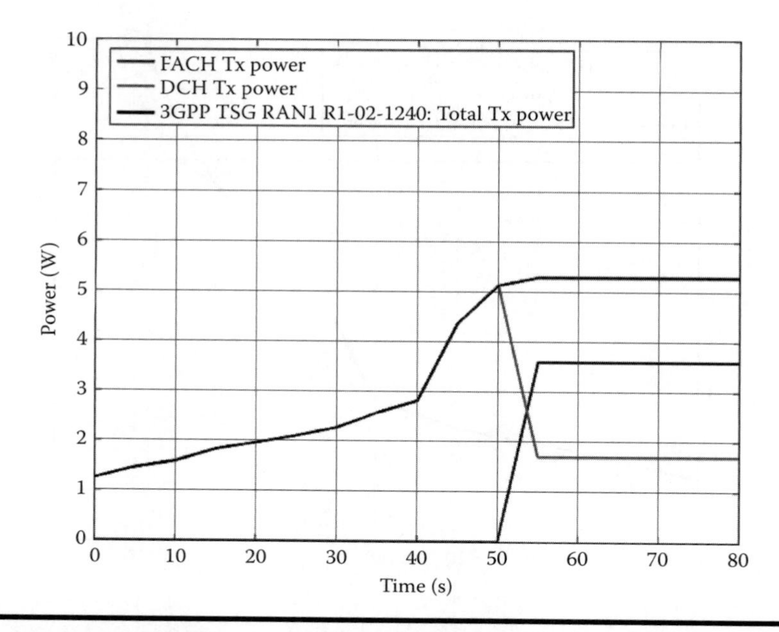

Figure 9.11 3GPP TSG RAN1 R1-02-1240 Tx power levels.

9.6 Proposed MBMS Mechanism

This section proposes an advanced version of the aforementioned mechanisms that performs optimal power allocation during MBMS transmissions. The proposed mechanism dynamically determines the optimal MBMS radio bearer, based on the required transmission power to serve a multicast group. The scheme takes advantage of the HSPA technology (including MIMO support) and contributes to RRM mechanisms of UMTS by adopting a novel framework for MBMS that efficiently utilizes power resources. The main research motivation is to reduce MBMS power consumption, which translates into improved capacity, thus enabling the mass delivery of rich multimedia services in UMTS networks.

More specifically, the mechanism selects for the delivery of the multicast traffic of the transport channel with the lowest power requirements. The fact that any changes in such dynamic environments are directly reflected to the base station transmission power makes the proposed mechanism highly adaptive. Furthermore, the proposed scheme incorporates the premier HS-DSCH transport channel used in HSPA, in contradiction to the MBMS counting mechanism that considers only Release'99 bearers (DCH and FACH). HS-DSCH in many cases is less power consuming, which combined with the power-based bearer switching criterion further improves MBMS power efficiency. However, even more power resources can be saved when MIMO technology is supported.

Next in this section, we present the architecture and the functionality of the proposed scheme, the block diagram of which is illustrated in Figure 9.12. More specifically, the mechanism comprises three distinct operation phases: the parameter retrieval phase, the power level computation phase, and the transport channel selection phase. Additionally, a periodic check is performed at regular time intervals. The RNC is the responsible node of the MBMS architecture for the operation of

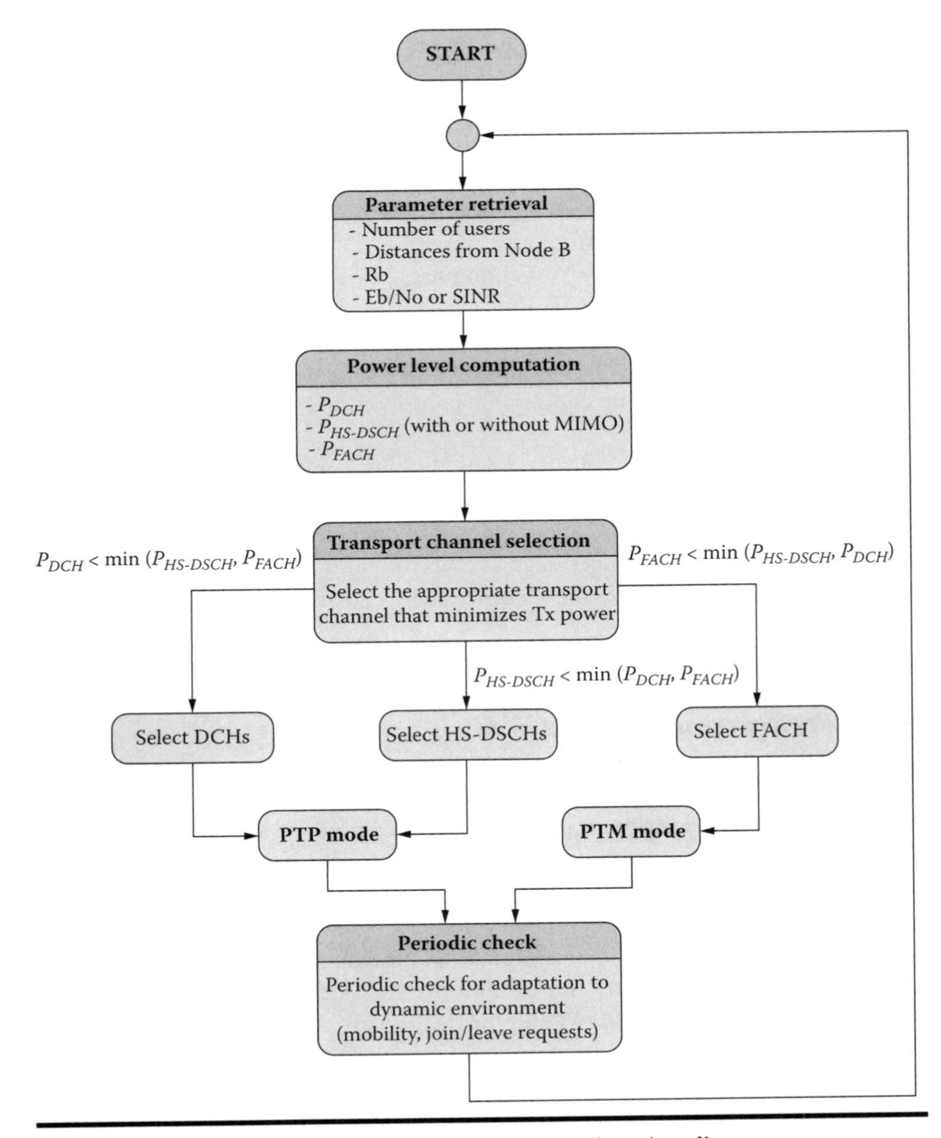

Figure 9.12 Power counting scheme with MIMO functionality.

this algorithm and the final decision on the most efficient transport channel for the delivery of MBMS multicast data.

During the parameter retrieval phase (Figure 9.12) the mechanism retrieves parameters of the existing MBMS users and services in each cell. User-related parameters, such as the number of users requesting a specific MBMS session, their distances from the base station, and their QoS requirements are received from the RNC through appropriate uplink channels. Moreover, the MBMS bit rate service is retrieved from the BM-SC node.

The power level computation phase substantially processes the information received from the parameter retrieval phase. In this phase, the required power to be allocated for MBMS session delivery in each cell is computed. The computation is based on the assumption that the transmission of the multicast data can be performed over multiple DCHs, HS-DSCHs, or over a single FACH. Consecutively, P_{DCH}, $P_{HS-DSCH}$ (with or without MIMO), and P_{FACH} power levels are computed, respectively, for each type of transport channel.

During the transport channel selection phase, the appropriate transport channel for the transmission of the MBMS multicast content is selected. P_{DCH}, $P_{HS-DSCH}$, and P_{FACH} values are compared in order to select the most power-efficient bearer for an MBMS session in a cell. The algorithm dynamically decides which case requires less power and, consequently, chooses the corresponding transport channel for the session.

Finally, the mechanism performs a periodic check and re-retrieves user and service parameters in order to adapt to any changes during the service provision. This periodic check is triggered at a predetermined frequency rate and ensures the mechanism is able to conceive changes, such as users' mobility, join/leave requests, or any fading phenomena, and configure its functionality so as to maintain high resource efficiency.

9.6.1 Performance Evaluation

9.6.1.1 Efficient MBMS Transport Channel Selection

This subsection presents performance results concerning the most critical aspect of the proposed scheme: the transport channel selection phase. This power efficient channel deployment is illustrated in Figures 9.13, 9.14, and 9.15, for 60%, 80%, and 100% cell coverage areas, respectively. In these figures, transmission power levels (overall output of the power level computation phase) for DCH, HS-DSCH, (with and without MIMO support) and FACH channels are depicted. The simulation scenario considers a 64-kbps MBMS session delivery in a cell, whose users are assumed to be in groups (of varying population), located at the bounds of the earlier coverage areas each time.

Regarding the 60% cell coverage case (Figure 9.13), we observe that for a multicast group with ten or fewer users, DCH is the optimal transport channel. For a multicast

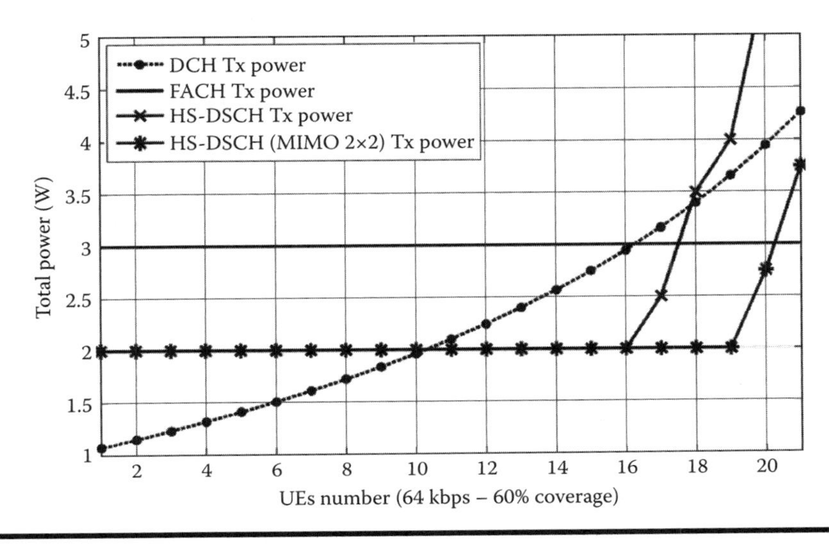

Figure 9.13 MBMS power allocation, 64 kbps, 60% coverage.

population of 10 to 17 users, HS-DSCH (without MIMO) is less power consuming and, thus, it should be preferred for MBMS content transmission (PTP mode). When MIMO 2 × 2 is supported, the above upper threshold is further increased to 20 users. For more than 17 users (or 20 users with MIMO support), FACH is more power efficient and should be deployed (PTM mode). Similar results can

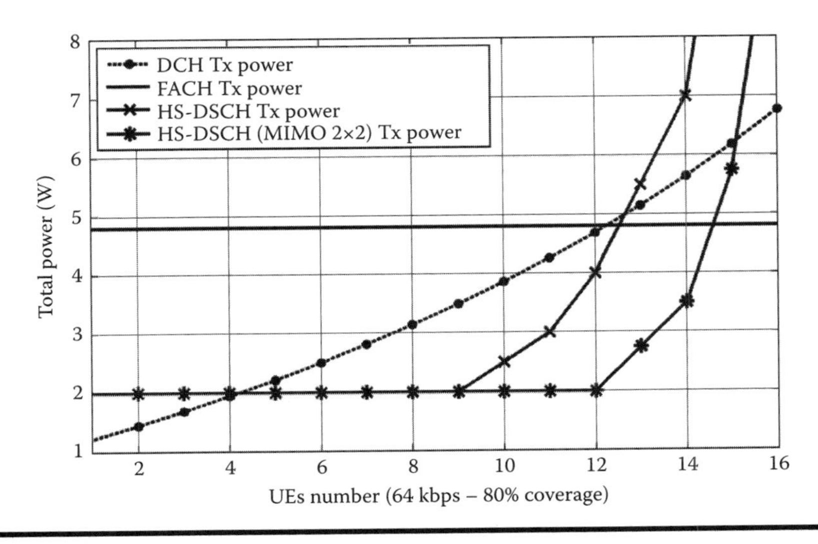

Figure 9.14 MBMS power allocation, 64 kbps, 80% coverage.

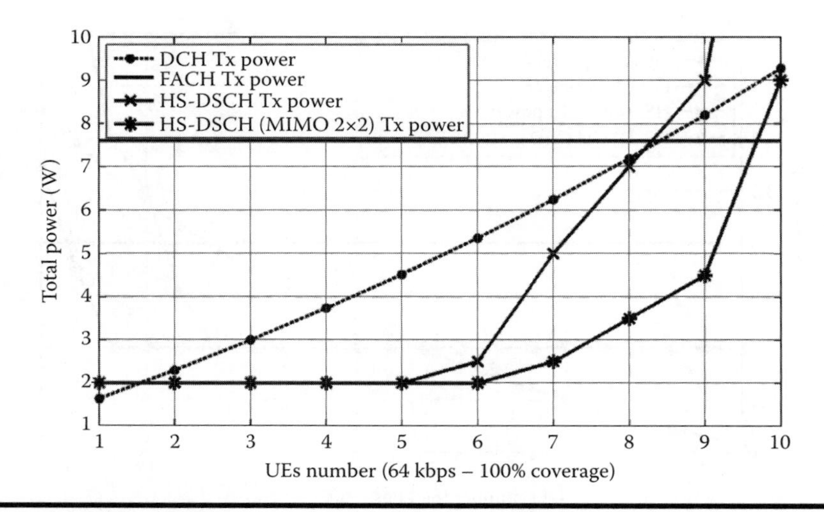

Figure 9.15 MBMS power allocation, 64 kbps, 100% coverage.

be extracted for the cases of 80% and 100% cell coverage from Figures 9.14 and 9.15, respectively. However, from these figures we may additionally conclude that for higher cell coverage areas HS-DSCH is prevailing over the DCH even for a small multicast group and should be exclusively used instead of DCH in PTP mode.

In general, in cases where the number of users is small, PTP transmissions are preferred, while PTM transmissions are favored for a large multicast population. However, the enhanced mechanism does not only decide to use PTP or PTM transmissions, it makes a further distinction between DCH and HS-DSCH in PTP mode. This is an important notice since HS-DSCH appears to use less power than DCH in most cases, especially when MIMO is supported. MIMO schemes significantly reduce MBMS power consumption compared to other radio bearers and further maximize power efficiency. This power gain, in turn, leads to a major gain in capacity and enables the provision of multimedia services to a greater number of MBMS users in future mobile networks.

9.6.1.2 Comparison with the MBMS Counting Mechanism

The superiority of the mechanism can be better illustrated if we compare the performance of the proposed approach with the most prevailing 3GPP approach, the MBMS counting mechanism or TS 25.346. For a more realistic performance comparison, both mobility issues and a varying number of served users are taken into consideration and investigated.

At this point, it should be remembered that the MBMS counting mechanism considers a static switching point between PTP and PTM modes (or else between DCH and FACH), based on the number of MBMS served users. Such a reasonable

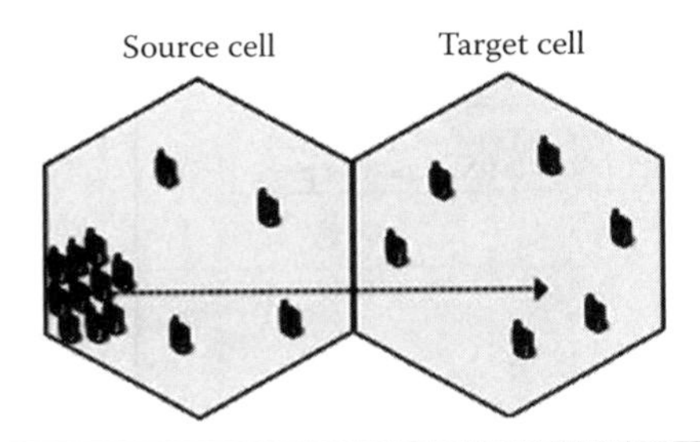

Figure 9.16 Simulation topology.

threshold for a macro cell environment would be eight multicast users. For less than eight users in PTP mode, multiple DCHs (and no HS-DSCH) would be transmitted, while for more than eight multicast users in PTM mode, a single FACH with such power as to provide full (100%) coverage would be deployed.

The simulation scenario considers the provision of a MBMS multicast session in a segment of a UMTS macrocellular environment. We examine the performance of both approaches for two neighboring cells (called source cell and target cell) as depicted in Figure 9.16. A 64-kbps MBMS session with 2000-s time duration is delivered in both cells.

Figures 9.17 and 9.18 depict the downlink power of the available transport channels, as extracted from the power level computation phase, in source and target cells, respectively. Figures 9.19 and 9.20 depict the transmission power of the transport channel that is actually deployed both for the proposed mechanism and

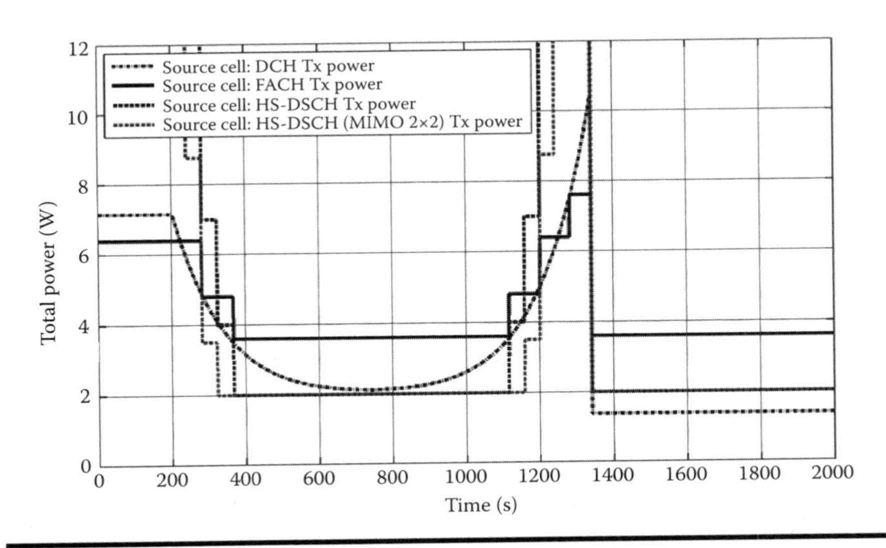

Figure 9.17 Source cell—output of power level computation phase.

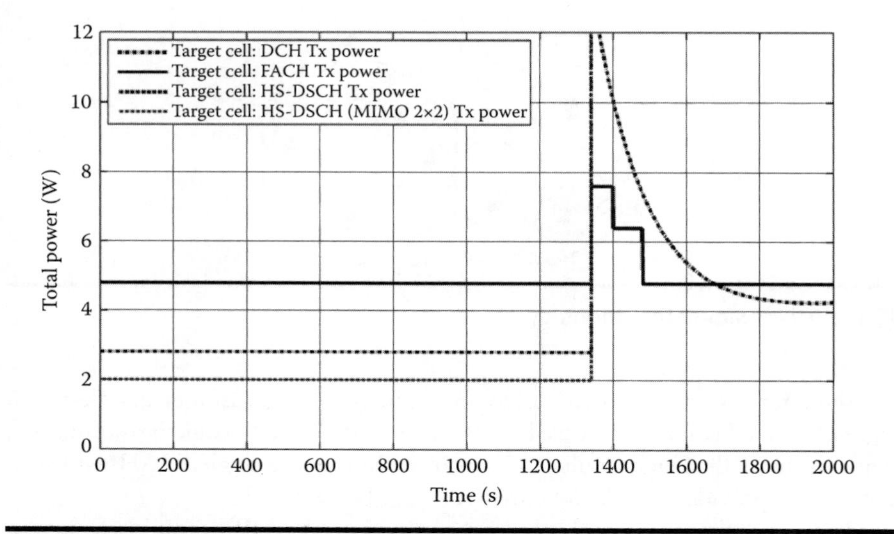

Figure 9.18 Target cell—output of power level computation phase.

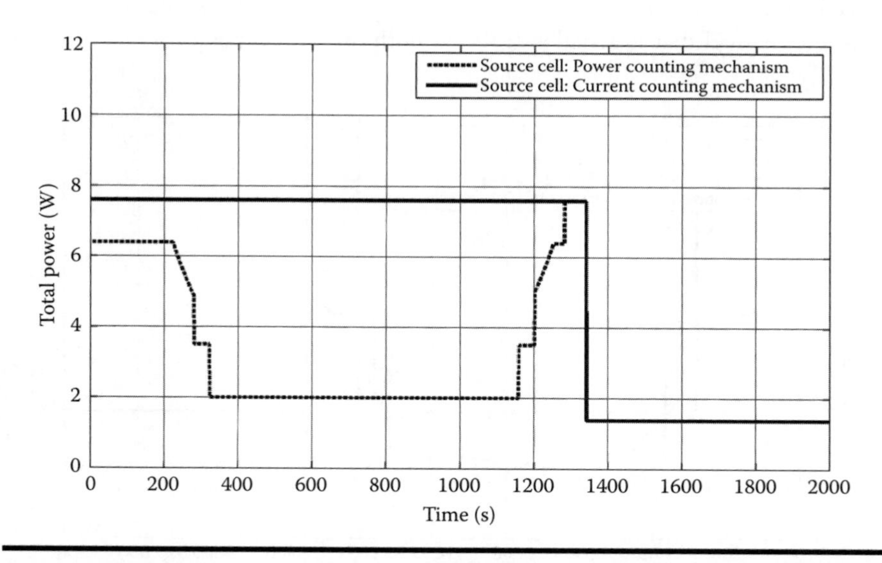

Figure 9.19 Source cell—proposed mechanism versus the MBMS counting mechanism.

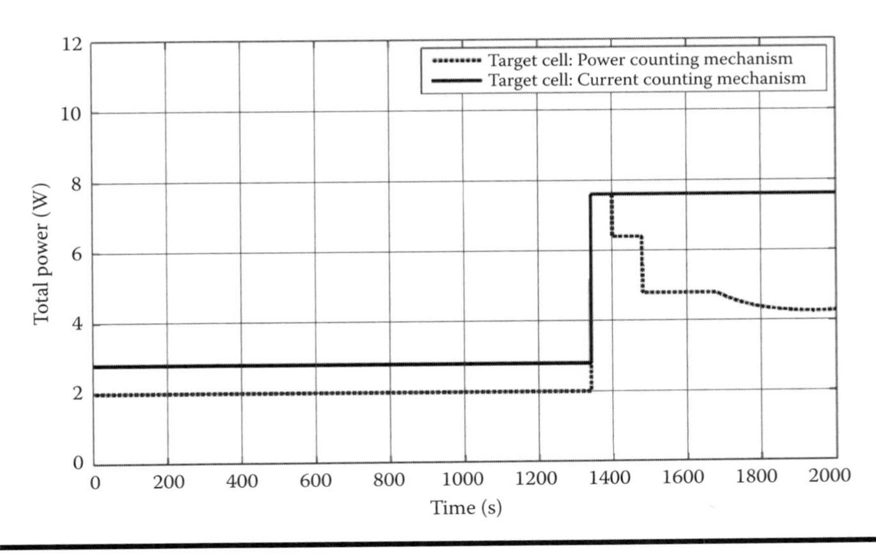

Figure 9.20 Target cell—proposed mechanism versus the MBMS counting mechanism.

the MBMS counting mechanism, in source and target cell, respectively. In the proposed approach, this transmission power level represents the power consumed by the channel selected in the transport channel selection phase. The selected channel for each cell can be easily extracted from Figures 9.17 and 9.18 (the one with less power requirements for each time instance). Regarding the MBMS counting mechanism, this power level is either the total DCH power for less than eight users, or the fixed FACH power, equal to 7.6 W, for more than eight users.

The source cell initially consists of 14 multicast users, while 6 users reside in the target cell. During the first 200 s of the simulation time, all users in both cells are static. In the source cell, the proposed mechanism favors the transmission of MBMS content over FACH with a power set to 6.4 W in order to serve users with the worst path loss, located at a distance of 90% cell coverage. On the other hand, the MBMS counting mechanism uses a FACH with power set to 7.6 W to provide full cell coverage, resulting in a power wasting of 1.2 W in the source cell (Figure 9.19). The target cell is a PTP cell, since it serves less than eight users. However, we observe that HS-DSCH has better performance than DCH, with almost a 1 Watt power saving (Figure 9.20). Thus, the proposed scheme performs better than the MBMS counting mechanism in the target cell, too.

A group of ten users in the source cell, which is located near the cell edge (90% cell coverage), starts moving at time instance 201 s toward the target cell, according to the trajectory depicted in Figure 9.16, while the rest of the users remain static. This group enters the target cell at time instance 1341 s. During the time period 201–1341 s, we can make the following observations for the source cell. The enhanced mechanism is able to track users' mobility and further improves power efficiency. When multicast users get close to the source cell's base station, PTP bearers (DCH and HS-DSCH) use less power than the PTM bearer (FACH) even for a large number of served users. Similarly, when users reside near the cell edge, FACH is more efficient. On the other hand, the MBMS counting mechanism fails to deal efficiently with users' mobility, in the absence of any adaptive procedure, and uses exclusively FACH since simultaneous users receiving the MBMS service exceed the threshold of eight users. As a result, we observe that a significant power budget, approaching 5.6 W, is wasted (Figure 9.19). Both mechanisms have identical performance (FACH deployment) only when moving users are on the cell border. Moreover, we observe that HS-DSCH with MIMO support requires less power compared to pure HS-DSCH for some time instances. The target cell still remains in PTP mode, with the same power gains arising from the proposed scheme during the first 200 s of simulation (Figure 9.20).

Finally, at time instance 1341 s, the group of 10 moving users enters the service area of the target cell. At this point, according to the MBMS counting mechanism, the source cell switches to PTP mode (multiple DCHs) since it serves only four users. The enhanced mechanism also uses DCHs and, thus, both approaches have similar performance. At the same time, the target cell switches to PTM mode (a single FACH) and serves 16 users. However, as the moving group reaches the base station, the proposed scheme appropriately adapts its functionality and results to better utilization of power resources in contradiction to the static FACH channel assignment of the MBMS counting mechanism's specifications. Power gains approach almost 3 W.

Conclusively, from Figures 9.19 and 9.20 it is obvious that the proposed approach is prevailing over the MBMS counting mechanism. The power-based criterion for switching between transport channels as well as the deployment of the HS-DSCH, especially when MIMO is supported, strongly optimizes resource allocation and enhances MBMS performance.

9.7 Open Issues

Regarding the operation of the proposed mechanism, several enhancements can be incorporated to further improve the MBMS performance. The steps that follow this work could be, at a first level, the evaluation of the mechanism through additional simulation scenarios. The scenarios could be simulated in the ns-2 simulator, in which the proposed mechanism could be implemented. In that way, except for the performance of the proposed mechanism, other parameters such as delays in UTRAN interfaces during MBMS transmissions could be measured.

Furthermore, several power saving techniques such as rate splitting and macro diversity combining could be integrated in the proposed mechanism. The use of these techniques will further improve the overall performance of the proposed mechanism, which in turn means that a better utilization of radio and network resources can be achieved.

Finally, it may be considered whether the multicast broadcast single frequency network (MBSFN) transmission mode, included in the evolved UTRAN technologies of the LTE, can be used as an alternative PTM transmission mode for MBMS. MBSFN tries to overcome the cell-edge problems of MBMS and to reduce the inter-cell interference. Therefore, MBSFN can be used in order to achieve very high receiver output SNR and significantly improve the overall spectral efficiency.

9.8 Conclusion

This chapter introduced the key concepts of MBMS services. The main target was to highlight the importance of power control and its commanding role during the delivery of MBMS multicast content for the overall efficiency of next-generation networks. To this direction, the power profiles of several transport channels, which could be employed for the transmission of MBMS services to the mobile users, were investigated. Moreover, the reader was introduced to certain problems that MBMS current specifications are facing and became familiar with techniques/solutions proposed to overcome such limitations.

Finally, this chapter proposed a novel mechanism for efficient transport channel selection during MBMS transmissions in UMTS networks. The proposed mechanism defines downlink power as the switching criterion between different radio bearers and is capable of conceiving any dynamic changes and, therefore, optimally adapting its functionality. Furthermore, the proposed mechanism conforms to next-generation mobile networks' requirements and takes advantages of MIMO antennas to further improve resource efficiency. Simulation results prove that the proposed scheme strongly outperforms the current counting mechanism of MBMS specifications, by maximizing power and capacity efficiency.

References

[1] 3rd Generation Partnership Project TR 23.846. (2003). Technical Specification Group Services and System Aspects; Multimedia Broadcast/Multicast Service; Architecture and functional description (Release 6). Version 6.1.0.

[2] 3rd Generation Partnership Project TS 22.146. (2008). Technical Specification Group Services and System Aspects; Multimedia Broadcast/Multicast Service; Stage 1 (Release 9). Version 9.0.0.

[3] Holma, H. and Toskala, A. (2007). *WCDMA for UMTS: HSPA Evolution and LTE (4th edition)*. The Atrium, Southern Gate, Chichester, England: John Wiley & Sons.

[4] Alexiou, A., Antonellis, D., Bouras, C., and Papazois, A. (2006, October). "An Efficient Multicast Packet Delivery Scheme for UMTS." Paper presented at the 9th ACM/IEEE International Symposium on Modeling, Analysis and Simulation of Wireless and Mobile Systems (MSWiM 2006), Torremolinos, Malaga, Spain.

[5] Boni, A., Launay, E., Mienville, T., and Stuckmann P. (2004, October). "Multimedia Broadcast Multicast Service—Technology Overview and Service Aspects." Paper presented at the 5th IEE International Conference on 3G Mobile Communication Technologies (3G 2004), London, UK.

[6] Holma, H. and Toskala, A. (2006). *HSDPA/HSUPA for UMTS: High Speed Radio Access for Mobile Communications.* The Atrium, Southern Gate, Chichester, England: John Wiley & Sons.

[7] 3rd Generation Partnership Project TR 101.102. (2002). Universal mobile telecommunications system (UMTS); Selection procedures for the choice of radio transmission technologies of the UMTS. Version 3.2.0.

[8] Zihuai, L., Sorensen, T. B., and Mogensen, P. E. (2007). *Downlink SINR Distribution of Linearly Precoded Multiuser MIMO Systems.* IEEE Communications Letters, 11(11), 850–852.

[9] Perez-Romero, J., Sallent, O., Agusti, R., and Diaz-Guerra, M. (2005). *Radio Resource Management Strategies in UMTS.* The Atrium, Southern Gate, Chichester, England: John Wiley & Sons.

[10] 3rd Generation Partnership Project TR 25.803. (2005). Technical Specification Group Radio Access Network; S-CCPCH performance for MBMS, (Release 6). Version 6.0.0.

[11] Parkvall, S., Englund, E., Lundevall, M., and Torsner, J. (2006). *Evolving 3G Mobile Systems: Broadband and Broadcast Services in WCDMA.* IEEE Communication Magazine, 44(2), 30–36.

[12] Chuah, P., Hu, T., and Luo, W. (2004, June). "UMTS Release 99/4 Airlink Enhancement for Supporting MBMS Services." Paper presented at the 2004 IEEE International Conference on Communications (ICC 2004), Paris, France.

[13] 3rd Generation Partnership Project R1-021239. (2002). MBMS Power Usage, Lucent Technologies. Lucent Technologies, TSG-RAN WG1#28.

[14] 3rd Generation Partnership Project TS 25.346. (2009). Technical Specification Group Radio Access Network; Introduction of the Multimedia Broadcast Multicast Service (MBMS) in the Radio Access Network (RAN); Stage 2, (Release 8). Version 8.3.0.

[15] 3rd Generation Partnership Project TR 25.922. (2007). Technical Specification Group Radio Access Network; Radio Resource Management Strategies (Release 7). Version 7.1.0.

[16] 3rd Generation Partnership Project R1-021240. (2002). Power Usage for Mixed FACH and DCH for MBMS. Lucent Technologies, TSG-RAN WG1#28.

Chapter 10

Managing Coverage and Interference in UMTS Femtocell Deployments

Jay A. Weitzen, Balaji Raghothaman, and Anand Srinivas

Contents

10.1 Introduction to Femtocell Technology

This chapter discusses the issues and challenges associated with the deployment and optimization of UMTS femtocell networks. UMTS femtocell access points (FAPs), also known as home NodeBs (HNBs), or just femtocells, are small, inexpensive, low-power NodeB base stations designed primarily for home use to provide high-quality indoor coverage. Additionally, they allow home-user traffic to be offloaded from the macrocell network. They generally have peak transmitter powers on the order of less than 13 to 17 dBm (which is less than the maximum power of a typical handset device), and generally can support between 4 and 10 radio access bearers (RABs). While physically small, femtocells must support the standard UMTS RABs: voice (e.g., AMR), data (HSDPA/HSUPA), and supplemental services such as SMS, caller ID, call forwarding, voice mail notification, etc. UMTS femtocells generally implement a flat architecture incorporating both the NodeB and RNC functionality. For their backhaul connection to the operator's packet-switched and circuit-switched core networks, they use home Internet connections such as DSL, cable, or fiber to the home. They generally connect to the operator's core through a concentrator via secure tunnels using either a variant of Iu or other protocols. Key to understanding the deployment of femtocells is having an understanding of the differences between an HNB (femtocell) and an access point technology such as Wi-Fi, as well as the differences compared to a fixed infrastructure such as macro, micro, and picocellular NodeB.

Wireless LAN technology embodied in IEEE 802.11 access points, also known as Wi-Fi, is a totally ad-hoc access technology in which the users purchase and own the access point units. Each Wi-Fi access point operates in an unlicensed spectrum, is self-managing, independent of all others, and does the best job it can with no intervention to minimize interference between it and other units. It has a MAC protocol based on a wireless optimized version of carrier sense multiple access, designed specifically for unlicensed wireless access points with small footprints.

UMTS NodeB macro networks often consist of hundreds, or at most a few thousand, NodeB macrocell, microcellular, and picocell base stations covering a typical market. They are totally planned—that is, the location and coverage footprint of each NodeB has been carefully selected, analyzed, and optimized in the context of all the other NodeB units that surround it. The network design remains basically static.

UMTS femtocells on the other hand, while they may be owned by an individual, operate in a licensed spectrum owned by a cellular/PCS operator often using the same frequency channel as the macro network, and are part of the operator's overall network. There could be hundreds of thousands of femtocells randomly located throughout the overall macro network deployment and potentially on the same channel with the macrocell network. Each day, hundreds of femtocells may be added, removed, or moved to different locations. The addition of each HNB femtocell potentially changes the dynamics of the network design.

Femtocells provide a great improvement in indoor coverage for voice calls in areas where macrocell coverage is weak or spotty. For high-speed mobile broadband data services, the advantages of deploying femtocells for indoor coverage are even more compelling than for pure voice services, both in terms of greatly improved coverage as well as the overall user experience [1]. A typical user can expect to achieve broadband data rates, throughout their house, of approximately three to five times that of the existing wireless macro network [2].

Femtocells deliver indoor coverage with high data rates and with extremely high reliability, a combination that is hard to achieve using macrocells alone. Measurements have demonstrated that femtocell data rates are higher and significantly more consistent than the macrocell rates, exceeding 3.5 to 5 Mbps in almost all locations of the home on a statistically reliable basis. The bottom line is that femtocells provide a dramatic improvement in indoor UMTS data performance when compared with using the existing macrocell network alone. These dramatic improvements are primarily due to:

- The proximity of the femtocell transmitter to the user device where the receiver is a few meters or at most tens of meters from the transmitter versus the receiver in the macro network, which is potentially thousands of meters away. The effects of spatial and temporal fading are magnified as the distance between transmitter and receiver increases.
- The absence of significant penetration losses that take place when radio signals travel through exterior walls, other structures, or geographical obstacles within the macro coverage area. Moreover, these obstacles that prevent good coverage from a macrocell are actually advantageous for an indoor user with a femtocell, since they provide isolation from macrocell and other femtocell interference.

Macrocell HSDPA data rates of 0.5 to 1.5 Mbps are adequate for text messaging and text-heavy applications like email. Higher rates possible with femtocells will better support broadband applications such as web browsing, video streaming, picture messaging, and music downloads. The higher rates will also improve the speed and responsiveness of email and text messaging.

Macrocell networks are effective at delivering broad coverage for voice and moderate-speed data services, especially outdoors in a mobility environment. However, these network topologies cannot efficiently serve the needs of emerging broadband data devices and applications, even after upgrades to fourth-generation air interface technologies. It is a matter of link budget, the accounting of all of the gains and losses between the transmitter and the receiver as constrained by distance, interference, and other factors, rather than air interface technology. Supporting broadband data rates indoors with high reliability would require the deployment of thousands of new cell sites, which would be cost-prohibitive [1]. Femtocells, as part of an operator's overall network architecture, complement the macrocell network by providing coverage and data performance in the home, a place that the macro network has difficulty reaching, but where the user highly values wireless services.

There is reason to believe the throughput advantages of femtocells will be greater than what is predicted by measurements based on the effective airlink data rate alone. Airlink data rate is an upper bound on throughput when there is only one user. HS-DPA is a shared medium and it is likely that in the macrocell there will be a number of users competing for the airlink bandwidth, as opposed to a very small number in a femtocell. Each macrocell serves many more users than each femtocell, so the actual data rate experienced by a user in a macrocell during busy periods will be much less than what is predicted by the effective airlink data rate when compared to a user of a femtocell.

Finally, over the next few years, femtocells will evolve from stand-alone units primarily serving the HNB function only, to integrated data portals, consisting of DSL/Cable modems, fixed routers, Wi-Fi routers, and HNBs. These integrated units will become generalized access portals for a plethora of new services as the vision of universal mobile broadband and fixed mobile convergence becomes a reality. There is also considerable interest in femtocells for fourth-generation technologies such as LTE[3] and WiMAX[4], which is driving a lot of activity in the industry forums and standardization bodies. We focus our attention on UMTS femtocells in this chapter.

While universal mobile broadband is the goal and there are numerous advantages to deploying a network of HNB UMTS femtocells, a number of significant issues and challenges exist in deploying a femtocell network as an underlay to the existing macro/micro network architecture[5–9]. We start with the "Prime Directive" from the operator's point of view, which is that the deployment of femtocells cannot noticeably degrade the performance of the macrocell network. Because UMTS uses 5-MHz channel allocations, and most operators have two or at most three UMTS carriers, deploying femtocells on the same frequency with the macrocell network is a fact of life. Therefore, minimizing and managing interference between femtocells and the macrocell network is of prime concern to carriers. As the number of femtocells grows into the millions, managing interference between femtocells will become a primary challenge, which means centralized planning and control will evolve into distributed interference management architectures. This must all be done in the context of a dynamic network with femtocells constantly entering and leaving the network.

This chapter discusses both the challenges and available solutions in managing the interference and performance of large femtocell networks. It is divided into five main sections plus the introduction.

- Section 10.2 discusses the primary deployment considerations for femtocells. The main topics discussed are frequency planning, cell selection, and access control.
- Section 10.3 describes techniques for managing downlink coverage and interference. Remote environment monitoring and automatic configuration and planning mechanisms are discussed, in which the femtocell senses both the presence and strength of macrocells and other femtocells and then sets forward link parameters based on the measurements to provide the desired level of coverage while minimizing interference. Results of measurements and simulations are presented.
- Section 10.4 is dedicated to interference control on the uplink in femtocell deployments. Techniques for combating uplink interference both at the femtocell and the macrocell are discussed. The differences between the receiver and transmitter requirements for femtocells versus macro NodeBs are discussed.
- Section 10.5 summarizes and discusses challenges related to the deployment of large femtocell networks in the future. These include incorporating advanced interference mitigation techniques into the femtocells, developing standards for femtocell interoperability, inter-femto communication for RF coverage and interference mitigation, and developing standards-based enhancements to the macrocell network to enable more efficient macrocell/femtocell interference management.

10.2 Deployment Considerations

10.2.1 Frequency Planning

One of the first choices any UMTS operator considering a femtocell deployment must make is the frequency channel allocation policy. Assuming a limited number of UMTS channels—for example, consider two—there are three possible deployment configurations that need to be considered. This is illustrated in Figure 10.1.

In Figure 10.1, the scenario represents a dedicated carrier deployment that provides separate macrocell and femtocell carriers. This scenario has the advantage of minimizing interactions between the two networks and has a number of other advantages if the spectrum is available. The dedicated carrier scenario might work in more rural areas where the load on the macrocell network would support only one carrier. Scenario c takes all available UMTS carriers and shares them between the macrocell network and the femtocell deployment. It has the advantage of more

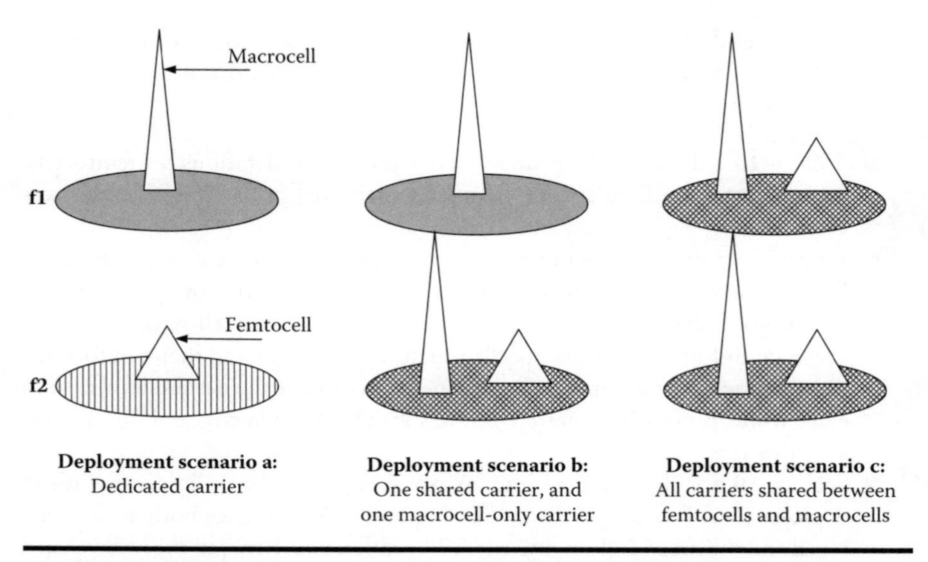

Figure 10.1 Deployment scenarios for sharing two UMTS carriers: F1 and F2.

degrees of freedom to manage interference between femtocells, especially in dense urban deployments. However, scenario c also requires the greatest degree of overall network management to ensure minimal degradation of the macrocell network due to the presence of co-channel femtocells. Scenario b represents a compromise between scenarios a and c in which there is one carrier shared between the macrocell network and femtocell network, and one carrier is reserved for macrocells only. Another option is to use the shared macrocell and femtocell network carrier primarily for macrocell data services (HSxPA), and share this with the voice and data femtocells so any interference introduced by the femtocells affects only macrocell data.

10.2.2 UE Selection of the Femtocell

When a user enters the femtocell home zone coverage region, the first function of the femtocell network is to attract the handset devices to the femtocell. Throughout the chapter, we will refer to a handset device as a user equipment (UE), which is the standard UMTS terminology. Cell selection and reselection refers to the process by which UEs select the cell/sector in which they camp when they are in idle mode. Camping on a cell refers to the fact that the UE can be paged through that cell. In idle state, the UE is asleep for most of the time, and wakes up periodically to listen to the paging channel. During this time, if the received pilot (also known as CPICH) strength of its camping cell is lower than a pre-set threshold, defined by the quantity called $S_{intrasearch}$, the UE searches for other stronger cells in the same frequency. Similarly, if the pilot strength is below another threshold called

$S_{intersearch}$, then the UE searches for better cells in other frequencies. The values of these thresholds are broadcast along with other system parameters in the broadcast channel of the cell. These parameters play a major role in determining how often a UE performs these cell searches. The settings of $S_{intrasearch}$ and $S_{intersearch}$ in the macrocell are one way of forcing a UE idling on the macrocell to look for a nearby femtocell at frequent intervals. Similarly, the parameters on the femtocell can be used to control the UE's behavior in searching for nearby femtocells and macrocells. While more frequent searches result in quicker detection and use of femtocells when entering the femtocell home zone, they may have some impact on the battery life of all users. This aspect is being studied in detail by the Femto forum, a consortium of companies interested in femtocell technology.

Another possible mechanism for enabling UEs to look for femtocells is to enable hierarchical cell selection (HCS), and to prioritize the femtocells to have a higher priority compared to the macrocells [10]. In this configuration, the UEs camped on the macrocell will constantly look for higher priority cells (in this case femtocells) to camp on. While this will improve the speed with which UEs find femtocells, there is a definite impact on the battery life of all UEs in the system, including those that are far away from any femtocell. Due to this reason, the use of HCS for femtocells has not met with widespread acceptance.

10.2.3 Access Control

Access control is an important consideration in a femtocell deployment. In the open access configuration, the femtocell can be used by any UE that is authorized to use the operator's network. While it is somewhat easier to deploy open-access femtocells, the owner of the femtocell must then accept the fact that any UE in the vicinity can use their Internet bandwidth and their femtocell airlink resources. The result is a possibility of a degraded user experience for the femtocell owner, including the possibility of call blocking when all of the femtocell's resources are used by other UEs in the vicinity.

Closed access deployment, on the other hand, restricts the use of the femtocell to a small set of UEs designated as authorized users by the femtocell owner. This configuration provides greater availability and higher throughput and coverage to the authorized users. Support for closed access in femtocells was provided in the recently concluded UMTS 3GPP Release 8 specifications in the form of the Closed Subscriber Group (CSG). The CSG-ID is an additional identity for the femtocell that can be provisioned to be globally or locally unique. The CSG-ID is broadcast by the femtocell, and the UEs can be provisioned at the point of sale, through over-the-air procedures or some other mechanism with the CSG-IDs of the femtocells that it is authorized to use. During the cell selection and reselection process, the UE decodes the CSG-ID from the broadcast channel and registers with the femtocell based on a CSG-ID match, and ignores the femtocell if there is no match.

However, a disadvantage of using closed access is that it may create a "black hole" for unauthorized users located close to the forbidden femtocell. These unauthorized macrocell users will not be able to connect to the femtocell (due to access control restrictions) nor the macrocell (due to the downlink interference caused by the femtocell). Preventing the "black hole" effect requires that unauthorized users select a different carrier if they are not allowed to camp on the femtocell. Other possibilities are also being explored, such as allowing an unauthorized UE to camp on the femtocell and then either redirecting them to the macrocell for actual calls, or providing them a lower grade of service at the femtocell itself in a mode called hybrid access.

10.3 RF Provisioning and Downlink Interference Mitigation

This section discusses the setting of femtocell RF parameters that define the downlink coverage zone as well as the key interference scenarios and techniques used to control and mitigate interference. The interactions and trade-offs between coverage and interference are elaborated upon. The presentation is based on theory, measurements, and simulations.

The optimization and deployment process for femtocell networks starts with setting the primary RF parameters that affect downlink coverage and interference. These are the downlink transmitter power, frequency channel (from a possible list of available channels in a multi-channel deployment), and the primary scrambling code (PSC).

For the task of RF parameter planning and optimization, there is more than one architectural choice. The first option is to provision the parameters centrally, incorporating global information about RF parameters and measurement reports from all relevant femtocells. In the second option, the parameters can be provisioned in a completely distributed manner in which they are set independently by the individual femtocells. The third option is an architecture that lies in between the centralized and decentralized approaches.

In a purely distributed RF planning architecture, each femtocell senses the RF environment including the presence and strength of both macrocells and other femtocells. They then independently set their RF parameters (CPICH transmitter power, PSC, channel, etc.) based on the detected local RF environment. By contrast, in a centralized planning architecture, a central entity collects sets of RF environment information from the individual femtocells, and makes the decision on the RF parameters based on this global information. In this way, a centralized algorithm can take the interference effect on the whole network (femtocell and macrocell) into account when provisioning the RF parameters for a given femtocell. The centralized approach has potential scalability issues for this automatic network planning function, both because of the large number of femtocells in a region (potentially hundreds of thousands), and the fact that new femtocells continuously enter and leave the network, changing the overall network topology. Within the centralized planning architecture is the choice of whether to re-optimize the entire network when a femtocell enters or leaves the network, or whether to incrementally optimize the new femtocell's RF parameters while keeping the RF parameters of already deployed femtocells fixed. Centralized planning is quite similar to the standard planning algorithms used in macrocell networks and considers each femtocell to be just a very small macrocell.

It is possible to consider an approach that is a hybrid between the centralized and the distributed approaches, in which the RF provisioning algorithm works on the basis of optimizing small neighborhoods of femtocells. This approach can work via a central entity or can use advanced inter-femtocell communication techniques where femtocells can learn the RF parameters and environment of their neighbors and take this into consideration when they set their parameters. Such an approach incorporates techniques from self-organizing or "ad hoc" networks, without the mobility.

Regardless of whether the planning is central or distributed, to properly provision RF parameters, each femtocell must detect its local RF environment. Two general techniques are used to form a picture of the RF environment around the individual femtocell: remote environment monitoring (also called "RF sniffing") and monitoring via collecting information from the UEs served by the femtocell.

For remote environment monitoring, at startup and possibly periodically, each femtocell scans a set of frequencies or channels. It detects the total received signal power on each channel as well as the received energy on any given scrambling code it detects. We call the received signal power of each channel h, $RSSI(h)$, and denote the signal energy of each scrambling code s on channel h as $E_c(h, s)$. This information represents the minimum requirement for the femtocell to approximate its surrounding interference profile to try to establish a target downlink coverage radius.

The femtocell gains additional information by decoding the broadcast channels of nearby macrocells and other femtocells. This provides the femtocell with important information about its neighborhood such as neighbor IDs, locations, transmit powers, and so on, which can be utilized as part of an overall RF provisioning scheme. This is required if femtocells are to be able to hand over either to each other or to the macrocell network.

To this point, environment monitoring has been limited to the location of the femtocell (i.e., because it is doing the "sniffing"). However, in general the RF characteristics can vary across the femtocell coverage region. For example, a femtocell located in the basement of the house may not be able to detect the presence of a macrocell, yet a UE upstairs and near a window can clearly see the macrocell. Thus, a femtocell can augment its own "sniffed" measurements by commanding active UEs connected to it to make measurements on various frequencies and send those reports back in the form of measurement reports. Additionally, the presence of unauthorized macrocell UEs can be noted when they unsuccessfully attempt to register on the femtocell. Thus, the femtocell can estimate the amount of interference it is causing to the macrocell network and react accordingly [6]. Note, however, that future macrocell UEs will likely recognize closed subscriber groups. If this is the case, then unauthorized macrocells UEs may not even attempt registrations on the femtocell and this source of UE feedback will not be available to the femtocell.

It should be noted that utilizing UE measurements to detect the RF environment and provision RF parameters, is non-trivial. One issue is that at startup the femtocell will not in general have access to an active UE connected to it and thus cannot rely on these measurements being available. Moreover, when a UE makes a measurement report, the femtocell may not know the UE's exact location relative to itself. This causes an ambiguity as to whether the UE is making a measurement from a location that should be covered, or maybe should not be. For example, it is reasonable to expect a femtocell UE located in an upstairs bedroom to receive good service from the femtocell, but perhaps not so reasonable to expect this good service when it is located down the street.

10.3.1 RF Provisioning Problem Formulation

At a high level, the RF parameter provisioning problem can be formulated as one in which the objective is to set RF parameters (transmit power, frequency, and scrambling code) for each femtocell such that (1) each femtocell's coverage meets (or comes as close as possible to meeting) a desired coverage goal, and (2) the minimum amount of interference is caused to the macrocell network and other femtocells.

To describe this problem mathematically, we start with elaborating upon the concept of femtocell "coverage." Conceptually, the coverage region refers to the area around the femtocell in which a user can expect a minimum level of "service" (e.g., the ability to make voice calls and/or high-rate data calls—5 Mbps). From an engineering perspective, managing coverage is difficult since RF propagation does not stop at property lines and the covered region is both irregular and depends on the location of the femtocell within the dwelling, the size, the layout, and the distribution of walls, etc. A useful first-order definition of coverage is based on the concept of path loss relative to the femtocell. Specifically, we say that a femtocell f has coverage equal to ρ_f dB if any "femtocell UE" (fUE) it serves sees a CPICH SINR (i.e., pilot strength) of at least Y dB whenever the path loss between the fUE and femtocell is less than or equal to ρ_f dB. This is shown in Figure 10.2, which shows an fUE at the femtocell "cell edge" (i.e., the path loss between the femtocell and UE is exactly ρ_f dB) attaining a CPICH SINR of exactly Y dB.

The idea behind this definition of coverage is that all fUEs that are less than or equal to ρ_f dB away from the femtocell will support a "minimum" CPICH SINR, Y. For example, achieving a 5-Mbps data rate in HSDPA requires a CPICH SINR of approximately $Y = -1$ dB*, and this definition of coverage would mean

* The calculation assumes the CPICH is allocated 10% and HS-DSCH allocated 80% of the total femtocell downlink transmit power.

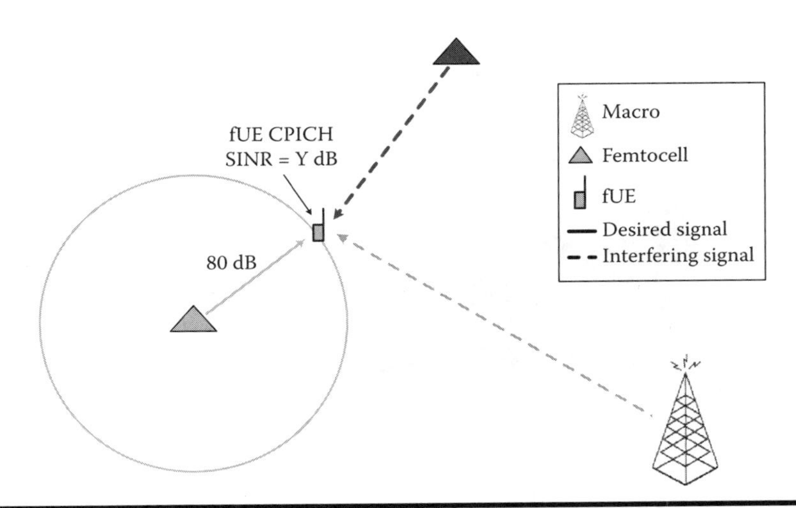

Figure 10.2 Illustration of femtocell coverage: The femtocell on the left has a coverage of $\rho_f = 80$ dB as defined by the fact that any "cell edge" UE has CPICH SINR of at least Y dB. Interference due to another femtocell and a macrocell is shown.

that an fUE that is less than or equal to ρ_f dB away from the femtocell can support a downlink data rate of 5 Mbps or more. Under this first-order model of coverage, an additional assumption is made that the average interference from other sources (such as macrocells or other femtocells) is uniform over the coverage region (averaging the effects of shadowing and micro-spatial fading). While this assumption is not justified in the presence of nearby interferers, it has two useful purposes along with simplifying the mathematics. The first is the fact the femtocell, at least at startup, can only measure interference at its location and cannot possibly know the exact interference distribution over its entire coverage region. In practice, we assume that at a given mean distance from the femtocell, the distribution of path loss will be log-normally distributed (due to central limit theorem) about the mean path loss value. The mean path loss can be predicted using standard indoor short range path prediction models such as the ITU indoor propagation model. An example of actual path loss measurements from multiple locations within 20 houses is shown in Figure 10.3, along with the best fit to a standard log-distance model.

We can mathematically summarize the first-order coverage model as follows. For a given femtocell f, let $P_{f\text{-cpich}}$ denote its CPICH transmit power and I_f the total "interference + noise" (both in linear units) seen at f. Thus the first-order coverage of f, ρ_f, is defined as:

$$\rho_f = [P_{f\text{-cpich}}/I_f]_{\text{dB}} - \text{Y} \tag{10.1}$$

where the notation $[]_{\text{dB}}$ is used to indicate a conversion to decibel units. Note that the first term on the right-hand side of Equation 10.1, $P_{f\text{-cpich}}/I_f$, expresses

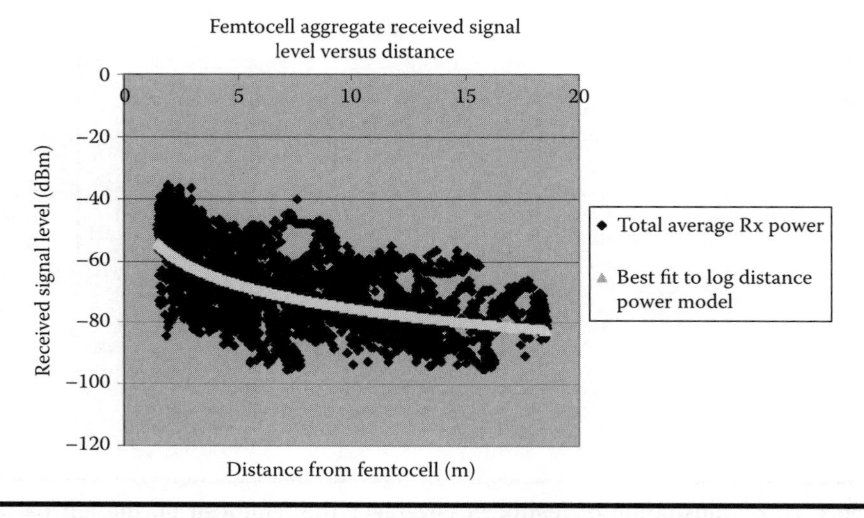

Figure 10.3 Mean path loss measurements for 0 dBm transmitter power and best fit to the log-distance model over an ensemble of 20 residences.

the CPICH SINR that a (theoretical) UE located 0 dB path loss away from the femtocell would see. Using the assumption that interference is constant, it therefore follows that a UE ρ_f dB from the femtocell would see CPICH SINR of $[P_{f\text{-cpich}}/I_f]_{\text{dB}} - \rho_f$.

When optimizing coverage, the operator provides a target value for the coverage radius for each femtocell f, $\rho_{f\text{-des}}$ (i.e., assuming a pre-specified constant value for minimum service SINR, Υ). This target radius can and should be different based on a customer profile. For example, users in a suburban house would want a larger radius than in an urban apartment.

At this point, it should be noted that no RF parameter allocation may exist that results in a solution in which every femtocell obtains its desired coverage of $\rho_{f\text{-des}}$. For example, such a situation can occur when a femtocell faces too much macrocell interference. In this case, it is a power saturation problem in which, due to hardware constraints, the femtocell cannot transmit at a high enough transmit power to overcome the macrocell interference. It is similarly problematic if two or more femtocells are closely deployed and there is only one available frequency channel. In this case, the desired coverage cannot be simultaneously achieved because if one femtocell increases its transmit power, it degrades the service of its nearby femtocells, causing them to increase their transmit power, and so on. If not careful, the end result could involve all of these femtocells transmitting at their maximum power—causing additional interference to the macrocell network—yet not achieving any discernable gains in coverage. This specific issue is explored in greater detail later in the section.

An intuitive and general way to formulate the RF parameter provisioning problem is to define a utility function $Z(\rho_f)$ that is a function of the actual coverage, ρ_f, that femtocell f obtains. The problem can then be formulated as one of maximizing the sum utility over all femtocells. Thus, the utility function should represent a mathematical description of the high-level optimization goals. To this end, consider the utility function depicted in Figure 10.4, where we have assumed a minimum acceptable coverage ρ_{min} specified by the operator. Note that the characteristics of the function include: (1) no increase in utility once coverage exceeds $\rho_{f\text{-des}}$, (2) a steep slope to the left of ρ_{min} to encourage every femtocell achieving at least this minimum coverage, and (3) a shallower curve to indicate the marginal utility gained from increasing coverage between ρ_{min} and $\rho_{f\text{-des}}$. Note the purpose of requirement (1) is to avoid unnecessary femtocell transmit power expenditure, and that of requirements (2) and (3) are to balance fairness with overall coverage optimization. In general, a generic utility function along the lines of that depicted in Figure 10.4 would have three segments. The segment on the left increases with a large slope to ensure each femtocell has a minimum coverage. The segment in the middle increases gradually to indicate increasing utility until a femtocell achieves its desired coverage. The segment on the right is flat to indicate no additional utility is gained through coverage that is larger than the desired coverage.

Finally, it should be noted that the utility function in Figure 10.4 does not explicitly take into account the interference caused to macrocell UEs. One consequence of this is that a solution in which femtocells transmit at high power but that has a slightly higher sum utility will be preferred to a much lower-power solution. To see the effect of this, consider the situation involving two neighboring femtocells separated by channel gain G (in linear units), shown in Figure 10.2. Assuming they both transmit at the same power P (assume only pilot transmitted, for simplicity),

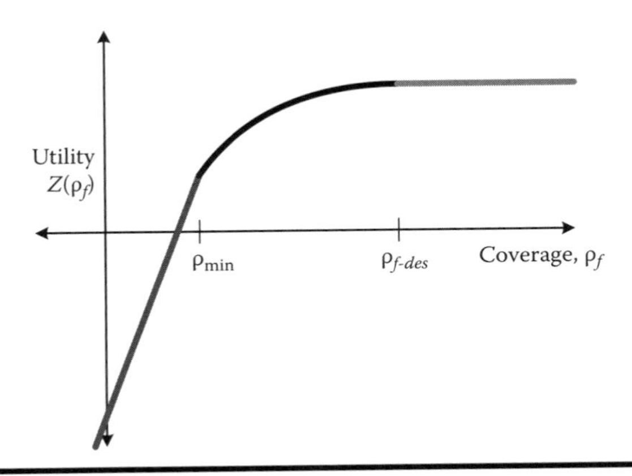

Figure 10.4 The generic utility function.

they will both obtain coverage equal to

$$\rho = [P/(PG + N_0)]_{\text{dB}} = [1/(G + N_0/P)]_{\text{dB}} \tag{10.2}$$

where N_0 includes interference and noise from sources other than the neighboring femtocell. Examining the preceding equation, illustrated in Figure 10.5a, we see that increasing P to infinity yields rapidly diminishing returns, and an asymptote in mutual coverage is reached at $\rho = [1/G]_{\text{dB}}$. Furthermore, if $\rho_{f\text{-des}} > [1/G]_{\text{dB}}$, then it is impossible for both femtocells to simultaneously achieve coverage of $\rho_{f\text{-des}}$. Given such a scenario, it is clear that a lower transmit power solution that yields operation close to the "knee" of the curve depicted in Figure 10.5b is preferable to a high transmit power solution that yields operation in the diminishing returns region. Thus a properly designed utility function should include other factors such as femtocell transmitter power. This modified utility function, $Z(\rho_f, P_f)$, can at least indirectly better account for interference caused to the macrocell network.

To summarize, the RF provisioning problem can be formulated as follows:

> We are given a set of femtocells F and for each $f \in F$, a range of possible pilot transmit powers $[P_{\text{f-min}}, P_{\text{f-max}}]$, a set of allowable transmit frequencies H_f, allowable scrambling codes SC_h for each allowable frequency $h \in H_f$, and a "desired coverage" $\rho_{f\text{-des}}$. The problem is to select a transmit power $P_f \in [P_{\text{f-min}}, P_{\text{f-max}}]$, frequency $h \in H_f$, and scrambling code $s \in SC_h$ for each femtocell f such that the sum utility [e.g., $\Sigma_f (Z\rho_f, P_f)$] is maximized.

Thus, the overall RF provisioning problem becomes a joint optimization problem involving transmit power, and frequency, as well as the scrambling code selection. We split the RF provisioning algorithms into two main categories: incremental and global provisioning. Incremental provisioning refers to provisioning the RF

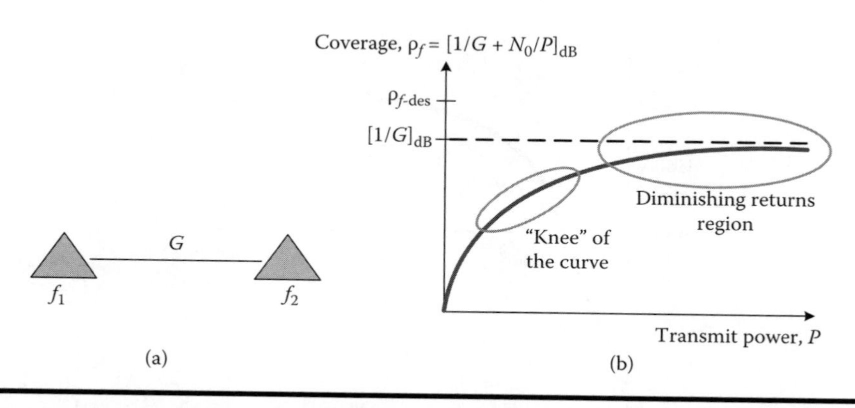

Figure 10.5 (a) Two femtocells separated by a channel gain equal to G. (b) Illustration of the curve showing coverage versus transmit power, when both femtocells transmit at the same transmit power P and when $\rho_{f\text{-des}} > [1/G]_{\text{dB}}$.

parameters of the Nth femtocell when it attempts to join the network, given the presence of $(N-1)$ previously deployed femtocells. Global provisioning refers to an offline global optimization of all deployed femtocells, and the subsequent update of their RF parameters.

For incremental provisioning, we formulate the problem as one of RF parameters of the "current" (e.g., N^{th}) femtocell treating the RF parameters of nearby femtocells as fixed. A more complicated algorithm allows changes to the RF parameters of nearby femtocells as well. However, allowing RF changes to neighboring femtocells has the disadvantage that changes in RF conditions propagating to femtocells that are the neighbors of the current femtocell's neighbors, and so on. This would be the case every time a new femtocell comes online. Moreover, a basic design principle for incremental provisioning should be that it is simple, efficient, near real time, and keeps changes as local as possible.

Given the earlier formulation, there are several possible approaches. For all of the approaches, the first step is for the femtocell to detect the total received energy on each frequency, and the received energies on the various frequency, scrambling-code, that the femtocell is provisioned by the operator to use. The simplest RF provisioning algorithm can simply set the current femtocell's frequency h as the one with the least amount of received energy, and the scrambling-code s as the one with the least energy among all (h, s). Next, the transmit power could either be set as: (1) a fixed value, (2) a function of total received power on the selected frequency h, or (3) a function of the total received power not including the contributions of other femtocells. The disadvantage of approach (1) is that a fixed power allocation cannot adapt to varying levels of external interference (e.g., from a nearby macrocell) to obtain the exact desired coverage $\rho_{f\text{-des}}$. The disadvantage of approach (2) is seen when we consider the situation that was depicted in Figure 10.5. Specifically, if two femtocells very close to each other iteratively perform this power-setting algorithm, each reacting to the additional interference caused by the other increasing its transmit power, then both femtocells will eventually race to their maximum powers; this is undesirable from the perspective of interference caused to the rest of the network. Finally, criteria (3) mitigates this issue somewhat since other femtocell interference is not used in setting transmit power. One disadvantage of approach (3) is that by failing to take into account interference from other femtocells, a suboptimal solution may result.

It should be noted that the simple incremental provisioning solutions discussed thus far can be implemented in a completely distributed manner. One minor complication that arises with approach (3) is that the femtocell should autonomously be able to tell whether a particular combination of frequency and scrambling code belongs to a macrocell or whether it corresponds to another femtocell. For this to work, additional information such as a database containing macrocell deployment information should be known a priori by the femtocell planning algorithm. This information can be either distributed from a centralized location or used via query during distributed planning. Information gleaned through decoding the broadcast channels can also aid in distinguishing neighboring macrocells and femtocells.

In terms of maximizing the sum utility, however, the preceding methods are suboptimal in that they do not consider the utility change to the current femtocell's neighbors due to the additional interference caused by the current femtocell itself. An "optimal" incremental provisioning algorithm jointly optimizes the selection of frequency, scrambling, code, and transmit power to maximize the sum utility of the current femtocell as well as its neighbors. Note that to make the optimal method work, the algorithm requires the knowledge of the current coverage of the neighboring femtocells as well as their transmit powers. Its implementation requires partial centralization, and/or advanced techniques such as inter-femtocell communication.

Finally, since global provisioning is the idea of re-provisioning several femtocells simultaneously, it has greater applicability in a more centralized system that can combine the RF information of all of the femtocells in a small region. The basic idea is to attempt to maximize the sum utility over the entire femtocell network. To do this, several approaches are possible, including optimization methods such as gradient search, simulated annealing, genetic algorithms, and others.

10.3.2 Downlink Interference Scenarios

10.3.2.1 Interference to Macrocell Mobiles from Femtocells

The first interference scenario we illustrate is the most common and occurs to some degree for every femtocell: interference caused to macrocell (or other femtocells) UEs by femtocells operating on the same or adjacent channel. This scenario is of greatest concern when the femtocell is located near the edge of coverage of the overlaid macrocell. An example of the interference impact to macrocell UEs due to the presence of femtocells is illustrated in Figure 10.6. The figure shows a macrocell UE being interfered with by two femtocells f_1 and f_2 operating on the same channel, and one femtocell f_3, operating on an adjacent channel. Note that f_3's coverage region is drawn smaller than that of f_1 and f_2 to indicate that its interference effect on the macrocell UE is discounted by 33 dB (the adjacent channel rejection) [11]. The combined effect of the three femtocells results in a fairly large zone in which macro UEs face considerable interference or may be totally excluded from service unless they move on their own, or are redirected to a different UMTS frequency or different radio access technology (e.g., GSM).

The effect of interference from femtocells can be such that surrounding each femtocell is a "dead zone" or "black hole" in which a UE, operating on the same frequency, that is restricted from registering or using the femtocell will not be able to see the macro network due to forward link interference from the femtocell. If the macrocell UE is on an active call as it enters the femtocell zone, it must be redirected either to 2G GSM service or to another UMTS carrier, and perform a hard hand-over. Otherwise, the call will drop when the macrocell SINR drops below the minimum needed

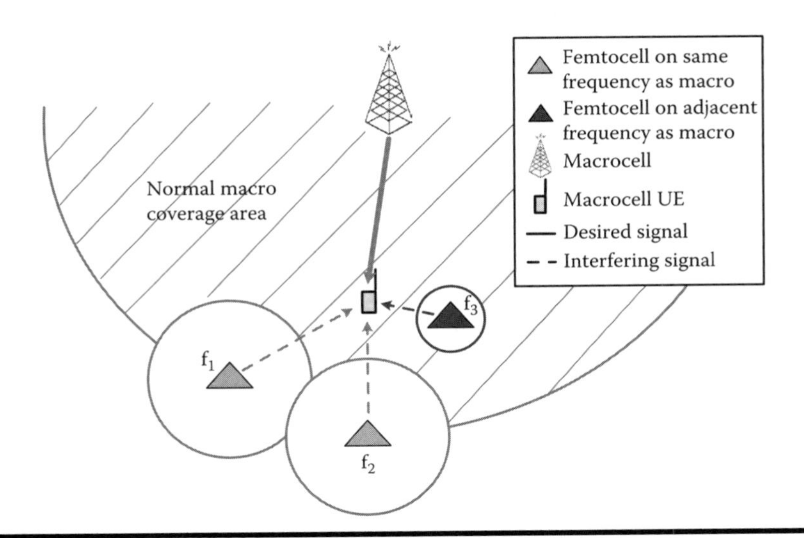

Figure 10.6 Example of interference to the macrocell by femtocells.

to sustain service. When an idle mobile approaches the femtocell zone, the situation is even more complicated and is dependent on the version of the UE device and on the access control mode of the femtocell. If the femtocell is operating using "open access," in which any user may register onto the femtocell while idle, then the problem is relatively simple. Idle users perform same frequency cell-reselection, register, and are granted service from the femtocell. In closed access, in which only a limited list of users (family and friends) may use the femtocell, unauthorized users will be rejected, and depending on the UE behavior will need to do an inter-frequency search for another UMTS carrier or move to GSM. A hybrid technique allows all users to register on the femtocell, but when they attempt either to originate or terminate a call, unauthorized users are redirected to the macrocell carrier. Making sure there is always one clean carrier available for this redirection tends to favor scenario c in Figure 10.1, the scenario in which all carriers are shared between femtocells and macrocells.

To characterize and measure the "dead-zone," conceptually illustrated in Figure 10.6, a series of experiments was conducted at approximately 20 homes. Specialized software communicated with UMTS handset devices and allowed users to enter waypoints as they walked around their houses measuring key parameters. As an example, consider the set of following (Figures 10.7, 10.8, and 10.9) in which the CPICH pilot strength was set to 0, −10, and −20 dBm, respectively. The nearest macrocell is located approximately 3 km from the house. In Figure 10.7, we see that at 0 dBm CPICH, over the entire house, there is total exclusion—that is, no macrocell pilot is even strong enough to be considered as a candidate (the add threshold is set to −12 dB). Figure 10.8 shows candidate set pilot signal strength (those that exceed the add threshold) as a function of location for −10 dBm CPICH. We observe that in

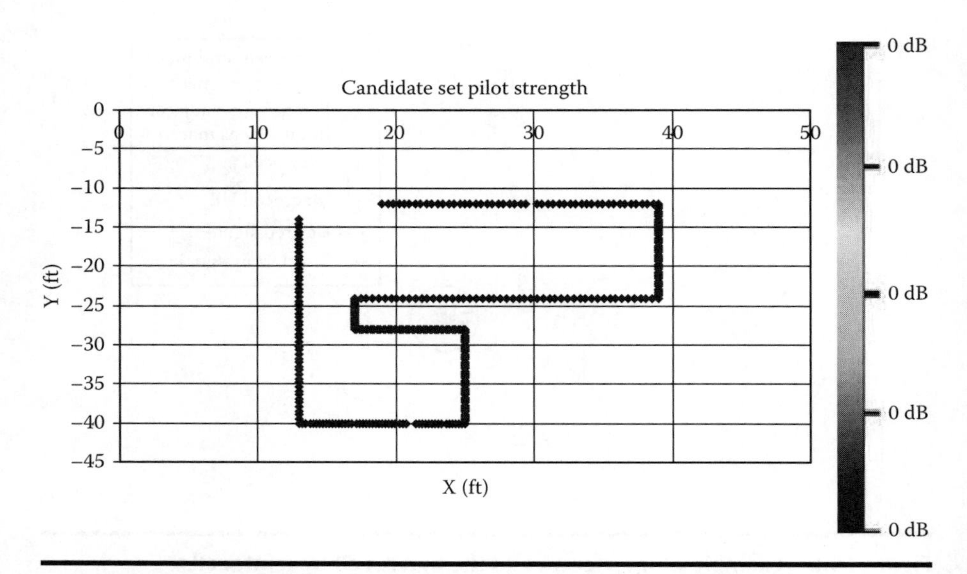

Figure 10.7 Candidate set pilot strength for 0 dBm femtocell transmitter power as measured by the macrocell UE. This figure shows a total dead zone within the house for the macrocell system.

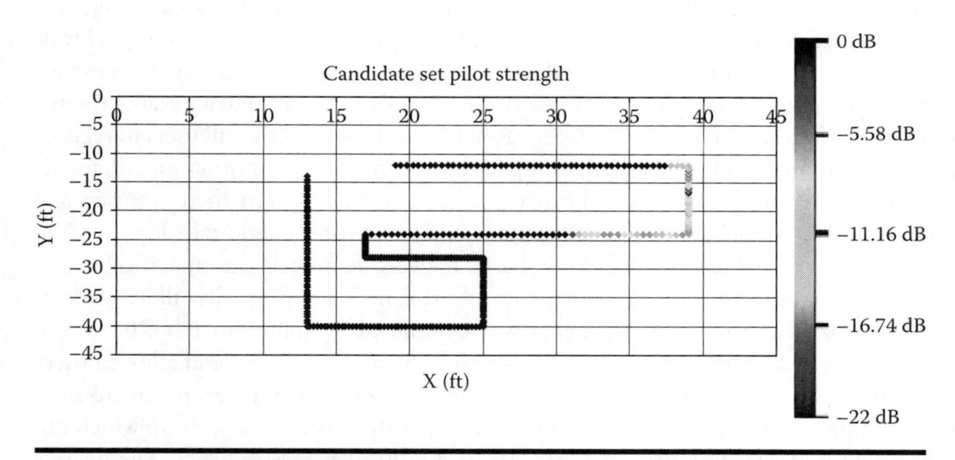

Figure 10.8 Candidate set pilot strength for −10 dBm femtocell transmitter power as measured by the macrocell UE. This figure shows weak coverage of the macrocell at the far corners of the house near a window only.

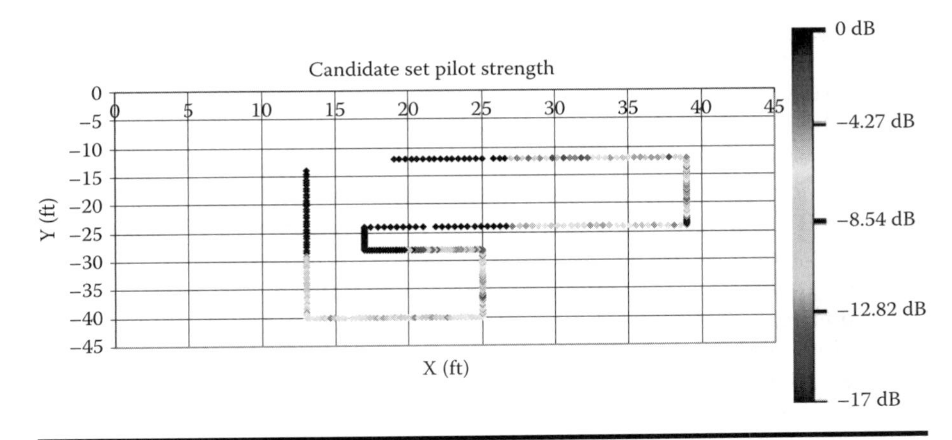

Figure 10.9 Candidate set pilot strength for −20 dBm femtocell transmitter power as measured by the macrocell UE. This figure shows weak coverage of the macrocell at the far corners of the house near a window only.

the corners of the house, near windows and farthest from the femtocell, a user would be able to keep a voice call, but everywhere else there is still a dead zone. Finally, at −20 dBm, we see how the "dead zone" continues to shrink. It is shown in Figure 10.9.

Thus, managing forward link interference poses a complicated set of trade-offs. For example, if a femtocell increases its transmitter power to improve data rates for its UEs, this directly causes an increase in interference to nearby macrocell UEs—thus, the need for the utility function described in Figure 10.5 which attempts to find an optimum point, making sure that all femtocells get at least a minimum level of service. It can also be scaled to take into account foreign registrations through the use of a weighting function.

10.3.2.2 Interference to Femtocell Mobiles from Macrocells and Other Femtocell

From the perspective of femtocell UEs, signals emanating from nearby macrocell base stations transmitting on the same or an adjacent frequency constitute downlink interference. This was illustrated in Figure 10.6. The scenario currently under discussion is of most concern when the femtocell is located close (in terms of propagation loss) to the macrocell and is also directly linked to the uplink interference scenario. Unlike interference from other femtocells, interference from macrocells is constant and is something the femtocell network has no control over. Thus, the only way femtocells can combat this interference is to increase their transmit power to satisfy the target coverage radius or to move to an unoccupied frequency, if available. This is one of the key functions of the automatic RF planning and provisioning function.

Assuming there is no soft handover between femtocells, interference to femtocell UEs from other femtocells can be treated similarly to downlink interference from macrocells. A key difference as mentioned earlier is that the operator of the network has the ability to set the downlink transmit parameters (e.g., transmit power and frequency) of the femtocells. In this sense, assuming a common or adjacent frequency situation, interference can be managed either by (1) a femtocell increasing its transmit power, or (2) reducing the transmit power of interfering femtocells. It is also possible that a femtocell is located so close to a macrocell that the target radius cannot be achieved without using the femtocell's maximum transmitter power. In this situation, the automatic RF planning and provisioning function may determine (i.e., based on the specific operator's policy) that it is better not to turn on the femtocell rather than create the additional interference or to accept a more limited service radius. In addition, for femtocells located near the macrocell, femtocell mobiles located on the femtocell cell edge, have the potential to create uplink degradation of the macrocell. This is an additional metric for the RF planning function to use when determining whether to allow a femtocell located very close to a macrocell, and operating especially co-channel, to be allowed to turn on or whether to reduce its target service radius.

10.4 Uplink Interference Scenarios and Mitigation Techniques

In this section, we discuss the uplink or reverse link interference issues in femtocell deployments. There are crucial differences between downlink and uplink interference scenarios. In particular, while the former is controlled in large part by proper power and frequency assignment to the femtocell and is performed in a quasi-static manner, the latter tends to be dynamic, based on the current location and mobility condition of UEs with respect to the femtocells and macrocells that are controlling their power. Thus, the techniques for controlling and mitigating uplink interference need to be dynamic and adaptive. The uplink interference issue can be divided into three scenarios, analogous to the downlink scenarios: (1) interference from macro UEs to the femtocells, (2) interference from the femtocell UEs to the macrocell, and (3) interference from femtocell UEs to neighboring femtocells. Most of these uplink interference scenarios are caused by the fact that mobiles are power-controlled by either the femtocell or the macrocell, but not both. Soft handover between femtocells and macrocells can alleviate this situation, but is not likely in the near future due to the complexity of implementation at the core network level.

10.4.1 Interference to Femtocell Reverse Link from Macrocell UEs

To study reverse link interference from a macrocell UE, it is useful to perform a detailed analysis of the situation. Consider the situation shown in Figure 10.10. In this figure, the UE connected to the macrocell is called the mUE, while the UE connected to the femtocell is called the fUE. The transmit powers of the fUE and mUE are denoted by P_{fUE} and P_{mUE}. The path gains (multiplicative inverse of path loss) between a UE and a cell is denoted as $G_{Gx,Y}$, where x is either mUE or fUE, and Y is equal to either M (for macro) or F (for femtocell). The received powers from the mUE at the macrocell and the femtocell are given by $P_{mUE}G_{mUE,M}$ and $P_{mUE}G_{mUE,F}$. Note that the values of parameters are expressed in dB and in linear terms in interchangeable fashion in the following analysis.

Let the sum of the total interference and noise levels at the macrocell be I_M. The received uplink power at the macrocell from the mUE is given by:

$$P_{mUE}G_{mUE,M} = I_M \kappa_{UL} \tag{10.3}$$

where κ_{UL} is the uplink SNR dependent on the data rate being transmitted by the UE, and could be within a wide range (e.g., from -18 dB to $+10$ dB)

The received power from the mUE at the femtocell is computed as:

$$P_{mUE}G_{mUE,F} = P_{mUE}G_{mUE,M}(G_{mUE,F}/G_{mUE,M}) = I_M \kappa_{UL}(G_{mUE,F}/G_{mUE,M}) \tag{10.4}$$

Using Equation 10.4, we judge the interference levels on the femtocell uplink by examining the mUE femtocell-to-macro path gain ratio given by $(G_{mUE,F}/G_{mUE,M})$. First, note that the path gain from the mUE to the femtocell, $G_{mUE,F}$, can be substantially higher than the path gain to the nearest macrocell, $G_{mUE,M}$. Figure 10.11 illustrates this using the distribution of path gain differences. We simulate a dense

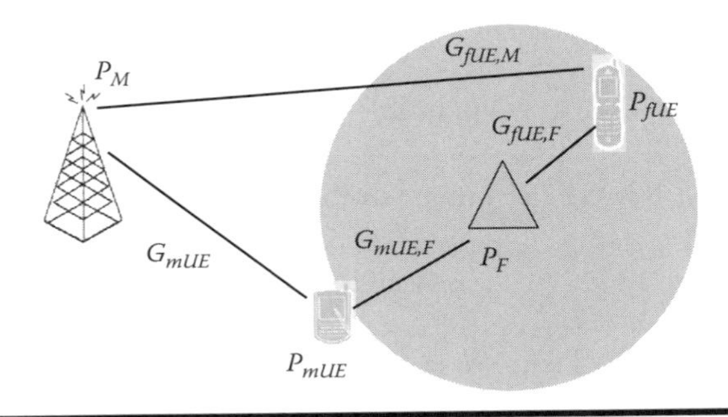

Figure 10.10 The geometry of reverse link interference.

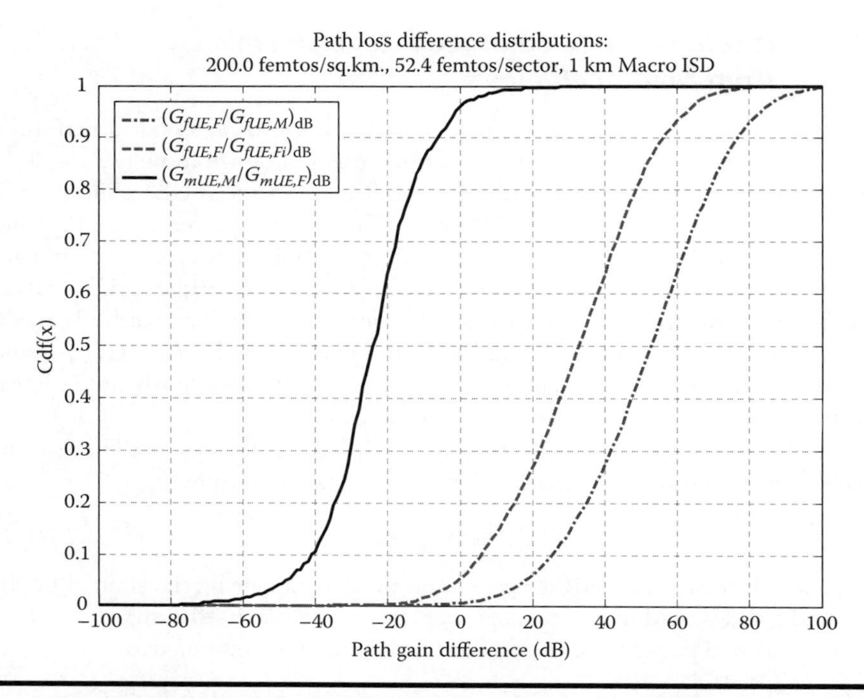

Figure 10.11 Cumulative distribution of path gain difference in a shared macrocell–femtocell deployment.

suburban femtocell deployment assuming 200 femtocells/sq.km in a suburban environment. The path gains are computed based on a 3GPP model for outdoor to indoor and a COST231-based indoor-to-indoor multiwall model [12] as appropriate:

$$\text{Macro to UE [3GPP model]} : -[15.3 + 37.6^{*}\log10(d) + \text{applicable wall loss] dB} \quad (10.5)$$

$$\text{Femto to UE [COST231 indoor-to-indoor]}: -[38.46 + 20^{*}\log10(d) + \text{wall loss] dB} \quad (10.6)$$

$$\text{Femto to other UE: max(A,B)} \quad (10.7)$$

The curve showing the difference between the path gains from an mUE to its serving macrocell and to the nearest femtocell is skewed well into the negative direction, from which we can conclude that the typical UE is much closer to a femtocell than it is to its nearest macrocell. It should also be noted that the downlink power settings influence the cell boundary, and hence the ratio $(G_{mUE,F}/G_{mUE,M})$. The reason for this is that a mUE can get closer (in path gain) to a femtocell with a smaller transmit power without having to either switchover to the femtocell, or switch to a different macrocell on a different channel. Thus the femtocell downlink power setting has an interesting interaction with uplink interference. A higher femtocell downlink transmit power causes higher downlink interference to co-channel macrocell UEs, but can reduce uplink interference from macrocell UEs to femtocells, as well as that from femtocell UEs to the macrocell, as will be seen later in the section. Using $(G_{mUE,F}/G_{mUE,M}) = 40$ dB and $\kappa_{UL} = 10$ dB, both of which are plausible values in a real deployment, we can see that the interference from the macrocell UE at the femtocell could be 50 dB above the interference and noise level at the macro. This leads to the necessity of using a different (and possibly dynamic) mechanism for operating the femtocell reverse link instead of a conventional macrocell technique of using a fixed operating point above a fixed thermal noise threshold.

An additional conclusion of the preceding analysis is that the dynamic range of the femtocell receiver needs to be much higher than that in the macrocell, a fact that has been recognized in 3GPP [8] as part of the requirements for home NodeBs.

A series of experiments was conducted to try to reproduce and understand the scenario of a mUE causing interference to the femtocell and to help understand the required femtocell response. A house that was about 3 km (about 2 miles) from the nearest macrocell was selected, close to the limit of its coverage. In the experiment, whose setup is illustrated in Figure 10.12, the femtocell was located at (0, 0) in a local coordinate system. The transmitter power of the femtocell was lowered to −20 dBm, to move the downlink "dead zone" region so that the mUE (a data card was used) could operate within the house without dropping its call. Several calibrated points

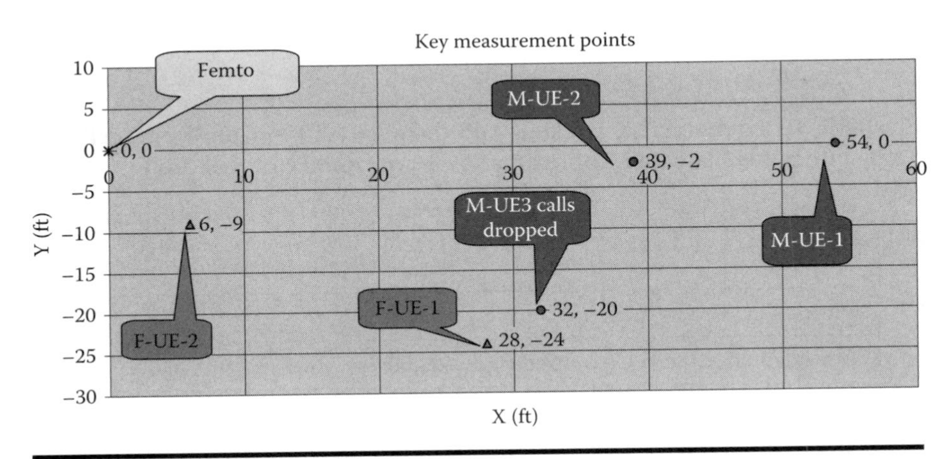

Figure 10.12 Locations of macro UEs and femtocell UEs relative to the femtocell.

within the house were identified, as shown in Figure 10.12. Locations were selected such that a sustained ftp upload on the macrocell network would not be interrupted due to femtocell forward link interference. Femtocell uploads were then performed, recording the overall upload throughput and transmitted upload power for the two cases of the macro UE (1) turned off and (2) performing a sustained ftp upload. Table 10.1 summarizes the overall user experience for one of the test points.

The table clearly shows that the femtocells see a significant level of interference from the mUE that is causing the femtocell UE to power up between 10 and 15 dB relative to the case where the mUE is turned off. Note that the throughputs at the femtocell are not severely degraded due to the increased interference levels from the active macro UE. This is due to the fact that the femtocell uplink MAC and power control was suitably adapted to perform under these high-interference conditions. A description of the possible methods to achieve this are provided in the next section.

10.4.1.1 Mitigating Uplink Interference to Femtocell Receivers

This section describes some possible techniques and methods to be employed in the design of femtocells to ensure they operate in a reliable manner in the presence of uplink interference from the macrocell UEs in their operating frequency.

The reverse link resource allocation for conventional macrocell systems operates based on maintaining uplink stability by controlling the total received in-cell power. Airlink resources are allocated in such a way that the noise rise, which is the difference between the total received power and the equivalent thermal noise level, is maintained at or below a predetermined threshold. This noise rise threshold (also known as the rise-over-thermal) is typically set between 5 and 10 dB. Variants of this technique, such as load control, are also employed, but the underlying spirit remains the same. This mechanism counters the near-far effect, and serves cell-edge UEs in a fair manner, while at the same time maintaining system stability. A completely controlled network is assumed in the macrocell system, wherein all the neighboring cells also follow the same policy.

Table 10.1 Macro UE Located Approximately 13 m from Femtocell. Femtocell UE Located Approximately 10 m from Femtocell Cell

	Avg Tx Total Power	*Fwd Link RSSI (Control)*	*F-UE Upload Speed (kbps)*
Femtocell UE, Macro UE off	−36.27	−67.9	440
Femtocell UE, Macro UE uploading	−23.08	−69.7	410
Change in value	13.62	1.8	30
Macro UE	19.18	−95	75

As has been discussed, femtocells that are overlaid within a macrocell network and deployed in the same frequency as the macrocell network are not able to control the total received power when a macrocell UE is nearby. It is important to ensure that the femtocell can operate under these raised interference conditions. Under consistently strong interference, it is possible to operate the femtocell uplink reliably by simply having the femtocell recalibrate its self-imposed limitation on noise rise, as long as the dynamic range of the receiver is not exceeded. However, the issue becomes more complicated when the interference is strong as well as bursty.

The problem of bursty uplink interference is demonstrated pictorially in Figure 10.13. Simulations show that the path gain from a mUE to the nearest femtocell could be 30 dB higher than its path gain to its serving macrocell, as shown in Figure 10.11. Thus, when a mUE tries, for example, to initially access the macrocell using access probes, if its received power level for these access bursts at the macro is ZdB, (where Z could be below the macrocell's equivalent noise level), then the received level at the femtocell could be > (Z + 30) dB (which could be well above the femtocell's noise level). Even accounting for the difference in noise figure between the femtocell and macrocell, the effect of this burst could be significant. This problem occurs during data transmission as well—for example, when a macrocell UE either transmits bursty data, or in the case of HSUPA, transmits data on some hybrid ARQ (HARQ) processes and not others. In these cases, we have the situation where the swing for the bursts is from the (pilot + control) to (pilot + control + traffic), which is still significant.

Under these bursty conditions, the typical implementation of the reverse link power control fails, as shown in Figure 10.13. At the beginning of every burst

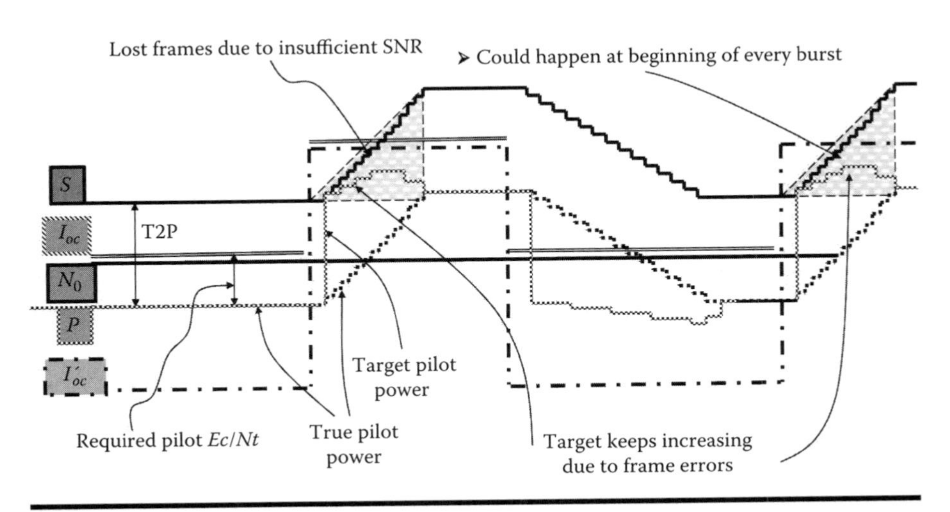

Figure 10.13 Bursty interference on the UMTS uplink.

of interference, the femtocell UE's pilot as well as the control/traffic channel SNRs (E_c/N_0) at the femtocell drop to a very low level, resulting in many packet errors. The fUE's transmit power rises only at a rate determined by the inner loop power control (ILPC), and the fUE cannot react any faster. Eventually (provided the connection is not lost), the fUE raises its power sufficiently to operate in the new interference condition. At the end of the burst, the fUE goes back to its pre-burst transmit power, and this cycle repeats for every burst. Thus there is a potential for the loss of a large number of packets in every burst of interference, which is unacceptable for the femtocell UE.

Consider the following example. Suppose the fUE is operating at its target DPCCH E_c/N_0 before the burst begins. Next, suppose there is a burst that increases the "out-of-cell" interference + noise, I_{oc}, from $(N_0 + N_{F,\text{femto}})$ to a level of $(N_0 + N_{F,\text{femto}} + 30\text{ dB})$, where N_0 stands for the thermal noise and $N_{F,\text{femto}}$ is the noise figure of the femtocell receiver. To counteract this interference, the UE must raise its transmit power by 30 dB. Suppose the inner loop power control step size used by the UE is 1 dB. The number of slots taken by the UE to raise its power by 30 dB is controlled by the power control frequency, which cannot be changed. Meanwhile, the traffic channel is also operating at 30 dB below its target SNR and hence the femtocell is highly likely to lose at least two consecutive frames in the process. It may seem that a two-frame loss is not significant, but consider the case where the interfering macro UE is transmitting on one of the four HARQ processes (for a 10-m TTI). Now, the swings of 30 dB are repeated every four TTIs. So during the three TTIs with no interference, the ILPC will again settle down at, or close to, its original value before the burst, only to have to respond again to another TTI of interference, which causes every fourth frame to be in error. Moreover, if a femtocell UE happened to be using the same HARQ process as the macrocell UE, every one of its data packet transmissions can be in error during this period. This situation will persist until the OLPC has raised the target high enough due to the frame errors.

S = Traffic, N_0 = Thermal noise, P = Pilot, I'_{oc} = Out-of-cell interference, $I_{oc} = I'_{oc}$+ noise

We now discuss two possible solutions for dealing with bursty uplink interference: adaptive attenuation and virtual attenuation.

10.4.1.1.1 Adaptive Attenuation—Adaptive attenuation uses an adjustable attenuator in the femtocell receiver to change the noise figure, and hence the equivalent thermal noise level in the femtocell receiver. The value of the attenuation is set to be a filtered version of the out-of-sector interference. The attenuator is designed so it reacts instantaneously to upswings in interference, whereas when there is a downswing, it reacts very slowly, as shown by the N_{adapt} line in Figure 10.14. In this manner, the pilot and traffic powers reduce slowly, and when the next burst occurs, the correction to be made is much smaller and more manageable. In this manner, the effects of bursty interference are accommodated.

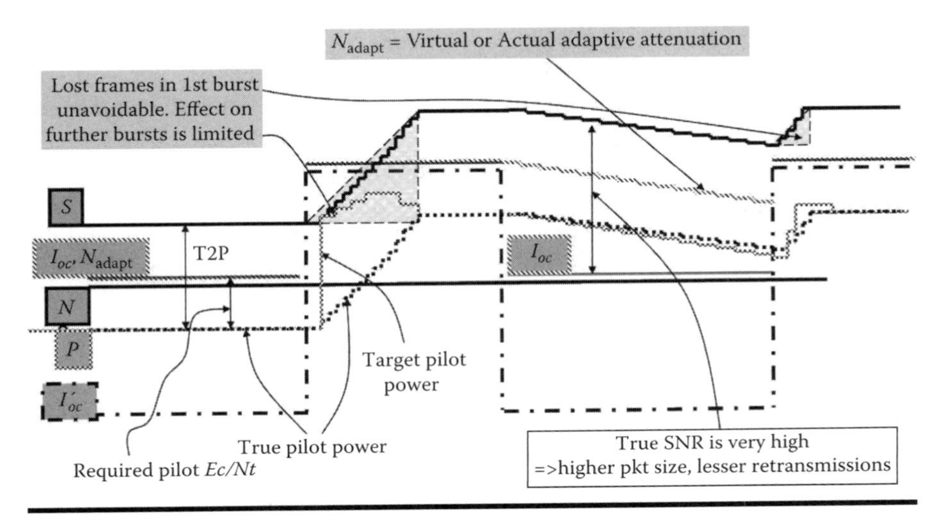

Figure 10.14 Management of bursty interference by adaptive/virtual attenuation.

One side effect to be noted, however, is that the average transmit power of the UE increases.

10.4.1.1.2 Virtual Attenuation—In the virtual attenuation solution, we use a smoothed filtered version of the out-of-cell interference, which can be called virtual Ioc, for inner loop power control. The filtering could be the same as described in the adaptive attenuation technique. A virtual pilot SNR is computed based on this virtual Ioc. In contrast with the adaptive attenuation method, in the period between bursts, the actual SNR is much greater (since there is no additional physical attenuation).

10.4.2 Interference to the Macrocell Reverse Link Due to Femtocell UEs

Interference caused to the macrocell due to femtocell UEs is by far the most serious uplink interference scenario. Though the occurrence of this situation is infrequent, it has the potential to cause severe degradation to the performance of the macro network, and hence must be prevented. In general, femtocells UEs radiate lower power compared to the macro UEs, because the propagation loss between the UE and femtocell is relatively small. This, coupled with the fact that the macrocell is usually at a much greater distance from the femtocell UE compared to the femtocell itself, ensures that in a vast majority of cases, there is no noticeable interference from the femtocell UE to the macrocell.

To analyze the scenario under which a femtocell UE can cause interference to the macro, it is important to first acknowledge that the interference level at a femtocell can be a lot higher than that at the macrocell, due to the possibility of the presence of macrocell UEs nearby, as explained in the previous section. When the femtocell is operating under such conditions, the UEs transmitting to the femtocells also must increase their power to be "heard" by the femtocell.

Let us first analyze the case of a femtocell UE in isolation, and then extend the analysis to the case where there is also a macrocell UE in the vicinity of the femtocell. The received power from the femtocell UE at the femtocell is given by:

$$P_{fUE} G_{fUE,F} = I_F \kappa_{fUE,UL} \tag{10.8}$$

where $\kappa_{fUE,UL}$ is the uplink SNR dependent on the data rate being transmitted by the fUE, and I_F is the sum of the total uplink interference and noise seen at the femtocell. The femtocell UE's received power at the macrocell (in other words, the interference it causes to the macro) is given by:

$$\begin{aligned} P_{fUE} G_{fUE,M} &= P_{fUE} G_{fUE,F} \times (G_{fUE,M}/G_{fUE,F}) \\ &= I_F \kappa_{fUE,UL} \times (G_{fUE,M}/G_{fUE,F}) \end{aligned} \tag{10.9}$$

using Equations 10.8 and 10.9, and also noting that $(G_{fUE,M}/G_{fUE,F})$ is likely to be a very small value as demonstrated by the path gain difference distribution in Figure 10.11, we can conclude that if the nearby macrocell UE is not transmitting, the femtocell UE interference at the macro will be well below the interference level seen by the femtocell on its uplink, I_F. If the nearby macrocell UE in Figure 10.10 is not transmitting, and there is no other source of interference, then I_F is near the equivalent thermal noise level, and hence the interference from a single femtocell UE to the macrocell is negligible. It would take a very large number of such UEs transmitting simultaneously (an unlikely scenario) to start affecting the macrocell. For voice service, the uplink power is so low that the probability of interference is negligible.

In relation to the analysis shown earlier and in the previous section, it is relevant to point out that a slightly reduced sensitivity (i.e., higher noise figure) at the femtocell does not hurt either the femtocell or the macrocell performance. A reduced sensitivity can be attained at a lower cost, and is thus desirable for femtocells. The 3GPP requirements study for femtocells has also made a recommendation to this effect [13].

Now consider the situation where the macrocell UE in Figure 10.10 is transmitting high-speed data using significantly higher transmitter powers than required for voice. As seen from Equation 10.4, the interference from the mUE at the femtocell is given by:

$$I_F \approx P_{mUE} G_{mUE,F} = I_M \kappa_{mUE,UL} (G_{mUE,F}/G_{mUE,M}) \tag{10.10}$$

Combining Equations 10.9 and 10.10, the femtocell UE interference at the macro is now given by:

$$P_{fUE}G_{fUE,M} = I_M \kappa_{mUE,UL} \kappa_{fUE,UL} (G_{mUE,F}/G_{mUE,M}) \times (G_{fUE,M}/G_{fUE,F}) \tag{10.11}$$

Observe that the ratios $(G_{mUE,F}/G_{mUE,M})$ and $(G_{fUE,M}/G_{fUE,F})$ more or less multiply to 1 if the mUE and fUEs are both located near the femtocell coverage boundary. Under these conditions, when both the macrocell UE and the femtocell UE are operating at high data rates ($\kappa_{mUE,UL}$, $\kappa_{fUE,UL} \geq 1$), the interference from this single femtocell UE at the macrocell can be near or even above the overall interference level seen from other sources at the macrocell. This can be a serious issue, since the macrocell network is typically designed to operate at a total reverse link interference level that is close to the noise level. Hence, even one or a few such femtocell UEs can cause a disruption at the macrocell on the uplink. Techniques that can be used to mitigate and manage this situation are described in section 10.4.3.

10.4.2.1 Measurements of Uplink Interference to Macrocell Base Stations

The previous interference scenario is more likely to occur when the femtocell is located very close to the macrocell base station and a femtocell UE is doing a high-speed (and therefore high-power) upload such as HSUPA. In this scenario, if the femtocell UE is located near the femtocell edge of coverage, there is the potential for the femtocell UE to create out-of-cell interference into the macro network, which can cause performance degradation of the macrocell.

To show how this happens, we used a macrocell test network and a femtocell located close to it (less than 50 m). We started an upload on the femtocell UE, and then an upload on the macrocell UE and watched the change in the transmitter power. Table 10.2 summarizes this experiment and shows that the upload on the femtocell caused the macro-UE to power up by close to 10 dB over what it would have done normally. Now if the macro-UE were near the cell edge, it might have dropped the call due to the large rise in out-of-cell interference.

10.4.3 Mitigating Uplink Interference from Femtocell UEs to Macrocell Base Stations

To prevent uplink interference from the femtocell users to the macrocell, we must limit the femtocell UE's transmit power. A blanket limit that places the same transmit power limit on all fUEs will be unnecessarily restrictive, and will reduce the achievable throughput in femtocells. A smart approach is to analyze the level of interference caused by each femtocell UE at the macrocells, and take appropriate action.

Table 10.2 Changes in Macro-UE Transmitter Power When the Femtocell UE Moves Close to the Macro Network

	Avg. Tx. Pilot Power (dBm)	*Avg. Tx. Total Power (dBm)*	*Fwd Link RSSI (dBm)*	*F-UE Upload Speed (kbps)*
Macro UE, femtocell UE off	−14.8	0.61	−76.1	544
Macro UE, femtocell UE uploading	−3.67	9.43	−76.9	319
Change in value	11.13	8.82	0.8	225
Femtocell UE	−14.35	−0.93	−65.04	

The maximum allowable addition to the received power at the macro, due to all femtocell UEs, $\Delta P_{I,\max}$, must be managed so it does not contribute significantly to the noise rise at the macro. That is to say

$$\Sigma_i P^i_{fUE} G^i_{fUE,M} \leq \Delta P_{I,\max} \tag{10.12}$$

where P^i_{fUE} is the transmit power of the ith fUE, and $G^i_{fUE,M}$ is the path gain from the ith fUE to the macrocell. Jointly optimizing the impact from all femtocell UEs to a macrocell involves a complex global approach involving instantaneous power of all the femtocells in the network. A viable approach of lesser complexity is to place a limit on each femtocell UE's incremental effect on the macrocell noise rise. From Equation 10.12, the limit for each UE is given by:

$$P^i_{fUE,\max} G^i_{fUE,M} \leq \Delta P_{I,\max}/N \tag{10.13}$$

where $1/N$ is a fraction dependent on the number of UEs that are susceptible to simultaneously cause substantial interference to a specific macrocell. This factor can either be periodically updated based on the automatic RF planning and provisioning function or it can be set to a conservative fixed value.

Once the limit on per-UE interference is fixed, the remaining task is to compute the path gain between the UE and the macrocell, so that the UE transmit power limit can then be fixed. Both HNB measurements as well as UE measurements can be used to compute path gain between the UE and the macro [13]. Note that the underlying assumption in both the methods is that the P-CPICH power of the macrocell base station can be ascertained from decoding the system parameters transmitted in the broadcast channel. One possible method is described next.

The value of $P_M G^i_{fUE,M}$ can be known from the UE's measurement report of CPICH RSCP of all significant macrocells. From a knowledge of P_M obtained by decoding the broadcast channel from the macrocell NB, we can now estimate $G^i_{fUE,M}$, the UE's path gain to the macrocell. The maximum of all the path gains (i.e., the path gain to the closest macro) is used to compute the most conservative power limit. Based on this path gain to the closest macro, and the configured limit in Equation 10.13, the power limit for the UE can be computed. The power limit is not directly applied, but is instead implemented through the scheduling algorithm. This per-UE power limit can be updated at a much slower rate compared to the slot duration. Figure 10.15 provides a pictorial representation of the algorithm described earlier to set the power limit of a UE with respect to the macro. In Figure 10.15, the received power at each UE (RSCP) from a nearby macrocell is divided by the estimated transmit power from that macrocell, to provide an estimate of the path gain between the UE and the macrocell. The first quantity is obtained from the UE's measurement reports received on the femtocell's uplink, while the second quantity is obtained from the femtocell's REM (RF sniffing) function, as described in section 10.3, by decoding the broadcast channel from the macrocell. The maximum of the path gains is then divided by the configured limit on the allowable per-UE interference to the macrocell, which provides us an absolute transmit power limit on the UE. This absolute power limit information is then used by the femtocell's uplink scheduler in an appropriate manner when determining the resource grant to the UE in question.

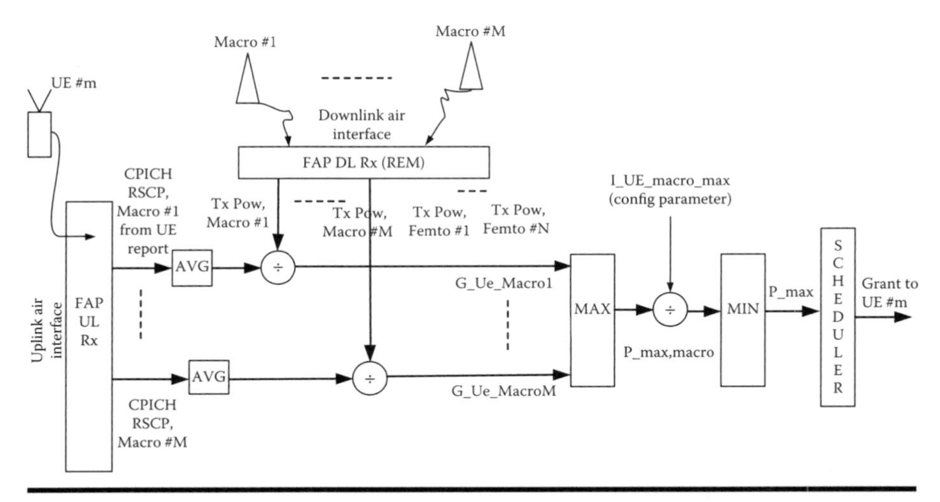

Figure 10.15 Power limit for mitigation of interference to macros.

10.4.4 Limiting Interference to Other Femtocells

The power limit technique was explained earlier with respect to the mitigation of femtocell UE interference to the macrocell network. The methods discussed are also applicable to the mitigation of the interference caused by a femtocell UE to an adjacent femtocell (Figure 10.16). The control parameter in this case is the amount of interference allowable to a neighboring femtocell. While the interference to the macro can be reliably based on the thermal noise level at the macro and the noise figure for wide area or medium range BS classes, the interference to femtocells should be based on the operating region of the femtocells themselves, which can be quite different from that of the macros, as discussed in previous sections. For example, if the limit of the allowable interference to a neighbor femtocell is set to the equivalent thermal noise level, while the neighbor femtocell is operating well above N_0, due to the interference it is seeing from a macro UE, then this limit becomes unnecessarily restrictive to the femtocell UE in question, and may hamper its throughput to its serving femtocell.

Two approaches will set the inter-femtocell interference limit:

1. Setting the interference limit to a specific fixed but high value above the equivalent thermal noise (say 30 dB).
2. Setting the ceiling on the interference from a UE to a neighboring femtocell to be a configurable value above the total interference seen at one's own femtocell. This technique is more efficient because it avoids the situation that could arise where the UEs operate at a higher power level even though there are no macro UEs around.

Either way, once the interference limit to a neighboring femtocell is determined, the rest of the procedure is as defined in Section 10.4.3.

10.5 Summary, Challenges, and Opportunities

Femtocell deployments offer the vision of universal mobile broadband, merging the bandwidth of the Internet with very high-quality coverage both indoors and outdoors. Femtocell deployments are in their infancy and as both the number and density of femtocells grows into the millions, techniques for managing the trade-offs between coverage and interference will evolve. Interference scenarios and mitigation techniques are fairly well understood at this point in time, and are summarized in the next section. The sophistication of the implementation will continue to grow and evolve.

Standards-based network architectures, to provide for more open network deployments such as 3GPP2 UMTS Release 8 and Release 9, are either completed or nearing completion. Advanced techniques for managing networks are under consideration and development.

This section briefly summarizes the challenges and opportunities for deploying very large networks of femtocells.

10.5.1 Summary of Interference Mitigation Techniques

Sections 10.3 and 10.4 describe a number of interference scenarios. In this section, we show that the problem is well understood and summarize the solutions that can mitigate and manage the four basic interference scenarios often discussed [5].

The first scenario occurs on the downlink when the femtocell, operating on the same channel as the macrocell, creates a dead zone around the femtocell preventing service to macrocell users not authorized to use the femtocell. This scenario can be mitigated by a combination of the following techniques:

- Active calls need to be redirected either to other UMTS carriers or to 2G GSM service.
- The automatic network planning function uses a combination of femtocell-based measurements and UE-based measurements to optimally set the femtocell radius to provide the desired coverage radius without spilling too much RF interference.
- Idle users who are rejected by femtocell use HCS or cell reselection to register on a different macrocell frequency.
- All users are allowed to register on the femtocell, but are redirected to macrocells when they attempt to originate or become active.
- Use unauthorized UE registration requests as an indicator that the serving radius of the femtocell is too large and reduce radius to maximize coverage while controlling unauthorized registrations.

The second key problem occurs when femtocells located close to macrocells operating on the same channel may not have a large enough coverage radius. The solutions for this scenario include:

- Use remote environment monitoring supplemented with UE-based measurements to determine the required power to provide the target level of coverage.
- Not allow the femtocell to turn on if it is too close to the macrocell.

The third scenario occurs when the femtocell is located near the macrocell cell edge and UEs far from macrocell doing high-speed data uplink create reverse link interference for femtocell users. The solutions include:

- Design the femtocell reverse power rate limiting algorithms so they are much more tolerant to large increases in out of cell interference (ROT).
- Use either adaptive attenuation or virtual attenuation to adjust the femtocell noise figure [14].
- Design the femtocell receiver so it has a very wide dynamic range.

The final scenario occurs when the femtocell and its UEs located close to a macrocell operating on the same frequency, have high-speed data on the uplink that creates reverse link interference into the macrocell. The solutions to prevent macrocell interference are:

- The femtocell attempts to measure downlink interference from neighboring macrocells and femtocells and reduces the uplink data rate (controls transmitter power) when DL interference indicates there may be a potential for UL interference.
- Reduce the femtocell target service radius. This moves the cell edge closer to the femtocell, reducing the transmitter power of mobiles at the cell edge and thus reducing interference to the macrocell.
- Not provide service for femtocells that are located extremely close to macrocells and that have the greatest potential to cause uplink interference.

10.5.2 Inter-Femtocell Communications

As the density of femtocell deployments grow, there will be increasing requirements for more advanced and scalable algorithms for managing the interference between femtocells, and in the process a migration from centrally planned RF provisioning algorithms to more distributed algorithms will occur. To facilitate this, femtocells will need to be able to directly communicate with each other in a completely distributed manner (e.g., somewhat similar to the LTE X2 interfaces for inter-eNodeB communication). The presence of such autonomous communication opens up several possibilities for "co-operative" interference management. One such possibility is a completely distributed implementation of the utility maximization method mentioned in section 10.3. More advanced interference mitigation techniques that leverage inter-femtocell communication include nearby femtocells coordinating their transmissions in a time-sharing manner, utilizing interference cancellation or multiuser detection, using coordinated beam-steering/forming to avoid interference, and other advanced signal processing technologies. Additional research needs to be done in this area.

10.5.3 Standardization in the Deployment of Femtocell Networks

UMTS femtocells, or home NodeBs (HNBs), are being developed by many companies, and commercial deployments are in their early stages. These initial deployments use pre-standard products, with each solution having a combination of standards based on proprietary interfaces between the femtocells, the core network, and the gateways. Simultaneously, standardization for femtocells is being carried out at a brisk pace by 3GPP. Furthermore, the Femto Forum was created as a means of promoting femtocells, bringing together companies interested in femtocell technology and focusing on solutions that overcome barriers to the adoption of this technology. Some of the major achievements of this forum include the publication of a white paper on interference issues and management for femtocells [5], and bringing about consensus among a large set of industry players on the overall femtocell architecture (including the IuH standard interface between femtocells and the femtocell network gateway function) described in UMTS 3GPP Release 8.

In the 3GPP standards body that governs UMTS; significant progress has been made in femtocell standardization by producing a comprehensive set of specifications, including the aspects of architecture, security, IuH interface, mobility management, radio aspects, and conformance testing. The first set of femtocell standards was published in Release 8 of the overall 3GPP standard. As part of this effort, a new class of BS, called the HNBs has been included, and the RF requirements for the HNB have been suitably modified [13]. 3GPP is also currently studying the interference mitigation aspects of the LTE femtocell as part of HeNB standardization [15, 16].

References

[1] Femto Forum Online White Paper, "Femtocell Business Case Presentation," Signals Research, February 2009.

[2] J. Weitzen, and T. Grosch, "*Advantages of Femtocells for Mobile Broadband Data Services,*" Airvana White Paper, www.airvana.com.

[3] "LTE Takes Shape," white paper from picoChip, http://www.picochip.com/downloads/lte_takes_shape.pdf.

[4] R. Y. Kim, J. S. Kwak, and K. Etemad, "WiMAX Femtocell: Requirements, Challenges, and Solutions," *IEEE Comm. Magazine*, vol. 74, no. 9, 2009.

[5] Femto Forum Online White Paper, "Interference Management in UMTS Femtocell Systems," December 2008.

[6] T. L. Ho, and H. Clasussen, "Effects of User-Deployed Co-channel Femtocells on the Call Drop Probability," Proceedings of 18th Annual International Symposium on Personal, Indoor, and Moble Radio Communications (PIMRC), September 2007.

[7] H. Claussen, "Performance of Macro-and Co-channel Femtocells in a Hierarchical Cell Structure," Proceedings of 18th Annual International Symposium on Personal, Indoor, and Moble Radio Communications (PIMRC), September 2007.

[8] V. Chandrasekhar, J. G. Andrews, and A. Gatherer, "Femtocell Networks: A Survey," *IEEE Comm. Magazine*, vol. 46, pp. 59–67, September 2008.

[9] Mehmet Yavuz et al., "Interference Management and Performance Analysis of UMTS/HSPA+ Femtocells," *IEEE Comm. Magazine*, vol. 74, no. 9, 2009.

[10] 3GPP TS 25.304: "UE Procedures in Idle mode and Cell Reselection Procedures in Connected mode."

[11] 3GPP TS 25.101: "User Equipment (UE) Radio Transmission and Reception (FDD)."

[12] "Digital Mobile Radio towards Future Generation Systems," COST231 Final Report.

[13] 3GPP TR 25.967: "FDD Home NodeB RF Requirements (FDD)."

[14] 3GPP R4-081597, "Impact of Uplink Co-channel Interference from an Uncoordinated UE on the Home NodeB," Airvana, Vodafone, IP.Access, 3GPP RAN4 #47, Munich, Germany.
[15] R4-091976, "LTE-FDD HeNB Interference Scenarios," Vodafone, AT&T, Alcatel Lucent, picoChip Designs, Qualcomm Europe.
[16] R4-092712, "HeNB to Macro eNB Cochannel Interference Simulations–Uplink," picochip Designs.

LTE PLANNING AND OPTIMIZATION

Chapter 11

RF Planning and Optimization for LTE Networks

Mohammad S. Sharawi

Contents

11.1 Introduction

Long-term evolution (LTE) is the next generation in cellular technology to follow the current universal mobile telecommunication system/high-speed packet access (UMTS/HSPA)*. The LTE standard targets higher peak data rates, higher spectral efficiency, lower latency, flexible channel bandwidths, and system cost compared to its predecessor. LTE is considered to be the fourth generation (4G) in mobile communications [1, 2]. It is referred to as mobile multimedia, anywhere anytime, with global mobility support, integrated wireless solution, and customized personal service (MAGIC) [1]. LTE will be internet protocol (IP) based, providing higher throughput, broader bandwidth, and better handoff while ensuring seamless services across covered areas with multimedia support.

Enabling technologies for LTE are adaptive modulation and coding (AMC), multiple-input multiple-output systems (MIMO), and adaptive antenna arrays. LTE spectral efficiency will have a theoretical peak of 300 Mbps/20 MHz = 15 bits/Hz (with the use of MIMO capability), which is six times higher than 3G-based networks that have 3.1 Mbps/1.25 MHz = 2.5 bits/Hz [i.e., evolution data only, (EV-DO)]. LTE will have a new air interface for its radio access network (RAN), which is based on orthogonal frequency division multiple access (OFDMA) [3].

This chapter focuses on the radio frequency (RF) planning and optimization of 4G LTE cellular networks, or the so-called evolved universal terrestrial radio access networks (E-UTRAN) and discusses the physical layer modes of operation for the user equipment (UE) as well as base stations (BS) or the so called evolved node B

* Estimated first commercial deployment is in 2011 (from Qualcomm Inc., February 2009).

(eNB) subsystem. Frequency division duplexing (FDD) and time division duplexing (TDD) modes of operation and their frequency bands are also discussed and illustrated according to the 3GPP specification release 8, 36 series, as of September/December 2008.

RF aspects of cell planning such as cell types, diversity, antenna arrays and MIMO system operation to be used within this architecture will be discussed. Various wireless propagation models used to predict the signal propagation, strength, coverage and link budget are to be explained. The main performance and post deployment parameters are then discussed to assess the RF network performance and coverage. Model tuning according to field measurements is discussed to optimize the network performance. These will follow the standard recommendation for mobile and stationary users. All these aspects are essential for the RF planning process.

11.2 LTE Architecture and the Physical Layer

11.2.1 LTE Network Architecture

The LTE network architecture is illustrated in Figure 11.1. The data are exchanged between the UE and the base station (eNB) through the air interface. The eNB is part of the E-UTRAN where all the functions and network services are conducted. Whether it is voice packets or data packets, the eNB will process the data and route it accordingly. The main components of such a network are [4]:

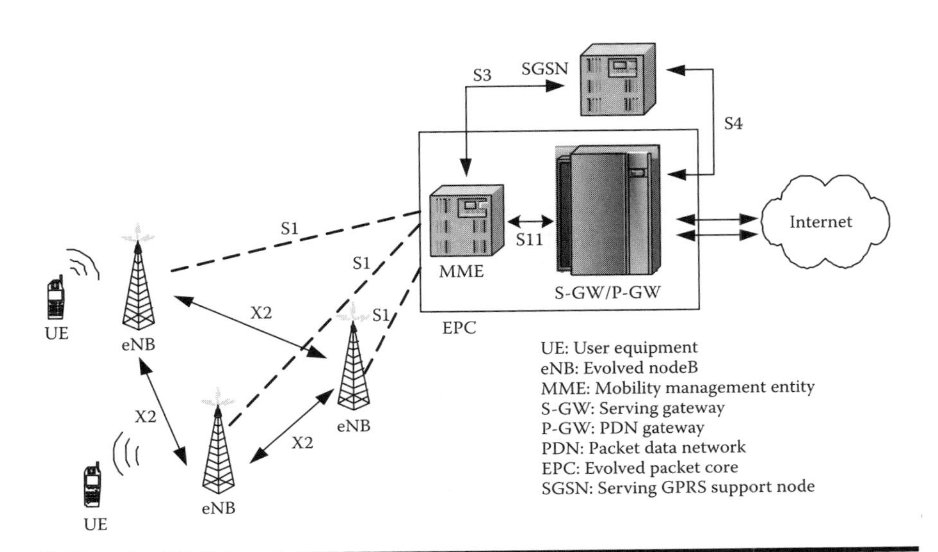

Figure 11.1 LTE network architecture.

- **User Equipment (UE):** This is the user device that is connected to the LTE network via the RF channel through the BS that is part of the eNB subsystem.
- **Evolved NodeB (eNB):** The eNB functionalities include radio resource management (RRM) for both uplink (UL) and downlink (DL), IP header compression and encryption of user data, routing of user data, selection of MME, paging, measurements, scheduling, and broadcasting.
- **Mobility Management Entity (MME):** This portion of the network is responsible for nonaccess stratum (NAS) signaling and security, tracking UE, handover selection with other MMEs, authentication, bearer management, core network (CN) node signaling, and packet data network (PDN) service and selection. The MME is connected to the S-GW via an S11 interface [5].
- **Serving Gateway (S-GW):** This gateway handles eNB handovers, packet data routing, quality of service (QoS), user UL/DL billing, lawful interception, and transport level packet marking. The S-GW is connected to the PDN gateway via an S5 interface.
- **PDN Gateway (P-GW):** This gateway is connected to the outside global network (Internet). This stage is responsible for IP address allocation, per-user packet filtering, and service level charging, gating, and rate enforcement.
- **Evolved Packet Core (EPC):** It includes the MME, the S-GW as well as the P-GW.

Logical, functional, and radio protocol layers are graphically illustrated in Figure 11.2. The logical nodes encompass the functional capabilities as well as radio protocols and interfaces. Interfaces S1–S11 as well as X2 are used to interconnect the various parts of the LTE network and are responsible for reliable packet routing and seamless integration. Details of such interfaces are discussed in the 3GPP specification and is discussed in this chapter. Radio protocol layers are the shaded ones in Figure 11.2. After a specific eNB is selected, a handover can take place based on measurements conducted at the UE and the eNB. The handover can take place between eNBs without changing the MME/SGW connection. After the handover is complete, the MME is notified about the new eNB connection. This is called an intra-MME/SGW handover. The exact procedures for this operation as well as inter-MME/SGW handover are discussed in detail in [4]. Handovers are conducted within layer-2 functionality (i.e., radio resource control (RRC)).

When comparing the new LTE standard release 8 to the currently deployed cellular systems in terms of maximum data rates, modulation schemes, multiplexing, among other system specific performance parameters, several improvements can be easily observed. Table 11.1 lists the major technologies and system performance for different networks evolved from 2.5G up to 4G. The North American system (based on CDMA) is shown in the shaded columns. The RF channel that connects the UE to the eNB is the focus of RF planning for LTE network design. The duplexing, multiplexing, modulation, and diversity are among the major aspects of the system

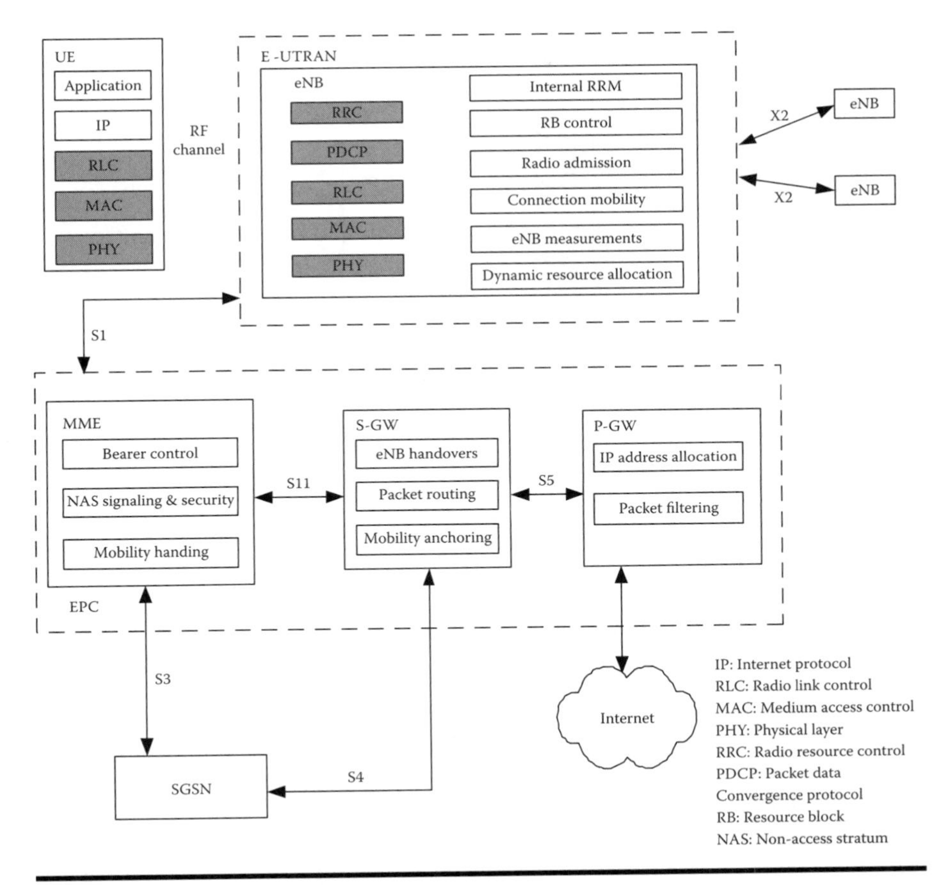

Figure 11.2 Logical, functional, and radio protocol layers for the LTE network.

architecture that affect the planning process. Also, the wireless propagation model, antenna types and number (LTE supports multiple antennas in the UE and eNB), and semiconductor technology used are key components in RF planning and design. The UE as well as the eNB (UL and DL) have to be designed, analyzed, deployed, and optimized in order achieve the system performance metrics defined within the standard.

11.3 Duplexing, Coding, and Modulation in LTE

In LTE, time division duplexing (TDD) and frequency division duplexing (FDD) are supported. If the cellular system is using two different carrier frequencies for the UL and DL, then the duplexing is called FDD. In this case, both the UE and the eNB can transmit at the same time. For FDD, a channel separation is needed to reduce the interference between the UL and DL traffic. Another precaution should

Table 11.1 Characteristics of Different Cellular Networks

	2.5G		*3G*		*3.5G*		*4G*
	EDGE	*cdma2000*	*UMTS*[1]	*EV-DO*[2]	*HSDPA*	*EV-DV*	*LTE*
Channel bandwidth (MHz)	0.2	1.25	5	1.25	5, 10	1.25, 3.75	5, 10, 15, 20
Duplexing	FDD	FDD	FDD	FDD	FDD	FDD	FDD/TDD
Multiplexing	TDMA	TDMA	WCDMA	TD-CDMA	WCDMA	TD-CDMA	OFDM/ SCFDMA
Modulation	GMSK/8PSK	GMSK/8PSK	QPSK	QPSK/8PSK /16QAM	QPSK/ 16QAM/	QPSK/8PSK /16QAM	QPSK/ 16QAM/ 64QAM
Coding	C	CTC	CTC	CTC	CTC	CTC	CTC
Maximum data rate	(UL) 0.04	(UL) 0.05	(UL) 0.14	(UL) 1.8	(UL) 2	(UL) 1	(UL) 50
(Mbps)	(DL) 0.18	(DL) 0.38	(DL) 0.38	(DL) 3.1	(DL) 7.2	(DL) 3-5	(DL) 100[3]

1: Universal Mobile Telecommunications Systems R99
2: Evolution data optimized (EV-DO) REV A
3: No MIMO
GMSK: Gaussian minimum shift keying
QPSK: Quadrature phase shift keying
QAM: Quadrature amplitude modulation
TD-CDMA: Time division-synchronous CDMA
OFDMA: Orthogonal frequency division multiple access
SC-FDMA: Single carrier frequency division multiple access
CTC: Convolutional/Turbo coding

be taken in the RF chain design that should provide enough out-of-band rejection in the transceiver. This is accomplished using high-quality RF filters.

In TDD-based systems, the communication between the UE and the eNB is made in a simplex fashion, where one terminal is sending data and the other is receiving. With a short enough delay time, the operation might seem as if it was a simultaneous process. The amount of spectrum required for FDD and TDD is the same. Although FDD uses two bands of frequencies separated by a guard band, TDD uses a single band of frequency, but it needs twice as much bandwidth. Because TDD sends and receives data at different time slots, the antenna will be connected to the transmitter at one time and to the receiver chain at another. The presence of a high-quality, fast-operating RF switch is thus essential.

LTE FDD supports both full-duplex and half-duplex transmission. Table 11.2 shows the LTE frequency bands for FDD. There are 14 bands shown (out of 15 defined in [6], band 17 is not shown). The DL as well as the UL bands are presented with their respective channel numbers. The channel numbers are also identified as the evolved absolute radio frequency channel numbers (EARFCN). The carrier frequency in the DL and UL is calculated based on the assigned EARFCN from the eNB. Equations 11.1 and 11.2 relate the EARFCN to the carrier frequency used in megahertz.

$$f_{DL} = f_{DL_{\text{Low}}} + 0.1(N_{DL} - N_{DL-\text{offset}}) \tag{11.1}$$

$$f_{UL} = f_{UL_{\text{Low}}} + 0.1(N_{UL} - N_{UL-\text{offset}}) \tag{11.2}$$

The offset value for the DL ($N_{DL-\text{offset}}$) and UL ($N_{UL-\text{offset}}$) are found from Tables 11.2 and 11.3 for FDD and TDD, respectively. The offset value is the starting value of the channel numbers for the specific band (i.e., for E-UTRA band 7, the $N_{DL-\text{offset}}$ is 2750). The nominal channel spacing between two adjacent carriers will depend on the channel bandwidths, the deployment scenario, and the size of the frequency block available. This is calculated using the following:

$$\text{Nominal channel spacing} = \frac{(BW_{\text{channel}-1} + BW_{\text{channel}-2})}{2} \tag{11.3}$$

where $BW_{\text{channel}-1}$ and $BW_{\text{channel}-2}$ are the channel bandwidths of the two adjacent carriers. The FDD mode utilizes the frame structure Type 1 [4]. The frame duration is $T_f = 307200 \times T_s = 10$ ms for both UL and DL. The sampling time (T_s) is given by

$$T_s = \frac{1}{15000 \times 2048}\,\text{s}$$

The denominator of T_s comes from the OFDMA subcarrier spacing (15 kHz) and the number of fast fourier transform (FFT) points. Each Type 1 frame is divided into 10 equally sized subframes, each of which is in turn equally divided into two slots. Each slot consists of 12 subcarriers with 6-7 OFDMA symbols (called a resource block). Figure 11.3 shows the structure of a Type 1 radio frame for LTE in FDD mode.

Table 11.2 LTE FDD Frequency Bands and Channel Numbers

E-UTRAN Band	Downlink (DL) (UE Receive, eNB Transmit)		Channel Numbers (N_{DL})	Uplink (UL) (UE Transmit, eNB Receive)		Channel Numbers (N_{UL})
	f_{DL_Low} (MHz)	f_{DL_High} (MHz)		f_{UL_Low} (MHz)	f_{UL_High} (MHz)	
1	2110	2170	0–599	1920	1980	13000–13599
2	1930	1990	600–1199	1850	1910	13600–14199
3	1805	1880	1200–1949	1710	1785	14200–14949
4	2110	2155	1950–2399	1710	1755	14950–15399
5	869	894	2400–2649	824	849	15400–15649
6	875	885	2650–2749	830	840	15650–15749
7	2620	2690	2750–3449	2500	2570	15750–16449
8	925	960	3450–3799	880	915	16450–16799
9	1844.9	1879.9	3800–4149	1749.9	1784.9	16800–17149
10	2110	2170	4150–4749	1710	1770	17150–17749
11	1475.9	1500.9	4750–4999	1427.9	1452.9	17750–17999
12	728	746	5000–5179	698	716	18000–18179
13	746	756	5180–5279	777	787	18180–18279
14	758	768	5280–5379	788	798	18280–18379

Table 11.3 LTE TDD Frequency Bands and Channel Numbers

E-UTRAN Band	*Downlink (DL) (UE Receive, eNB Transmit)*		*Channel Numbers (N_{DL})*	*Uplink (UL) (UE Transmit, eNB Receive)*		*Channel Numbers (N_{UL})*
	f_{DL_Low} *(MHz)*	f_{DL_High} *(MHz)*		f_{UL_Low} *(MHz)*	f_{UL_High} *(MHz)*	
33	1900	1920	26000–26199	1900	1920	26000–26199
34	2010	2025	26200–26349	2010	2025	26200–26349
35	1850	1910	26350–26949	1850	1910	26350–26949
36	1930	1990	26950–27549	1930	1990	26950–27549
37	1910	1930	27550–27749	1910	1930	27550–27749
38	2570	2620	27750–28249	2570	2620	27750–28249
39	1880	1920	28250–28649	1880	1920	28250–28649
40	2300	2400	28650–29649	2300	2400	28650–29649

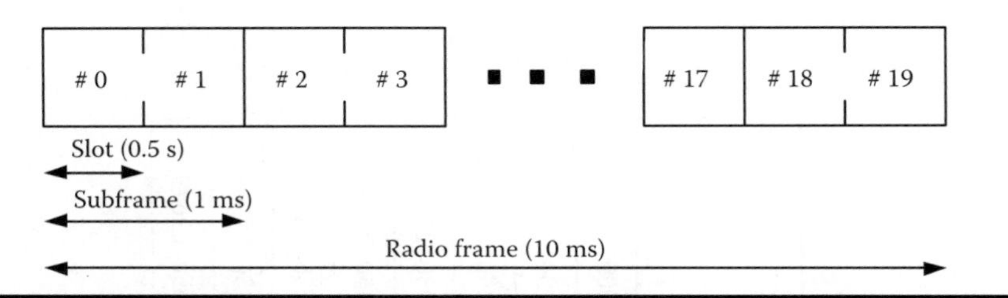

Figure 11.3 FDD frame type 1 structure.

For the TDD mode of operation, the frequency bands used and their respective channel numbers are illustrated in Table 11.3. There are 8 bands in TDD mode. Frame structure Type 2 is used for TDD duplexing. Each radio frame is 10 ms long. The frame consists of two 5-ms subframes. Each subframe is in turn divided into eight 1.5 ms slots and three special fields: downlink pilot time slot (DwPTS), guard period (GP), and uplink pilot time slot (UpPTS). The length of the combined three fields is 1 ms. Figure 11.4 shows the structure of a Type 2 radio frame.

11.3.1 LTE Physical Channels

The following physical channels are used in the LTE architecture [4]:

- **Physical Broadcast Channel (PBCH):** The transport blocks are mapped into four subframes within a 40-ms interval and then decoded with no special signaling. This channel is used for correcting mobile frequencies, control channel structure, frame synchronization, and the like.
- **Physical Control Format Indicator Channel (PCFICH):** This channel is transmitted in every subframe and indicates the number of OFDMA symbols used for the PDCCH.
- **Physical Downlink Control Channel (PDCCH):** This channel carries the uplink scheduling information and informs the UE about resource

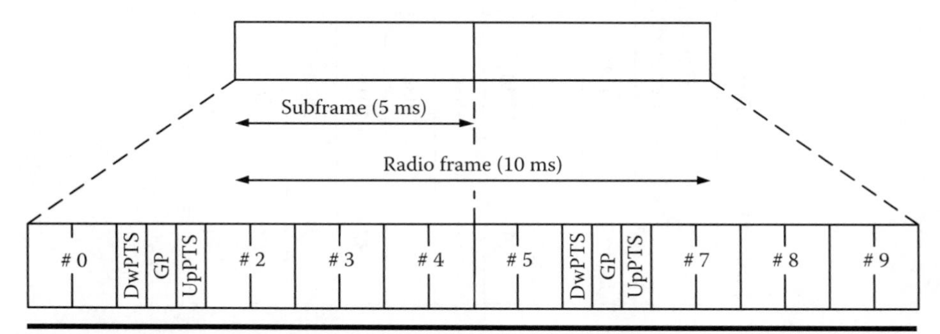

Figure 11.4 TDD frame type 2 structure.

allocation and hybrid automatic repeat request (HARQ) for the paging channel (PCH) and the downlink synchronization channel (DL-SCH).
- **Physical Hybrid ARQ Indicator Channel (PHICH):** This channel carries the HARQ of acknowledge/not-acknowledge (ACK/NACK) for the uplink transmissions.
- **Physical Downlink/Uplink Shared Channel (PDSCH/PUSCH):** This channel carries the DL synchronization channel (SCH) and UL-SCH as well as PCH information.
- **Physical Multicast Channel (PMCH):** This channel carries the multicast information.
- **Physical Uplink Control Channel (PUCCH):** This channel carries HARQ for the downlink transmissions, as well as scheduling requests and channel quality indicator (CQI) reports.
- **Physical Random Access Channel (PRACH):** This channel carries the random access preamble.

Three main procedures are given to the physical layer: cell search, power control, and link adaptation. In cell search, the DL synchronization and reference signals are tracked to acquire frequency and time synchronization, as well as the cell ID. Power control is used to control the output power spectral density from the UE. Finally, link adaptation will decide which modulation and coding schemes (bearers) are to be used based on the performed channel measurements and data coming from the eNB.

The physical layer maps the physical channels to transport channels for layer-2 processing. Layer-2 is divided into three sublayers: medium access control (MAC), radio link control (RLC), and packet data convergence protocol (PDCP). The MAC sublayer is responsible for mapping between transport and logical channels, scheduling reporting, error correction, priority handling, and padding. The RLC sublayer performs ARQ error correction, RLC reestablishment, passing protocol data units (PDUs) to upper layers and concatenation, segmentation, and reassembly of service data units (SDUs). The PDCP is responsible for header compression, ciphering and deciphering, data transfer, and PDU/SDU delivery [4].

11.3.2 Coding, Modulation, and Multiplexing

There are two levels of coding: source coding and channel coding. The former is used for data compression and the latter is used to minimize channel effects on transmitted symbols. Convolutional turbo coding (CTC) channel coding is used in LTE equipment. It has proven to enhance data transmission in complex fading channels. The coded blocks are then scrambled using a length 31 gold code (GC) sequence. Initialization on the sequence generator is made at the beginning of every frame. A GC sequence generator can be implemented using a linear feedback shift register (LFSR). A block of bits to be transmitted on the PBCH consists of 1920 bits with normal cyclic prefix (CP) is denoted by b. This block has to be scrambled with a cell-specific sequence prior to modulation. Let the GC sequence for this cell be denoted by C. The scrambling process is made according to Equation 11.4 [7,8]:

$$\tilde{b}(i) = b(i) \oplus C(i) \tag{11.4}$$

where $\oplus$ is the module-2 addition operator. To combat bursty channel errors in a multipath environment, the coded data is interleaved in such a way that the bursty channel is transformed into a channel with independent errors. Interleaving will spread the bursty errors in time so that they would appear independent of each other and then conventional error correcting mechanisms can be used to correct such errors. For more on interleaving and coding, please refer to [1,9].

Linear modulation schemes are used in LTE. M-ary digital modulation is utilized since the bit rate used is higher than the channel bandwidth assigned (300 Mbps data rate in a 20-MHz bandwidth). The modulation scheme is determined depending on the channel characteristics. In bad channel conditions, a low constellation modulation scheme is used, which is QPSK. Here, two bits are encoded into a single word (phase) for transmission. The 16-QAM and 64-QAM modulation schemes are used in better channel conditions, and the data are mapped into both phase and amplitude changes on the carrier frequency. The signal constellation of a QAM modulation consists of a square grid. The modulated signals contain a level based on the number of bits used. For 16-QAM, every 4 bits are given a signal value from the 16-level constellation. Figure 11.5 shows the difference between the QPSK and the 16-QAM signal constellations [7]. A 64-QAM modulation scheme follows that of the 16-QAM but instead encodes 6-bits into one signal level/phase compared to 4-bits in 16-QAM.

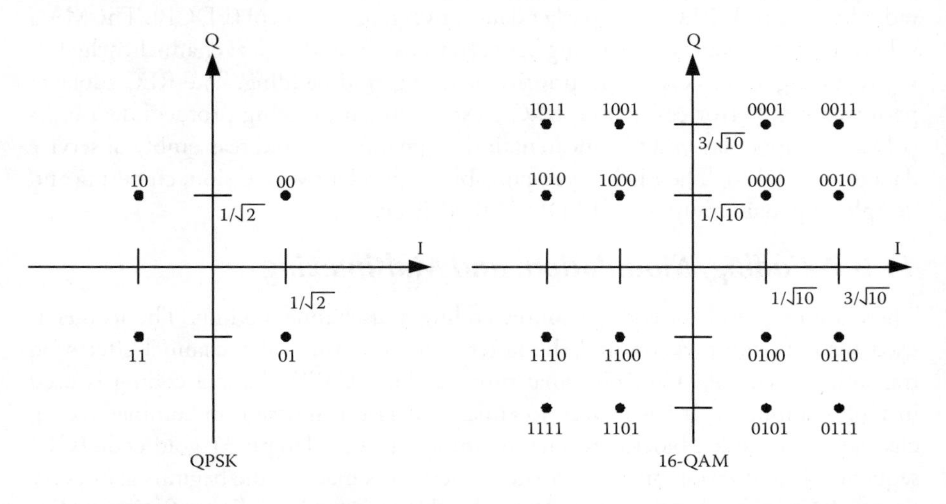

Figure 11.5 QPSK and 16-QAM signal constellations, gray coded. (From 3GPP TS 36.211, V8.4.0, E-UTRA "Physical Channels and Modulation," September 2008.) © 2008. 3GPP™.

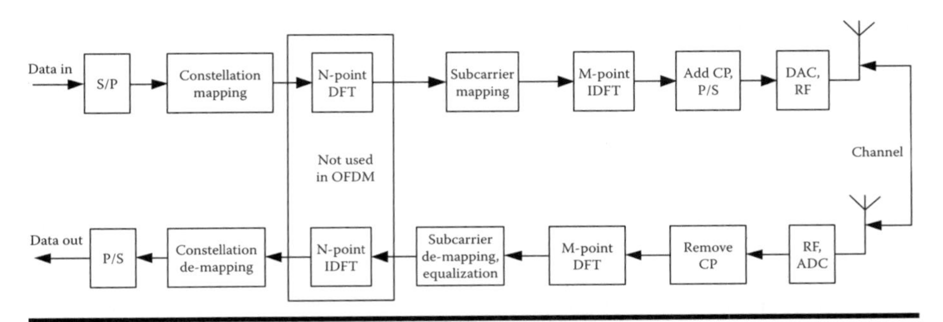

Figure 11.6 SC-FDMA/OFDMA tranceiver block diagram (DL/UL).

In LTE, adaptive modulation and coding (AMC) is implemented on the UL/DL streams according to channel conditions. Thus, the modulation scheme as well as the coding scheme are changed automatically for best transmission performance for the given channel conditions. Multicarrier multiplexing is attractive because a frequency-selective channel would appear as a flat-fading one for individual orthogonal carriers (subcarriers). Thus compensation for channel impairments would become easier to realize in hardware [1,9]. Two multiplexing schemes are used in the LTE architecture. OFDMA is used for the downlink and SC-FDMA is used for the uplink. Figure 11.6 shows a block diagram for a SC-OFDMA transceiver and the modification required to obtain that of an OFDMA [7].

CP is needed to overcome inter-symbol interference (ISI) that is introduced on the data by the wireless channel. The cyclic bit extension (adding a copy of the last data portion of the OFDMA symbol to the beginning of it instead of a guard band) to the FFT output adds a guard interval to the data to be transmitted.

In a typical OFDMA system block diagram (Figure 11.6), the serial input binary data is converted into a parallel (S/P) stream that is mapped into a complex constellation (modulation and coding) before being formatted for subcarrier mapping through an IFFT operation. This process is followed by the addition of a CP before being passed to a digital-to-analog converter (DAC). The data are then passed to the RF part and the antenna elements (in case if more than 1 is used, i.e., MIMO). In the receiver, the opposite sequence of operations is followed with the use of an FFT processor. A SC-FDMA system includes an extra FFT/IFFT operation in the transmitter/receiver, respectively. The size of the former FFT/IFFT processor is less than the latter one ($M > N$). This change in the signal chain of the system diagram gives several advantages of SC-FDMA over OFDMA. The major advantage of using SC-FDMA over OFDMA is the lower peak-to-average power ratio (PAPR) that minimizes problems to power amplifiers within the terminals. Thus, although the SC-FDMA will entail more signal processing complexity (which can be handled with today's DSP), it will allow the creation of cheaper UE (RF portion is still relatively expensive) and a better link budget because a lower PARP is achieved [10]. OFDMA is discussed in detail in [9,11].

11.4 Cell Planning

The aim of the cell planning engineer is to establish the proper radio network in terms of service coverage, QoS, capacity, cost, frequency use, equipment deployment, and performance. In order to plan a cellular radio network, the designer has to identify specifications, study the area under consideration and create a database with geographic information (GIS), analyze the population in the service area, create models (i.e., cell types, IDs, locations, etc.), and perform simulations and analysis using proper propagation scenarios and tools. Afterward, simulation and coverage results are analyzed, followed by cell deployment and drive testing. The results of field measurements are compared against the simulation model results, and the model is tuned for performance optimization. Each of the aforementioned stages in turn consist of a number of steps that need to be performed.

11.4.1 Coverage

Coverage planning is an important step in deploying a cellular network. This process includes the selection of the proper propagation model based on the area's terrain, clutter, and population. Propagation models (empirical models) are too simplistic to predict the signal propagation behavior in an accurate fashion; they provide us with some relatively good accuracy of how things would behave. Field measurements are the most accurate in predicting radio coverage in a certain area. For example, in buildings coverage will add about 16 to 20 dB of extra signal loss, and inside vehicle ones can increase the loss by an extra 3 to 6 dB [1].

Engineers rely on prediction tools to study and analyze the performance of the network for a geographic area via its coverage. In LTE, the air interface and radio signal electronics are going to be different than those already deployed (in terms of multiplexing, AMC, and MIMO capability for both the UE and eNB). Modeling and simulation using some current RF planning tools (i.e., Atoll [12]) for LTE cells will give a good idea about the coverage performance of a certain grid within a specific area. Based on the simulations made, the planning engineer would change eNB locations, add more towers, replace antenna types, add more sectors to some towers, and so on. Most cells are designed to be hexagonal in theory; in reality, this is not the case, as several factors affect the location selection decision (political, humanitarian, economical). Figure 11.7 shows simulated signal power levels (color coded) of four LTE RF cells in downtown Brussels, Belgium. Note the description of one of the sites where the frequency band and bandwidth are shown along with the RF equipment characteristics: antenna parameters, tower-mounted amplifier (TMA) characteristics, and feeder loss. This information is to be used in the link budget as well (Section 11.5.4).

11.4.2 Cell IDs

For LTE cells, the eNB antenna is 45 m tall in rural areas and 30 m tall in urban areas. Typically eNBs (or sites) in a macrocellular deployment are placed on a hexagonal

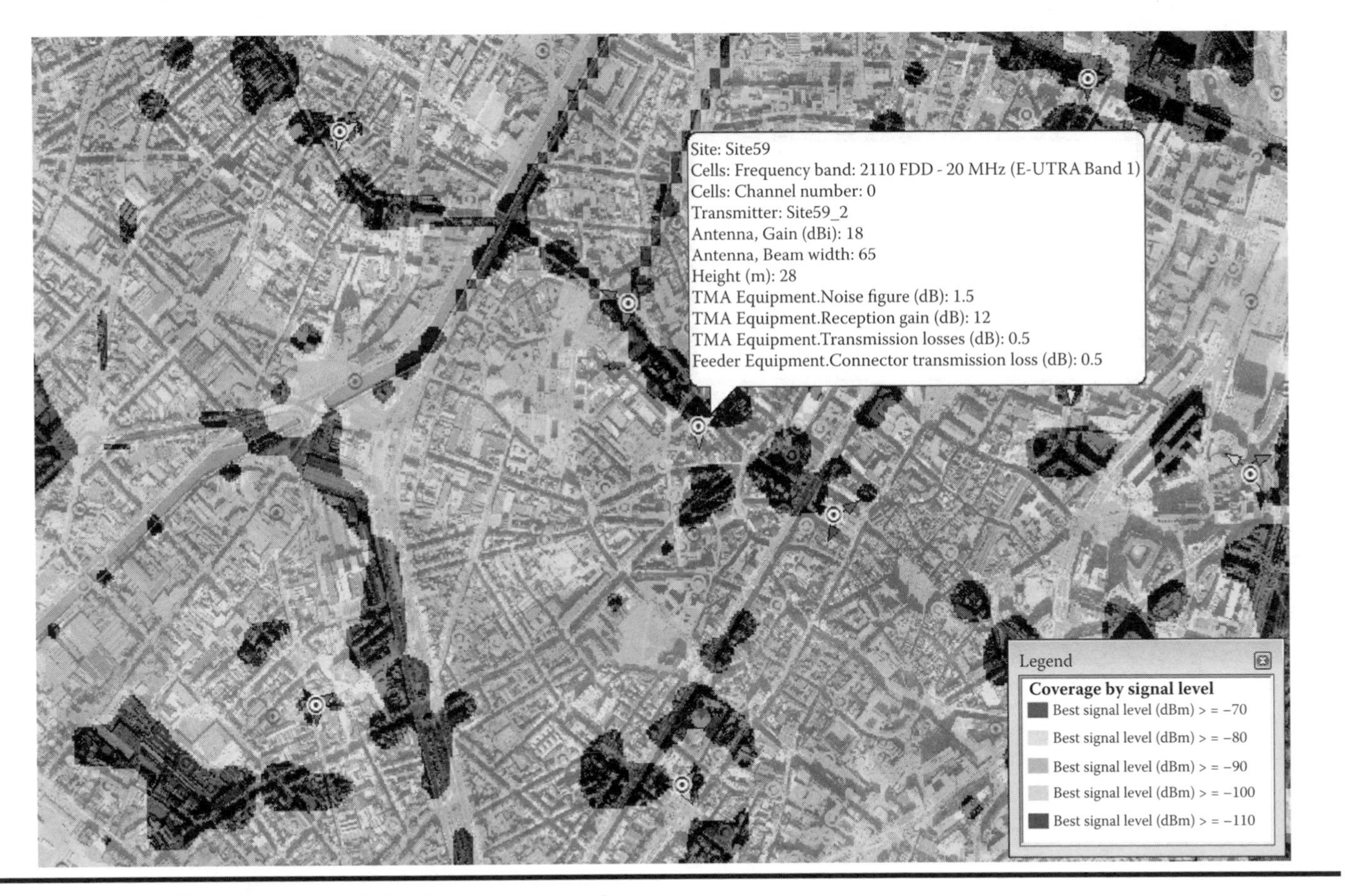

Site: Site59
Cells: Frequency band: 2110 FDD - 20 MHz (E-UTRA Band 1)
Cells: Channel number: 0
Transmitter: Site59_2
Antenna, Gain (dBi): 18
Antenna, Beam width: 65
Height (m): 28
TMA Equipment.Noise figure (dB): 1.5
TMA Equipment.Reception gain (dB): 12
TMA Equipment.Transmission losses (dB): 0.5
Feeder Equipment.Connector transmission loss (dB): 0.5
Legend
Coverage by signal level
Best signal level (dBm) > = −70
Best signal level (dBm) > = −80
Best signal level (dBm) > = −90
Best signal level (dBm) > = −100
Best signal level (dBm) > = −110

Figure 11.7 Atoll coverage example in downtown Brussels.

grid with an intersite distance of $3 \times R$, where $R = 500$ m is the cell radius. Each eNB has three sectors with an antenna placed at each sector. In a multioperator cellular layout, identical cell layouts for each network shall be applied, with second network sites located at first network cell edges [13].

In an LTE system, the same carrier frequency is used, and thus the system relies on scrambling and pseudo-noise (PN) codes to distinguish between users and sites as well as to establish synchronization between the UE and eNB. A cell ID and scrambling code is to be given to each site. There are 504 unique cell IDs that can be used within the LTE physical layer. These IDs are grouped into three 168 groups, each group contains three identities. The cell ID is found from:

$$N_{\text{cell}} = 3N_G + N_{ID} \tag{11.5}$$

where $N_G \in [0, 167]$ is the physical layer cell group ID, and $N_{ID} \in [0, 2]$ is the identification number within the group. N_{ID} is also used to pick one of the 64 Zandoff-Chu scrambling codes used for the primary and secondary synchronization channels (reference channels). A Zandoff-Chu sequence is a complex-orthogonal sequence that is used to give unique signatures to radio signals. Orthogonal codes are used to distinguish between radio transmissions and thus distinguish between surrounding eNBs. In UMTS, Walsh codes were used for this purpose. In LTE, Zandoff-Chu sequences are used. These give rise to constant amplitude radio signal after the scrambling process. A root Zandoff-Chu sequence can be found using:

$$d_u(n) = e^{-j}\frac{\pi un(n+1)}{N_{zc}} \tag{11.6}$$

where, $N_{zc} = 63$ in LTE, and $0 \leq n \leq N_{zc}$; the root index u is related to N_{ID}. In the UE, GC sequences with different shifts are used based on the subscriber identity and the physical channel type [8].

For cell ID and scrambling code planning, several strategies exist based on minimum reuse distance, domain constraints, minimum E_c/I_0 levels, number of codes per cluster, etc. Several automatic scrambling code planning algorithms exist within RF planning packages that can be used as well. The fact that there are plenty of cell IDs that can be used allows for a large pool of sequences and thus a larger area between similar reused sequences. Some of these strategies are [12, 14]:

- **Cluster Reuse-Based Method**: This method assigns code sets according to a code set reuse pattern that is predefined (i.e, 13 cell clusters). Then, based on the propagation loss exponent and the processing gain of the radio scheme, the minimum reuse distance is found.
- **Graph Optimization Technique**: In this method, heuristic algorithms are used to assign cell IDs and scrambling codes by minimizing the number of sets to be used based on an optimization criteria. The algorithm first finds the inter-cell distances, and then starts automatic code assignments based on the optimization criteria and their priorities.

- **Distributed Per-cell/Per-site**: In the per-cell strategy, the pool of codes is distributed among as many cells as possible, thus increasing the minimum reuse distance. The distribution per site allocates a group of different codes to adjacent sites, and from these groups, one code per transmitter is assigned.

11.4.3 Cell Types

Third-generation cellular networks utilize three cell types: macro, micro, and pico based on their coverage area and user capacity [15]. In LTE as well as WiMAX, a fourth type is introduced to serve a single household—femtocell. According to [16], these four cell types are defined as:

- **Macrocells**: The largest cell types that cover areas in kilometers. These eNBs can serve thousands of users simultaneously. They are very expensive due to their high installation costs (cabinet, feeders, large antennas, 30–50 m towers, etc). The cells have three sectors and constitute the heart of the cellular network. Their transmitting power levels are very high (5–40 W).
- **Microcells**: Provide a smaller coverage area than macrocells, and are added to improve coverage in dense urban areas. They serve hundreds of users and have lower installation costs than macrocells. You can find them on the roofs of buildings, and they can have three sectors as well, but without the tower structure. They transmit several watts of power.
- **Picocells**: Used to provide enhanced coverage in an office like environment. They can serve tens of users and provide higher data rates for the covered area. The 3G networks use picocells to provide the anticipated high data rates. They have a much smaller form factor than microcells and are even cheaper. Their power levels are in the range of 20 to 30 dBm.
- **Femtocells**: Introduced for use with 4G systems (LTE and WiMAX). They are extremely cheap and serve a single house/small office. Their serving capacity does not exceed 10 users, with power levels less than 20 dBm. A femtocell will provide a very high DL and UL data rates, and thus provide multi-Mbps per user, thus accomplishing MAGIC (see Section 11.1, Introduction).

11.4.4 Multiple-Input Multiple-Output Systems (MIMO)

MIMO systems are one of the major enabling technologies for LTE. They will allow higher data rate transmission through the use of multiple antennas at the receiver/transmitter. Let the number of transmitting antennas be M_T and the number of receiving antennas be N_R where $N_R \geq M_T$.

In a single-input single-output system (SISO)—used in current cellular systems, 3G and 3.5G—the maximum channel capacity is given by the Shannon-Hartley relationship:

$$C \approx B \times \log_2(1 + SNR_{\text{avg}}) \tag{11.7}$$

where C is the channel capacity in bits per second (bps), B is the channel bandwidth in Hz, and SNR_{avg} is the average signal-to-noise ratio at the receiver. In the SISO case, $M_T = N_R = 1$. In a MIMO case, the channel capacity becomes [1]:

$$C \approx B \times \log_2(1 + M_T \times N_R \times SNR_{\text{avg}}) \tag{11.8}$$

thus obtaining an $M_T N_T$ fold increase in the average SNR and increasing channel capacity. If $N_R \geq M_T$, we can send different signals within the same bandwidth and still decode them correctly at their corresponding receivers. For each channel (transmitter-receiver), its capacity will be given as:

$$C_{\text{single}-CH} \approx B \times \log_2\left(1 + \frac{N_R}{M_T} \times SNR_{\text{avg}}\right) \tag{11.9}$$

In the same bandwidth, we will have M_T dedicated channels (M_T transmitting antennas), resulting in an M_T-fold increase in capacity:

$$C \approx M_T B \times \log_2\left(1 + \frac{N_R}{M_T} \times SNR_{\text{avg}}\right) \tag{11.10}$$

Hence, with respect to the transmitting antennas we obtain a linear increase in the system capacity. It is very important in a MIMO system to identify the correlation matrices between the transmit/receive antennas, as well as the channel propagation conditions. Based on the transmit (M_T) and receive (N_R) number of antennas, a correlation matrix of $M_T \times N_R$ will be obtained. As an example, the spatial correlation matrix, R_{spatial} for a 2 × 2 MIMO system is given by:

$$R_{\text{spatial}} = R_{eNB} \bigotimes R_{UE} = \begin{bmatrix} 1 & \alpha \\ \alpha^* & 1 \end{bmatrix} \bigotimes \begin{bmatrix} 1 & \beta \\ \beta^* & 1 \end{bmatrix} = \begin{bmatrix} 1 & \beta & \alpha & \alpha\beta \\ \beta^* & 1 & \alpha\beta^* & \alpha \\ \alpha^* & \alpha^*\beta & 1 & \beta \\ \alpha^*\beta^* & \alpha^* & \beta^* & 1 \end{bmatrix} \tag{11.11}$$

where $\bigotimes$ is the Kronecker product operator, and R_{eNB} and R_{UE} are the eNB and UE correlation matrices, respectively. The other combinations of 1×2, 2×4, and 4×4 are listed in [6]. Values for α and β depend on the environment under consideration as defined by the standard. α can take one of the values in (0, 0.3, 0.9), whereas β can be one of the values (0, 0.9, 0.9) for the slow-, medium- and high-delay spread environments, respectively. Figure 11.8 shows a simplified MIMO system block diagram.

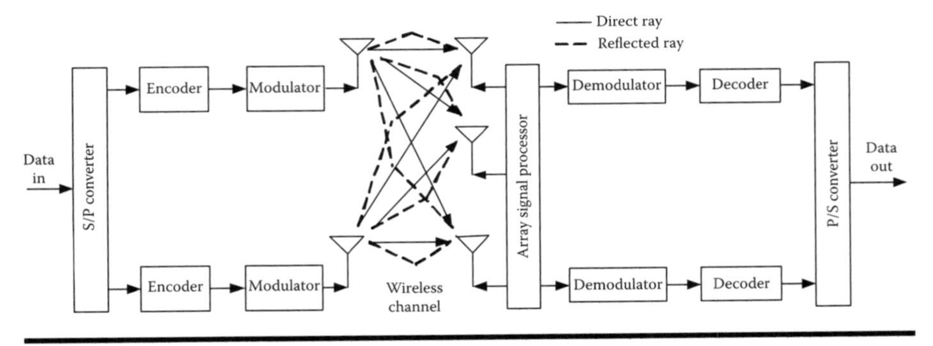

Figure 11.8 A MIMO system block diagram (simplified).

11.4.5 Diversity

For MIMO-based systems, different kinds of diversity techniques are used. MIMO-based diversity systems can be described as follows [12]:

- **Transmit Diversity**: The signal to be transmitted is forwarded and sent over all antennas, the same signal that is sent on all transmit antennas reaches the receiver, and the combined signal level will be higher if only one transmit antenna was used, making it more interference resistant. Transmit diversity will increase the carrier-to-interference plus noise ratio (CINR) level, and is used at cell locations with low CINR (i.e., further from eNB toward the cell edges).
- **Spatial Multiplexing**: Different signals are passed to different transmit antennas in this diversity technique. If the transmit terminal has M antennas and the receive terminal had N antennas, the throughput through the transmit-receiver link can be increased by [min(M, N)]. This diversity technique will increase the channel throughput provided that good CINR levels exist.
- **Adaptive MIMO Switching**: This technique allows switching between transmit diversity and spatial multiplexing based on the environment conditions. If the CINR exceeds a certain threshold, spatial multiplexing is chosen to provide the user with higher throughput. On the other hand, if CINR is below the defined threshold, transmit diversity is picked to improve user reception by choosing to operate at a lower throughput.

11.4.6 Antenna Arrays

Antenna arrays are used to provide directional radiation characteristics and higher gain to the transmitted/received signal. The outputs of individual antenna elements within the array are combined to provide a certain desired radiation pattern and gain. The more directional the antenna array, the narrower the half power beam width becomes. Relationships between these parameters are of importance to the antenna design engineer and is found in [17]. Another important parameter for cellular antennas is their polarity. Vertical polarization is used in cellular systems.

The eNB antenna gain within a macrocell in urban and rural areas is to be between 12 and 15 dBi, including the feeder losses within the bands of operation. These gain values are important in formulating the RF link budget of the system and in determining the coverage power levels. The antenna radiation pattern to be used for each sector within the three-sector cell site is given by [13]

$$A(\theta) = -\min\left[12\left(\frac{\theta}{\theta_{3dB}}\right)^2, A_m\right] \tag{11.12}$$

where θ_{3dB} is the 3-dB (half power) beam width; in this case, it is 65°, $-180^\circ \leq \theta \leq 180^\circ$, and A_m is the maximum attenuation of 20 dB between the main lobe and the highest side lobe in the antenna gain pattern. The UE does not have such a directional pattern and will use an omnidirectional antenna for UL, whereas it will have the option of using one, two, or four antennas for downlinking. It is believed that the first UE prototypes are to have two antennas. Such antennas will have omnidirectional radiation patterns to receive as much signal power as possible. Its gain is assumed to be 0 dBi.

With proper use of adaptive techniques, the weighting of signal levels coming out of each antenna element with optimized coefficients gives better signal-to-noise ratio (SNR), interference reduction, and source signal tracking. This is called antenna beam forming (BF). If BF is utilized, the eNB antenna array can keep the main lobe in the direction of the UE, thus providing maximum antenna gain. This means higher SNR (CINR) levels, and thus better throughput and higher data transmission rates. Five predefined angles for BF have been suggested [0°, 30°, 45°, 60°, 70°] along with their image (negative) angles. The weights for the antenna array can be stored in a lookup table in the antenna electronics, or can be achieved using an RF butler matrix.

11.5 Propagation Modeling

In wireless communications, a multipath channel is the one that describes the medium between the UE and the eNB. A multipath channel is characterized by the delay profile that is characterized by the RMS delay spread and the maximum delay spanned by the tapped-delay-line taps, along with the Doppler spread. Four environments are defined for LTE: extended pedestrian (low delay profile), extended vehicular (medium delay profile), extended typical urban (high delay spread), and the high-speed train (nonfading). This section presents the four propagation scenarios supported in the LTE standard, followed by the propagation channel models used. Statistical and deterministic channel modeling is presented. Creation of the LTE link budget is discussed as well.

11.5.1 Propagation Environments

Multipath channel characteristics can be described by a combination of a delay spread profile, the Doppler spectrum, and the effect of multiple antennas in a MIMO system through the use of correlation matrices. The delay spread profile can be modeled as a tapped delay line with predefined delay elements and relative power contributions. There are four propagation scenarios in LTE [6]:

1. **Extended Pedestrian A Model**: This model covers walking users with speeds up to 3 km/h. The tapped delay line model consists of 7-taps with delays ∈ [0, 30, 70, 90, 110, 190, 410] ns, and relative power ∈ [0.0, −1.0, −2.0, −3.0, −8.0, −17.2, −20.8] dB. The maximum Doppler shift is 5 Hz.
2. **Extended Vehicular A Model**: This model covers moving vehicles with speeds up to 50 km/h. The model consists of 9-taps with delays ∈ [0, 30, 150, 310, 370, 710, 1090, 1730, 2510] ns and relative power ∈ [0.0, −1.5, −1.4, −3.6, −0.6, −9.1, −7.0, −12.0, −16.9] dB. The maximum Doppler shift is between 5 and 70 Hz.
3. **Extended Typical Urban Model**: This model covers moving vehicles with speeds up to 90 km/h. The model consists of 9-taps with delays ∈ [0, 50, 120, 200, 230, 500, 1600, 2300, 5000] ns and relative power ∈ [−1.0, −1.0, −1.0, 0.0, 0.0, 0.0, −3.0, −5.0, −7.0] dB. The maximum Doppler shift is between 70 and 300 Hz.
4. **High-Speed Train**: This model covers train users with speeds of 300 km/h. This is considered a nonfading model with a 1-tap delay line. The maximum Doppler shift is 750 Hz.

11.5.2 Empirical/Statistical Path Loss Models

Path loss models are important in the RF planning phase to be able to predict coverage and link budget among other important performance parameters. These models are based on the frequency band, type of deployment area (urban, rural, suburban, etc.), and type of application. Two path loss models for macro/microcell propagation are listed in [13] that are accurate if used beyond 100 m distances from the site for both urban and rural areas. Table 11.4 lists the most widely used propagation models in current cellular systems. Most of these models are a fusion of empirical formulas extracted from field measurements and some statistical prediction models. Three of the listed models that will be used in LTE are discussed in detail in the rest of this section.

11.5.2.1 Okumura-Hata

The Okumura-Hata model is a widely used wireless cellular propagation model that can predict channel behavior in the 150 to 2200 MHz range. It covers distances from 1 to 20 km. The model has three environment formulas:

Table 11.4 Commonly used Wireless Channel Propagation Models

Model	*Frequency (MHz)*	*Recommended use*
COST-231	800–2000	$0.02 < d < 5$ km, UMTS, GSM1800, LTE
Erceg-Greenstein	1900–6000	$0.1 < d < 8$ km, Fixed WiMAX
IMT-2000	800–2800	Indoor office, vehicular, outdoor to indoor
ITU-526	30–1000	Fixed receivers
ITU-529	300–1500	$1 < d < 100$ km, GSM900, CDMA2000, LTE
Okumura-Hata	150–2200	$1 < d < 20$ km, GSM900, CDMA2000, LTE
WLL	30–10000	Fixed receivers, Microwave Links, WiMAX

IMT: International Mobile Telecommunication
ITU: International Telecommunication Union
WLL: Wireless Local Loop

- **Typical Urban**

$$L_{P(\text{urban})} = 69.55 + 26.16\log(f_c) + [44.9 - 6.55\log(h_{BS})]Log(d) - 13.82\log(h_{BS}) - a(h_{UE}) \quad \text{dB} \tag{11.13}$$

where f_c is the carrier frequency $400 \leq f_c \leq 2200$ MHz, $h_{BS} \in [30, 200]$ m and $h_{UE} \in [1, 10]$ m are the base station and mobile station heights, respectively, $d \in [1, 20]$ km is the distance between the BS and UE, and $a(h_{UE})$ is the UE antenna height correction factor. For $f_c \geq 400$ MHz, $a(h_{UE})$ given by:

$$a(h_{UE}) = \begin{cases} 3.2[\log(11.75h_{UE})]^2 - 4.97 & \text{Dense urban} \\ [1.1\log(f_c) - 0.7]h_{UE} - [1.56\log(f_c) - 0.8] & \text{Urban} \end{cases} \tag{11.14}$$

- **Typical Suburban:**

$$L_{P(\text{suburban})} = L_{P(\text{urban})} - 2\left[\log\left(\frac{f_c}{28}\right)^2 - 5.4\right] \quad \text{dB} \tag{11.15}$$

- **Rural:**

$$L_{P(\text{rural})} = L_{P(\text{urban})} + 18.33\log(f_c) - 4.78[\log(f_c)]^2 - 40.94 \quad \text{dB} \tag{11.16}$$

11.5.2.2 COST-231

Cooperation in science and technology (COST-231) is one of the models anticipated to be used for LTE channel prediction. It covers frequencies from 800 to 2000 MHz, and distances from the BS starting at 20 m and up to 5 km. It is widely used in Europe for the GSM 1800-MHz system. The model is valid for $h_{BS} \in [4, 50]$ m and $h_{UE} \in [1, 3]$ m. The path loss formula is given by:

$$L_P = 32.4 + 20\log(d) + 20\log(f_c) + L_{rts} + L_m \quad \text{dB} \tag{11.17}$$

where L_{rts} and L_m are the rooftop to street diffraction and scatter factor and multi-screen loss, respectively. The formulas for these two factors are given by:

$$L_{rts} = 10\log(f_c) + 20\log(h_r - h_{UE}) + L_\phi - 10\log(W) - 16.9 \quad \text{dB} \tag{11.18}$$

$$L_m = L_{BS2B} + K_a + K_d\log(d) + K_f\log(f_c) - 9\log(b) \tag{11.19}$$

where, h_r is the average building height, W is the street width, b is the distance between adjacent buildings, L_ϕ is the loss due to the incident angle relative to the street, and L_{BS2B} is the loss factor due to the difference between the BS and average building height. The various relationships are given by (all in dB)

$$L_\phi = \begin{cases} -10 + 0.354\phi, & 0 \le \phi \le 35^\circ; \\ 2.5 + 0.075(\phi - 35), & 35 \le \phi \le 55^\circ; \\ 4 - 0.114(\phi - 55), & 55 \le \phi \le 90^\circ. \end{cases} \tag{11.20}$$

$$L_{BS2B} = \begin{cases} -18\log(11 + h_{BS} - h_r), & h_{BS} \ge h_r; \\ 0, & h_{BS} < h_r. \end{cases} \tag{11.21}$$

$$K_a = \begin{cases} 54, & h_{BS} > h_r; \\ 54 - 0.8h_{BS}, & d \ge 500m,\ h_{BS} \le h_r; \\ 54 - 0.8h_{BS}\left(\frac{d}{500}\right), & d < 500m,\ h_{BS} \le h_r. \end{cases} \tag{11.22}$$

$$K_d = \begin{cases} 18, & h_{BS} < h_r; \\ 18 - \frac{15(h_{BS} - h_r)}{h_{UE} - h_r}, & h_{BS} \ge h_r. \end{cases} \tag{11.23}$$

$$K_f = \begin{cases} 4 + 0.7\left(\frac{f_c}{925} - 1\right), & \text{midsize city/suburban;} \\ 4 + 1.5\left(\frac{f_c}{925} - 1\right), & \text{metro area.} \end{cases} \tag{11.24}$$

11.5.2.3 IMT-2000

International mobile telecommunications (IMT-2000) is the standard that includes the system requirements for 3G-based cellular systems from which UMTS is derived. This standard has the following several propagating environment models for outdoor and indoor channels [1]:

- **Indoor Environment**: This model covers indoor scenarios with small cells and low transmit power levels. It is suitable for RMS delay spread values of 35 to 460 ns. It uses a log-normal shadowing with a 12-dB standard deviation. The path loss is given by:

$$L_{P(\text{indoor})} = 37 + 30\log(d) + 18.3n^{\left[\frac{n+2}{n+1} - 0.46\right]} \quad \text{dB} \tag{11.25}$$

  where d is the distance between the transmitter and receiver stations, and n is the number of floors.
- **Pedestrian and Outdoor-to-Indoor Environment:** The model uses small cells, with low transmit power levels, and RMS delay spread of 100 to 800 ns. It covers only nonline of sight (NLOS) scenarios, and utilizes a log-normal shadowing with a 10-dB standard deviation. The path loss is given by:

$$L_{P(\text{ped-out2in})} = 40\log(d) + 30\log(f_c) + 49 \quad \text{dB} \tag{11.26}$$

- **Vehicular Environment:** The model covers large cells and higher transmit power levels, with an RMS delay spread of 4 to 12 μs. A log-normal shadowing with a 10-dB standard deviation is used. The path loss formula is given by:

$$\begin{aligned} L_{P(\text{vehicle})} &= 40\left(1 - 4 \times 10^{-2}\Delta h_{BS}\right)\log(d) - 18\log(\Delta h_{BS}) \\ &\quad + 21\log(f_c) + 80 \quad \text{dB} \end{aligned} \tag{11.27}$$

  where Δh_{BS} is the BS antenna height measured from the average rooftop level of the vehicle in meters.

11.5.3 Deterministic Path Loss Models

The previous section discussed three of the most widely used empirical/statistical path loss models used in 3G models that will also be used in LTE. These models are derived from extensive measurement scenarios from which the wireless channel is described by probability functions of statistical parameters. Empirical/statistical models provide general results. Another group is based on deterministic channel modeling. The channel characteristics are obtained by tracing the reflected, diffracted, and scattered rays based on a specific geometry with a database what includes the sizes of the physical objects and their material properties. Deterministic models have the advantage of providing very accurate, site-specific results that are reproducible. However, they suffer from the need of more model specific data and computation time [18–20].

Three-dimensional (3D) ray tracing is a deterministic channel modeling method that has proven to give good accuracy for indoor MIMO channels [19]. It is based on the combination of geometrical optics and the uniform theory of diffraction (GO/UTD). After specifying transmitter and receiver locations, a shouting-and-bouncing algorithm is used to obtain the electric field $\overrightarrow{E}_i$ (amplitude, phase, polarization, direction of departure (DOD), direction of arrival (DOA), delay spread, etc.) from the ith transmitter antenna to the jth receiver antenna. The path loss can be found [20] as follows:

$$L_j = 20\log\left(\frac{\lambda}{4\pi}\frac{|\overrightarrow{E}_{\mathbf{T}}|}{|\overrightarrow{E}_{\mathbf{0}}|}\right) \tag{11.28}$$

$$\overrightarrow{E}_{\mathbf{T}} = \sqrt{\sum_{i=1}^{n} \overrightarrow{E}_{\mathbf{i}}^{\mathbf{2}}} \tag{11.29}$$

where $\overrightarrow{E}_{\mathbf{T}}$ is the total received electric field, n is the number of received rays (paths), $\overrightarrow{E}_{\mathbf{0}}$ is the transmitted electric field, and L_j is the jth receiver antenna. Because 3D ray tracing will provide all the parameter values of the propagating signal for a transmitter/receiver path, the channel impulse response can be constructed and used to predict other channel effects on the transmitted waveform.

11.5.4 Link Budget

The fact that the LTE radio channel is adaptive according to channel variations as well as having the option of using MIMO at the UE and eNB makes the link budget formulation dynamic as well. The channel bandwidth and the measurements made at the UE and eNB also vary the noise density. A link budget is formed by specifying the power levels from the output of the transmitter module right before the antenna feeders, through the antenna, passing the wireless channel (fading, diffraction, shadowing, interference, noise), to the receiving antenna, feeders, and finally the input point of the receiver module.

In a spread-spectrum based communication system, the spreading of the data introduces an extra gain called the processing gain. The value of the processing is gain is given by $G_P = \frac{R_{\text{code}}}{R_{\text{data}}}$ where R_{code} is the code sequence chipping rate, and R_{data} is the data rate of the transmitted signal. The processing gain should be included in the link budget calculation. The geometry factor (G-factor), which describes the desired signal levels to inter-/intra-cell interference plus noise, should also be accounted for in the link budget (same as the CINR). Table 11.5 presents the minimum specified transmit and receive power levels for the UE and eNB in FDD mode of operation. These levels depend on the E-UTRA band of operation as

Table 11.5 Power Levels for Link Budget, UE, and BS in FDD Mode

		UE	*BS*	*Level*
Transmit	Maximum Transmit Power	23	43	dBm
	Cable/Interconnect Losses	0.5–2	0.5–2	dB
	Noise Figure	9	5	dB
	Antenna Again	0	12–15	dB
Receive	Antenna Gain	0	12–15	dB
	Noise Figure	9	5	dB
	Cable/Interconnect Losses	0.5–2	0.5–2	dB
	Sensitivity (min, BW = 10 MHz)	−94	−101.5	dBm

well as the operating BW and modulation scheme. The reference sensitivity is the minimum mean power applied to the antenna ports at which throughput shall meet the minimum requirements for the specified channel. For QPSK transmission, sensitivity is based on ≥95% throughput level from the absolute maximum. No MIMO gain is shown in the table, and the output powers are based on the active ON state of the UE/eNB. For other levels and modes of operation, refer to [6,13,21]. Table 11.6 show a simple link budget calculation for a 64 Kbps UL with the use

Table 11.6 Sample UL Budget for Four Resource Blocks with 720 KHz BW and 128 Kbps Operation, FDD Model

UE	a. Max. transmit power	23	dBm
	b. Cable losses	0.5–2	dB
	c. Body loss	0	dB
	d. Antenna again	0	dBi
	e. EIRP	21	dBm, (a − b − c + d)
eNB	f. Antenna gain	15	dBi
	g. Feeder loss	2	dB
	h. Noise figure (NF)	5	dB
	i. Thermal noise	−110.4	dBm, (KTB)
	j. Receiver noise floor	−105.4	dBm, (h + i)
	k. CINR	−5	dB, From simulations
	l. Receiver sensitivity	−106.8	dBm, or (j + k)
	m. Max. channel loss	140.8	dB (e + f − g − l)

Note: 1. KTB = Boltzmann's Constant × Temperature in degrees Kelvin (290) × Bandwith in hertz.

2. Receiver sensitivity (i) was used from [21]. Also it can be calculated from j + k in Table 11.6. The value used is based on the minimum sensitivity value required for 1.4 MHz BW in [21].

Table 11.7 Sample DL Budget for 10 MHz BW and 1 Mbps Operation, FDD Mode

eNB	a. Max. transmit power	43	dBm
	b. Feeder losses	2	dB
	c. Antenna again	15	dBi
	d. EIRP	66	dBm, (a − b + c)
UE	e. Antenna gain	0	dBi
	f. Body loss	0	dB
	g. Cable loss	2	dB
	h. Noise figure (NF)	9	dB
	i. Thermal noise	−104	dBm (KTB)
	j. Receiver noise floor	−95	dBm (h + i)
	k. CINR	−5	From simulations
	l. Receiver sensitivity	−91	dBm, or (j + k)
	m. Max. channel loss	155	dB (d + e − f − g − l)

Note: 1. KTB = Boltzmann's constant × temperature in degrees Kelvin (290) × Bandwdth in hertz.
2. Receiver sensitivity (i) was used from Table 11.5. It can also be calculated from j + k in Table 11.7. The value used is based on the minimum sensitivity requirement value listed in [6] for 10 MHz BW.

of two resource blocks. While Table 11.7 shows a simple DL budget with 1 Mbps, and a bandwidth of 10 MHz. The CINR values depend on the levels of interference and the modulation and coding combination used. These can be extracted from cite specific simulations. The effective isotropic radiated power (EIRP) is the power level exiting the UE/eNB when transmitting. The maximum channel loss calculated at the end of the link budget is the maximum level of power loss through the channel below which the receiver will not be able to capture the received signal, and a call will be dropped. Any of the channel models presented in the previous sections can be used to determine the maximum distance between the UE and the eNB for a certain configuration.

The presence of neighboring cells, users and other networks increases the interference and noise levels. There are certain threshold levels for the CINR at the UE and eNB below which the service cannot be granted. The coexistence of LTE with other 2G and 3G networks dictates the need for careful design procedures and tighter interference requirements. The LTE standard identifies the adjacent channel interference ratio (ACIR) and the adjacent channel leakage ratio (ACLR) as two bandwidth dependent parameters that are monitored and describe the amount of interference and its impact on the DL and UL throughput. Several simulation examples and scenarios can be found in [15].

11.5.5 CW Testing

Continuous wave (CW) testing, also called CW drive testing, is essential to the RF planning process and deployment of cellular networks. A CW test should be conducted to examine the signal levels in the area of interest: indoor, outdoor, and in vehicle. There are two types of drive tests:

1. **CW Drive**: A CW drive test is conducted through different routes in the area to be covered before the network is deployed. A transmit antenna is placed in the location of interest (future site), and is configured to transmit an unmodulated carrier at the frequency channel of choice. A vehicle with receiver equipment is used to collect and log the received signal levels.
2. **Optimization Drive**: This drive test is conducted after the cellular network is in operation (different call durations, data uploads, and data downloads). Thus, the modulated data signal is transmitted and then collected by the on-vehicle receiver equipment, then the data are analyzed for different performance parameters like reference channels (similar to the pilot in 3G systems), power measurements, scrambling codes, block error rates, and error vector magnitudes.

11.5.6 Model Tuning

Model tuning is the step that follows CW testing. The logged CW data are used to come up with a tuning factor for the initially picked propagation model used for the area under investigation. Propagation model optimization/tuning is performed using various curve fitting and optimization algorithms that are proprietary to the planning tool, and after the process is complete, statistical performance measures are obtained to illustrate the effect of optimization on the model behavior in terms of the mean, standard deviation, and RMS error. This process will provide a mode accurate channel model.

11.6 Network Performance Parameters

11.6.1 Performance Parameters

Several types of parameter measurements are made at the UE or the eNB. These measurements are used to quantify the network performance and thus will aid in the adaptation of the appropriate coding/modulation as well as the link/cell traffic and capacity. In idle mode, eNB broadcasts the measurements within messages in the frame protocol. To initiate a specific measurement from the UE, the eNB transmits an "RRC connection configuration message" to the UE, along with the measurement type and ID, objects, command, quantity, and reporting criteria. The UE performs the measurement and responds to the eNB request with the measurement ID and results via a "measurement report message" [21–23]. Some of the most common performance metrics in LTE are:

- **Received Signal Strength Indicator (RSSI)**: This measures the wideband received power within the specified channel bandwidth. This measurement is performed on the broadcast control channel (BCCH) carrier. The measurement reference point is the UE antenna connector. This measurement is easy to perform, as it does not need any data decoding, rather it shows whether a strong signal is present or not. It does not give any details about the channel or signal structure.
- **Received Signal Code Power (RSCP)**: measures the received power on one code on the primary common-pilot channel (CPICH). If the measurement is made while the equipment is in spatial multiplexing, the measured code power from each antenna is recorded, and then all are summed together. If transmit diversity is chosen, the largest measurement from all antennas is picked. The measurement reference point is the UE antenna connector.
- $\boldsymbol{E_c/N_0(E_c/I_0)}$: This is the received energy per chip divided by the noise power density (E_c/N_0) (interference power density E_c/I_0) in the band. When spatial multiplexing is used, the individual received energy per chip is measured for each antenna, and then summed together. The sum is divided by the noise power density in the band of operation. If transmit diversity is used, the measured E_c/N_0 for antenna i should not be lower than the corresponding RSCP level. The measurement reference point is the UE antenna connector. Usually the E_c/I_0 level is indicated as the interference levels are more profound and affect signal quality than noise levels (i.e., thermal noise).
- **Block Error Rate (BLER)**: This is used to measure error blocks within a specific channel transmission as a measure of transmission quality. This is performed on the transport and dedicated channels (TCH, DCH).
- **Carrier–Interference Plus Noise Ratio Power Level [CINR ($\boldsymbol{C/(I+N)}$)]**: The CINR is measured in both the UE and eNB to determine the radio bearer to be used based on some predefined set of thresholds. The radio bearer defines which modulation and coding scheme to use for the data to be transmitted. The higher the CINR, the higher the spectrum efficiency by using a higher constellation modulation and coding scheme. The calculation of CINR is more involved than the RSSI, and it provides a better indication on the channel and signal qualities. CINR is sometimes referred to as the *G*-factor.
- **Error Vector Magnitude (EVM)**: It measures of the difference between the measured symbol coming out of the equalizer to that of the reference. The square root ratio of the mean error vector power to the mean power of the reference symbol is defined as EVM. The required EVM percentage over all bandwidths of operation performed over all the resource blocks and subframes for LTE is based on the modulation scheme used. Thus, for QPSK, 16-QAM, and 64-QAM modulation is given by 17.5%, 12.5%, and 8%, respectively.

11.6.2 Traffic

Traffic intensity is a measure of the average number of calls taking place at a specific time interval. The traffic intensity (I) is usually measured in Erlangs. One Erlang represents a call with an average duration of 1 hour.

$$I = \frac{\sum_{i=1}^{N_c} t_i}{T} = \frac{N_c \bar{t}_i}{T} \quad (11.30)$$

where N_c is the total number of calls, t_i is the holding time for user i, T is the monitoring time interval, and $\bar{t}_i$ is the average holding time for user i. A call that cannot be completed because the connecting equipments are busy (fully utilized) is termed as a blocked call. Several probability distribution models (formulas) are used based on the way calls are handled. In the USA, the Poisson's formula is used, while in Europe and Asia the Erlang-B formula is used. Each formula has its own assumptions based on the way the calls originate and processed [1]. The Poisson's formula results in higher blocking rates than the Erlang-B one for a given traffic load.

The actual cellular network performance in terms of traffic and capacity depend on the eNB transceivers capacity as well as the average Erlangs per subscriber. The user might be blocked from service if its CINR level is below the minimum threshold as well [21]. In LTE, dynamic scheduling for the UL and DL is included in the MAC of the eNB. It takes into account the traffic volume and the QoS of each connected UE. Only service granted UE are allowed to transmit. Resources are allocated based on the radio channel condition measurements (CQI) and layer-2 measurements. These measurements will assign scheduling, load balancing and transmission priorities per traffic class [4, 5]. Layer-2 measurements are discussed in detail in [24]. The QoS class identifier (QCI) and the allocation and retention priority (ARP) are two parameters that control node specific parameters and control bearer level packet forwarding, scheduling weights, queue management, and priority levels to establish/modify such requests and whether to grant or decline them in the presence of resource limitations.

11.6.3 Measurement Types

The LTE is still in its early stages, as no actual deployments of the system has been reported in any part of the world. Several experiments are taking place as we speak. WiMAX, on the other hand, has been deployed in several regions around the world. The radio interface for LTE and WiMAX are similar in several aspects, the main difference being the use of SC-FDMA for the UL radio interface compared to an OFDMA interface for WiMAX. The measurement of signal levels and performance parameters in a complex system with MIMO capability is not a trivial task.

Although RF giants like Agilen, and Keithley, among others in the measurement arena, provide complete integrated solutions for RF planning engineers to test their networks, the engineer should know what to look for and how to read, interpret, and analyze the measurements conducted with proper setup procedures. There are generally two types of measurements involved: measurements of the prototype and measurements of the deployed network. Although the former is essential to make sure that the designed equipment satisfy the specification requirements in the laboratory environment, the latter provides an actual field characterization of the developed equipment.

- **Predeployment/Prototyping**: The measurements performed in this stage are aimed to show compliance of the developed equipment (UE/eNB) with the technology requirements. For LTE, laboratory measurements should reflect as much of the actual field environment as possible. Thus, several RF equipment vendors are providing MIMO channel emulators that can give close estimates of the RF propagation channel to be encountered by the LTE terminals. Vector signal generators (VSG) are used to generate LTE modulated signals, and vector signal analyzers (VSA) analyze the signals and provide all necessary information and performance measures at the receiver [25].
- **Postdeployment**: The measurements performed in this stage are made in two steps, as mentioned in Section 11.5.5. A CW test drive is made to check the coverage scenario, then a detailed test with active network traffic is performed to log performance metrics such as the EVM, subcarriers, spreading codes, signal strength, I/Q imbalance and errors, frequency shifts, cell ID, timing offset, and the like [26].

11.7 Postdeployment Optimization and Open Issues

11.7.1 Postdeployment Optimization

As with all currently deployed cellular networks, whether it be a 2G GSM network or even a 3G UMTS one, an LTE network will have to be optimized after deployment to provide better coverage, throughput, lower latency and seamless integration as the specification asks for. The optimization process contains several steps. It starts with data drive testing as mentioned in Section 11.5.5, where all performance parameters are tested and logged in the field after the network is active. This test should also include the different coverage/propagation scenarios along with their respective models (e.g., pedestrian, vehicular, indoor). The field data will then be used to tune the models for better network performance and coverage.

Based on the collected data, RF planning engineers analyze the performance and maybe decide to add more eNBs for coverage, mainly pico and femtocells, in the areas that show degraded power levels or data throughput. Femtocells will be used in LTE, as they will provide service for households and small businesses. Usually, the optimization process is an iterative one with no specific steps involved, rather than a set of consistent procedures that characterizes network performance and coverage in a certain area; actions are taken accordingly.

11.7.2 Open Issues

There are several open issues that original equipment manufacturers (OEMs) has to take into account when designing LTE terminals and equipment. Some of the issues are being addressed, whereas others are still under extensive investigation. Here, we identify some of these open issues in two categories: UE and eNB.

11.7.2.1 UE

There are several challenges that has to be overcome in implementing LTE UE. The use of MIMO technology dictates the use of highly reliable and complex equalization techniques. In a worse-case scenario, and using a minimum-mean-square-error (MMSE) technique, the equalization might consume 1500 MIPS (million instructions per second) performed on 600 subcarriers. This poses a challenge in performing parallel computations, minimizing power consumption and silicon area. Memory requirements for coding and decoding is also a challenge that needs to be overcome [27,28].

11.7.2.2 eNB

Although designers always try to minimize power consumption and silicon area in their designs, there are less stringent requirements at the eNB side. The challenges with complexities of hardware also exist within the eNB equipment. However, there are other challenging aspects that have to be solved such as using BF to improve DL performance. BF needs the use of antenna arrays, which require the use of adaptive algorithms and electronics to be able to operate in real time and automatically. The fact that BF will coexist within MIMO system is also a challenge. The coexistence with legacy systems like 2G and 3G networks in the vicinity is another obstacle to be overcome (4G-3G, 4G-2G). This coexistence will increase interference levels and raises the thesholds of noise and interference. The LTE specification specifies strict intermodulation levels due to this network coexistence. There are stringent requirements within it that details the compliance levels within legacy systems bands that OEMs should pay attention to.

11.8 Conclusion

LTE-based cellular networks are anticipated to be deployed in 2011. Such complex networks need careful design and planning. This chapter touched on the physical layer (air interface) of such a network and discussed the RF planning process. Adaptive coding and modulation schemes as well as OFDMA multiplexing were presented to explain the physical layer interface and operation. The planning engineer should be aware of the channel models, MIMO system operation, and performance parameters and metrics to analyze and study the network behavior. Three statistical propagation channel models were discussed in detail (Okumura-Hata, Cost-231, and IMT-2000), as was the deterministic ray tracing method to identify the main parameters used for modeling the LTE channel. A coverage example was provided, using a simulation tool that incorporates MIMO systems. Measurement types and methods for pre- and postnetwork deployment were highlighted. The four propagation scenarios listed in the LTE specification were presented. Finally, some of the challenges on the UE and eNB sides were discussed.

Acknowledgments

I thank Mr. Michael Dial from FORSK Inc. for providing open access to the ATOLL simulation tool for LTE cellular network planning. I also thank Mr. Ghaith Abu-Sleiman (consultant at T-Mobile) for his practical points of view on the manuscript and its flow of ideas. Finally, I thank Dr. Tareq Y. Al-Naffouri (KFUPM) and the anonymous reviewers for their valuable comments that improved the content of this chapter.

References

[1] V.K. Garg, *Wireless Communications and Networking*, Elsevier Morgan Kaufmann, San Fransisco, California, USA, 2007.
[2] M. Rumney, "What next for mobile telephony," *Agilent Measurements Journal*, No. 3, pp. 33–37, 2007.
[3] E. Boch, "Backhaul for WiMAX and LTE: High-bandwidth ethernet radio systems," *Microwave Journal* (supplement), pp. 22–26, November 2008.
[4] 3GPP TS 36.300, V8.6.0, "UTRAN and E-UTRAN overall description, stage 2," September 2008.
[5] 3GPP TS 23.401, V8.4.1, "GPRS enhancements for E-UTRAN access," December 2008.
[6] 3GPP TS 36.101, V8.3.0, "E-UTRA user equipment radio transmission and reception," September 2008.
[7] 3GPP TS 36.211, V8.4.0, "E-UTRA physical channels and modulation," September 2008.
[8] 3GPP TS 36.212, V8.5.0, "E-UTRA multiplexing and channel coding," December 2008.
[9] S. Haykin and M. Moher, *Modern Wireless Communications*, Prentice Hall, Englewood Cliffs, NJ, 2005.
[10] M. Rumney, "Introducing Single-Carrier FDMA," *Agilent Measurement Journal*, no. 1, pp. 1–10, 2008.
[11] A. Goldsmith, *Wireless Communications,* Cambridge University Press, New York, 2005.
[12] Atoll, Global RF planning solution, Forsk Inc., Blagnac, France, 2008.
[13] 3GPP TR 36.942, V8.0.0, "EUTRA radio frequency system scenarios," September 2008.
[14] Y. Jung and Y. Lee, "Scrambling code planning for 3GPP W-CDMA systems," *IEEE Vehicular Technology Conference* (VTC), vol. 4, pp. 2431–2434, May 2001.
[15] 3GPP TR 25.942 V.8.0.0, "Radio frequency (RF) system scenarios," December 2008.

[16] http://www.picochip.com
[17] C.A. Balanis, *Antenna Theory: Analysis and Design*, 3rd Ed., John Wiley, Hobsken, NJ, 2005.
[18] R. Hoppe et. al., "Comparison of MIMO channel characteristics computed by 3D ray tracing and statistical models," *IEEE Second European Conference on Antennas and Propagation*, pp. 1–5, November 2007.
[19] S. Loredo, A. Rodriguez-Alonso, and R.P. Torres, "Indoor MIMO channel modeling by rigorous GO/UTD-based ray tracing," *IEEE Transactions on Vehicular Technology*, vol. 57, no. 2, pp. 680–692, March 2008.
[20] H. Zare and A. Mohammadi, "A fast ray tracing algorithm for propagation prediction in broadband wireless systems," *IEEE 8th International Conference on Communication Systems* (ICCS), vol. 1, pp. 6–10, November 2002.
[21] 3GPP TS 36.104, V8.4.0, "EUTRA base station radio transmission and reception," December 2008.
[22] 3GPP TS 36.214, V8.4.0, "EUTRA physical layer—Measurements," September 2008.
[23] 3GPP TS 36.331, V8.3.0, "EUTRA radio resource control," September 2008.
[24] 3GPP TS 36.314, V8.0.1, "Evolved universal terrestrial radio access (E-UTRA); layer 2—measurements," January 2009.
[25] N9080A, "LTE measurement application," *Application Note*, Agilent Technologies, September 2008.
[26] M.G. Sanchez, A.V. Alejos, and I. Cuinas, "Urban wide-band measurement of the UMTS electromagnetic environment," *IEEE Transactions on Vehicular Technology*, vol. 53, no. 4, pp. 1014–1022, July 2004.
[27] J. Berkmann et. al., "On 3G LTE terminal implementation—Standard, algorithms, complexities and challenges," *IEEE International Wireless Communications and Mobile Computing Conference* (IWCMC), pp. 970–975, August 2008.
[28] A. Ghosh et. al., "Multi-antenna system design for 3GPP LTE," *IEEE International Symposium on Wireless Communication Systems*, pp. 478–482, October 2008.

Chapter 12

Advanced Radio Access Networks for LTE and Beyond

Petar Djukic, Mahmudur Rahman,
Halim Yanikomeroglu, and Jietao Zhang

Contents

12.1 Introduction

Current state-of-the-art standardization activities of the 3rd Generation Partnership Project (3GPP) long-term evolution (LTE) [1] and worldwide interoperability for microwave access (WiMAX) [2] have resulted in cellular standards with high data rates, close to the IMT-advanced spectral efficiency requirements of 15 bits/sec/Hz peak downlink and 6.75 bits/sec/Hz uplink [3]. Due to a spectrum limitation of 20 MHz, these standards fall short of the IMT-advanced data rate requirements of 600 Mbps peak downlink and 270 Mbps peak uplink at 40 MHz bandwidth. Current standardization activities are aiming for even higher data rates of 1 Gbps for downlink and 500 Mbps for uplink [4], as originally proposed in IMT-advanced [5]. Although it is still early for standardization bodies to consider beyond-4G data rates (tens of gigabits per second on the downlink), this is clearly a major research topic due to the exponential growth of user traffic on existing networks.

Even though the standards allow for very high spectral efficiency transmissions, the laws of physics, combined with the Shannon's capacity bound, show that high spectral efficiency would only be available when the distance between the transmitter and the receiver is small. Taking the approach of scaling the cellular radio access network (RAN) architecture to decrease the distance between the users and the base station (BS) is not practical from the cost perspective. Ubiquitous very high data rate coverage is also an extremely challenging problem with the conventional radio resource management (RRM) approaches, as the rates decrease substantially at the periphery of BS coverage regions (the well-known cell-edge coverage problem). The conventional cellular design also uses fixed (a priori) radio resource allocations and assignments, which are inefficient; this inefficiency becomes even worse in a dense network due to the increased interference. It is, therefore, necessary to examine new RAN architectures, which can cost effectively increase radio port density in the RAN coverage area, and related RRM optimization techniques, which effectively manage the interference.

This chapter provides an overview of current 4G RAN architectures and RRM optimization techniques and the current consensus in the community about the elements of future RANs and associated advanced RRM optimization techniques. For readers who are familiar with the RAN architecture concepts, the chapter is an easy introduction into the area. For readers who are familiar with the RAN architecture concepts, the chapter gives a perspective on the evolution of 4G RANs, summarizes the current consensus in the community on architectures beyond-4G, and introduces network management concepts from data networks, which will be necessary in beyond-4G RANs.

First, we review the recently standardized, 3rd Generation Partnership Project (3GPP) long-term evolution (LTE) [1], and LTE-advanced, 4th generation (4G) RAN architectures currently undergoing standardization, and related RRM optimization problems. LTE and LTE-advanced use orthogonal frequency division multiple access (OFDMA) technology, which allows flexible spectrum usage. OFDMA's flexible spectrum is already allowing more efficient RRM techniques, such as fractional frequency assignment, which only assigns portions of channels to cells [6, 7]. These new RRM optimization techniques are enabled by OFDMA's flexibility and were not previously possible with time-division multiple access (TDMA) and frequency division multiple access (FDMA) physical layers. We focus on the evolution of the LTE RAN architecture, which shows trends expected in future RAN architecture development. For example, relay elements are currently under 3GPP standardization discussions [8] for inclusion into LTE-advanced. Thus, we expect future advanced RANs to contain various types of relays.

Second, we provide an overview of advanced RAN elements, such as distributed antenna ports, femto-BSs and various forms of relays, and coordinated multipoint (CoMP) transmission and reception techniques, which increase over the network's coverage area. These elements can cost effectively increase radio port density, bring the user closer to the source of the wireless signal, and effectively manage the interference. For example, recent standardization activities are adding a variety of advanced devices to the network, such as distributed antenna ports [8], femto-BSs [9, 10], various forms of relays [8, 11], and CoMP transmission and reception techniques. As many of the new elements are currently the subject of advanced standardization efforts, the discussion of advanced RAN architecture is timely. The advanced architecture uses OFDMA and provides network component integration and signaling required to implement centralized and distributed user-centric RRM techniques. This approach is enabled by OFDMA and requires extensive support from RAN to coordinate the resource assignment.

Third, we review several open issues in RRM optimization for the advanced RAN architecture. We argue that to achieve the full potential of OFDMA, it is necessary to investigate new user-centric RRM techniques, which strive to provide ubiquitous high-rate coverage when and where it is needed, depending on user location and needs. We propose user-centric utility optimization of radio resources, which was recently pursued in the wired community, as a potential RRM optimization framework for advanced RANs. For example, the transmission control protocol (TCP) was shown to be a user-centric distributed utility optimization of network resources [12, 13]. A nice feature of user-centric utility optimizations is that it lends itself to top-down protocol development, which we believe will be useful in future RAN standardization efforts.

The rest of the chapter is organized as follows: we review the 3G generation universal mobile telecommunications system (UMTS) terrestrial radio access network (UTRAN), its OFDMA-based 4G RAN architecture—the evolved-UTRAN (E-UTRAN)—also known as LTE, and its successor, LTE-advanced, in Section 12.2; we review RRM techniques for OFDMA-based RANs in Section 12.3; and provide an overview of the advanced RAN architecture based on OFDMA and open issues RRM optimization for advanced RANs in Section 12.4.

12.2 Evolution of 4G OFDMA-based RANs

We now review the evolution of 4G RANs toward LTE and LTE-advanced. Multiuser access technology for all 4G RANs is based on OFDMA, which uses orthogonal frequency-division multiplexing (OFDM) in the physical layer. OFDMA allows relatively easy assignment of radio resources in time and frequency, which is an improvement over the existing 2G and 3G technologies. Because LTE shares many of the feature of its predecessors, we start with a short overview of UTRAN, which is a precursor for 4G E-UTRAN. Then we give an overview of E-UTRAN, and its successor LTE-advanced. We note that, in addition to higher bandwidth and spectral efficiency, the overall trend in the evolution of RANs is also toward decentralization.

12.2.1 UMTS Radio Access Network

The 2nd generation (2G) wireless network provided support for low-rate service such as voice traffic and short messaging service (SMS). Two prevalent 2G systems are the TDMA-based global system for mobile (GSM) and the CDMA-based IS-95 (cdmaOne). More evolved TDMA-based systems, such as the general packet radio service (GPRS) and enhanced data rate for GSM evolution (EGDE), emerged to provide a two- to threefold gain in rates by exploiting advanced modulation and encoding techniques and enabled services beyond SMS such as Internet access. However, recent demand for high-speed wireless Internet and video telephony have driven the move toward 3G technologies that can provide peak data rates of 384 kbps under mobile conditions and 2 Mbps under stationary and low-speed mobility conditions.

Almost all well-accepted 3G standards—wideband CDMA (WCDMA), CDMA2000, and time-division synchronous CDMA (TD-SCDMA)—are based on CDMA, which is fundamentally different from its predecessor 2G/2G+ technologies, such as TDMA-based GSM/GPRS. However, 3G RAN has been built on a 2G core network in order to facilitate the coexistence of TDMA-based GSM/GPRS services. Figure 12.1 shows the RAN architecture of 3G networks that coexist with the GSM evolved radio access network (GERAN). Although GERAN supports both packet and circuit switched services, UTRAN is developed toward "all IP services" through a packet switched part of the core network. The dual-mode GERAN/UTRAN core network provides necessary interfaces to support both packet and circuit switched services.

The primary operating mode of UTRAN is WCDMA with frequency-division duplexing (FDD), whereas the other variant is time-division duplex (TDD)-based TD-SCMDA. TD-SCDMA is a Chinese home-grown technology in collaboration with major industry players around the world, which is based on 3GPP specifications. It has the advantage of dynamic spectrum usage due to TDD duplexing. Narrowband 3G (CDMA2000) operates on FDD, which was developed by 3GPP. Here we focus our discussion on UTRAN.

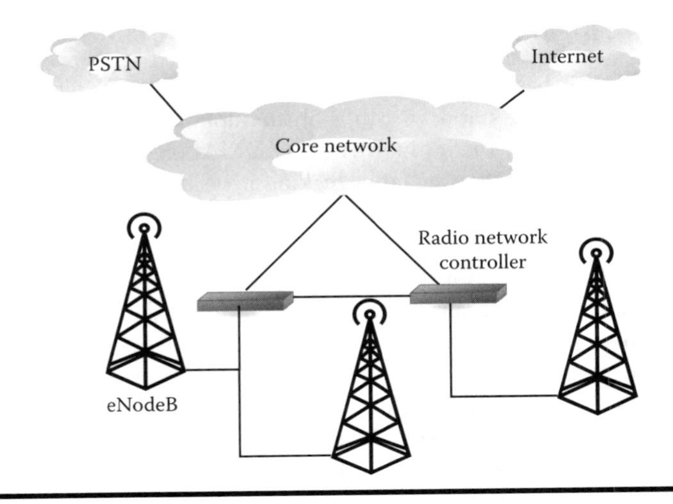

Figure 12.1 3G RAN architecture.

Although there are differences in the implementation of FDD WCDMA and TDD TD-SCDMA, the radio network architecture is quite similar (Figure 12.1). RAN consists of base stations (NodeB) and a radio network controller (RNC), which together make the radio network subsystem (RNS) [14]. UTRAN comprises a number of RNSs connected to the core network (CN), which bridges the public service telephone network (PSTN) and the Internet with the RAN. NodeBs communicate with user terminals. The RNC is responsible for major RRM decisions such as handover and admission control, which may require control signaling among user terminals, NodeBs, or other RNCs.

The main purpose of the RNC is to aid macrodiversity, which uses multiple radio signal streams through multiple NodeBs to communicate with the mobile terminals. Because multiple radio streams may go through the same RNC, the RNC must perform link-layer functionality. RNC also aids in power control, which is vital in the WCDMA systems. The dynamic inner-loop power control performed on a short time scale is done at NodeB and is controlled by the outer-loop power control overseen by the RNC. Scheduling of data is performed by the RNC.

3GPP has also released a version of WCDMA for beyond 3G: high-speed downlink packet access (HSDPA), high-speed uplink packet access (HSUPA) in Release-5 specifications [15]. These standards are based on UMTS WCDMA and provide peak data rates of 14.4 Mbps on the downlink and 5.76 Mbps on the uplink. Advanced modulation and coding, fast packet scheduling, and hybrid automatic repeat-request (ARQ) are among the added features behind these increased rates. Further improvements, such as MIMO, have been provided in HSPA+ (also called as evolved-HSPA) specifications (in Release-7 and Release-8). With these enhancements, peak rates are 42 Mbps on the downlink and 22 Mbps on the uplink.

12.2.2 Long-Term Evolution RAN (E-UTRAN)

LTE is an OFDMA-based cellular system that can achieve peak data rates of 100 Mbps on the downlink and 50 Mbps on the uplink [4]. LTE uses OFDMA in the downlink and single-carrier FDMA (SC-FDMA) on the uplink. SC-FDMA reduces the peak-to-average power ratio, making it easier to implement it on user terminals [16]. LTE has spectral efficiency three to four times higher than UTRAN and supports a scalable bandwidth from 1.4 to 20 MHz. It also uses MIMO configurations (4 × 2 and 1 × 2 for the downlink and uplink, respectively). Although the main motivation for LTE air interface is an improved data rate, it also focuses on removing shortcoming experienced in the UMTS system, such as nonscalable bandwidth, latency, and poor cell-edge performance. LTE system is optimized for low mobility while it can obtain high performance at speeds of up to 100 km/hr, and it also supports mobility of up to 350 km/hr. LTE also supports coexistence and internetworking with GERAN, UMTS, HSxPA, and WiMAX access technologies.

The major differences between UTRAN and LTE systems are the OFDMA-based air interface in LTE, which can achieve high spectral efficiency, and omission of the RNC in the RAN to obtain reduced latencies. The radio access part of the LTE system is termed the evolved-UTRAN (E-UTRAN) which consists of evolved-NodeB (eNodeB) and UE. RRM functionalities, which resided in the RNC in the 3G system, have been implemented in the eNodeBs in the LTE. The LTE system is termed an evolved packet system (EPS), which comprises an E-UTRAN radio access and an evolved packet core (EPC) network [17], as shown in Figure 12.2. LTE RAN does not include relays.

An eNodeB has all of the functionality required for the RRM operations such as radio bearer control, call admission, mobility managements and scheduling. Multiple

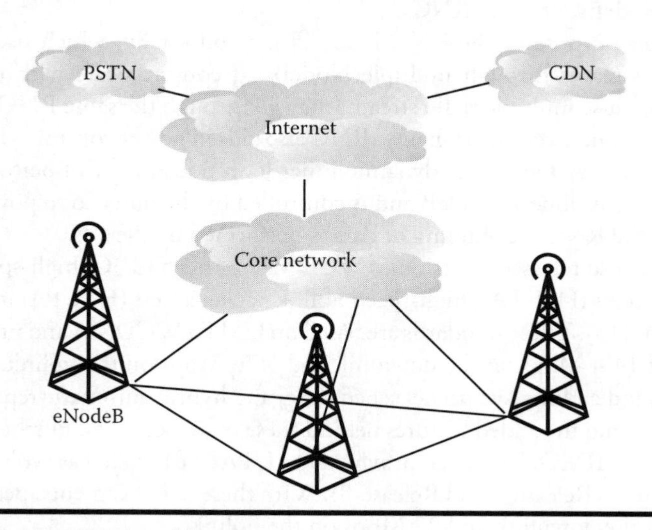

Figure 12.2 LTE RAN architecture.

eNodeBs are connected and signal each other to facilitate distributed RRM functionality, such as handoff, inter-cell interference coordination, and load balancing. For example, the handover mechanism is performed by signaling between the source and target eNodeBs. Decisions are made at the eNodeB level, and the multicell mobility entity (MME) and the serving gateway (S-GW) are notified about the new association so that packets can be forwarded to the new eNodeB on completion of the handover process. Unlike UMTS systems where handover is performed by the RNC, handover process in LTE is performed by direct signaling between eNodeBs, greatly reducing signaling latencies.

12.2.3 LTE-Advanced RAN

In response to 4G system requirements set out by ITU [3], the 3GPP has initiated development of LTE-advanced specification [4]. Although LTE-advanced will inherit much of the LTE features (it is going to be back-compatible with LTE), there are many envisioned improvements, most notably inclusion of relay-based deployment, cooperative diversity, bandwidth expansion, and higher order MIMO antenna configurations. The LTE-advanced targets support of peak data rates of 1 Gbps in the downlink and 500 Mbps in the uplink for low-mobility scenarios [18]. However, more importantly than simply advertising high peak data rates, LTE-advanced is also moving to supporting these rates in a greater part of the cell with the use of devices such as relays.

The LTE RAN is capable of reaching rates close to the Shannon's limit with the clever use adaptive modulation and coding. OFDMA flexibility allows for flexible bandwidth use from 1.25 MHz to around 20 MHz, for peak rates of about 300 Mbps. In order to reach IMT-advanced requirements of 1 Gbps, LTE-advanced increases the transmission bandwidth to the maximum of 100 MHz, which may be used in noncontiguous frequency blocks [18]. Noncontiguous blocks are necessary, as the current frequency allocations do not always have 100 MHz frequency blocks.

If the current LTE spatial multiplexing is used, 100 MHz transmission bandwidth would allow for peak data rates of about 1.5 Gbps. Some discussions regarding LTE-advanced standardization are moving to using even more spatial multiplexing layers, however, the well-known paradox of spatial multiplexing is that gains are available at higher signal-to-noise ratios (SNRs), which are available close to the base station where there are not that many spatial channels available. Thus, further gains using spatial multiplexing may not be that great.

In terms of cell throughput, it is more promising to evolve the network toward a higher density of radio ports, rather than to increase spatial multiplexing. We reflect on this more in Section 12.4 when we discuss advanced, beyond-4G RANs. For now, we explain how LTE-advanced is increasing the number of radio ports with relays.

LTE-advanced standardization is considering relays for extending the base-station coverage and increasing port density in the cell [8]. Relays are already in the 802.16j standard [11] and are the basis of 802.11-based mesh networks. A relay acts as an intermediary between the base station and the user terminal by receiving data intended for the terminal and then retransmitting it to the user terminal. Because the relay is closer to the base station and the user terminal, than the user terminal and the base station are to each other, there is potential for high spectral efficiency transmissions [19]. Broadly speaking, if the relay simply amplifies the signal, it is the amplify-and-forward (AF) type, whereas if the relay also decodes and reencodes the data it is the decode-and-forward (DF) type. Relays in 802.16j and 802.11 multi-hop networks are the DF type. The current discussion in LTE-advanced is to decide which type will prevail in that RAN.

12.3 4G Radio Resource Management

So far, we have seen that 2G systems were based on TDMA, whereas 3G systems were based on CDMA. On the other hand, 4G, LTE, and WiMAX systems use the flexible OFDMA physical layer. We now review RRM techniques for OFDMA, applicable to both LTE and LTE-advanced. All of our examples follow LTE standard parameters. We start by a short system overview of OFDMA RRM and then provide a more detailed overview of various RRM techniques.

12.3.1 Overview of OFDMA RRM

Available RRM techniques depend on the multiple access technology used to share the radio channel. In TDMA, users are multiplexed in time. Time is divided into fixed size frames, and each user is allocated a portion of the frame for exclusive use. In FDMA, the users are always on and are multiplexed in frequency, thus a user is assigned a portion of the available bandwidth. In CDMA, users are always on, use the entire frequency space, and are multiplexed in an orthogonal code space. OFDMA is the most flexible scheme, which combines TDMA and FDMA and allows assignment of either portions of time or frequency to users.

Figure 12.3 shows the available radio resources where the time and the frequency/code are shown as a grid. RRM involves techniques necessary to assign to users full columns in the grid (in the case of TDMA), full rows (in the case of FDMA and CDMA), or planes in the case of multiuser multiple-input multiple-output (MU-MIMO). The number of assigned rows, columns, and planes depends on the maximum rate the user can achieve and the rate the user has requested. Ideally, all users should get the rate they requested; however, this may not always be possible.

In TDMA and CDMA, it is generally easy to assign resources to users in a flexible way. A TDMA user gets a higher rate with more time, whereas a CDMA user gets a higher rate with a larger portion of the available code space [20]. The two approaches can also be combined. For example, in HSPA the users get flexible rate assignment in code and in time [21]. In addition to allowing for better sharing of resources,

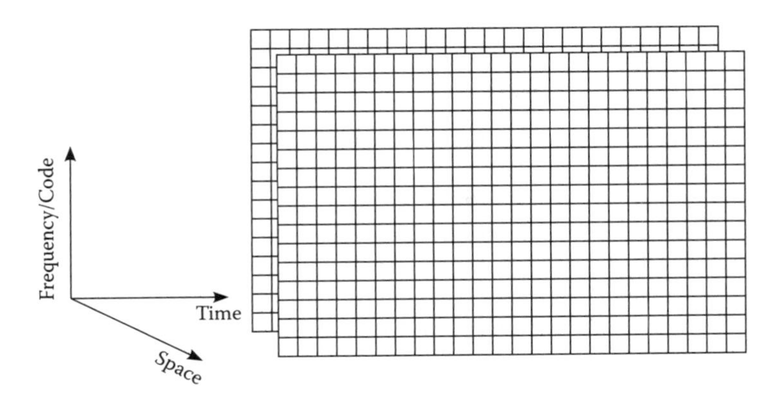

Figure 12.3 Available radio resources.

flexible time assignment has the advantage of potentially taking advantage of time diversity if the channel state information (CSI) is available. RRM is generally less efficient with FDMA due to the lack of granularity. However, with OFDMA it is easy to map data to frequencies, as in OFDM, each transmission is divided into multiple parallel transmission on distinct subcarriers. The total bandwidth taken by a transmission depends on the number of subcarriers used in the transmission, and the total time for transmitting one symbol depends on the bandwidth required by the subcarrier.

12.3.2 Transmission Scheduling in Time and Frequency

Residing in the MAC layer, scheduling function is responsible for efficient short-term allocation of available shared radio resources taking into account users' QoS considerations such as delay, end-to-end errors, and rate requirements. Additionally, an optimal scheduling scheme should consider channel condition, generally termed as channel quality indicator (CQI) available from PHY layer measurement and feedback to exploit channel variations inherent to any wireless system. The periodicity of the scheduling operation is defined by the radio resource controller (RRC).

As mentioned earlier, OFDMA scheduling takes advantage of channel variations in both time and frequency. The channel variability in time depends on the Doppler shift dominated primarily by the speed of the terminal, whereas the frequency correlation among subcarriers are dependent on the environment. In order to best exploit these variations, CQI measured on both time and frequency are necessary at the transmitter. However, measurement and reporting of CQI on each subcarrier require excessive signaling bandwidth as well as complexities, which makes it inefficient if not impossible. Instead, radio resource is partitioned into a number of subchannels in the frequency dimension; these subchannels within a specified time duration are typically the scheduling granularity of resources. For example, in LTE

12 subcarriers over seven or six OFDM symbols, depending on the length of cyclic prefix, form a scheduling resource granularity and is termed as physical resource block (PRB) [1].

Cell-specific reference signals that consist of known OFDM symbols are inserted into different specific portion of the downlink PRB and transmitted to users. Interpolation in time and frequency are done by the user terminal to estimate the channel in the other part of the PRB to prepare CQI values that represent the channel status required by the scheduler. CQI can be periodic or aperiodic and can be in various forms such as wideband and multiband and supports MIMO operation.

LTE supports a variety of scheduling disciplines appropriate for different service types. Figure 12.4 shows an example of scheduling scheme that is based on the maximum signal-to-interference plus noise (SINR) for a simple two-user case. In this example, 50 PRBs are considered and allocated between two users based on their SINRs. Allocation has been shown for a 100 PRB duration, and the base of the figure shows the share of resources between these two users. Being only channel dependent, a maximum SINR scheduler can provide throughput benefit, it seriously lacks fairness. A proportional fair (PF) scheduler, on the other hand, has attracted attention as a fair scheduler that takes both channel condition and user rates into consideration.

Unlike time-slot based scheduling, OFDMA scheduler works with two-dimensional resources (i.e., in time and frequency). By exploiting time and frequency variations, OFDMA allocation can achieve multiuser as well as frequency diversities. However, modifications of scheduling principles that are designed for slot-based are required. For example, a proportional fair scheduler [22], such as that used in a CDMA system, is not directly applicable to an OFDMA system. Such a slot-based

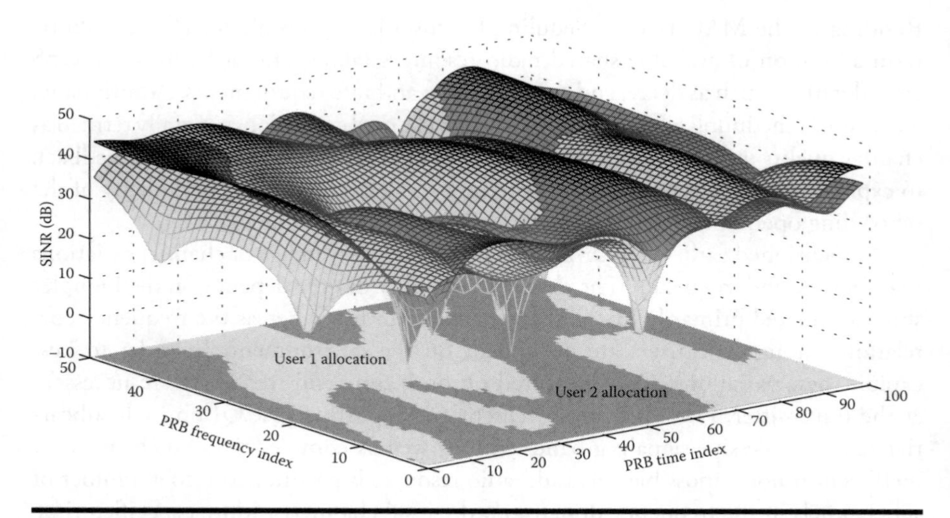

Figure 12.4 Maximum SINR time-frequency scheduling example with two users.

PF scheduler works as follows. At a particular scheduling instant t, transmission opportunity is given to user k^* based on the maximum $R_k(t)/T_k(t)$, where $R_k(t)$ is the instantaneous achievable rate at time t and $T_k(t)$ is the filtered average throughput over a past time-window t_c for user k and t_c is a tuning parameter that determines the trade-off between fairness and throughput. The average throughput is updated after each scheduling instant as follows:

$$T_k(t+1) = \begin{cases} \left(1 - \frac{1}{t_c}\right) T_k(t) + \frac{1}{t_c} R_k(t), & k = k^* \\ \\ \left(1 - \frac{1}{t_c}\right) T_k(t) & k \neq k^* \end{cases} \tag{12.1}$$

For the OFDMA system, user k will be given PRB n based on the following:

$$\arg\max_k \frac{R_{k,n}(t)}{T_k(t)} \tag{12.2}$$

In this case, user terminal throughput is updated after all PRBs are allocated. Alternatively, PRBs can be partitioned into a number of equal segments and throughput can be updated after allocating each of these segments [23].

12.3.3 Adaptive Modulation and Coding

Adaptive modulation and coding (AMC) is an effective way to enhance the spectral efficiency of the wireless channel. The basic idea of AMC is to use a high constellation modulation scheme with less redundant coding to achieve high throughput when the channel has a high SINR and to use a lower level modulation with more redundant coding scheme when the channel has a low SINR. The quality of the received signal depends on a number of factors such as the distance between the transmitter and the receiver, the path-loss exponent, log-normal shadowing, multipath fading, and noise. This implies that the SINR a receiver experiences varies over time, frequency, and space. The decision of selecting appropriate modulation and coding scheme is performed at the transmitter, which is based on CQI measured at the receiver side and fed back to the transmitter. Clearly, the performance of an adaptive modulation scheme is dependent on the accuracy of the channel estimation by the receiver and the reliability of the feedback path.

In LTE, quadrature phase-shift keying (QPSK), 16-quadrature amplitude modulation (QAM), and 64-QAM modulation modes are used for data channels, whereas only the more robust binary phase-shift keying (BPSK) and QPSK are specified for control channels [1]. Turbo and convolutional codes are specified for data, whereas additional coding schemes such as block and repetition codes are used for control channels [24]. Similar modulation and channel coding schemes are also used in WiMAX [11].

12.3.4 Power Control

LTE defines different power control procedures for both downlink and uplink transmissions [25]. Energy per resource element is determined by downlink power control. Average powers to be used on different physical uplink channels are determined by the uplink power control. Both open-loop and closed-loop power control schemes are supported in LTE to combat against deep fading, the near-far effect, and multiuser and inter-cell interference. Power control is an effective mechanism to ensure a certain level of bit error rate (BER) regardless of channel conditions.

12.3.5 Interference Avoidance

Target high data rates in the beyond 3G cellular systems require dense reuse of frequency with the obvious pitfall of having high inter-cell interference. Therefore, to realize the full potential of the OFDMA in a dense reuse environment, appropriate interference mitigation technique(s) has to be used. To that end, interference mitigation is regarded as one of the major issues to be investigated by different standardization bodies and forums focusing beyond 3G cellular systems.

Interference mitigation techniques are generally classified into three major categories such as interference cancelation, interference averaging, and interference avoidance. The benefits of these techniques are mutually exclusive, hence a combination of these approaches is likely to be used in the system. Interference avoidance is an RRM issue where restrictions in resource usage in terms of resource partitioning and power allocation are imposed [26, 27]. We provide a brief description of some methods of interference avoidance available in the literature in this section.

Interference avoidance using classical clustering technique [28], for example, a reuse of 3, may have been good enough for early networks focusing primarily voice service, however, they are not applicable to future systems envisioned to support ranges of high data rate applications. Recently, fractional frequency reuse (FFR) schemes have attracted enormous attention from the researchers in different standardization bodies and forums. A common example of FFR for a network with trisector base stations (BSs) is a blend of reuse factors of 1 and 3 in the cell-center and the cell-edge areas, respectively. In most of these schemes, higher power is allocated to the resources used for cell-edge user terminals (UTs). Partial frequency reuse (PFR) [29] and soft frequency reuse (SFR) [30] are two popular variations of the FFR schemes.

In SFR for three-sector cell sites, the cell-edge band, termed as a major band, uses one-third of the available spectrum, which is orthogonal to those in the neighboring cells, and forms a structure of a cluster size of 3. The cell-center band (i.e., the minor band) in any sector is the collection of frequencies used in the outer zones of neighboring sectors. These bands are assigned transmission powers, depending on the desired effective reuse factor while keeping the total transmission power fixed. Let us assume that $P^{(T)}$ is the total transmit power per sector, N is the total number of available PRBs, and α is a power amplification factor applied to the cell-edge band whose value is greater than 1. Then, the power per PRB is $P^{(T)}/N$ in the case of reuse factor of 1 without coordination and it is $\alpha P^{(T)}/N$ for the cell-edge band of the SFR scheme. For constant sector power, the power per PRB in the cell-center band of the SFR would have to be $P^{(T)}(3-\alpha)/2N$, giving a ratio of powers of minor to major bands as $(3-\alpha)/2\alpha$. The cell-edge band can be used in the cell center as well if it is unoccupied by the cell-edge UTs; however, the cell-center band is available to cell-center users only. This scheduling restriction implies that increasing the power ratio from 0 to 1 effectively moves the reuse factor from 3 to 1. Hence, SFR is a compromise between reuses 1 and 3 in a network with trisector BSs. Users have to be categorized into cell edge and cell center based on user geometry determined by the received signal power (averaged over multipath fading) taking into account the large-scale pathloss, shadowing, and antenna gains.

The idea of PFR, as first presented in [31], is to restrict some resources so that a portion of available frequencies is not used in some sectors at all. The PFR and some of its variants are studied in the 3GPP and WINNER projects (see, e.g., [26, 29]). The effective reuse factor of this scheme depends on the amount of unused frequency. Let us assume available system bandwidth to be β, which is divided into inner and outer zones with β_1 and β_2, respectively. Here, β_1 and β_2 are used with reuse factors of 1 and 3 in the inner and the outer zones, respectively. Then, the effective frequency reuse factor is expressed by $\beta/[\beta_1 + (\beta_2/3)]$. Power used on PRBs in the outer zone is usually amplified.

In the literature, most FFR schemes rely on static or semistatic coordination among BSs; it is seen that such static or semistatic FFR schemes do not provide much gain, as the cell-edge throughput can only be improved with severe penalty to the system throughput [26]. In addition, such schemes requiring frequency planning cannot be applied to networks using femto BSs [32], as femto BSs are expected to be placed at the end user locations in an ad hoc manner, which makes prior frequency planning difficult. Dynamic coordination-based schemes, on the other hand, do not require frequency planning and are based on dynamic interference information from surrounding transmitters. As a result, dynamic avoidance schemes are not only more effective to avoid interference in macrocell-macrocell scenario, they are capable of handling interference from macrocells if applied to femto BSs. Dynamic inter-cell coordination-based schemes can best exploit channel dynamics to achieve maximum interference avoidance gain; however, only a few such studies can be found in the literature [33, 34]. It has been shown that the dynamic interference avoidance schemes provide enhanced cell-edge throughput without impacting overall cell throughput.

12.3.6 RRM Techniques for Multihop OFDMA Networks

So far, we have discussed RRM techniques for single-hop wireless networks (i.e., scenarios where the user terminals connect directly to the base station). However, because relays make the 4G networks multihop networks, it is also necessary to examine multihop RRM techniques. In the multihop wireless networks, the RRM needs to consider network load balancing and end-to-end delay, which were are not issues in single-hop wireless networks. Both load balancing and delay are decided with multihop OFDMA scheduling.

We note that the scheduling access part of the network is different from the backhaul part of the network in terms of the wireless channel and the offered traffic. The wireless channel in the backhaul varies more slowly than the wireless channel in the access network. Traffic patterns also change more slowly in the backhaul than in the access network, due to the static nature of relays. Thus, backhaul RRM algorithms and the resulting RAN protocols can be more accurate, although they may be slower to converge, than in the access part of the network.

In general, multihop OFDMA scheduling is closely related to graph coloring, which is a computationally hard problem. Relationship to graph coloring is also present in cellular frequency assignment, where spatial reuse is required [35]. Nevertheless, if the end-to-end scheduling delay is fixed, finding multihop OFDMA schedules takes polynomial time [36] and can be easily distributed [37]. Scheduling delay occurs when packets arriving on an inbound link must wait to be transmitted on the outbound link and can be large on multihop paths. Because high data rate OFDMA-based MACs are scheduled, the end-to-end delay depends on scheduling only. Without getting into details of schedule OFDMA networks, they are "stop-and-go" queuing networks [38], thus traffic delay can be controlled at the ingress part of the network and does not vary with competing end-to-end traffic.

The scheduled operation over multiple hops also means that hop-by-hop load balancing is achieved implicitly by simply forwarding traffic. Hop-by-hop load balancing is required for multiple path routing, which simplifies network management. The lack of multipath routing in wired networks is a major reason why many wired network traffic management problems are difficult [39]. In wired networks, network traffic management optimization must, in addition to optimizing end-to-end traffic, ensure that all traffic only traverses one path between the source and the destination. The requirement on the solution to only use one path makes the optimization a more difficult "unsplittable flow" problem [40]. Due to the use of scheduled MACs in the advanced RAN, a networkwide RRM can be simplified with implicit load balancing, which allows multiple path routing. In a wireless network, a network layer solution using multiple paths also benefits from using multiple radio ports, thus increasing diversity.

Thus, one can consider optimization problems, which result in multipath routed solutions such as cross-layer optimization techniques [41, 42]. Formally, a cross-layer RRM optimization is:

$$\max_{\substack{x_1, \ldots, x_m \in \mathcal{S} \\ x_1, \ldots, x_m \in \mathcal{N}}} \sum_{l=1}^{m} U_l(x_l) \tag{12.3}$$

where $x_1, \ldots, x_m$ are the rates of the m users in the network, $U_l(\cdot)$ is the utility of user l, and the optimization maximizes the total system utility subject to the existence of user rates $x_1, \ldots, x_m \in \mathcal{S}$ and $x_1, \ldots, x_m \in \mathcal{N}$ where $\mathcal{S}$ is the set of all m-tuples of "schedulable" end-to-end rates and $\mathcal{N}$ is the set of all m-tuples of "routable" end-to-end rates. Only the schedulability constraint is encountered in single-hop RANs. The networking constraint is required for multi-hop RANs.

The utility function is chosen to represent the "satisfaction" of each user with the service (rate) he is getting. There are many utility functions that correspond to different types of user satisfaction with the network. With a proper choice of utility functions [43], one may have an optimization that maximizes the total "weighted proportional fairness," a game theoretic optimum, "max-min fairness," which eliminates starvation, or simply "maximum total throughput." The utility function can also be chosen to represent the satisfaction of the network operator with the rates "maximum area spectral efficiency" or "maximum profit" may be one utility for the system are examples of such utilities. It is also possible to have utility functions that take the combination of traffic and profit into account [44].

Because cross-layer optimizations for 4G networks combine the areas of classical network research and classical wireless research, they are currently a "hot-topic" in the wireless network research.

12.4 Advanced RANs for Beyond-4G Networks

We discussed the architectures of 3G and 4G RANs in Section 12.2, and we now discuss advanced RAN architectures, which are becoming the community consensus as RANs for beyond-4G wireless networks. This section describes the elements of the advanced RAN architectures, which can cost effectively implement dense radio port coverage, and to differentiate these elements from the elements of the classical cellular RAN. We first motivate the need for a new RAN architecture by showing that high data rates can only be achieved by decreasing the distance between the transmitter and the receiver. Then, we propose a new RAN architecture, which provides cost-effective dense radio port coverage. Finally, we discuss some RRM techniques and open issues for advanced RAN architectures.

Our motivation for proposing a new RAN architecture comes from the fundamental laws of wireless transmission, under which wireless signals attenuate with distance from the transmitter. As the receiver distances from the transmitter, it has a lower received signal power, which lowers its peak data rate. The only way to solve this problem in a conventional cellular network is to increase the density of the base stations in the system's coverage area, which decreases the distance between the base stations and the mobile terminals. However, this approach is not cost effective, so a new RAN architecture is needed.

The well-known Shannon capacity formula, adjusted for spatial multiplexing, shows that the achievable rate is limited by the number of antennas available at the transmitter and the receiver, signal-to-noise ratio (SNR) of the received signal, and the bandwidth used for the transmission

$$R \leq C = n^{(a)} W \log_2 (1 + \text{SNR}) = n^{(a)} W \log_2 \left(1 + \frac{g P^{(T)}}{N^{(T)} N_0 W}\right) \tag{12.4}$$

where

$$\text{SNR} \triangleq \frac{g P^{(T)}}{N^{(T)} N_0 W} \tag{12.5}$$

where SNR is measured at the receiver, R is the user's rate, C is the upper limit on the rate (the Shannon capacity), $n^{(a)} = \min\{N^{(T)}, N^{(R)}\}$ is the minimum of the antennas available at the transmitter $[N^{(T)}]$ and at the receiver $[N^{(R)}]$, W is the bandwidth used by the signal, g is attenuation of the signal transmitted with power $P^{(T)}$, and N_0 is the noise power spectral density at the receiver. Here, we refer to $n^{(a)}$ as spatial multiplexing gain, as it comes from the use of MIMO techniques. We note that SNR has a one-to-one correspondence with capacity when the spatial multiplexing gain, $n^{(a)}$, and the signal bandwidth, W, are fixed.

At first glance, it appears that there are many ways to increase the achievable data rate: the number of antennas can be increased with the corresponding increase in the spatial multiplexing gain, the bandwidth can be increased, or the transmitter power can be increased. However, neither of these methods is very effective. The spatial multiplexing gain is limited by the size of the user device and at the writing of this report is limited with $N_T = 8$ and $N_R = 4$ [3]. The number of transmitter antennas is also is in the argument of the logarithm function, decreasing the energy available to transmission. Increasing the bandwidth, W, is also ineffective because it decreases the SNR in the logarithmic part of the equation. In practice, bandwidth is also limited by licensing issues [45, pp. 15–18]. For example, the next generation of LTE—LTE-advanced—limits the bandwidth up to 100 MHz [4, 18]. Similar to the limits on bandwidth, maximum transmit power is also regulated. Nevertheless, increasing power is not effective due to the logarithmic relationship and cost of amplifier design for high signal powers and terminal battery limitations.

We conclude that, for ubiquitous high data rate coverage needed for beyond-4G networks, it is important to decrease the distance between the base station and the user terminals. However, using the approach of scaling the cells is not cost effective, requiring alternative, cost-effective approaches. These approaches lead to advanced RAN architectures, which we discuss next.

12.4.1 Advanced RANs for Beyond 4G

In classical cellular RAN architecture, there is essentially one network element—the base station. As we have shown, increasing the density of radio ports by increasing the number of base stations is not practical. A consensus is currently forming in the community about the next generation of advanced RAN architecture, which contains many other network elements, such as distributed antenna elements, femto BSs, and relays. Indeed relays are already part of the WiMAX 4G standard and are considered for addition into the LTE-advanced 4G standard. The new elements are to provide a high density of radio ports to (1) decrease the distance to the receivers and (2) enable new coordinated multipoint (CoMP) transmission and reception techniques, which promise high data rates. Here, we refer to the classical base station as a full base station (full BS) to distinguish it from a femto BS.

In advanced RAN [46, 47], elements other than the base station either do not implement all of the functionality of the base station, or are not directly connected to the Internet. The elements in the new RAN work together to provide dense radio port coverage (Figure 12.5). The radio ports are attached to the various elements used throughout the RAN: full BSs, femto BSs, and relays. A full BS is a gateway to the Internet for multiple RAN elements, whereas relays connect to the full-BS station with wireless connections. Femto BSs connect to the RAN through the Internet and provide indoor coverage.

The base station (RAN anchor) is an important element of the advanced RAN. It manages multiple radio ports and has a wired connection to the Internet. RAN anchors do not require radio resources to provide backhaul services. We distinguish two types of RAN anchors: full base station (full BS) and femto base station (femto BS). A full BS is a gateway to the Internet for multiple RAN elements, whereas a femto BS is a gateway to the Internet for indoor elements.

In addition to the various types of base stations, RAN also uses many types of relays. Unlike base stations, which are directly connected to the Internet, relays

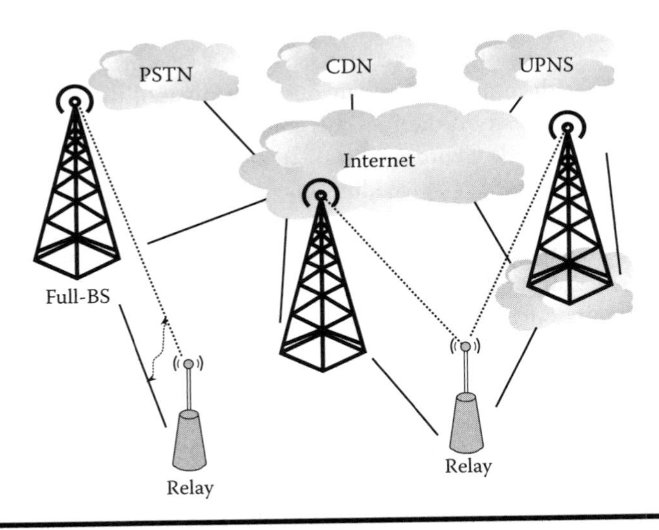

Figure 12.5 Potential beyond-4G RAN architecture.

connect to the Internet through direct wireless connections to RAN anchors or through multihop wireless connections over other relays, which connect directly to RAN anchors. A relay may have multiple radio ports attached to it, as a base station would and may have to participate in hand-off and other RRM procedures, as a base station would. However, a relay is not connected to the RAN with a wired connection—it must at least connect to a base station to get to the Internet and it may connect to the base station in the network layer by using multihop transmissions through other relays.

The advanced RAN contains various types of relays, which vary in complexity. For example, a relay may be a fairly simple amplify-and-forward relay, which does not examine the data flow, or a much more complex decode-and-forward relay, which examines and forwards packets. Typically, a relay is also expected to have a shorter range than a base station, so it requires a lower power amplification and thus cheaper power amplifier than a base station. Cheaper power-amplifier circuitry also makes relays cheaper than a base station, from an engineering point of view.

The essential part of the proposed RAN are radio ports, which perform the radio transmission. The radio ports are available densely throughout the RAN coverage area, so that the distance between the terminal and a radio port is always small. Because radio ports are deployed densely throughout the RAN coverage area, it may be possible for the user terminals to simultaneously send (and receive) radio signals to (and from) multiple radio ports. Similar technologies are already proposed for LTE-advanced is OFDMA macrodiversity, also known as coordinate multipoint transmission (CoMP) [18]. However, CoMP has to be back compatible to the existing LTE standard, so it may not be able to take full advantage of the high port density.

In advanced RAN, the port density will have to be much higher to achieve rates in the range of tens of gigabytes per second, so new precoding technologies will be necessary. With multiple simultaneous transmission, the terminal can take advantage of spatial diversity if the transmissions and receptions through the multiple points is coordinated. Coordination of transmissions and receptions leads to potentially higher rates [48], as precoding may be done to take spatial properties channel of the distributed channel. We note that precoding takes the concept of macrodiversity one step further, beyond simple signal combining. The concept of CoMP transmission and reception has many names in the literature, such as "distributed antenna ports" [49, 50], and in the standardization process "multicell MIMO," "network MIMO," and "network cooperative MIMO" [8], to mention a few.

The proposed RAN is a mesh of RAN elements, where any one element can connect to any other element. Due to the flat hierarchy in the RAN, RRM does not belong to any given RAN element. In the proposed RAN, RRM is a networkwide set of protocols and algorithms that allow network elements with different capabilities to negotiate assignments of radio resources to users.

12.4.2 Open Issues in RRM Optimization in Advanced RANs

So far, we have shown that advanced RAN architectures cost effectively facilitate dense deployments of radio ports in the RAN coverage area. To take full advantage of the radio port density, advanced RAN needs advanced RRM algorithms and related protocols. We now discuss issues, that should be considered in optimizing RRM of advanced RANs [47]:

- **Need for User-Centric RRM**: Ubiquitously high data rates require moving away from the system-centric radio resource management (RRM), used in cellular RAN architecture, and toward advanced user-centric RRM. System-centric RRM uses a divide-and-conquer approach, which assigns resources to cells first and to users second. On the other hand, user-centric RRM assigns resources to users first and then finds radio ports to provide these resources. By first assigning resources to users, user-centric RRM solves a fundamental inefficiency with the system-centric RRM—in system-centric RRM, resources are *assigned* to cells, while they are *used* by users. Because the user may be multiple radio hops away from the core network, user-centric RRM needs to consider cross-layer RRM techniques between the network and lower layers in the architecture.
- **Top-Down Protocol Design**: Recent research in top-down optimizable protocol design can also be used to devise user-centric RRM algorithms and RAN protocols, which implement them. Top-down protocol design starts with a global utility maximization problem, which optimizes the rates of individual users. The objective function of optimization is chosen so that, at the optimum (equilibrium) point, user rates satisfy some criteria specified by the network operator. Optimization is subject to the availability of radio resources, which can support the optimum user rates. Then, one uses mathematical decomposition to devise a distributed algorithm and a corresponding protocol, which solves the problem. We believe that this approach can be successfully used for design of advanced RRM algorithms and suitable RAN protocols for the advanced RANs.
- **Restrictions on OFDMA Schedules**: In general, OFDMA scheduling is closely related to graph coloring, which is a computationally hard problem. However, under some circumstances, finding schedules takes polynomial time [36] and can be easily distributed [37]. Nevertheless, to take advantage of distributed scheduling protocols, one should design RAN medium access control (MAC) protocols that allow easy OFDMA scheduling.

12.5 Summary

In this chapter, we first reviewed 3G and 4G OFDMA-based architectures. We concentrated on the evolution of LTE and LTE-advanced RAN architecture. We showed that the trend in architecture development is toward decentralized RAN architectures, where little information is exchanged through a central point in the RAN. Decentralization of the LTE RAN architecture is required by the OFDMA-based physical layer, which allows flexible frequency and time assignment in the cells. Without decentralization, the network would be too slow to adjust to varying user demands. We have also reviewed RRM algorithms for OFDMA-based 4G RANs and state-of-the art approaches to assign radio resources to the users.

Then, we argued that a new RAN architecture is required for beyond-4G RANs. Although current OFDMA-based RANs provide peak data rates in the order of hundreds of megabytes per second, we expect beyond-4G architectures to provide data rates in the tens of gigabytes per second. The fundamental problem with current RAN architecture is that it is based on the cellular RAN concept, which requires scaling down the RAN by introducing more base stations into the network. We reviewed an advanced RAN architecture, which seems to be the current consensus in the community. Advanced RAN uses advanced RAN elements such as distributed antenna ports, femto BSs, and relays and uses CoMP transmission and reception techniques. This advanced RAN architecture needs advanced RRM algorithms and protocols. We also reviewed open issues in advanced RRM optimization for advanced RAN architectures.

References

[1] 3GPP TS 36.211, "Evolved universal terrestrial radio access (E-UTRA); physical channels and modulation," 2008.

[2] IEEE P802.16Rev2/D7, "IEEE draft standard for local and metropolitan area networks part 16: Air interface for fixed broadband wireless access systems," 2008.

[3] Report ITU-R M.2134, "Requirements related to technical performance for imt-advanced radio interface(s)," 2008.

[4] 3GPP TR 36.913, "Requirements for further advancements for E-UTRA (LTE-advanced)," 2008.

[5] Recommendation ITU-R M.1645, "Framework and overall objectives of the future development of IMT-2000 and Systems Beyond IMT-2000," 2003.

[6] Y. Zhou and N. Zein, "Simulation study of fractional frequency reuse for mobile WiMAX," *Proceedings of the IEEE Vehicular Technology Conference*, pp. 2592–2595, 2008.

[7] M. Rahman, H. Yanikomeroglu, and W. Wong, "Interference avoidance with dynamic inter-cell coordination for downlink LTE systems," Proceedings of the 2009 IEEE conference on Wireless Communications and Networking Conference, pp. 1238–1243, April 2009.

[8] Ericsson Research, "3GPP TSG-RAN WG1 ♯53 R1-082024: A discussion on some technology components of LTE-advanced," Kansas City, MO, May 2008.

[9] V. Chandrasekhar, J.G. Andrews, and A. Gatherer, "Femtocell networks: A survey," *IEEE Communications Magazine*, vol. 46, no. 9, pp. 59–67, September 2008.

[10] S.-P. Y. S. Talwar, S.-C. Lee, and H. Kim, "WiMAX femtocells: A perspective on network architecture, capacity, and coverage," *IEEE Communications Magazine*, vol. 46, no. 10, pp. 58–65, October 2008.

[11] IEEE Draft Standard P802.16j/D5, part 16, "Air interface for fixed and mobile broadband wireless access systems—multihop relay specification," May 2008.
[12] F.P. Kelly, A. Maulloo, and D. Tan, "Rate control in communication networks: shadow prices, proportional fairness and stability," *Journal of the Operational Research Society*, vol. 49, no. 3, pp. 237–252, March 1998.
[13] F.P. Kelly, "Fairness and stability of end-to-end congestion control," *European Journal of Control*, vol. 9, pp. 159–176, 2003.
[14] 3GPP TS 25.401 V8.2.0, "UTRAN overall description (Release 8)," 2008.
[15] E. Dahlman, S. Parkvall, J. Sköld, and P. Beming, *3G Evolution HSPA and LTE for Mobile Broadband Phase-Locked Loops: Theory, Design and Applications*, 2nd ed. Academic Press, London, 2008.
[16] H.G. Myung, J. Lim, and D.J. Goodman, "Single carrier FDMA for uplink wireless transmission," *IEEE Vehicular Technology Magazine*, vol. 1, no. 3, pp. 30–38, September 2006.
[17] 3GPP TS 36.300 V8.7.0, "E-UTRAN overall description stage 2 (Release 8)," 2008.
[18] S. Parkvall, E. Dahlman, A. Furuskär, Y. Jading, M. Olsson, S. Wänstedt, and K. Zangi, "LTE-advanced—Evolving LTE towards IMT-advanced," *Proceedings of the 68th Vehicular Technology conference*, VTC 2008-Fall. IEEE 68th, pp. 1–5, September 2008.
[19] J. N. Laneman, D.N. Tse, and G.W. Wornell, "Cooperative diversity in wireless networks: Efficient protocols and outage behaviour," *IEEE Transactions on Infinity Theory*, vol. 50, no. 12, pp. 3062–3080, December 2003.
[20] F. Adachi, M. Sawahashi, and K. Okawa, "Tree-structured generation of orthogonal spreading codes with different lengths for forward link of DS-CDMA mobile radio," *Electronics Letters*, vol. 33, no. 1, pp. 27–28, January 1997.
[21] 3GPP TR 25.855 V5.0.0, "Technical report 3rd generation partnership project; technical specification group radio access network; high speed downlink packet access; overall UTRAN description (Release 5)," 2001.
[22] P. Viswanath, D. N. C. Tse, and R. Laroia, "Opportunistic beamforming using dumb antennas," *IEEE Transactions in Infinity Theory*, vol. 48, no. 6, pp. 1277–1294, June 2002.
[23] R. Srinivasan, J. Zhuang, L. Jalloul, R. Novak, and J. Park, "IEEE 802.16m-07/037r2: Draft IEEE 802.16m evaluation methodology," 2007.
[24] 3GPP TS 36.212, "Evolved universal terrestrial radio access (E-UTRA); multiplexing and channel coding," 2008.
[25] 3GPP TS 36.213, "Evolved universal terrestrial radio access (E-UTRA); physical layer procedures," 2008.
[26] *Interference Avoidance Concepts*, WINNER II Deliverable D4.7.2, June 2007. [Online]. Retrieved from http://www.ist-winner.org.on.
[27] *Physical Layer Aspects for Evolved Universal Terrestrial Radio Access (UTRA) (Release 7)*, 3GPP Std. TR 25.814 V7.1.0, October 2006. [Online]. Retrieved from http://www.3gpp.org. (retriwved on April 20, 2010).
[28] V. MacDonald, "The cellular concept," *Bell System Technical Journal*, vol. 58, pp. 15–41, January 1979.
[29] *OFDMA Downlink Inter-Cell Interference Mitigation*, 3GPP Project Document R1-060 291, February 2006. [Online]. Retrieved from http://www.3gpp.org.
[30] *Soft Frequency Reuse Scheme for UTRAN LTE*, 3GPP Project Document R1-050 507, May 2005. [Online]. Retrieved from http://www.3gpp.org.
[31] M. Sternad, T. Ottosson, A. Ahlen, and A. Svensson, "Attaining both coverage and high spectral efficiency with adaptive OFDM downlinks," *Proceedings of the IEEE Vehicular Technology Conference*, pp. 2486–2490, October 2003.

[32] V. Chandrasekhar, J. Andrews, and A. Gatherer, "Femtocell networks: a survey," *IEEE Communications Magazine*, pp. 59–67, September 2008.

[33] M. Rahman and H. Yanikomeroglu, "Interference avoidance through dynamic downlink OFDMA subchannel allocation using inter-cell coordination," in *Proceedings of the IEEE Vehicular Technonogy Conference*, pp. 1630–1635, May 2008.

[34] *Further Discussion on Adaptive Fractional Frequency Reuse*, 3GPP Project Document R1-072762, June 2007. [Online]. Retrieved from http://www.3gpp.org. (retrieved on April 20, 2010).

[35] I. Katzela and M. Naghshineh, "Channel assignment schemes for cellular mobile telecommunication systems: A comprehensive survey," *IEEE Personal Communications Magazine*, vol. 3, no. 3, pp. 10–31, June 1996.

[36] P. Djukic and S. Valaee, "Link scheduling for minimum delay in spatial re-use TDMA," in *26th Proceedings of the international conference on computer communications*, pp. 28–36, May 2007.

[37] P. Djukic and S. Valaee, "Distributed link scheduling for TDMA mesh networks," in *Proceedings of the IEEE International conference on Communications*, 2007.

[38] S. J. Golestani, "A framing strategy for congestion management," *IEEE Journal on Selected Areas in Communications*, vol. 9, no. 7, pp. 1064–1077, September 1991.

[39] J. He, J. Rexford, and M. Chiang, "Don't optimize existing protocols, design optimizable protocols," *ACM SIGCOMM Computer Communication Review*, vol. 37, no. 3, pp. 53–58, 2007.

[40] Y. Dinitz, N. Garg, and M.X. Goemans, "On the single-source unsplittable flow problem," Combinatorica, vol. 19, no. 1, pp. 17–41, April, 1999.

[41] X. Lin, N.B. Shroff, and R. Srikant, "A tutorial on cross-layer optimization in wireless networks," *IEEE Journal on Selected Areas in Communications*, vol. 24, no. 8, pp. 1452–1463, August 2006.

[42] M. Chiang, S. H. Low, A. R. Calderbank, and J. C. Doyle, "Layering as optimization decomposition: A mathematical theory of network architectures," *Proceedings of the IEEE*, vol. 95, no. 1, pp. 255–312, 2007.

[43] J. Mo and J. Walrand, "Fair end-to-end window-based congestion control," *IEEE/ACM Transactions on Networking*, vol. 8, no. 5, pp. 556–567, 2000.

[44] A. Elwalid, D. Mitra, and Q. Wang, "Distributed nonlinear integer optimization for data-optical internetworking," *IEEE Journal on Selected Areas in Communications*, vol. 24, no. 8, pp. 1502–1513, August 2006.

[45] "3GPP TS 25.101 V8.4.0: 3rd generation partnership project; technical specification group radio access network; user equipment (UE) radio transmission and reception (FDD)," 2008.

[46] H. Yanikomeroglu and J. Zhang, "Beyond-4G cellular networks: Advanced radio access network (RAN) architectures, advanced radio resource management (RRM) techniques, and other enabling technologies," *Wireless World Research Forum Meeting 21*, October 2008.

[47] P. Djukic, H. Yanikomeroglu, and J. Zhang, "User-centric RRM and optimizable protocol design for beyond-4G rans," *Wireless World Research Forum Meeting*, 22 May 2009.

[48] Ericsson Research, "3GPP TSG-RAN WG1 ♯53bis R1-082469 LTE-Advanced–coordinated multipoint transmission/reception," Warsaw, Poland, 2008.

[49] L. Dai, *Distributed Antenna Systems: Open Architecture for Future Wireless Communications.* CRC Press, 2007, ch. "Optimal Resource Allocation of DAS," pp. 169–200.

[50] X.-H. Yu, G. Chen, M. Chen, and X. Gao, "The Future project in China," *IEEE Communications Magazine*, Piscataway, NJ, USA, vol. 43, no. 1, pp. 70–75, January 2005.

Chapter 13

Physical Uplink Shared Channel (PUSCH) Closed-Loop Power Control for 3G LTE

Bilal Muhammad and Abbas Mohammed

Contents

13.1 LTE Physical Layer

13.1.1 Multiplexing Schemes

The capabilities of the evolved NodeB (eNodeB) and user equipment (UE) are obviously quite different; thus, the LTE (long-term evolution) physical layer (PHY), downlink (DL) and uplink (UL) are different.

13.1.1.1 Downlink

Orthogonal frequency division multiplexing (OFDM) is selected as the basic modulation scheme because of its robustness in the presence of severe multipath fading. Orthogonal frequency division multiple access (OFDMA) is used as the multiplexing scheme in the downlink.

13.1.1.2 Uplink

LTE uplink requirements differ from downlink in several ways. Power consumption is a key consideration for UE terminals. High peak-to-average power ratio (PAPR) and related loss of efficiency with OFDM signaling are major concerns. As a result, an alternative to OFDMA was sought for use in the LTE uplink.

The LTE PHY uses single-carrier frequency division multiple access (SC-FDMA) as the basic transmission scheme for the uplink. SC-FDMA is a modified form of OFDMA. SC-FDMA has a similar throughput performance and an overall

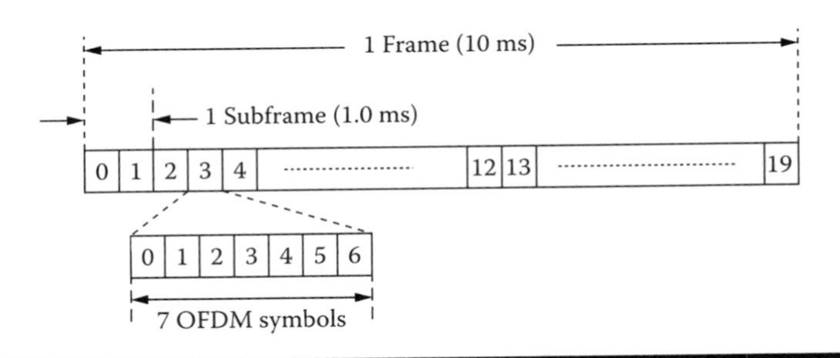

Figure 13.1 LTE generic frame structure.

complexity as OFDMA. The principle advantage of SC-FDMA is a lower PAPR than OFDMA.

13.1.2 Frame Structure

In an uplink, LTE transmissions are segmented into frames [1]; a frame consists of 10 subframes and a subframe is formed by two slots each 0.5 ms long and consists of seven SC-FDMA symbols. The generic frame structure is shown in Figure 13.1.

13.1.3 Physical Resource Block

In LTE, the physical resource block (PRB) is the smallest element of resource allocation assigned by the base station scheduler. A PRB is defined as a resource of 180 kHz in the frequency domain and 0.5 ms (one time slot) in the time domain. Because the subcarrier spacing is 15 kHz, each PRB consists of 12 subcarriers in the frequency domain, as shown in Figure 13.2.

13.1.4 Reference Signals

In contrast to packet-oriented networks, LTE does not use a PHY preamble to facilitate carrier offset estimate, channel estimation, or timing synchronization. Instead, special reference signals are embedded in the PRBs, as shown in Figure 13.3.

Reference signals are transmitted during the first and fifth OFDM symbols of each time slot and in every sixth subcarrier of each subframe. Reference signal is also used to estimate the path loss using reference symbol received power (RSRP).

13.2 Introduction to Power Control

The uplink transmitter power control is a key radio resource management feature in cellular communication systems. It is usually used to provide an adequate transmit power to the desired signals to achieve the necessary quality, minimizing interference

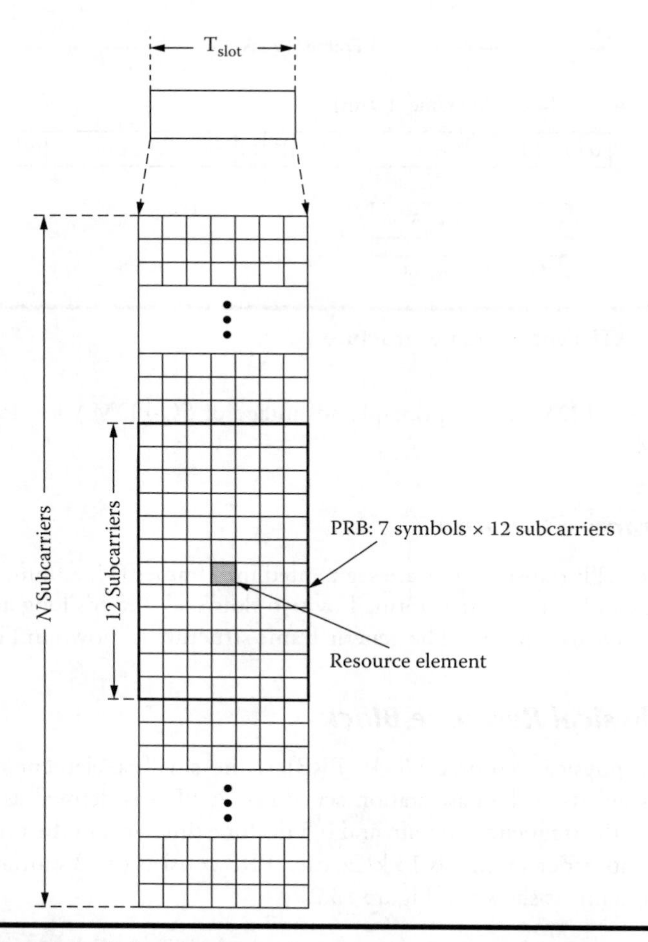

Figure 13.2 LTE uplink resource grid.

to other users in the system and maximizing the battery life of the mobile terminal. To achieve these goals, uplink power control has to adapt to the radio propagation channel conditions, including path loss, shadowing, and fast fading fluctuations while limiting the interference effects from other users, within the cell, and from neighboring cells.

LTE is a project within the Third-Generation Partnership Project (3GPP) that aims at improving the current 3G universal mobile telecommunications system (UMTS) standard to cope with future requirements and to maintain competitiveness in the long term. Hence, the important goals of LTE include achieving higher data rates, reducing latency, improving efficiency, enhancing services, exploiting new spectrum opportunities, improving system capacity and coverage, lowering costs, and better integration with other standards.

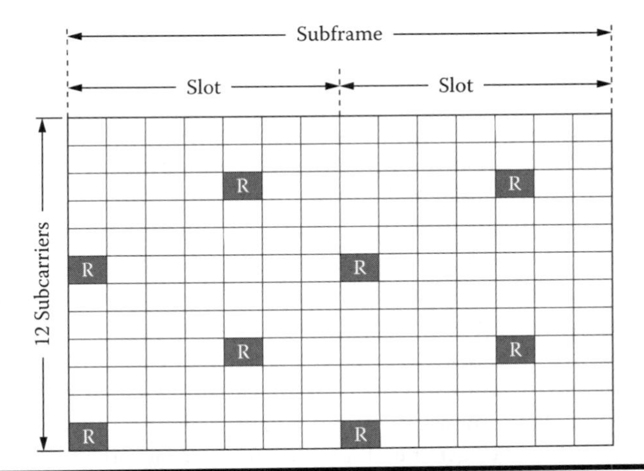

Figure 13.3 Resource element mapping of the reference signals for a single antenna only.

In order to meet some of these requirements, SC-FDMA has been chosen as the uplink radio access technology in LTE [2]. SC-FDMA has a low PAPR, which leads to lesser power consumption at the UE.

In a multiuser environment, a number of users share the same radio resources. A consequence of the limited availability of radio channels in the network is that the same channel has to be assigned to many users. Thus, a signal intended for a certain user will reach other users, introduce interference to their connection, and degrade the system quality. A user with very good quality may consider using low power and still having acceptable quality. The advantage is that it will disturb other users less, thereby improving their quality. Power control is essentially doing the same task but in a controlled manner.

Using an orthogonal transmission scheme eliminates mutual interference between users in the same cell (intra-cell interference) and the near-far problem typically encountered in CDMA systems. However, because transmission in the neighboring cell is not orthogonal, there is interference between users in neighboring cells (inter-cell interference), which ultimately limits system performance and capacity.

To maximize the spectral efficiency in LTE, a frequency reuse of 1 is selected for both the downlink and the uplink [2], which means that all cells in the network use the same frequency band. Thus, both data and control channels are sensitive to inter-cell interference. Cell-edge performance and capacity of a cell site can be limited by the inter-cell interference. Therefore, the role of a closed-loop power control becomes decisive because it ensures that the required SINR is maintained at an acceptable level of communication between the eNB and the UE while controlling interference caused to neighboring cells.

In addition, in the coming years, many portable devices (e.g., notebooks, ultraportables, gaming devices, and video cameras) are also expected to operate over mobile broadband technologies such as LTE. The battery power is an important and scarce resource in these devices. Thus, the application of an efficient power control mechanism is crucial in order to minimize the consumption of battery power and use the available resources efficiently.

The foregoing requirements are used in the LTE physical uplink shared channel (PUSCH) power control scheme [2], which is a combination of an open-loop and closed-loop mechanism. The scheme allows for full or partial compensation (of path loss and shadowing) as opposed to the conventional uplink power control scheme (full compensation) in which all users receive the same SINR [3, 4]. The open-loop component compensates for the slow channel variations based on signal strength measurements performed by the terminal, which reduces the power to cell-edge users since they are likely to generate higher interference to others. The closed-loop component, on the other hand, directly controls the terminals power using explicit transmit power control (TPC) commands in the downlink to optimize the system performance. The potential benefit of fractional path loss compensation is a relatively lower transmitted power for terminals closer to the cell border, implying less interference to other cells. However, this also leads to reduce data rates for those terminals.

This chapter presents a novel closed-loop power control algorithm with fractional path loss compensation for the PUSCH for a 3GPP LTE system. In contrast to conventional closed-loop power control, the proposed scheme sets a SINR target for all users based on their path loss, which allows users with good radio conditions to achieve better SINR and, at the same time, providing a better cell-edge bit rate. Different values of the fractional path loss compensation factor (in the range 0.7–1.0) are tested, and an optimal value that provides the best cell-edge performance for a given SINR target is selected for further investigation. Realistic simulation scenarios are modeled by taking into account the mobility, delay, error, and power headroom reporting and performance results compared with the ideal case [5]. Simulation results show that closed-loop power control with a fractional path loss compensation factor is advantageous compared to closed-loop power control with full path loss compensation. Using a simple upload traffic model, the closed-loop power control with a fractional path loss compensation factor improved system performance in terms of mean bit rate by 68% in the ideal case and 63% in the realistic case. In addition, the proposed algorithm provides a better cell-edge bit rate and better battery life performance.

The chapter is organized as follows. In Section 13.3 PUSCH power control formula and basic PUSCH power control signaling is presented. In Section 13.4 we present LTE power control schemes and their comparison based on power spectral density. In Section 13.5, we present a proposed closed-loop power control algorithm, traffic models, and realistic simulation environments. Simulation results are presented in Section 13.6. Finally, Section 13.7 concludes the chapter.

13.3 LTE PUSCH Uplink Power Control

The power control scheme for the physical uplink shared channel (PUSCH) is the combination of an open-loop power control (OLPC) and closed-loop power control (CLPC). The 3GPP specifications [2] define the setting of the UE transmit power for PUSCH by the following equation:

$$P_{\text{PUSCH}} = \min\{P_{\max}, 10 \cdot \log_{10} M + P_0 + \alpha \times \text{PL} + \delta_{\text{mcs}} + f(\Delta_i)\} \text{ [dBm]} \tag{13.1}$$

where

$P_{\max}$ is the maximum allowed transmit power, which depends on the UE power class;
M is the number of physical resource blocks (PRB);
P_0 is a cell/UE-specific parameter signaled by the radio resource control (RRC). However, we assume that P_0 is cell specific;
α is the path loss compensation factor. It is a three-bit cell specific parameter in the range [0 1] signaled by the RRC;
PL is the downlink path loss estimate and is calculated in the UE based on the RSRP;
δ_{mcs} is a cell/UE-specific modulation and coding scheme parameter defined in the 3GPP specifications for LTE. The setting of δ_{mcs} is beyond the scope of this chapter;
$f(\Delta_i)$ is UE specific. Δ_i is a closed loop correction value, and f is a function that permits us to use an accumulated or an absolute correction value.

The parameter P_0 is calculated [6] as follows:

$$P_0 = \alpha \cdot (SNR_0 + P_n) + (1 - \alpha)(P_{\max} - 10 \cdot \log_{10} M_0) \text{ [dBm]} \tag{13.2}$$

where

SNR_0 is the open-loop target SNR (signal-to-noise ratio);
P_n is the noise power per PRB;
M_0 defines the number of PRBs for which the SNR target is reached with full power. It is set to 1 for simplicity.

13.3.1 PUSCH Power Control Signaling

Many of the parameters listed in Equation 13.1 are broadcasted by the eNB toward the UEs; that is, they are the same for all the users in that specific cell. Figure 13.4 shows the UE parameters (e.g., Δ_i, α, P_0, and δ_{mcs}) received from the eNB. In the

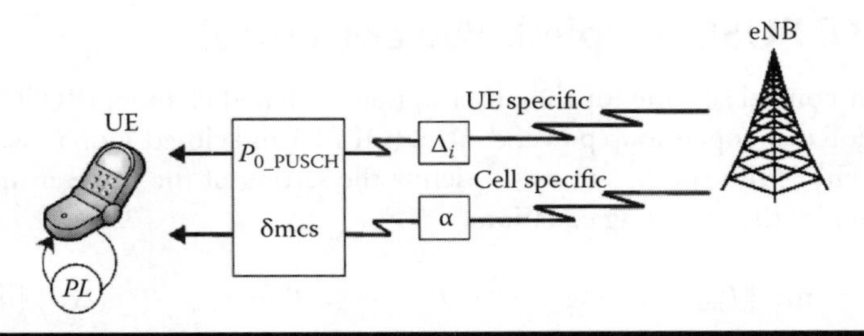

Figure 13.4 PUSCH power control parameters broadcast by the eNB toward UEs.

figure, PL is the estimate of the path loss calculated at the UE based on the reference symbol received power (RSRP). The figure also indicates that few parameters are cell-specific, implying that they vary from one UE to another. Cell-specific parameters indicate that they are same for all the UEs in that specific cell.

13.4 LTE Power Control Schemes

In this section, LTE power control schemes are discussed and different ways of categorizing them are presented.

13.4.1 Power Spectral Density (PSD)

The UE sets its initial transmission power based on parameters received from the eNB and the path loss calculated by the UE. It is worthwhile to note that Δ_i is signaled by the eNB to any UE after it sets its initial transmit power; that is, Δ_i has no contribution in the initial setting of the UE transmit power. The expression on which a UE sets its initial power can be obtained from Equation 13.1 by ignoring ∂_{mcs} and the closed-loop correction, whereas power limitation can be neglected because we assume that the UE has to take the power limitation into account, and is given by the following:

$$P_{PUSCH} = 10 \cdot \log_{10} M + P_0 + \alpha \times \text{PL [dBm]} \tag{13.3}$$

where M is the total number of PRB scheduled by the eNB. The power assignment for transmission at the UE is performed on the basis of PRB, and each PRB contains an equal amount of power. Thus, by neglecting M, the expression used by the UE to assign power to each PRB is given by

$$PSD_{T_x} = P_0 + \alpha \cdot \text{PL [dBm/PRB]} \tag{13.4}$$

Equation 13.4 represents the transmit power spectral density (PSD) of a PRB expressed in dBm/PRB. PSD_{T_x} is a helpful way to explain the basic difference between conventional and fractional power control.

The power control scheme can be categorized based on the value of α in Equation 13.4 as follows:

$\alpha = 1$ (full compensation of path loss), which is the well-known conventional power control scheme,
$0 < \alpha < 1$ (fractional compensation of path loss) turns to fractional power control,
$\alpha = 0$ (no compensation of path loss) leads to no power control; that is, all users will use the maximum allowed transmission power (P_{max}).

13.4.2 Conventional Power Control Scheme

If full compensation of path loss is used ($\alpha = 1$), then P_0 is given as

$$P_0 = SNR_0 + P_n \text{ [dBm]} \tag{13.5}$$

The PSD_{T_x} is thus defined as follows:

$$PSD_{T_x} = P_0 + PL = SNR_0 P_n + PL \text{ [dBm/PRB]} \tag{13.6}$$

Taking the path loss into account, the received PSD at the eNB is then given by

$$PSD_{R_x} = (SNR_0 + P_n) = P_0 \text{ [dBm/PRB]} \tag{13.7}$$

It is clear from Equation 13.7 that the received PSD at the eNB is equal to P_0, thus, this equation illustrates that the conventional power control scheme steers all users with equal power spectral density. This scheme is widely used in cellular systems that are not using orthogonal transmission scheme in the uplink, such as conventional CDMA-based systems. One of the advantages of this power control scheme is that it removes the near-far problem typically experienced by CDMA systems, as it equalizes power of all UEs received at the base station. Figure 13.5 shows the received PSD for users as a function of path loss. It can be clearly seen that, for a given SNR target, the received PSD is same for all users independent of their path loss. It is worthwhile to note that the "knee point" indicates the power limited region where users at this point and beyond will start to use P_{max}; in other words, it shows the maximum path loss that results in uplink power equal to P_{max} by the user. The knee point drifts to the left by increasing the SNR target (SNR_0); this means that users will be power limited more quickly. High SNR_0 mostly favors users close to the eNB, whereas a lower SNR_0 favors users at the cells' edge.

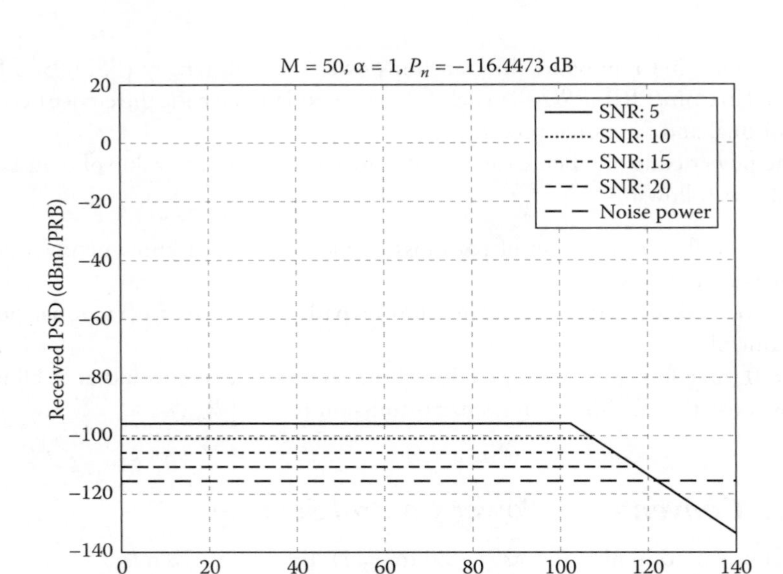

Figure 13.5 Plot of the received PSD for a conventional power control scheme.

13.4.3 Fractional Power Control Scheme

The fractional power control (FPC) scheme allows users to receive variable PSDs, depending on their path loss; that is, the user with good radio conditions will receive high PSD and vice versa. Using $0 < \alpha < 1$, the PSD is given by

$$\mathrm{PSD}_{T_x} = P_0 + \alpha PL = \alpha \cdot (SNR_0 + P_n) + (1 - \alpha)(P_{\max}) + \alpha PL \ [\mathrm{dBm/PRB}] \tag{13.8}$$

In contrast to conventional power control, which allows full compensation of path loss, FPC compensates only for a fraction of the path loss, hence the name. The PSD received can be found by taking the path loss in to account and is given by

$$PSD_{R_x} = P_0 + PL(\alpha - 1) \ [\mathrm{dBm/PRB}] \tag{13.9}$$

Attention is drawn here by comparing Equations 13.7 and 13.9 where the received PSD in a conventional power control scheme results in P_0, whereas in case of the FPC scheme it also has an additional term $PL(\alpha - 1)$. Because both P_0 and α are cell-specific parameters broadcast toward UEs by the eNB (same for all the UEs), then PL is the key factor in Equation 13.9 that allows users to be received with different power spectral densities. This observation can be explained by plotting Equation 13.9 as a function of the path loss (Figure 13.6). This figure clearly shows that, for the FPC scheme ($0 < \alpha < 1$), the users receive variable PSDs,

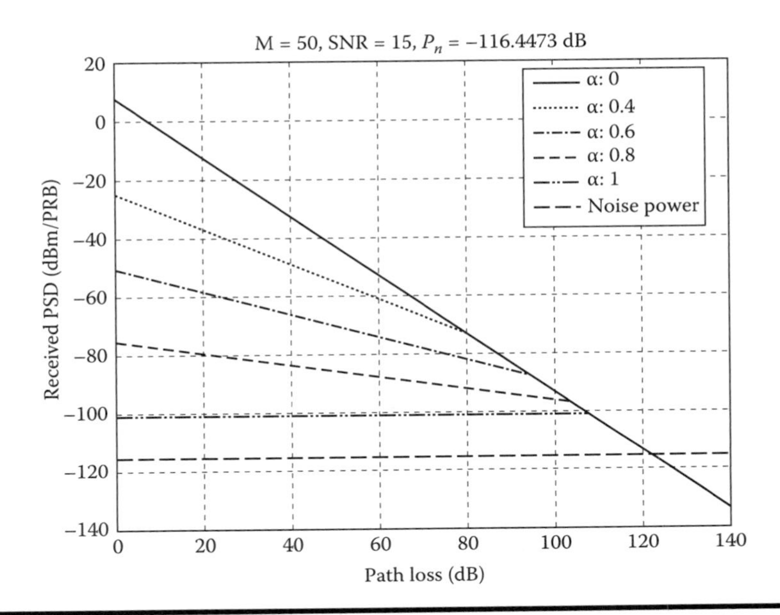

Figure 13.6 Plot of the received PSD for a fractional power control scheme.

depending on their path loss. The conventional or full compensation ($\alpha = 1$) and no compensation ($\alpha = 0$) power control schemes are also shown as benchmarks in this figure. Also notice that the knee point drifts toward the left by decreasing α and/or increasing SNR_0.

In Figure 13.6, $\alpha = 1$ (full compensation) and $\alpha = 0$ (no compensation) shows the conventional and no power control scheme, respectively; for $0 < \alpha < 1$ is the fractional setting where users are received with variable PSD depending on their path loss. The "knee point" drifts towards left by decreasing α and/or increasing SNR_0.

13.4.4 Slope of the Received PSD

In Section 13.4.1, the power control schemes are categorized based on the path loss compensation factor. However, the slope of the received PSD is another way of categorizing the different power control algorithms.

Figure 13.7 shows different slopes of the received PSD for various values of α ($0 \leq \alpha \leq 1$). For FPC ($0 < \alpha < 1$), the slope is $\alpha - 1$. If $\alpha = 0$, the slope is -1, which implies no power control, whereas if $\alpha = 1$ results in slope $= 0$, which leads to conventional open-loop power control. The figure also defines the region where users start using the maximum allowed power as identified by the knee point. At the knee point and beyond, users will experience maximum path loss, which results in using maximum power. The received PSD for the fractional power control is shown

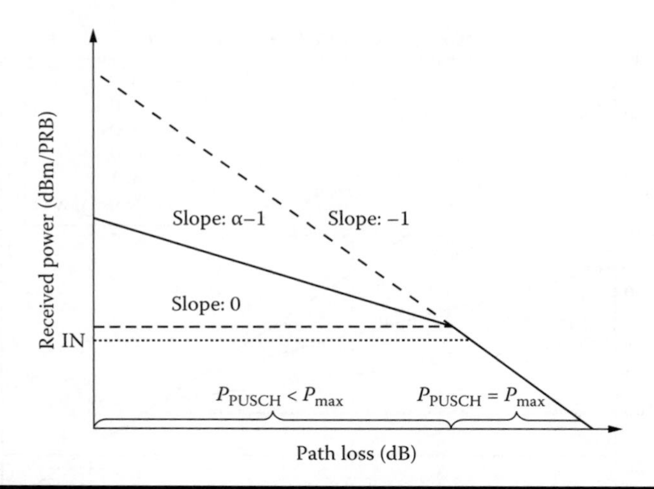

Figure 13.7 Illustration of the relationship between the received PSD and the path loss, describing the slopes for different power control algorithms.

by the solid line, and the received PSD for the conventional and no-power control schemes are shown by the dashed lines. For conventional open-loop power control, the received PSD is same for all the UEs, independent of their path loss as shown in Figure 13.7.

It is worthwhile to note that, conventional and fractional power control indicates the choice of value for α, whereas open-loop and closed-loop power control indicates the method of setting the UEs' transmission power.

13.4.5 Open-Loop Power Control

Open-loop power control is the means by which the UE transmitter is able to set its uplink transmit power to a specified value suitable for receiver operation. This setting, discussed in Section 13.4.1 is based on Equation 13.3, thus the uplink power (P_{OL}) set by the open-loop power controller can be written as follows:

$$P_{\mathrm{OL}} = \min(P_{\max}, 10 \cdot \log_{10} M + P_0 + \alpha \cdot \mathrm{PL}) \text{ [dBm]} \qquad (13.10)$$

The choice of α depends on whether a conventional or FPC scheme is used. Using $\alpha = 1$ leads to a conventional open-loop power control scheme, whereas $0 < \alpha < 1$ leads to fractional open-loop power control.

Figure 13.8 is a block diagram of the steps involved in setting the uplink transmit power using open-loop power control. An estimate of the path loss is obtained after measuring the reference symbol received power (RSRP). The calculation for transmission power is performed using Equation 13.10. The transmit block in the eNB represents the broadcast of parameters (P_0 and α) used in Equation 13.10.

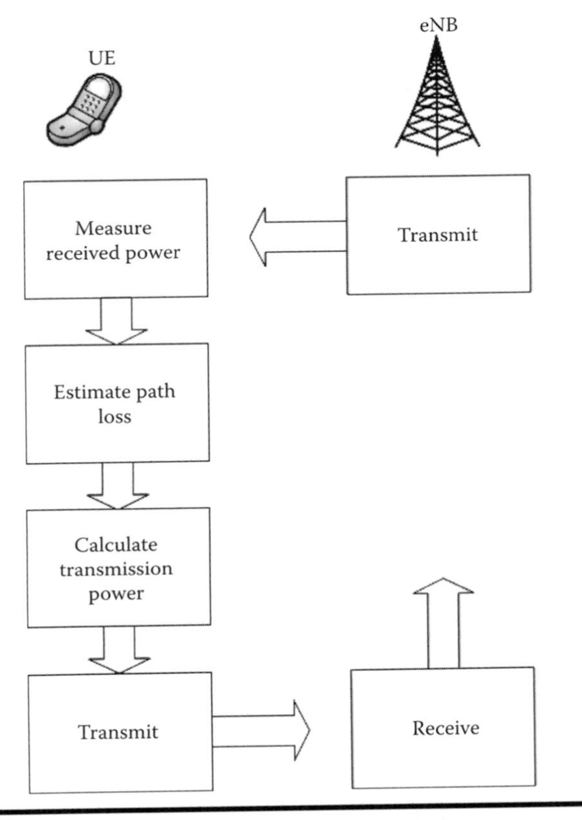

Figure 13.8 Block diagram of steps involved in setting the uplink power using the open-loop power control.

13.4.6 Closed-Loop Power Control

Closed-loop power control is the ability of the UE to adjust the uplink transmit power in accordance with the closed-loop correction values, which is also known as transmit power control (TPC) commands. The TPC commands transmitted by the eNB toward the UE are based on the closed-loop SINR target and the measured received SINR.

In an LTE closed-loop power control system, the uplink receiver at the eNB estimates the SINR of the received signal and compares it with the desired SINR target value. If the received SINR is below the SINR target, a TPC command is transmitted to the UE to request for an increase in the transmitter power. Otherwise, the TPC command will request for a decrease in transmitter power.

The LTE closed-loop power control mechanism operates around an open-loop point of operation. As discussed in Section 13.4.1, the UE adjusts its uplink transmission power based on the TPC commands it receives from the eNB when the uplink power setting is performed at the UE using open-loop power control. The closed-loop power control mechanism around open-loop point of operation is presented in Figure 13.9. The shaded blocks indicate the closed-loop power control components. It can be seen in Figure 13.9 that the closed-loop correction value is applied after calculating the transmission power using the open-loop power control. The PUSCH closed-loop power control expression is given by:

$$P_{\text{PUSCH}} = \min\{P_{\max}, P_{\text{OL}} + f(\Delta_i)\} \text{ [dBm]} \tag{13.11}$$

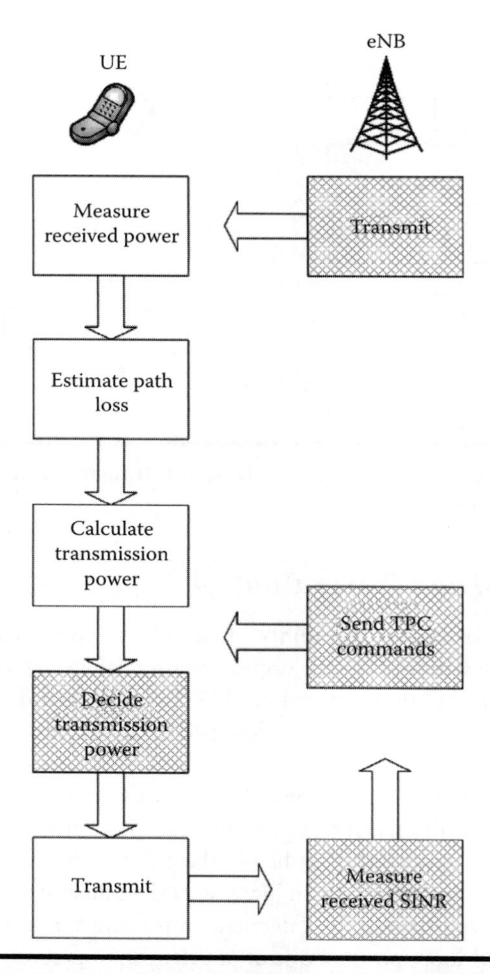

Figure 13.9 Block diagram of steps involved in adjusting the open-loop point of operation using the closed-loop power control.

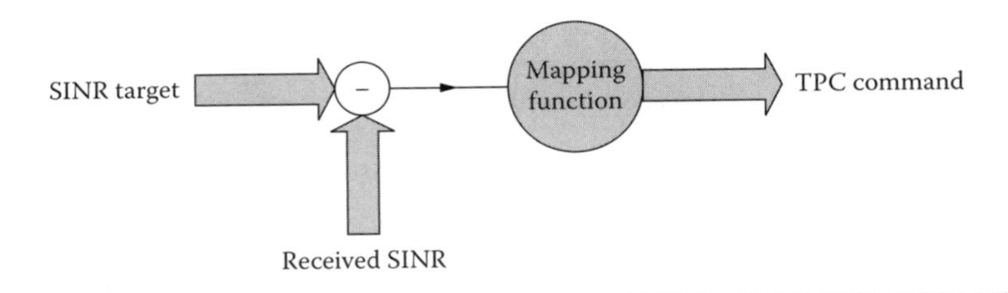

Figure 13.10 Generation of the TPC command at the eNB.

It is worthwhile to note that, if P_{PUSCH} is set using the closed-loop power control, power limitation is neglected in Equation 13.10 and is applied by Equation 13.11. In case of the conventional closed-loop power control, the open-loop component uses $\alpha = 1$. $f(\Delta_i)$ is the closed-loop adjustment to the open-loop point of operation and is defined as [2]

$$f(i) = f(i-1) + \delta_{\text{PUSCH}}(i - K_{\text{PUSCH}}) \text{ [dB]} \tag{13.12}$$

where δ_{PUSCH} is the UE specific correction value, which is also referred as the TPC command. TPC commands $[-1, 0, 1, 3]$ dB are used during the simulations. Here, $f(^*)$ represents accumulation and $f(0) = 0$ and $K_{\text{PUSCH}} = 4$ TTIs (transmission time intervals). K_{PUSCH} includes both processing and round-trip propagation time delay; this issue is discussed in Section 13.5.3. TPC commands are generated based on the difference between the SINR target and the received SINR, as shown in Figure 13.10.

The mapping function maps the resulting difference to one of the accumulated TPC commands. The generated TPC commands are transmitted by the eNB toward the UE. In this chapter, the conventional closed-loop SINR target is used in generating TPC commands, which is referred to as the baseline SINR target.

13.5 Proposed Closed-Loop Power Control Algorithm

In this section, a novel fractional closed-loop power control algorithm is presented and analyzed using different traffic models and realistic simulation scenarios. In addition, its performance is compared to that of a conventional closed-loop power control and open-loop power control algorithms.

The conventional closed-loop power control scheme steers all users to achieve equal received SINR, as seen in Figure 13.11. Consequently, users with good radio conditions (i.e., good channel quality) that possibly can achieve high received SINR are affected by this setting, thus resulting in a lower mean user bit rate. On the other hand, if we look at the open-loop power control performance, we see that users experiencing good radio conditions achieve high received SINR, but the cell-edge performance is worse compared to that of the conventional closed loop.

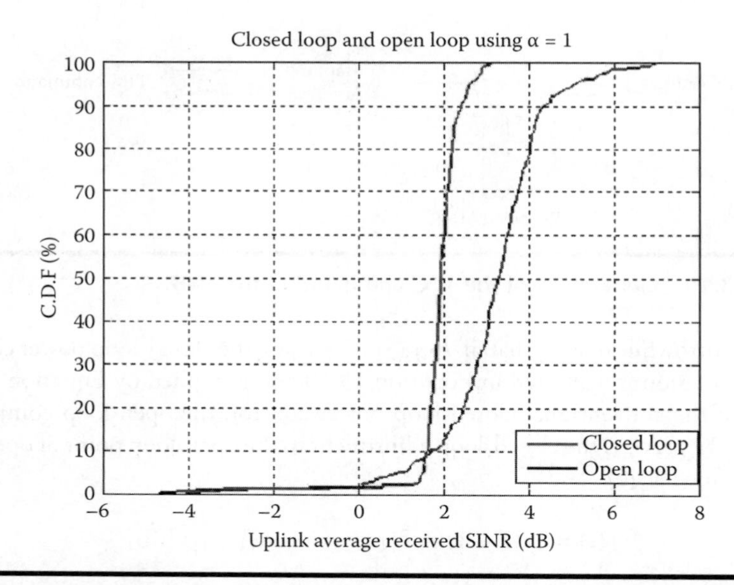

Figure 13.11 CDF plot of the uplink average received SINR for the open- and closed-loop power control using $\alpha = 1$.

In conventional closed-loop power control, the SINR target setting is the same for all users and is a trade-off between the cell-edge rate and the mean bit rate; that is, a high SINR target results in high mean user bit rate but a lower cell-edge bit rate, whereas a lower SINR target leads to a low mean and a high cell-edge bit rate.

It is desirable to design a closed-loop power control scheme that can provide a reasonable cell-edge bit rate and, at the same time, allow users with good radio conditions to achieve high received SINR, and subsequently attaining high mean user bit rate. In the following sections, we present a novel fractional closed-loop power control scheme that achieves these goals. In contrast to conventional closed-loop power control, the proposed scheme sets an SINR target for each user based on their path loss, which allows users with good radio conditions to achieve better SINR and, at the same time, provide a better cell-edge bit rate. In Section 13.5.2, setting an SINR target based on users' path loss is discussed and its mathematical expression is derived.

13.5.1 Power Headroom Report

Power headroom (P_h) is a mechanism by which the mobile terminal is configured to provide regular reports on its power to the network. The power headroom report can be used by the eNB to calculate the path loss of the users which is then used in setting of SINR target. Power headroom report is sent by the UE to the eNB which indicates how much power UE is left with to start using full power. In other words, it is the difference between the UE transmit power and the maximum UE transmit power and is given by

$$P_h = P_{\max} - P_{\text{PUSCH}} \text{ [dBm]} \tag{13.13}$$

The following triggers [7] should apply to power headroom reporting:

- The path loss has changed by a threshold value, as the last power headroom report is sent. The threshold value can be [1, 3, 6, inf] dB.
- The time elapsed from previous power headroom report is more than Y transmission time intervals (TTIs). The parameter Y can take values [10, 20, 50, 200, 1000, inf] TTIs.
- A power headroom report can only be sent when the UE has an UL grant. If one or several triggers are fulfilled when the UE does not have a grant, the UE should send the report when it has a grant again.

13.5.2 Mathematical Model for Setting an SINR Target Based on Path Loss of the Users

A mathematical expression needs to be derived to set the closed-loop SINR target based on the path loss of users while keeping the baseline SINR target for those users that are using full power ($P_{\max}$). Therefore, a relation is formed between the received SINR and path loss (PL) of the users with the aid of an illustration shown in Figure 13.12. In the figure, $PL_{\max}$ is the maximum path loss, at which users start using $P_{\text{PUSCH}} = P_{\max}$. $PL < PL_{\max}$ is the path loss of any arbitrary user, and $SINRtarget$ is the closed-loop baseline SINR target to start with. $SINRtarget'$ is the SINR target based on the path loss and α is the path loss compensation factor, whereas m is the slope and is given by $\alpha - 1$, and IN is the interference and noise power in dBm. The knee point in Figure 13.12 is denoted by $PL_{\max}$ where users use $P_{\max}$ at this point and beyond. The users at $PL < PL_{\max}$ are experiencing relatively better radio conditions than users at $PL \geq PL_{\max}$, and it is desirable that users should take advantage of good radio conditions.

Using Figure 13.12 and the information discussed above, the required mathematical equation that provides an SINR target based on the path loss of users can be obtained using the slope of the line and is given by

$$m = \frac{\Delta_Y}{\Delta_X} \tag{13.14}$$

where

$$\Delta_Y = SINR\,target' - SINR\,target \text{ [dB]} \tag{13.15}$$

$$\Delta_X = PL - PL_{\max} \text{ [dB]} \tag{13.16}$$

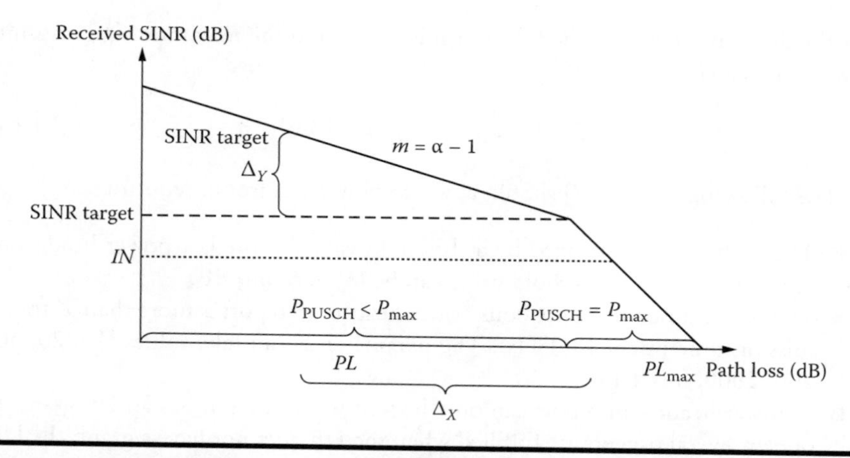

Figure 13.12 Illustration of setting the SINR target based on the path loss of users.

By using Equation 13.1, PL can be defined as

$$PL = \frac{1}{\alpha \cdot \{P_{\text{PUSCH}} - 10 \cdot \log_{10} M - P_0 - f(\Delta_i)\}} \text{ [dB]} \tag{13.17}$$

PL involves P_{PUSCH}, as can be clearly seen in Equation 13.17. In the real world, however, the eNB can use the power headroom report (P_h) received by the eNB from the UE to find the path loss of each user. Thus, PL can be rewritten as

$$PL = \frac{1}{\alpha \cdot \{P_h - 10 \cdot \log_{10} M - P_0 - f(\Delta_i)\}} \text{ [dB]}$$

$$P_h = P_{\text{PUSCH}} = P_{\max} \tag{13.18}$$

when $PL = PL_{\max}$.

By using Equations 13.13 through 13.17, the SINR target based on the path loss is given by

$$SINR\ target' = \begin{cases} (\alpha - 1) \cdot (PL - PL_{\max}) + SINR\ target & PL < PL_{\max} \\ SINR\ target & PL \geq PL_{\max} \end{cases} \text{ [dB]} \tag{13.19}$$

In Equation 13.18, users at $PL \geq PL_{\max}$ will use $SINRtarget' = SINRtarget$, indicating that there is no increase in the SINR target for users that are already using $P_{\text{PUSCH}} = P_{\max}$. Furthermore, $\alpha = 1$ turns the designed closed loop scheme into conventional closed-loop power control implying that the SINRtarget setting is independent of the path loss.

In the case of a closed loop with fractional path loss compensation factor, TPC commands will be generated based on the *SINRtarget′* and the received SINR. The closed-loop power control combined with fractional path loss compensation factor can be implemented using Equation 13.11 which defines the basic expression for the closed-loop power control. However, in contrast to conventional closed-loop implementation, which uses a baseline SINR target (same for all users) and $\alpha = 1$, a closed loop with fractional path loss compensation factor implementation involves an SINR target setting based on the path loss of the users and $0 < \alpha < 1$.

13.5.3 Processing and Round-Trip Time Delay Model

The eNodeB issues a TPC command to adjust the power at the UE. However, the adjustment takes place after some delay. This delay is typically propagation round-trip time (RTT) and processing time at the UE and the eNB. The RTT delay is due to the wave propagation, while the processing delay at the eNodeB occurs due to measuring the received SINR and issue of TPC command based on SINR target and received SINR. The processing delay at the UE occurs due to measuring the RSRP, calculating the *PL*, calculating the transmit power, and applying the adjustment based on the received TPC command. The total time delay used during simulations is of 4 ms. The time delay is demonstrated with the help of Figure 13.13. In the figure, $T_{\text{UL}} - T_{\text{DL}}$ is the difference between the duration of the UL and DL subframe. T_p is the propagation time, and the round-trip propagation time is $2 \times T_p$. We assume that the eNB received the first transmission from the UE; the TPC command is generated based on the difference between the SINR target and the received SINR measured at the eNB.

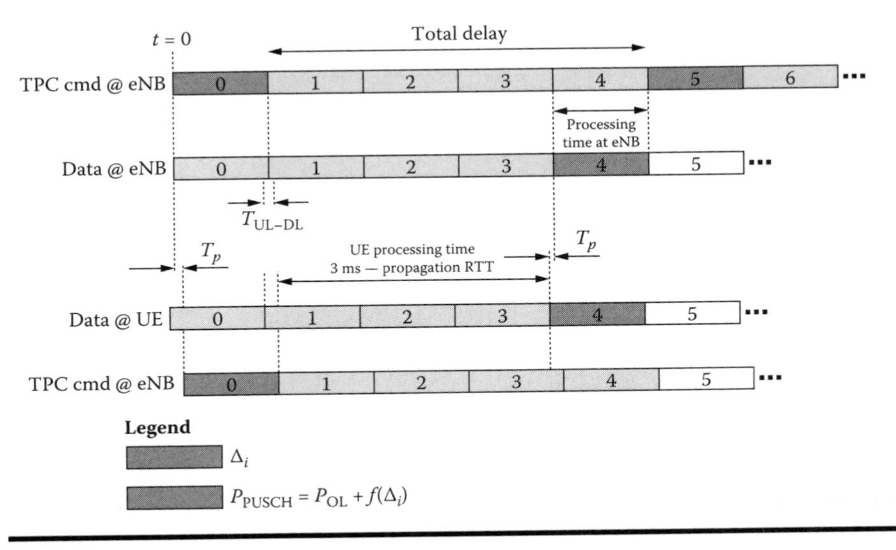

Figure 13.13 Illustration of time delay.

At time instant $t = 0$, the eNB sent the TPC command that was received by the UE after a delay equal to half of the propagation RTT (i.e., T_p). Processing time at the UE is 3 ms–propagation RTT [8]. Thus, the UE have to adjust the power and transmit half of the propagation time before $t = 4$. The power-adjusted transmission is received by the eNB at $t = 4$. The received transmission is decoded and a new TPC command is generated at $t = 5$. This makes a total time delay of 4 TTI.

As a consequence of four time instant delays, the eNB must take into account previous TPC commands while issuing a new TPC command. Let us say that the UE power needs to be adjusted by a single TPC command of +1 dB and is issued by the eNB at time instant $t = 0$. Now, until time instant $t = 5$, the eNB will receive a transmission without closed-loop adjustment; if the eNB continues issuing TPC commands independent of previously issued TPC commands, UE will end up at +4 dB more in P_{PUSCH}. In order to prevent this phenomenon, the sum of previously issued TPC commands are fed back to the mapping function, as illustrated in Figure 13.14. Thus, new TPC commands will be generated based on the SINR target, the estimated received SINR, and previously issued TPC commands. Now, as previously, issued commands are taken into account by the eNB between $t = 1$ and $t = 5$. The TPC command of 0 dB will be transmitted toward the UE.

13.5.4 Absolute Open-Loop Error Model

Open-loop power control errors are usually the result of several factors, such as the accuracy of measurements of the RSRP at the UE and inaccuracies in the radio parts such as temperature sensitivity and tolerances in the standard. The open-loop error is identified as a slowly varying component and varies between manufactures of UEs. Sources of open-loop power control error are illustrated in Figure 13.15.

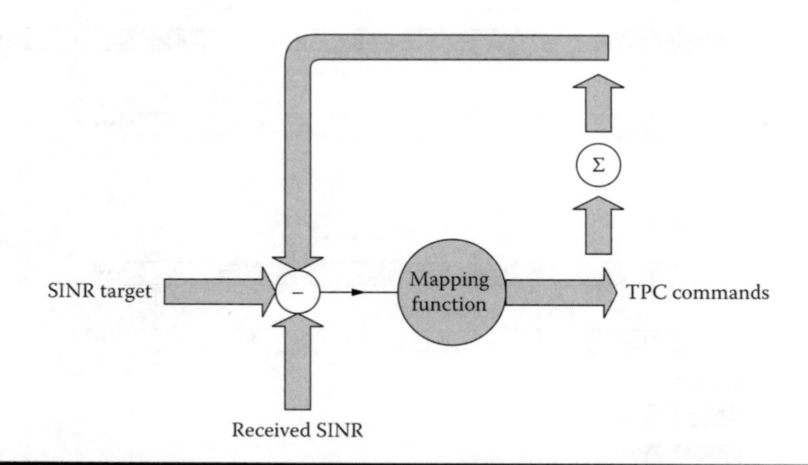

Figure 13.14 Generation of TPC commands at the eNB, taking into account previously issued TPC commands.

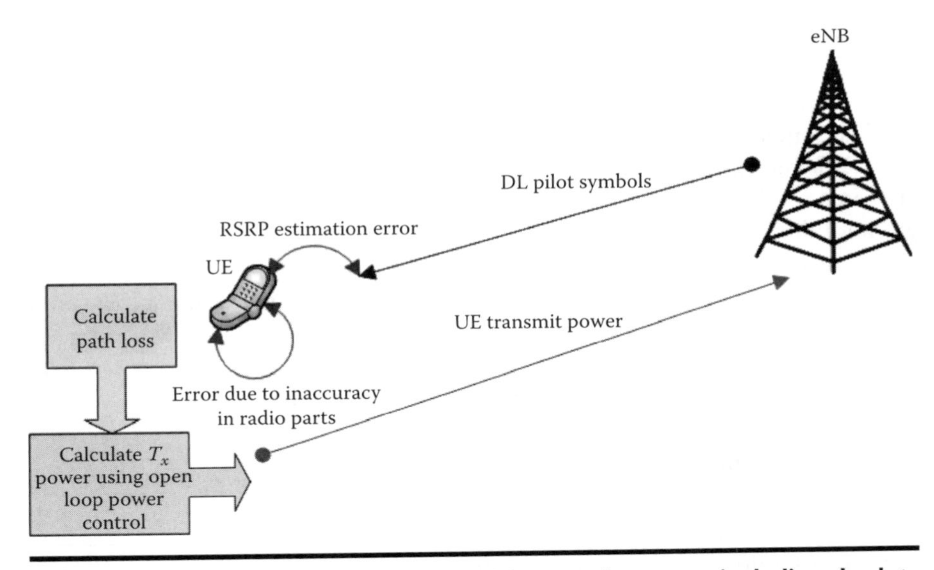

Figure 13.15 Illustration of the sources of the open-loop error including absolute error due to inaccuracy in radio parts.

At the time of writing, 3GPP had not yet standardized a tolerance in the standard, but because LTE RF components are the same as those used in wideband code division multiple access (WCDMA), the tolerance described in technical specification [4] can be used for first approximation. A tolerance of ±9 dB is required; however, a batch of UEs can handle ±4 dB. Thus, the absolute value of ±4dB with a uniform distribution is considered to be an open-loop error to evaluate the effect of the closed-loop power correction using TPC commands. By taking the absolute open-loop error into account, the expression for the closed-loop power control combined with fraction path loss compensation can be written as

$$P_{\text{PUSCH}} = \min\{P_{\max}, P_{\text{OL}} + \text{abserr} + f(\Delta_i)\} \text{ [dBm]} \quad (13.20)$$

where abserr is the absolute open-loop error.

13.6 Simulation Environment and Results

In this section, we present the simulation environment, the traffic models used, and the associated simulation results. The simulations are carried out using a multicell radio network dynamic simulator implemented in MATLAB. The simulator includes enhanced traffic and hybrid ARQ (HARQ) models. In the simulator, network performance is simulated for a certain period of time, which then includes events like arrival of new users, departure of users (whose calls are finished), and users' movements. The simulator also include a set of radio resource management (RRM) algorithms such as cell selection, scheduling, link adaptation, and transmit beamforming. The default simulation parameters are given in Table 13.1.

The simulator supports a variety of traffic models. In this chapter, we discuss two traffic models: full buffer and simple upload traffic models. The traffic models are based on Poisson processes of user arrivals. Different arrival intensities may be given for different services.

In the full buffer model, neither user leaves due to hang-up, nor does a new user arrive, as each user buffer is filled with infinite data, and the user will not leave until and unless it transmits all the data.

The simple upload traffic model is designed in such a way that users can have limited data in their buffers; thus, a user leaves when it transmits the data and new

Table 13.1 Default Simulation Parameters

Traffic Models	
User distribution	Uniform
Terminal speed	3 km/h
Data generation	Full buffer, simple upload traffic model
Radio Network Models	
Distance attenuation	$L = 35.3 + 37.6^* \log(d)$, $d =$ distance in meters
Shadow fading	Log-normal, 8 dB standard deviation
Multipath fading	Spatial channel model (SCM) [7], suburban macro
Cell layout	Hexagonal grid, 3-sector sites, 21 sectors in total
Cell radius	167 m (500 m intersite distance)
System Models	
Spectrum allocation	10 MHz (50 resource blocks) 180 kHz (1 resource block)
Max UE output power	250 mW into antenna
Max antenna gain	15 dBi
Modulation and coding schemes	QPSK and 16QAM, turbo coding
Scheduling algorithm	Round robin
Receiver	MMSE [10] with two-branch receive diversity

users are added to the system. This model provides the ease to define the user upload file size and mean bearer bit rate. The mean bearer bit rate, along with the offered cell throughput, defines the total number of users in the system. Moreover, the simple upload buffer model also allows inclusion of the effect of queuing delay when calculating the user bit rate. The queuing delay reflects more realistic results and provides a better scale for performance comparison in choosing the optimal value of α. For different values of α, the 5th percentile and mean user throughput are calculated by taking the effect of queuing delay into account.

The level of interference varies in both traffic models. However, the interferers are the same in the entire simulation in the case of the full buffer model as opposed to the simple upload traffic model because users neither leave nor arrive in the full buffer traffic model. Thus, the full buffer traffic model does not reflect a realistic scenario; however, it is used to compare the performance of the different power control schemes. In the following, simulation results are based on the simple upload traffic model unless otherwise stated.

13.6.1 Investigation for the Optimal Value of the Path Loss Compensation Factor

First, we investigate the performance of the fractional closed-loop power control for different values of α in terms of cell edge and mean bit rate using the full buffer and the simple upload traffic models, respectively. In addition, the value of α, which gives the best cell-edge performance for a given SINR target, will be selected for further investigation.

Figure 13.16 shows the performance of the closed-loop power control system using the full buffer traffic model. This figure shows that using $\alpha = 1$ results in a high cell-edge but a relatively low mean bit rate as compared to other values, as the system is less loaded in terms of bits. On the other hand, when using $\alpha = 0.7$ and 0.8, the user mean bit rate is relatively high, implying that users transmit more bits making the system more loaded, which may lead to a rise in the interference level and a lower cell-edge bit rate than $\alpha = 1$ is achieved. We can also see in Figure 13.16 that the optimal value for α cannot be chosen using the full buffer.

Using a realistic traffic model (i.e., the simple upload traffic model in our scenario), results in a high mean bit rate, which is advantageous, as users can upload (transmit) their data more quickly than a system with low mean bit rate, and users can leave the system early.

In contrast to the full buffer traffic model, Figure 13.17 shows that $\alpha = 1$ results in a low cell-edge bit rate as compared to $\alpha = 0.8$ in the simple upload traffic model. This is because of the relatively high interference, as more users are staying in the system for a longer time. Users stay longer because they cannot transmit their data quickly due to the low mean bit rate. It is evident from Figure 13.17 that the closed loop with full compensation results in a lower mean bit rate. The longer the users

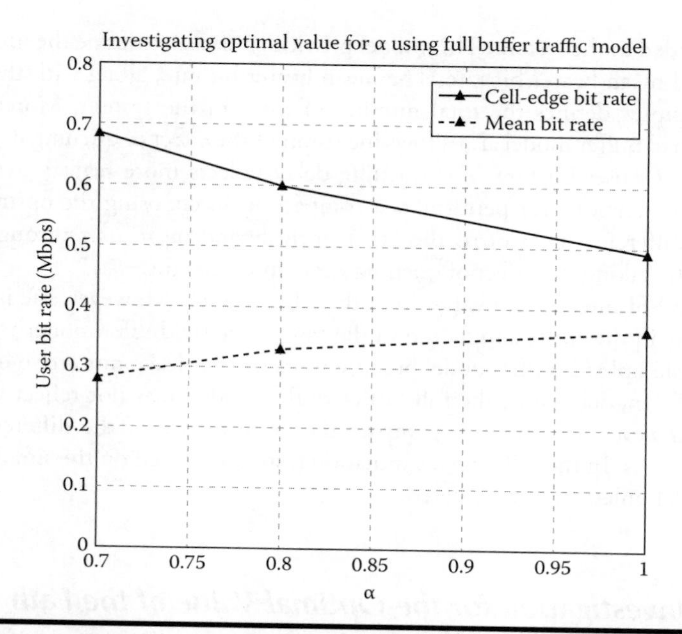

Figure 13.16 Investigating cell-edge and mean bit rate for different values of α for the closed-loop power control using full buffer model.

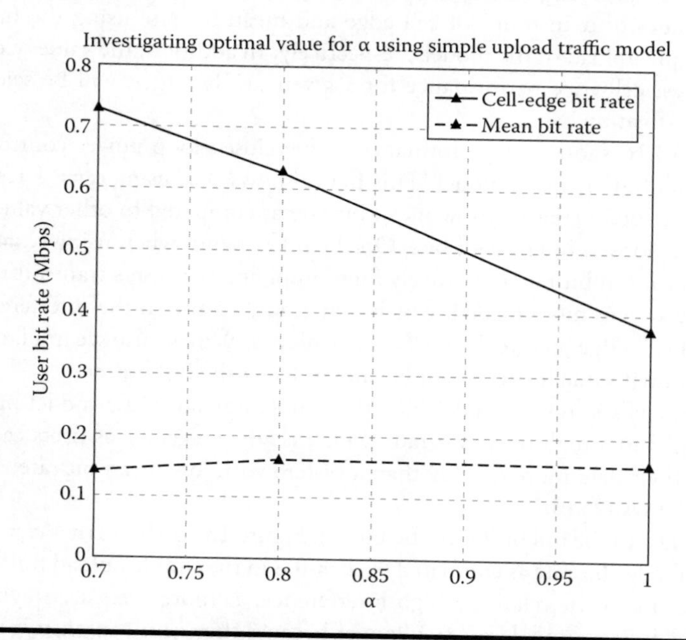

Figure 13.17 Investigating cell-edge and mean bit rate for different values of α for the closed-loop power control using the simple upload traffic model.

stay means that the system is more loaded (i.e., more users are in the system), which leads to a high queuing delay and, ultimately, a rise in the interference level.

It is worthwhile to note that, at the end of the simulation, the number of active users using $\alpha = 1$ was 2.6% more than the active users using $\alpha = 0.8$, keeping in view that for each value of α the simulation started with an equal number of users and lasted for an equal amount of time. This implies that due to the high queuing delay using $\alpha = 1$, 2.6% more users could not transmit all the data in their buffer.

Reiterating, and referring to Figure 13.17, compare the performance of $\alpha = 0.8$, 0.7, and 1, we can see that $\alpha = 0.8$ results in a high bit rate both in the cell edge and the mean. Therefore, $\alpha = 0.8$ is a better value for a fractional path loss compensation factor, as it results in the best cell edge and better mean bit rate.

13.6.2 Performance Analysis of the Closed-Loop PC Using $\alpha = 0.8$

In this section, we investigate the performance of the fractional closed-loop power control with the optimal value $\alpha = 0.8$, with the help of the cumulative distribution function (CDF) plots of the user bit rate and uplink-received SINR. The performance evaluation is carried out for both the ideal and realistic scenarios. In the realistic scenario, the time delay, absolute error, and power headroom reporting are taken into account, whereas in the ideal case they are not considered. The full compensation power control algorithm is also shown as a benchmark for comparison purposes.

13.6.3 Ideal Case

It is worthwhile to note that, in this study, calculating PL using Equation 13.16, assuming that $K_{\text{PUSCH}} = 0$ TTI, and taking into account abserr $= 0$, leads to an ideal study of the closed-loop power control with fractional path loss compensation factor. For a realistic study, PL is calculated using Equation 13.17, which involves power headroom reporting, $K_{\text{PUSCH}} = 4$ TTI is assumed, and an abserr of ± 4 dB with uniform distribution is considered.

Figure 13.18 shows the performance gain of the closed-loop power control using $\alpha = 0.8$ in both the mean and cell-edge bit rates. The mean bit rate is improved by 68% and provides better cell-edge performance than $\alpha = 1$ at the same time.

Figure 13.19 shows CDF plots of the uplink average-received SINR using the simple upload traffic model. We can see from this figure that the closed-loop power control with full compensation steers all users to achieve an equal uplink average-received SINR, as seen in both 5th percentile users and users close to the base station (i.e., users with good radio conditions) where all users get equal received SINR, as they all aim to achieve the same baseline SINR target. In contrast to full compensation, the closed-loop power control with fractional compensation keeps the baseline SINR target for the worst users and, at the same time, increases the baseline SINR target based for users with good radio conditions based on their path loss where low path

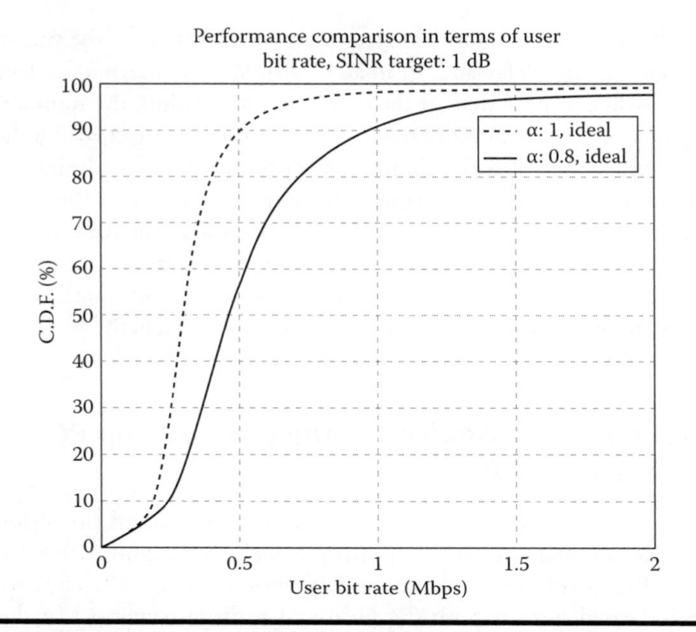

Figure 13.18 CDF plot of the user bit rate comparing $\alpha = 0.8$ and 1.

loss results in a high increase in the SINR target. The effect of SINR target setting based on path loss of the users is clearly evident from Figure 13.19, which shows that the better the radio conditions, the higher is the average received SINR.

13.6.4 Performance Analysis with Absolute Error and TPC Delay

Here, the individual and combined effects of the open-loop error and TPC delay are investigated. Figure 13.20 shows that the performance of the user bit rate is improved for the users with good radio conditions when taking the open-loop error in to account. This is because of the increase of the uplink power due to the open-loop error for the number of UEs, which results in a high received SINR. Performance in terms of user bit rate is slightly degraded for users in the low-CDF region, as the number of UE cannot satisfy the required SINR due to open-loop error. However, the performance change in terms of bit rate due to absolute error is just the initial phenomenon at the start of simulations (i.e., short simulation time), and will not be visible when simulated for longer time since the closed-loop power control compensates, for the open-loop error, using the TPC commands.

The TPC delay introduces an initial delay of only 4 TTI before the UE starts to use the TPC command it received from the eNB to correct its uplink power. It is worth noting that with round robin scheduling, it takes only 14 TTI before all

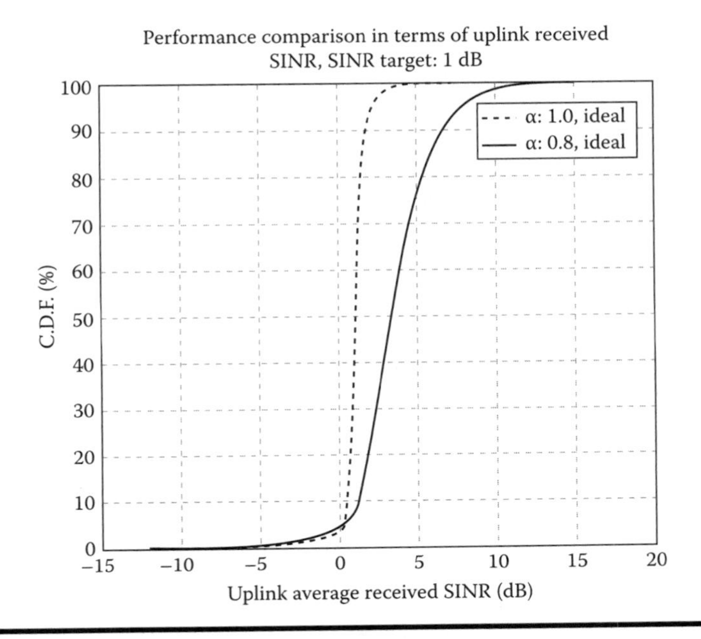

Figure 13.19 CDF plot of the uplink average received SINR.

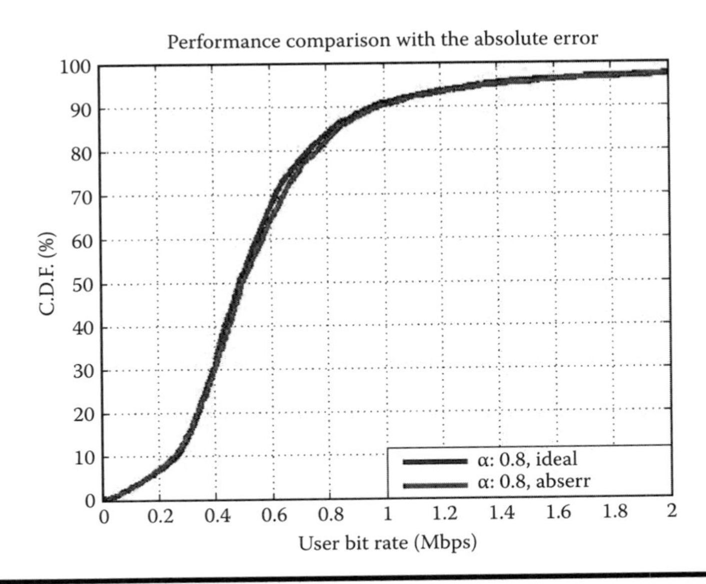

Figure 13.20 Performance comparison in terms of the user bit rate when the absolute error is taken into account.

users start to correct their uplink power using the TPC command. Thus, the effect of the TPC delay is not visible, as Figure 13.21 shows, where the closed-loop power control with TPC delay shows the same performance in terms of cell-edge and mean bit rate as that of the closed loop without TPC delay.

Finally, Figure 13.22 shows the combined effect of both the open-loop error and TPC delay. However, because there is no noticeable effect of TPC delay as discussed above, we can conclude that the effect on the user bit rate shown is due to the open-loop error only. It is also evident from the fact that the results in this figure show a similar trend to those in Figure 13.20.

13.6.5 Performance Analysis with the Power Headroom Report

Performance of the closed-loop power control with power headroom reports is analyzed in this section. It is worth mentioning that the power headroom is triggered at both the periodic intervals and change in the path loss by a threshold value. In this simulation, the power headroom report is trig(gered) at periodic intervals only.

A performance comparison in terms of user bit rate of the closed-loop power control with full compensation and $\alpha = 0.8$ with or without the power headroom report is shown in Figure 13.23. It can be seen from this figure that with power headroom reports, the user bit rate is degraded for the users with good radio conditions. The reason for the degradation in the mean bit rate is because the SINR target setting is based on an outdated path loss. The SINR target setting based on path loss aims

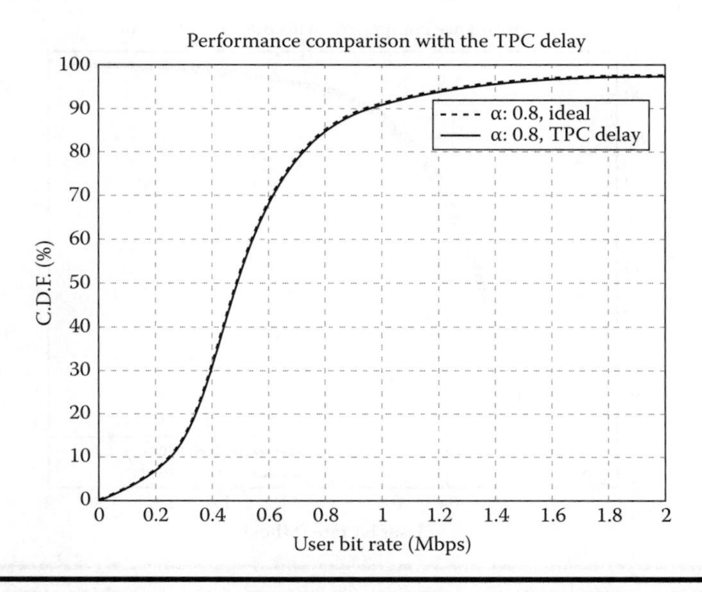

Figure 13.21 CDF plot of user bit rate showing both the closed-loop power control with or without the TPC delay. Total simulation time is 200 ms.

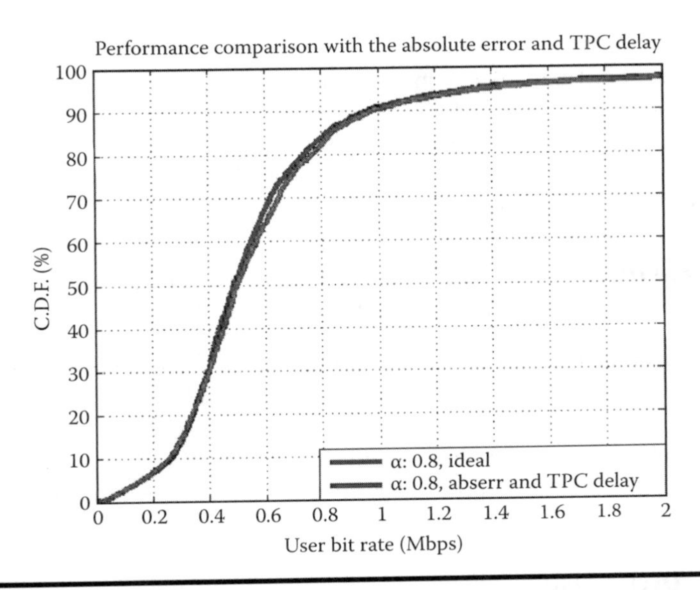

Figure 13.22 Performance comparison in terms of the user bit rate when both the absolute error and the TPC delay is taken into account.

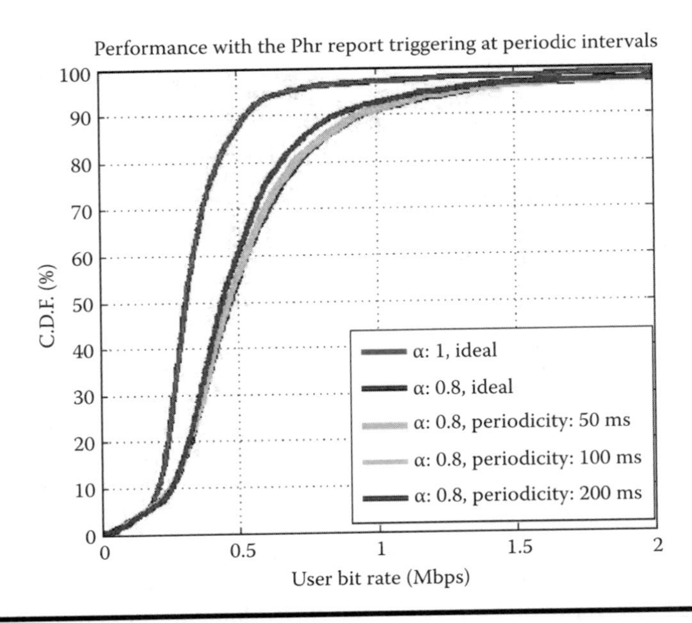

Figure 13.23 CDF plot of the user bit rate. The power headroom report triggers after 50, 100, and 200 TTIs.

to improve performance in terms of bit rate for users with good radio conditions. Thus, the more outdated the path loss, the more will be the degradation in the mean bit rate, as Figure 13.23 shows for high periodicity values. It is worthwhile to note that, for a longer simulation time and setting the power headroom periodicity to infinity, the performance of the SINR target setting based on path loss will be more like that of the absolute target setting.

13.6.6 Performance Analysis with Power Headroom Report, TPC Delay, and Absolute Error

Figure 13.24 shows performance in terms of user bit rate when taking into account the combined effects of absolute error, time delay, and power headroom report triggering at periodic intervals of 200 ms. The figure shows that the closed-loop power control using $\alpha = 0.8$ shows performance gain in both mean and cell-edge bit rate. The mean bit rate is improved by 63% and, at the same time, provides better cell-edge performance compared to $\alpha = 1$.

13.6.7 Power Use

The difference of the mean of the UEs power consumption results in zero, which implies that the overall power utilization is roughly the same using the closed-loop power control with full and fractional compensation, as can be seen in Figure 13.25.

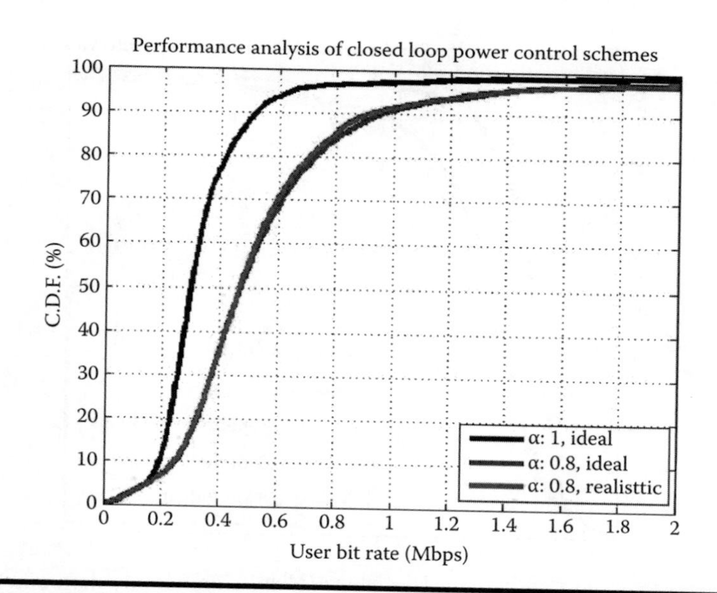

Figure 13.24 Performance analysis of the closed-loop power control schemes in terms of the user bit rate, taking into account the power headroom report, absolute error, and time delay.

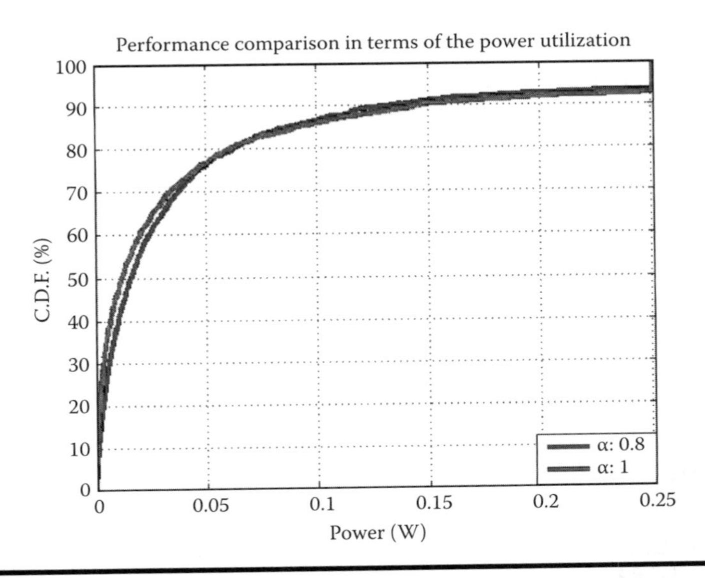

Figure 13.25 UEs power use of the closed-loop power control with full and fractional compensation.

However, the closed-loop power control using $\alpha = 0.8$ uses battery power more efficiently, as it provides better system performance in terms of cell edge and mean bit rate than the closed-loop power control with full compensation.

13.7 Conclusions

In this chapter, a novel fractional closed-loop power control algorithm for LTE system was proposed. The performance evaluation of the proposed algorithm was carried out using a dynamic radio network simulator. Both the ideal and realistic cases were investigated. The realistic case included the performance evaluation by simulating the effects of absolute error of ±4 dB, time delay, and power headroom report. The path loss compensation factor was investigated for the values in the range 0.7 to 1 as proposed by the standards. The closed-loop power control with full compensation was used as a reference for performance comparison. Simulation results have shown that the conventional closed-loop power control can be replaced by the closed-loop power control with fractional compensation, thus improving system performance in terms of the mean and cell-edge bit rate.

In the ideal case, $\alpha = 0.8$ has shown performance gain over $\alpha = 1$ by improving the mean bit rate by 68% and simultaneously providing the same cell-edge bit rate for a given SINR target. The realistic results using $\alpha = 0.8$ have shown that delay has no effect on the performance of a closed-loop power control. Absolute error has shown performance gain in terms of mean bit rate because of the initial phenomenon of users arriving with high uplink power. However, improvement due to the initial phenomenon will not be prominent when simulated for a longer time. Performance of the closed-loop power control using $\alpha = 0.8$ was simulated with a power headroom report triggering at periodic intervals, which showed performance degradation in the mean bit rate, as an outdated path loss was used in setting the SINR target. In the realistic case, performance in terms of the mean bit rate improved by 63% for a given SINR target and, at the same time, better cell-edge performance was achieved. This shows that a conventional closed-loop power control can be replaced by the proposed closed-loop power control combined with the fractional path loss compensation factor.

References

[1] 3rd Generation Partnership Project; "3GPP, TR 25.814 v7.1.0, Technical Specification Group Radio Access Network Physical layer aspects for evolved universal terrestrial radio access (UTRA) (Release 7)," 2006.

[2] 3rd Generation Partnership Project, "3GPP TS 36.213 V8.2.0, technical specification group radio access network; evolved universal terrestrial radio access (E-UTRA); Physical layer procedures (Release 8)," 2008.

[3] Ericsson Internal, "LTE uplink power control, overview and parameter setting."

[4] A. Simonsson and A. Furuskar, "Uplink Power Control in LTE—Overview and performance: Principles and Benefits of Utilizing Rather than Compensating for SINR Variations," *Proceedings of the 68th IEEE Vehicular Technology Conference*, 2008.

[5] Ericsson, R1-073036, "Intra-cell uplink power control for E-UTRA—Evaluation of fractional path loss compensation," 3GPP TSG RAN WG1 Meeting #49, Orlando, US, June 5–9, 2007.

[6] Ericsson, R1-074850, "Uplink power control for E-UTRA—Range and representation of P0," 3GPP TSG RAN WG1 Meeting #51, Jeju, Korea, November 5–9, 2007.

[7] 3rd Generation Partnership Project, R4-081162 "LS on power headroom reporting," 3GPP TSG RAN WG4 Meeting #47, Kansas City, MO, May 5–9, 2008.

[8] 3rd Generation Partnership Project, "3GPP TS 25.101 v8.3.0, universal mobile telecommunications system (UMTS); user equipment (UE), radio transmission and reception (FDD) (Release 8)," 2008.

[9] 3rd Generation Partnership Project, "Technical specification group radio access network; Spatial channel model for Multiple Input Multiple Output (MIMO) simulations," 3GPP TR 25.996, V6.1.0., 2004.

[10] J.H. Winters, "Optimum combining in digital mobile radio with co-channel interference," *IEEE Journal on Selected Areas in Communications*, vol. SAC-2, no. 4, July 1984.

Chapter 14

Key Technologies and Network Planning in TD-LTE Systems

Mugen Peng, Changqing Yang, Bin Han, Li Li, and Hsiao Hwa Chen

Contents

14.1 Introduction

As one of the commercial International Mobile Communication-2000 (IMT-2000) standards, time division-synchronous code division multiple access (TD-SCDMA) is being consummated dramatically in terms of key technologies, standard and products [1]. The key issues concerning the successful operation of TD-SCDMA for operators are [2]: (1) how to ensure the future profitability of TD-SCDMA and how to attract as many users as possible through providing all kinds of services, and (2) how to make TD-SCDMA have a smooth migration to IMT-advanced. Actually, the specialized trial for TD-SCDMA industrialization wound up in the middle of 2006 year has collected the evidence that TD-SCDMA is capable of building a full-coverage large-scale network and providing a good opportunity for some cutting-edge key technologies based on time division duplex (TDD) to be firmly validated.

The rapid development of wideband code division multiple access (WCDMA) and cdma2000 markets accelerated their evolution paces, and the short-term evolution (STE) technologies were presented for supporting the high bit rate transmission for packet services. For example, high-speed packet access (HSPA) is regarded as the STE in the 3G Partner Project (3GPP) for WCDMA. TD-SCDMA faces the same challenges of the evolution to HSPA, which has become an important factor for operators when they choose among the IMT-2000 standards to deploy their 3G networks. Consequently, the first evolution phase for TD-SCDMA is specified as TD-STE, in which single-carrier and multicarrier TD-HSDPA/TD-HSUPA (unified as TD-HSPA), TD-multimedia broadcast multicast service (TD-MBMS), and TD-HSPA evolution (TD-HSPA+) are incorporated. Note that the technologies for TD-STE systems are still based on the CDMA [3].

3GPP initiates the research on long-term evolution (LTE) with a view of keeping the competitive edge and the dominance of the cellular communication technologies in the market of mobile communications. The LTE program of TD-SCDMA (called TD-LTE) was undertaken in both 3GPP and the China Communications Standards

Association (CCSA). Based on TD-SCDMA mature industry ecosystem, TD-LTE is strongly supported by the TD-SCDMA industry chain. The target of the TD-LTE is to enhance the capabilities of coverage, service providing, and mobility supporting of TD-SCDMA. To save the investment and make full use of the network infrastructure available, the design of TD-LTE should take into account the features of TD-SCDMA and keep TD-LTE compatible to TD-SCDMA and TD-STE systems back forward to ensure a smooth migration.

14.2 Overview of TD-LTE Principles and Standards

The LTE systems will work at both (frequency division duplex) FDD and TDD modes. LTE TDD and FDD modes have been greatly harmonized in the sense that both modes share the same underlying framework, including radio access schemes: orthogonal frequency division multiple access (OFDMA) in downlink and single carrier frequency division multiplex access (SC-FDMA) in uplink, basic subframe formats, configuration protocols, etc [4]. In terms of architecture, there are no differences and the very few differences in the media access control (MAC) and higher layer protocols relate to TDD-specific physical layer parameters. Thus there will be high implementation synergies between the two modes allowing for efficient support of both TDD and FDD in the same network or user device. It is noted that TDD mode should be compatible with TD-SCDMA, which is known also as TD-LTE. The joint LTE trials organized by CMCC, Verizon Wireless, and Vodafone show that the operators want two LTE systems in a single device in order to achieve global interoperability [5]. The joint trials also made it clear that FDD-LTE and TD-LTE will evolve from a Universal Mobile Telecommunications System (UMTS) and TD-SCDMA, respectively.

The difference in the physical layer is due to two types of frame structures: one is for FDD LTE systems and is named as type 1, and the other is a special frame structure for TD-LTE systems to be aligned with TD-SCDMA smoothly and is named type 2, which is shown in Figure 14.1 [4]. In type 2, each radio

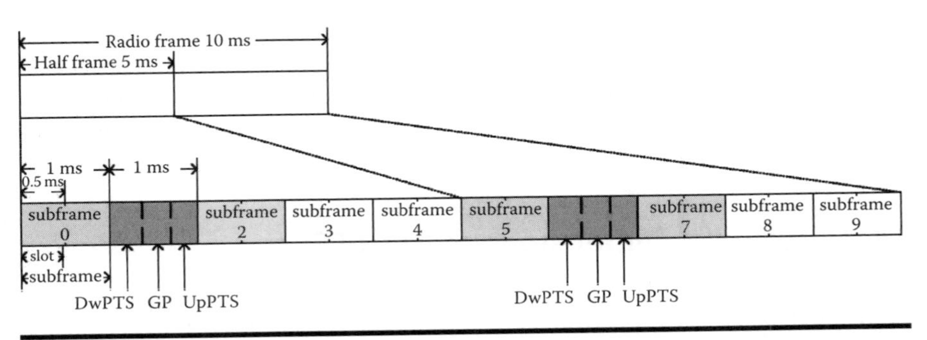

Figure 14.1 Frame structure type 2 for TD-LTE (for 5-ms switch-point periodicity).

frame consists of two half-frames with the length $T_f = 153600 \cdot T_s = 5$ ms. Each half-frame consists of eight slots and three special fields, downlink pilot time slot (DwPTS), guard period (GP), and uplink pilot time slot (UpPTS). The length of slot in the half-frame is $T_{\text{slot}} = 15360 \cdot T_s = 0.5$ ms. To be compatible with type 1, two slots are combined as one subframe, that is, the subframe i consists of slots $2i$ and $2i + 1$. Subframes 0, 5 and DwPTS are always reserved for downlink transmission.

In order to be compatible with both TD-SCDMA and WCDMA systems, two types of switch-point periodicities are defined in type 2: 5 ms and 10 ms. When the switch-point periodicity is 5 ms, UpPTS, subframes 2 and 7 are reserved for the uplink transmission. However, when the switch-point periodicity is 10 ms, DwPTS exists in both half-frames, whereas GP and UpPTS only exist in the first half-frame. Furthermore, UpPTS and subframe 2 are reserved for the uplink transmission and subframes 7 to 9 are reserved for the downlink transmission.

Due to the TDD features, uplink and downlink radio channels are reciprocated, and hence some open-loop mechanisms can be used, such as the open-loop transmit diversity and spatial multiplexing (including so-named "open-loop precoding") schemes can be deployed in TD-LTE systems. The reciprocity is a real advantage for TD-LTE but is not universal: (1) antenna configuration is different between evolved NodeB (eNB) and user equipment (UE); (2) the radio frequency (RF) impairment for uplink and downlink; and (3) fading and interference is time varying. Some multiple antenna configurations, such as single-input single-output (SISO), single-input multiple-output (SIMO), multiple-input multiple-output (MIMO), can be used in LTE systems. For SISO, a matched filter receiver is used, whereas the maximum ratio combining (MRC) is implemented at the receiver for SIMO. For the MIMO scheme, there are three types, space-frequency block coding (SFBC), spatial multiplexing (SM) with open-loop (OL) precoding and closed-loop (CL) precoding. Open-loop precoding is to allocate a codeword to each resource block (RB) in a transmission time interval (TTI) according to a predefined order known by both the transmitter and receiver, so it is unnecessary to feed back precoding matrix indicator (PMI) to the eNB. Closed-loop precoding is designed to allocate the codeword to a specific user depending on its channel state information (CSI) on one RB and it has to feed back PMI to the transmitter. Both open-loop and closed-loop precoding are specific to FDD-LTE systems. Whereas for TD-LTE systems, the CSI of downlink can be estimated effectively according to the uplink sounding reference signal (SRS) due to the channel reciprocity, especially when the UEs are operating in a low-mobility environment. With the estimated CSI, non-codebook-based precoding is possible for TD-LTE systems.

14.3 Capacity Dimensions of TD-LTE

Assume that the number of subchannels is N. Let K denote the frequency reuse factor. A single-cell has access to $M = N/K$ subchannels. Each UE will occupy one minimum resource unit with L subchannels in the frequency domain. Let Q denote the active UEs (or traffic channels) per cell. If $QL \leq M$, then the intra-cell interference can be completely avoided by assigning a different set of subchannels to each UE. Without loss of generality, we assume that one eNB transmits data to UE i. The signal-to-interference-plus-noise ratio (SINR) at the UE i is:

$$\gamma_i = \frac{S_i}{I_{\text{inter},i} + N} \tag{14.1}$$

where S_i denotes the received power, $I_{\text{inter},i}$ is the inter-cell cochannel interference, and N is the noise power. Let P denote the eNB power. The information sequence of each UE is transmitted over L parallel subchannels. The transmission power of eNB is equally divided among subchannels and UEs. Hence, the transmitted power allocated to a single UE on a subchannel is P/QL.

Let $g_{ij} = \bar{g}_{ij}\varsigma_{ij}$ denote the aggregate link gain [6] between UE i (receiver) and eNB j (transmitter). The link gain depends on the distance-based attenuation part $\bar{g}_{ij}$ and fast fading part ς_{ij}. The latter is a function of the complex channel responses $|h_{ii,l}|^2$ on the utilized subchannel i. In the case where $i = j$, ς_{ij} has the simple form [6]:

$$\varsigma_{ii} = \frac{1}{L}\sum_{l=1}^{L} |h_{ii,l}|^2 \tag{14.2}$$

Without loss of generality, we assume that subchannels 1 to L have been assigned to UE i. We assume that the UE receiver utilizes maximum ratio combining (MRC). Then, the received signal power at UE i can be written as:

$$S_i = \bar{g}_{ii}\sum_{l=1}^{L} |h_{ii,l}|^2 \frac{P}{QL} = g_{ii}\frac{P}{Q} \tag{14.3}$$

We assume that different cells are separately using pseudo-random scrambling codes. Furthermore, we assume that the interfering eNBs are evenly dividing their transmit power among the M subchannels available. Due to the frequency-domain subchannel number (L) and the full load in each cell, a fraction L/M of the eNB power ends up interfering with UE i. In that case, the inter-cell interference power can be written as:

$$I_{\text{inter},i} = \sum_{\substack{j=1 \\ j\neq i}}^{T} g_{ij}\frac{LP}{M} \tag{14.4}$$

where T denotes the number of eNBs that share the same L subchannels. Combining the above models into the SINR (see [7]) yields:

$$\gamma_i = \frac{S_i}{I_{\text{inter},i} + N} = \frac{g_{ii}\frac{P}{Q}}{\sum\limits_{\substack{j=1 \\ j\neq i}}^{T} g_{ij}\frac{LP}{M} + \nu \mathrm{L}} \tag{14.5}$$

where ν denotes the noise power per subchannel. We assume that the subchannels separation is greater than the coherence bandwidth so that they experience independent fading. Furthermore, we assume that the complex channel response $h_{ij,l}$ in a link between transmitter j and receiver i on subchannel l follows a Rayleigh distribution with unit mean. Thus, $|h_{ij,l}|^2$ follows an exponential distribution. Let us define $\xi = L\zeta_{ii} = \sum_{l=1}^{L} |h_{ii,l}|^2$ as a sum of channel power gains. It is well known that the probability distribution of the sum of L independent identically distributed exponential random variables follows the Erlang-L distribution [6]:

$$\Pr\{\xi_{ii} \leq x\} = 1 - \sum_{l=0}^{L-1} \frac{x^l}{l!} e^{-x} \stackrel{def}{=} F_{\xi}(x, L) \tag{14.6}$$

14.3.1 Outage Probability for Single-Cell Cluster

For the single-cell cluster, there is only the thermal noise, and $I_{\text{inter},i} = 0$. The outage probability that an UE receives an SINR lower than its target threshold $\gamma \geq 0$ can be expressed as:

$$F_{\gamma_i}(\gamma; L) = \Pr\{\gamma_i \leq \gamma\} = F_{\xi}\left(\frac{\gamma QL}{\bar{g}_{ii} P} \nu \mathrm{L}; L\right) = 1 - \sum_{l=0}^{L-1} \frac{\left(\frac{\gamma QL}{\bar{g}_{ii} P} \nu \mathrm{L}\right)^l}{l!} e^{-\left(\frac{\gamma QL}{\bar{g}_{ii} P} \nu \mathrm{L}\right)} \tag{14.7}$$

Take a single-cell downlink scenario into consideration, where 20 MHz system bandwidth is divided into $N = 100$ subchannels, out of which $M = 33$ are used per cell ($K = 3$). The noise power spectral density is -175 dBm/Hz, and the transmit power of the eNB is 30 dBm. The cell radius is assumed to be 1 km. All UEs are situated on the edge of the cell and have identical target SINR threshold.

Figure 14.2 shows the impact of subchannel number on the outage probability with SINR target varying from -15 dB to -5 dB. If L is invariant, the outage probability would increase when a higher target SNR is required by UE i. If we fix the SINR target, then the more the number of the subchannels UE_i used, the smaller outage probability is. Multipath diversity decreases the interference, leading to better performance.

14.3.2 Outage Probability for Multiple-Cell Clusters

We note that $|h_{ii,l}|^2$ and $|h_{ij,l}|^2$ are uncorrelated, that is:

$$E\left[\left(|h_{ii,l}|^2 - 1\right)\left(|h_{ij,l}|^2 - 1\right)\right] = 0 \tag{14.8}$$

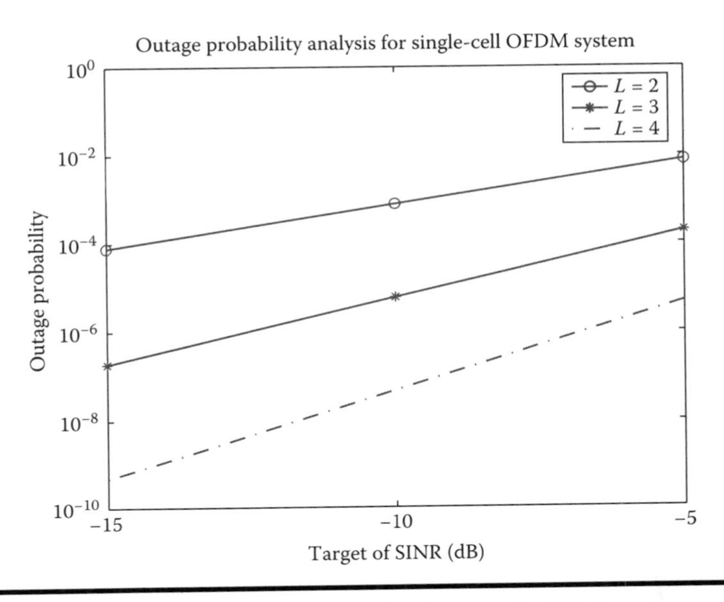

Figure 14.2 Outage probability analysis for single-cell OFDM system.

It thus follows that $I_{\text{inter},i} \propto \sum_{j \neq i} \bar{g}_{ij} \varsigma_{ij}$ and $g_{ii} = \bar{g}_{ii} \varsigma_{ii}$ are also uncorrelated. The central limit theorem implies that $I_{\text{inter},i}$ starts approaching a Gaussian distribution as the number of interfering eNBs grows ($T \to \infty$). The expected value and variance of $I_{\text{inter},i}$ (T) are given, respectively, by:

$$\bar{I}_{\text{inter},i} = \sum_{\substack{j=1 \\ j \neq i}}^{T} \bar{g}_{ij} \frac{LP}{M} \tag{14.9}$$

$$\sigma^2_{I_{\text{inter},i}} = \sum_{\substack{j=1 \\ j \neq i}}^{T} \bar{g}_{ij}^2 \left(\frac{LP}{M} \right)^2 \tag{14.10}$$

Let us define a random variable:

$$z_i = \partial_i \left(I_{\text{inter},i} + vL \right) \tag{14.11}$$

where $\partial_i = \frac{\gamma QL}{\bar{g}_{ii} P}$. Assume that $\gamma < \gamma_{\max}$. Now, we can rewrite Equation 14.7 as

$$F_{\gamma_i}(\gamma; L|z_i) = F_{\xi}(z_i; L) = 1 - \sum_{l=0}^{L-1} \frac{z_i^l}{l!} e^{-z_i} \tag{14.12}$$

If $I_{\text{inter},i}$ follows a Gaussian distribution, so does z_i. Denoting $_i$ and $_i$ as the mean and standard deviation of z_i respectively, we have:

$$\mu_i = \partial_i \left(\bar{I}_{\text{inter},i} + vL \right) \tag{14.13}$$

$$\sigma_i^2 = \partial_i^2 \sigma_{\text{inter},i}^2 \tag{14.14}$$

The moment-generating function (MGF) $M_{z_i}(t) = E\{e^{tz}\}$ of a Gaussian distributed random variable is known to be:

$$M_{z_i}(t) = E(e^{tz}) = \int_{-\infty}^{+\infty} e^{tz} \frac{1}{\sqrt{2\pi}\sigma} e^{-\frac{(z-\mu)^2}{2\sigma^2}} dz = \int_{-\infty}^{+\infty} \frac{1}{\sqrt{2\pi}\sigma} e^{-\frac{(z-\mu)^2}{2\sigma^2}+tz} dz$$

$$= \int_{-\infty}^{+\infty} \frac{1}{\sqrt{2\pi}\sigma} e^{-\frac{z^2-(2\mu+2\sigma^2 t)z+\mu^2+(2\mu+2\sigma^2 t)^2-(2\mu+2\sigma^2 t)^2}{2\sigma^2}} dz$$

$$= \int_{-\infty}^{+\infty} \frac{1}{\sqrt{2\pi}\sigma} e^{-\frac{[z-(\mu+\sigma^2 t)]^2-2\mu\sigma^2 t-\sigma^4 t^2}{2\sigma^2}} dz$$

$$= e^{\mu_i t+\frac{1}{2}\sigma_i^2 t^2} \int_{-\infty}^{+\infty} \frac{1}{\sqrt{2\pi}\sigma} e^{-\frac{[z-(\mu+\sigma^2 t)]^2}{2\sigma^2}} dz = e^{\mu_i t+\frac{1}{2}\sigma_i^2 t^2} \tag{14.15}$$

The outage probability $F_{\gamma_i}(\gamma; L)$ can be obtained by taking the expectation of Equation 14.12 with respect to z_i. We suggest the following simple iteration to solve $M_{z_i}^l(t) = E\{z^l e^{tz}\}$:

$$M_{z_i}^{(0)}(-1) = e^{-\mu_i+\frac{1}{2}\sigma_i^2} \tag{14.16}$$

$$M_{z_i}^{(1)}(-1) = (\mu_i - \sigma_i^2) M_{z_i}^{(0)}(-1) \tag{14.17}$$

$$M_{z_i}^{(n)}(-1) = \left(\mu_i - \sigma_i^2\right) M_{z_i}^{(n-1)}(-1) + (n-1)\sigma_i^2 M_{z_i}^{(n-2)}(-1),$$

$$n = 2, 3, \ldots, L-1 \tag{14.18}$$

Given the moment-generating function of z_i, we can write:

$$F_{\gamma_i}(\gamma; L) = E[F_{\gamma_i}(\gamma; L|z_i)] = E[F_{\xi}(z_i; L)] = E\left(1 - \sum_{l=0}^{L-1} \frac{z_i^l}{l!} e^{-z_i}\right)$$

$$= 1 - \sum_{l=0}^{L-1} \frac{1}{l!} E\left(z_i^l e^{-z_i}\right) = 1 - \sum_{l=0}^{L-1} \frac{1}{l!} M_{z_i}^l(-1) \tag{14.19}$$

Considering a hexagonal structure, we focus on the worstcase that UEs are located on the edge of the reference cell. Let R denote the cell radius, and D denote the distance from the eNB to the first tier of interferers. In addition, let us normalize the link gains so that $\bar{g}_{ii} = 1$, in which case the noise is multiplied by the factor R^{∂}. Now, the expected interference can be written as:

$$\bar{I}_{\text{inter},i} = \frac{LP}{M}\frac{6}{\sqrt{(3K)^{\partial}}}\sum_{\substack{m\geq 0,n\geq 0\\ m+n>0}}\frac{1}{\sqrt{(m^2+n^2+mn)^{\partial}}} \tag{14.20}$$

which is convergent for $\partial > 2$. The variance becomes

$$\sigma^2_{I_{\text{inter},i}} = \left(\frac{LP}{M}\right)^2\frac{6}{(3K)^{\partial}}\sum_{\substack{m\geq 0,n\geq 0\\ m+n>0}}\frac{1}{\sqrt{(m^2+n^2+mn)^{\partial}}} \tag{14.21}$$

which is convergent for $\partial > 1$.

We use the same parameter settings as those in Figure 14.2 for the hexagonal cellular system. Figures 14.3 through 14.5 illustrate the outage probability as a function of L for three different frequency reuse factor K (i.e., $K = 1$, $K = 3$, $K = 12$). We set different UEs for different K and make sure all the subchannels are used (i.e., $Q = M/L$ UEs per cell).

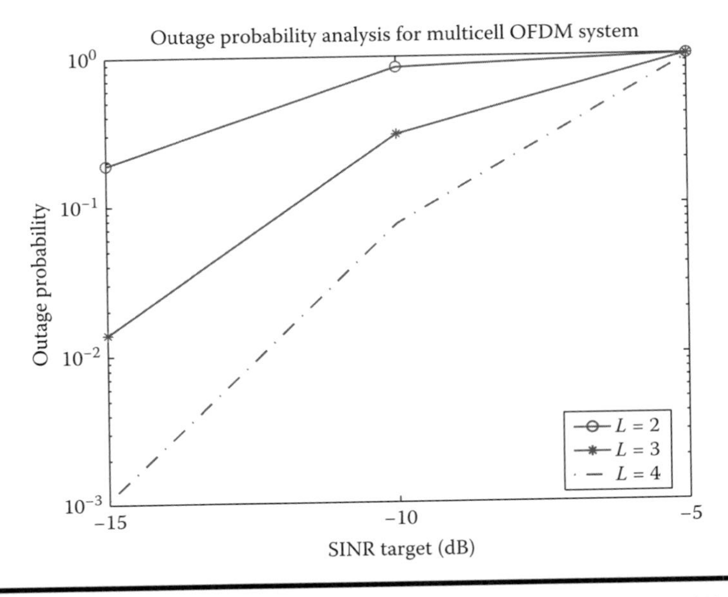

Figure 14.3 Outage probability analysis for multicell OFDM system ($K = 1$).

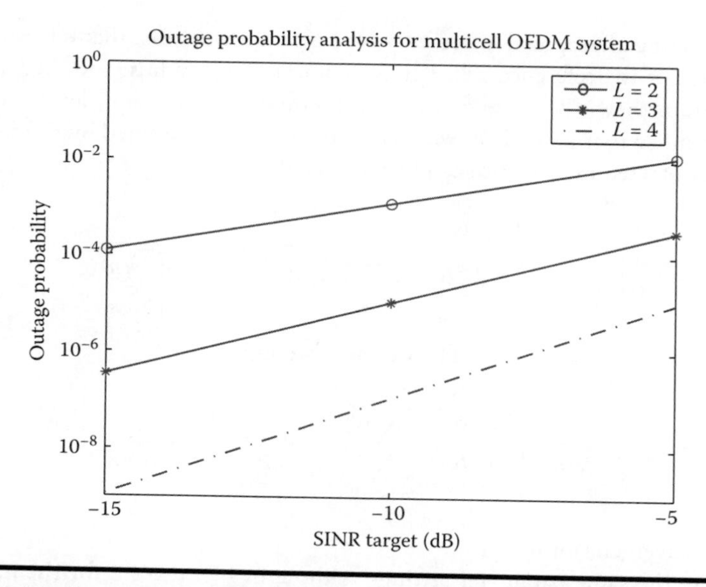

Figure 14.4 Outage probability analysis for multicell OFDM system ($K = 3$).

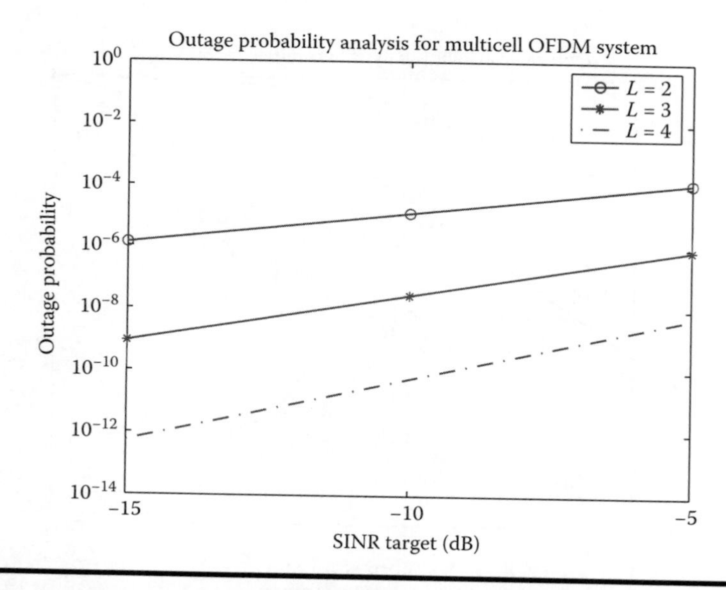

Figure 14.5 Outage probability analysis for multicell OFDM system ($K = 12$).

For a multicell system, whereas inter-cell interference becomes dominant, the frequency reuse factor K has a significant impact on the outage performance. For a bigger K, the reuse distance of the same subchannel is farther, and thus the inter-cell interference becomes less. Similarly, for a smaller K, the situation is reversed. In the three cases as shown in Figures 14.3 through 14.5, the outage probability is decreasing as the frequency reuse factor K increases. For example, when $K = 1$ and SINR target is -5 dBM, the outage probability is almost above 0.1 in Figure 14.3. However, when $K > 1$, the value of the outage probability is below 0.01 as shown in Figure 14.4.

Given an invariable SINR target and K, outage probability closely relates with L, and a smaller outage probability will be obtained as L increases. For instance, when SINR target is -5 dBM and $K = 3$ as in Figure 14.4, the value of the outage probability quickly declines from 0.01 to 0.0001. The positive effect of multipath diversity overcomes the drawback of fading and interference. Thus, for different SINR targets, we can choose appropriate K and L to ensure that all the active UEs have good service. In a word, our numerical results suggest that there is an optimal value L with frequency reuse factor K for different outage probability requirements.

14.4 Key Techniques in TD-LTE

14.4.1 Beamforming Technique

In TD-LTE systems, the eight-antenna uniform linear array (ULA) separated by 0.5 wavelengths is usually built for the outdoor macro and micro scenarios in eNB to enhance the UE receiving power [i.e., beamforming (BF) diversity gain, and mitigate interuser interferences, especially inter-cell interferences and increase cell-edge throughput and coverage]. Based on the deployed ULA smart antenna configurations in eNB, the single-user-based BF (SU-BF) scheme can be utilized when two antennas are deployed in the UE. On the eNB side, data flow are demultiplexed into two data streams. Different modulation and coding schemes (MCSs) can be applied to the two data streams, depending on the instantaneous channel condition of each transmit antenna, which is signaled by the UE. The two data streams are precoded, if needed, and then mapped to the two transmit antennas after spreading and scrambling operations. The UE can separate different data streams through MIMO detector and measure the channel quality of each transmit antenna. According to the receiving channel conditions in UL, the CSI for downlink is estimated at eNB.

In TDD system, since the uplink and downlink channel reciprocity can be used to obtain channel state information, this is beneficial to use beamforming scheme to realize space division multiple access (SDMA). The eNB can easily obtain the multiuser channel information from uplink channel information and effectively grouping the UEs for implementing SDMA. The CSI is obtained by utilizing the channel reciprocity of uplink and downlink. According to the DOA etc. information, eNB pair user groups that satisfy some certain constraint condition. The beam weights for each user group are generated based on particular algorithms, for example, the null-widening method. By now, the beam corresponding to the weight of a certain user has the following feature: the main lobe is formed on the DOA direction of the target user, and the widening null steering is formed on the DOA directions of other users which are in the same group with the target user. Therefore, the interuser interference is reduced. Moreover, the transmission weights of different users in the same group can also be orthogonalized to further eliminate the interference. Finally, the eNB schedules the optimal user group according to a certain criterion and multiply the data stream of each user in the user group by the corresponding beamforming weight to transmit through antennas. Downlink beamforming is completed then. The system model for MU-BF is described in Figure 14.6 [8].

Considering a fading channel, the received signal vector at UE is:

$$Y = HWS + N \tag{14.22}$$

where H is the DL MIMO channel, S is the transmission data vector precoded by W at eNB, and N is the additive white Gaussian noise.

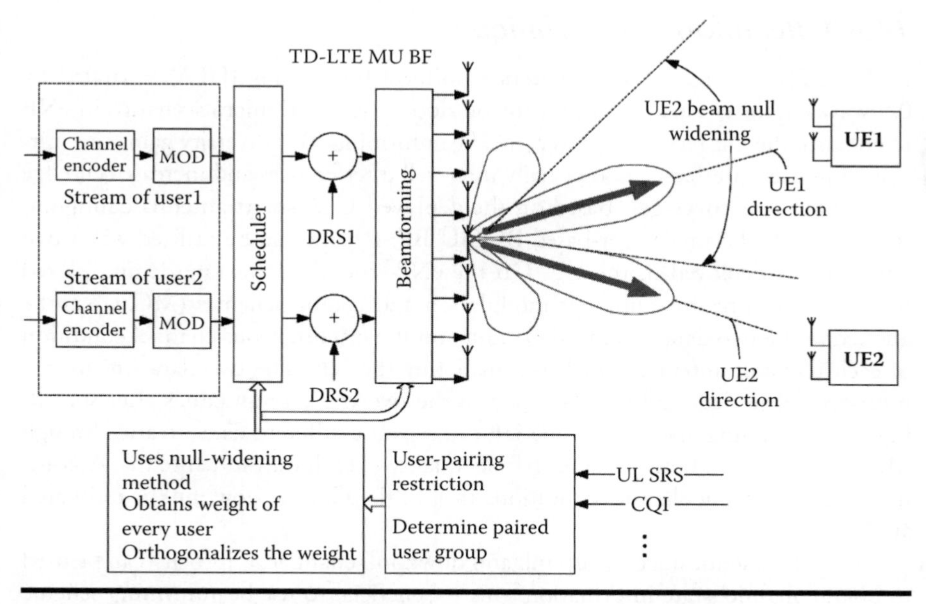

Figure 14.6 Block diagram of MU BF in a TD-LTE system.

In single-layer beamforming, the optimal precoding vector is:

$$W_o = \arg\max_{W}\{W^H H^H H W\} \tag{14.23}$$

That is, the precoding vector W_o is the eigenvector of matrix $H^H H$ corresponding to the largest eigenvalue.

In multilayer beamforming, one criteria to select the precoding matrix is:

$$W_o = \arg\max_{W}\{trace(W^H H^H H W)\} \tag{14.24}$$

That is, the precoding matrix W_o is composed of eigenvectors corresponding to the two largest eigenvalues of matrix $H^H H$.

The spectral efficiency in beamforming mode is evaluated with system level simulation in the following context. Up to eight polarized antennas are used in beamforming transmission with [$+45^a$, -45^a] polarization direction as shown in Figure 14.7. Two polarized antennas are assumed at UE.

Uplink-downlink configuration 1 is considered in which case each half radio-frame consists of two downlink subframes, one special subframe, and two uplink subframes. It should also be noted that downlink channel quality indicators (CQIs) calculated at UE for single-layer beamforming and two layers beamforming are based on an ideal channel estimation. Moreover, 2 codewords (CWs) beamforming with rank adaptation is used.

Cumulative distribution function (CDF) of user throughput (bps/Hz) is shown in Figure 14.8. The evaluation results of mean spectral efficiency and cell-edge spectral efficiency per cell are shown in Table 14.1. It can be inferred that nearly 26% mean spectral efficiency gain can be obtained with two layers beamforming over single-layer beamforming.

14.4.2 Inter-cell Coordination

Interference mitigation techniques in LTE systems can be classified into three major categories, such as interference cancelation through receiver processing, interference randomization by frequency hopping, and interference coordination achieved by

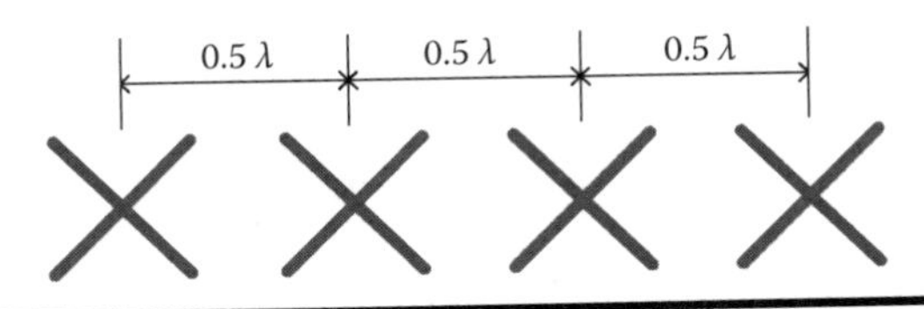

Figure 14.7 Configuration of eight dual polarized antennas.

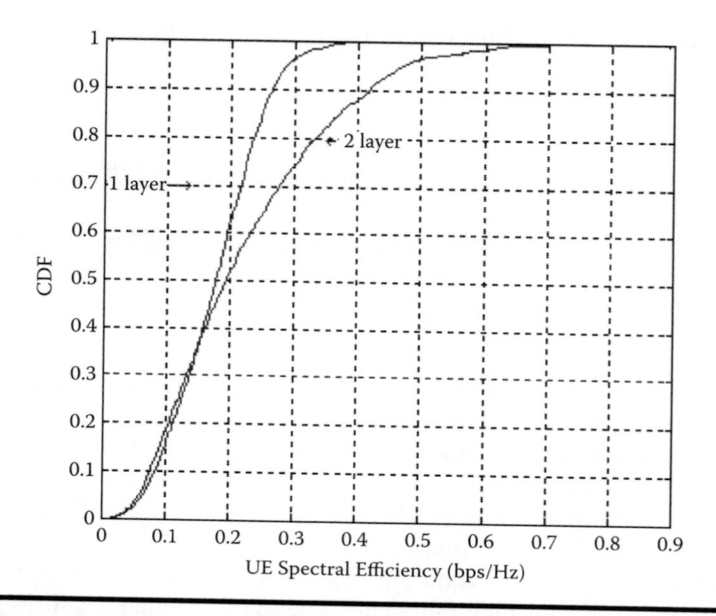

Figure 14.8 CDF of user throughput (bps/Hz).

restrictions imposed in resource usage in terms of resource partitioning and power allocation. For the interference coordination, the partial frequency reuse (PFR) [9] and soft frequency reuse (SFR) [10] are two variations. As shown in Figure 14.9 [11], for both PFR and SFR, only a part of the total resources is used for transmission to/from cell-edge user, therefore reducing the interference experienced by these users (at the cost of a reduced bandwidth). Note that for either downlink or uplink transmission, the orthogonal resource utilization of cell-edge users can reduce the inter-cell interference only when the proper power control mechanism is implemented, otherwise no coordination gain is available. Take the downlink transmission, for instance, if no power control mechanism is carried out, the resource not allocated to the cell-edge users may be used by the cell-center users, and it makes no difference for the interference suffered by the cell-edge users. For the uplink transmission scenario, the situation is almost the same.

Table 14.1 DL Spectral Efficiency in TD-LTE

Index	*Mean Spectral Efficiency (bps/Hz)*	*Cell-Edge Spectral Efficiency (bps/Hz)*
TDD LTE(single-layer beamforming)	1.77	0.066
TDD LTE(multi-layer beamforming)	2.23	0.06

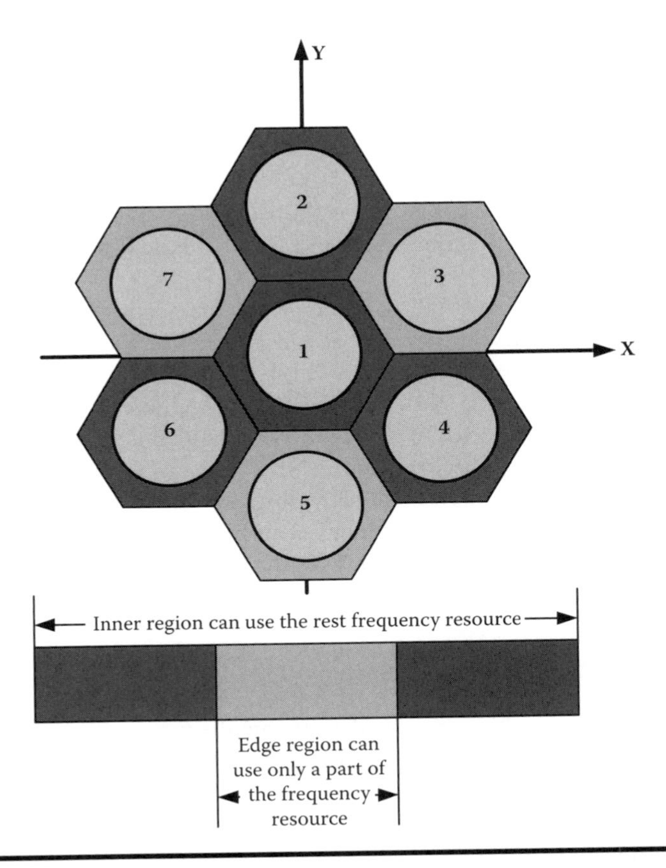

Figure 14.9 Illustration of ICIC.

Together with slow power control, inter-cell interference coordination can improve the throughput of the cell-edge users significantly. Because the radio resources used for transmission to/from UEs in a cell are controlled by the scheduler in the eNB, it makes sense to implement the coordination as part of the scheduling decision. In this way, coordination for downlink or uplink can simply be seen as constraints to the scheduler. The constraints can either be configured semistatically by a higher node [e.g., the radio network controller (RNC)] or derived and continuously updated by the eNB using an adaptive algorithm.

14.4.3 Scheduling and Link Adaptation

A scheduling algorithm is a method by which data flows are given access to system resources (e.g., transmission time, bandwidth). This is usually done to load balance a system effectively or achieve a target quality of service. The need for a scheduling algorithm arises from the requirement for most modern systems to perform multiplexing (transmit multiple flows simultaneously). Generally, there are three scheduling algorithms such as round-robin (RR), proportional-fair (PF), and maximum carrier-to-interference ratio (max *C/I*).

The RR algorithm allocates RB to users sequentially in rotation and can provide fairness. The max-*C/I* algorithm allocates RB to the user whose channel can support the highest MCS, but the user fairness is the worst. The PF algorithm can make a trade-off between system performance and user fairness. It allocates RB to user according to the PF factor.

Adaptive modulation and coding is one of the link adaptive techniques that is commonly used to enhance the spectral efficiency and user throughput in current and next-generation wireless system. In 3GPP LTE, quadrature phase-shift-keying (QPSK), 16-quadrature amplitude modulation (16-QAM) and 64-QAM with a wide range of coding rates are selected to adapt time-varying channel conditions. The UE can be configured to report CQI to assist the eNB in selecting an appropriate MCS for downlink transmissions. A simple method by which an UE can choose an appropriate CQI value could be based on a set of block error rate (BLER) thresholds. The UE would report the CQI value corresponding to the MCS that ensures $BLER \leq 10^{-1}$ based on the measured received signal quality [12].

$$CQI_n = \min\left[\text{MCS}\left(SINR \geq SINR_{\text{target}} \mid BLER \leq 10^{-1}\right)\right] \qquad (14.25)$$

The CQI can be used not only to adapt the modulation and coding rate to the channel conditions, but also for the optimization of the time/frequency selective scheduling.

As shown in Figure 14.10, generally the CQI feedback model consist of four steps: measuring signal to SINR, calibrating SINR according to a link adaptation

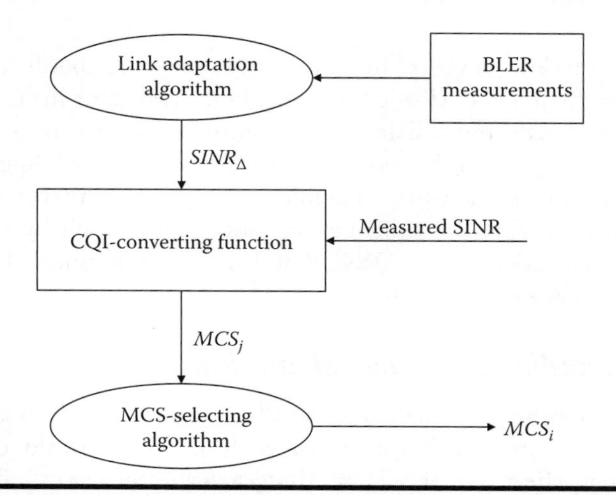

Figure 14.10 AMC procedure including CQI feedback and MCS selection.

algorithm, converting SINR values into discrete MCS index, and reporting CQI by a certain CQI feedback scheme. When receiving the CQI, the eNB decides which MCS is used on RBs that schedules to the given UE. All RBs allocated to this UE in a TTI should use the same MCS by the MCS-selecting algorithm.

Let M_j be the constellation size of MCS_j, r_j be the code rate associated with MCS_j and T_{slot} be the duration of a time slot. Then, the bit rate T_{slot}, which corresponds to a single RB is presented by:

$$R_j = \frac{r_j \log_2 M_j}{T_{slot}} N_{symbol} N_{subcarrier} \tag{14.26}$$

From Equations 14.25 and 14.26, it is known that R_j is a monotone increasing discrete function with the factor of SINR. So we can use MCS to stand for the channel quality and capacity directly.

Let 0, 1, 2, ... , J be the full MCS set. The set increases monotonically with the corresponding target SINR in Equation 14.25. Assume that U is the number of simultaneous users and UE_i ($i = 1, 2, \ldots, U$) feeds back the highest N_i CQI values of all RBs and the rest of the RBs who are not reported are assigned to MCS_0, the lowest MCS in the full MCS set. Of course, when N_i equals to N_{RB}, the number of the RBs at the whole system bandwidth, UE_i reports the full CQI values. MCS_j is the N_i CQI index set.

In a non-MIMO configuration system, all RBs allocated to a given UE in a TTI must use the same MCS. If the UE uses MCS_j, then only certain RBs whose channel quality supports the MCS_j can be scheduled to this user. For example, suppose $N_i = 5$, and

$$\begin{aligned} 0 &\le MCS_1(RB_3) < MCS_2(RB_5) < MCS_3(RB_1) \\ &< MCS_4(RB_4) < MCS_5(RB_2) \le J \end{aligned} \tag{14.27}$$

If MCS_3 is used, then only RB_1, RB_2, and RB_4 can be assigned to UE_i because only these RBs have good enough channel qualities to support the MCS_3 or higher. If RB_3 and RB_5 use MCS_3, it would result in unacceptable error rates for these RBs. In other words, all five RBs can be selected when using MCS_1 but it decreases the transmission rate on RB_1, RB_2, RB_4, and RB_5 and the total bit rate for UE_i might not be the maxim. This suggests that there is an optimal MCS_j which can be the throughput of UE_i.

If there are many users, the problem becomes to select the optimal MCSs for all users by which the system can get the maximal total throughput. Let U be the number of simultaneous users and N be the total number of RBs. The problem can be formulated as [13]:

$$\max_{A,B} \sum_{i=1}^{U} \sum_{n=1}^{N} a_{i,n} \sum_{j=1}^{J} b_{i,j} R_j \tag{14.28}$$

where J is the maximal value in MCS_j. In a non-MIMO system, each RB can be allocated to, at most, one user, and all allocated RBs for a user in a TTI must use the common MCS. In order to ensure these, we use the constraints as follows:

$$\sum_{i=1}^{U} a_{i,n} = 1 \quad \text{and} \quad \sum_{j=1}^{J} b_{i,j} = 1 \tag{14.29}$$

$$a_{i,n}, b_{i,j} \in \{0, 1\} \forall i, j \tag{14.30}$$

The objective in Equation 14.28 is to find optimal values in $A = \{a_{i,n}\}$ and $B = \{b_{i,j}\}$ to maximize the total bit rate. Although it can be solved by optimization techniques such as integer linear programming techniques, the complexity could not be neglected. Hence, a suboptimal algorithm is introduced.

The idea of reducing the computational complexity is that changing a joint multi-user optimization scheduling into U parallel single-user optimization problem. It decouples selection of MCSs and RBs. In the first stage, each RB is preallocated to the user according to the scheduling algorithm (e.g. Max *C/I* scheduling algorithm is used), the RB is assigned to the user whose channel quantity can support the highest MCS. In the second stage, the best MCS for each user is selected. In the last stage, the "rest" RBs that are not qualified for the determined MCS are reallocated. The procedure is illustrated in Figure 14.11.

Let $\Psi_{i,n}$ be the index of the RBs assigned for UE_i and $\nu_{i,j}$ the MCS value of the UE_i on these RBs. The size of $\Psi_{i,n}$ is N' while that of $\nu_{i,j}$ is J', and $J' \leq N'$ is due to the probable repetition of MCS. Let the MCS vector c_i for UE_i be:

$$c_i = [c_{i,1}, c_{i,2}, \ldots, c_{i,J'}] \tag{14.31}$$

Similar to Equation 14.28, the suboptimal algorithm is presented as:

$$\max_{c_i} \sum_{n=1}^{N'} \sum_{j=1}^{J'} c_{i,j} R_j \tag{14.32}$$

s.t.

$$\sum_{j=1}^{J'} c_{i,j} = 1, \; c_{i,j} \in \{0, 1\} \; \forall i, j \tag{14.33}$$

The objective in Equation 14.32 is to choose an optimal value in $c_i = \{c_{i,j}\}$ to maximize the bit rate of UE_i. Compared to Equation 14.28, it is much easier to be solved.

We simulate the MCS selection algorithm with the following CQI reporting schemes. The system is an OFDM based downlink system with TDD mode.

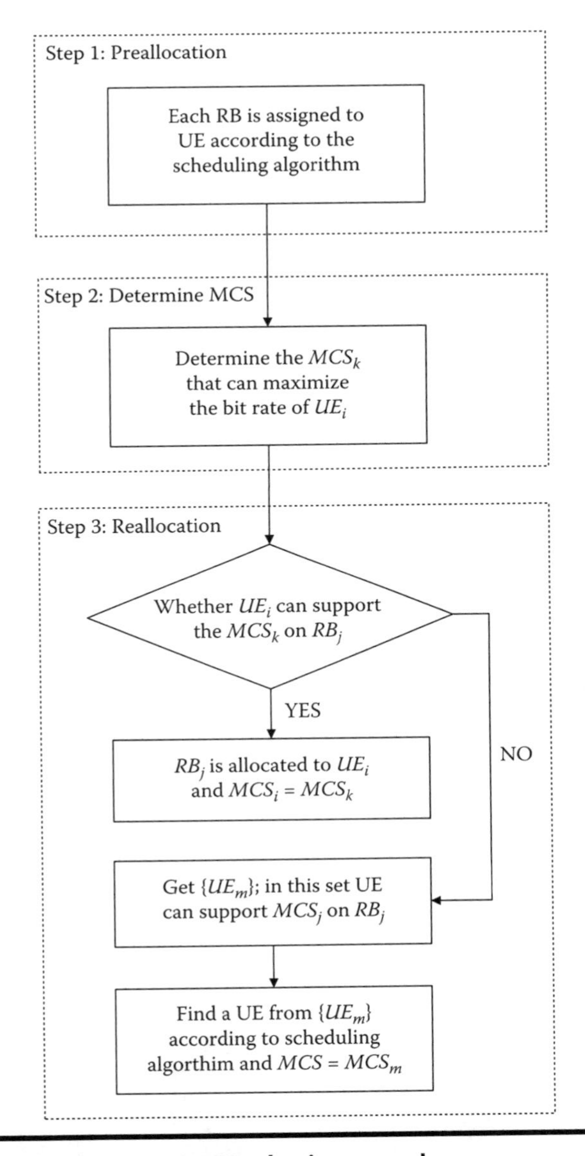

Figure 14.11 The three-step MCS selection procedure.

The simulation parameters and assumptions listed as Table 14.14 shows, and we use three downlink time slots.

- Full set scheme: The UE reports CQI values on all RBs or subbands
- Best M scheme I: The UE reports CQI values on each RB of the best M
- Best M scheme II: The UE reports CQI values on each subband of the best M

First, the impact of common MCS when transmitting data in downlink is investigated. Figure 14.12 shows the average cell throughput in four cases. Cases 1 and 3 adapt special MSCs, which means the RBs assigned to an UE can use different MCS, while in Cases 2 and 4, a common MCS is used. The average cell throughput is decreased by approximately 16% with 50 RB CQI feedback and 13% with 10 RB-group CQI feedback, respectively, when using a common MCS. It is a waste of the control signaling that eNB notices users that MCS adapted on a given RB, however. The results indicate that the common MCS selection procedure is effective. Then, we use the common MCS to evaluate the performance of AMC with limited CQI.

Next, the CQI reporting schemes are studied. Figures 14.13 and 14.14 illustrate the effect of the feedback number M on the average cell throughput and the uplink overhead when the UE reports the CQI value with best M scheme I. With the increase of M, the performance gets better, but the more overhead is used. If $M = 30$, the average cell throughput with the best M reporting scheme is approximate to that with the full set reporting scheme, but the feedback overhead decreases by about 40%. As more users are competing, it is more possible that recourses will be distributed to users with good channel quality. Therefore, the case that of 20 UEs per cell is better than that of 10 UEs per cell when the same M CQI values are reported, as Figure 14.13 shows.

Figures 14.15 and 14.16 show that the effect of feedback number M on the average cell throughput and the uplink overhead when the UE reports the CQI

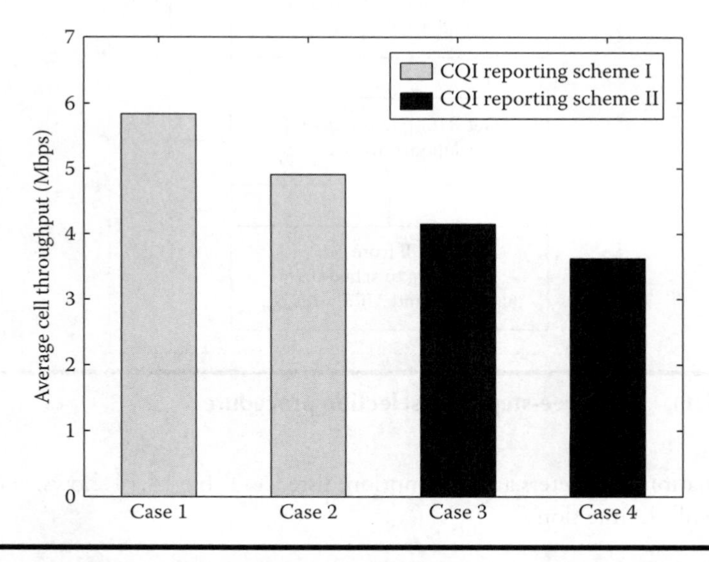

Figure 14.12 Average cell throughput for four different cases. Each UE feeds back CQI on 50 RBs in Cases 1 and 2, respectively; each UE feeds back CQI on 10 subbands (each subband consists of 5 RBs) in Cases 3 and 4, respectively.

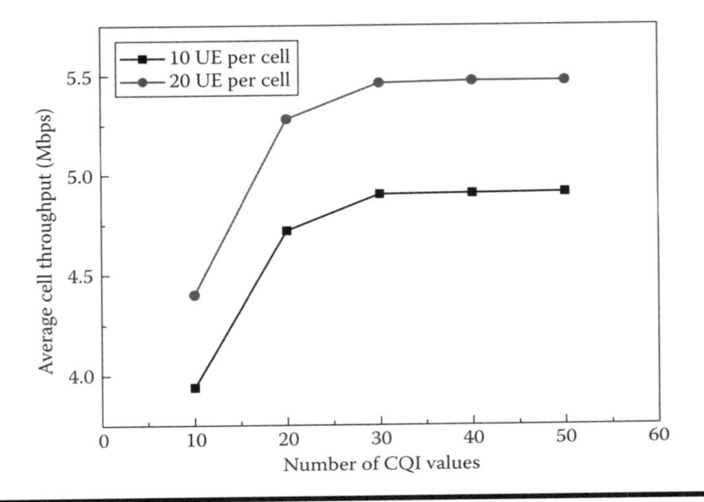

Figure 14.13 Average cell throughput with limited CQI feedback. The UE reports a CQI value with the best M scheme I ($M = 10, 20, 30, 40, 50$). The speed of the UE is 3 km/h.

value with best M scheme II. At the same time, the performance of the system also improves with the increasing M. The number of users has the same effect with CQI on RB. When $M = 5$, the average cell throughput with the best M reporting scheme is approximate to that with the full set reporting scheme, but the feedback overhead will decrease by 35%. The performance of best M scheme I will be better than that

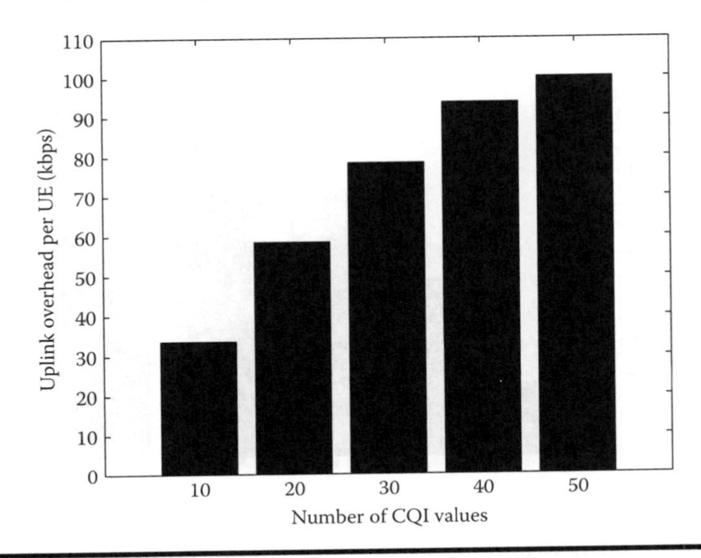

Figure 14.14 The uplink overhead per UE when UE reports CQI with the best M scheme I ($M = 10, 20, 30, 40, 50$).

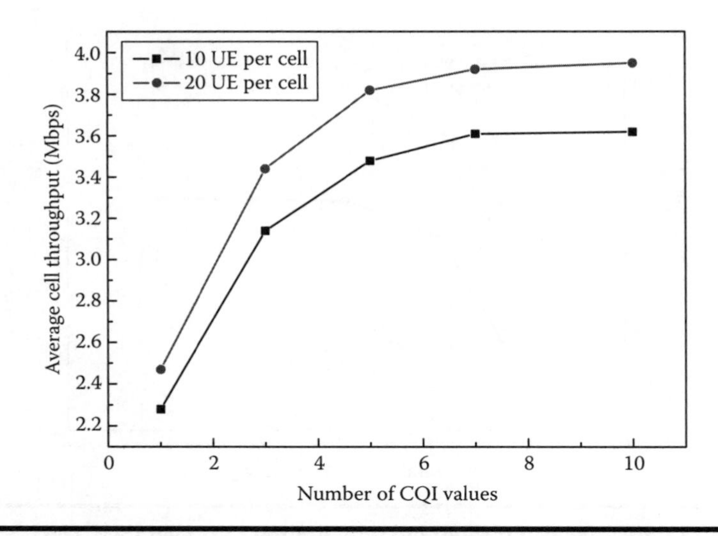

Figure 14.15 The average cell throughput with limited CQI feedback. The UE reports a CQI value with the best M scheme II ($M = 1, 3, 5, 7, 10$). The speed of the UE is 3 km/h.

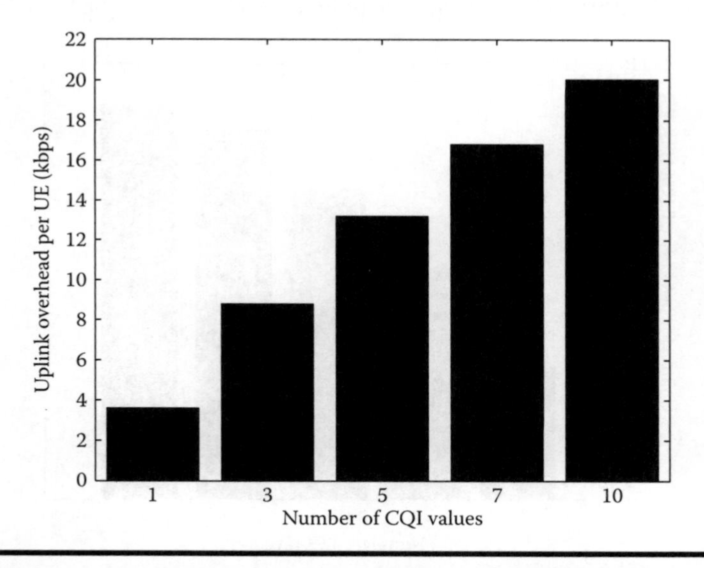

Figure 14.16 Uplink overhead per UE when UE reports CQI with the best M scheme I ($M = 1, 3, 5, 7, 10$).

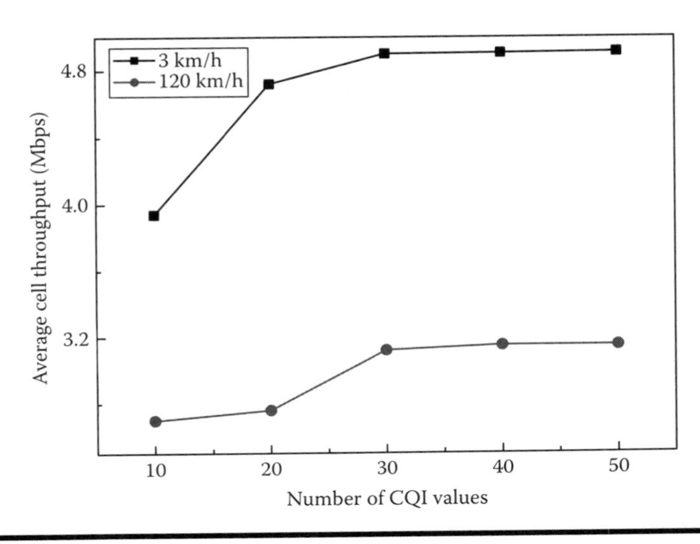

Figure 14.17 Average cell throughput when UE reports CQI with best *M* scheme I.

of best *M* scheme II which limits users with good channel quality to use higher MCS according to the average SINR on the subbands.

Finally, Figures 14.17 and 14.18 show the impact of the speed of UE on the average cell throughput and the uplink overhead. With best *M* scheme I, throughput decreases by approximately 36%, and it descends by 13% with best *M* scheme II. When the UE moves at a speed of 120 km/h, the performance is similar with the two different schemes, but the uplink overhead is more for best *M* scheme I. Hence, in a high-speed scenario, best *M* scheme II is preferred.

14.5 Link Budget of TD-LTE

To obtain link budget results for TD-LTE systems, performance of some basic physical channels should be evaluated, which are mostly from link level simulations.

14.5.1 Link Level Simulation

14.5.1.1 Physical Uplink Shared Channel (PUSCH)

An overview of the simulation process for the physical uplink shared channel (PUSCH) is illustrated in Figure 14.19 according to [4]. The number of RBs considered in this simulation is denoted by N_{RB}, with each RB containing 144 user data symbols. In particular, the 4th and 11th OFDM symbols in each subframe are occupied by a dedicated reference signal (DRS). We assume that there is a

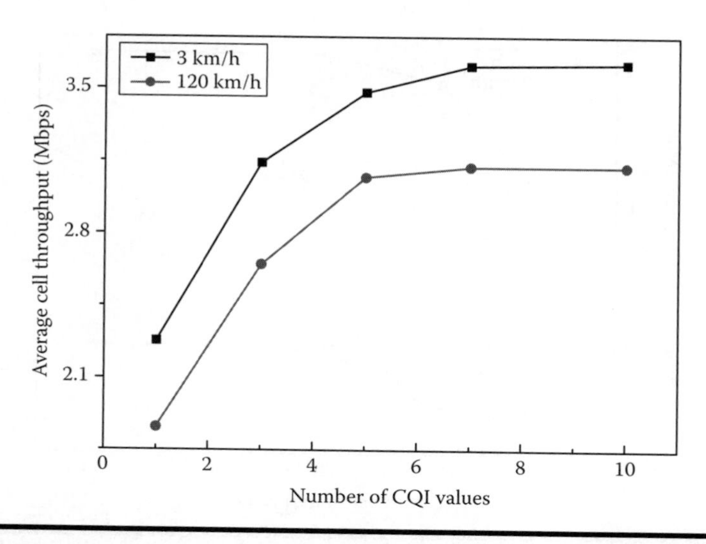

Figure 14.18 Average cell throughput when UE reports CQI with best *M* scheme II.

antenna in the transmitter, and multiple antennas performing MRC at the receiving end. Other simulation parameters for PUSCH are also given in Table 14.2.

In Figures 14.20 and 14.21, BLER and bit error rate (BER) performance for PUSCH with two receiving antennas are given, respectively. As expected, QPSK-1/2 outperforms the other two modulation and coding schemes in aspects of BLER and BER, and 64-QAM 3/4 demonstrates worst performance in both cases.

14.5.1.2 Physical Downlink Shared Channel (PDSCH)

The processing procedure of the physical downlink shared channel (PDSCH) in our link level simulation is similar to PUSCH, except for the multiple antenna technique BF applied to transmitter and minimum mean square error (MMSE) for the receiver to yield diversity gain. Table 14.3 lists some other primary simulation parameters for PDSCH simulation.

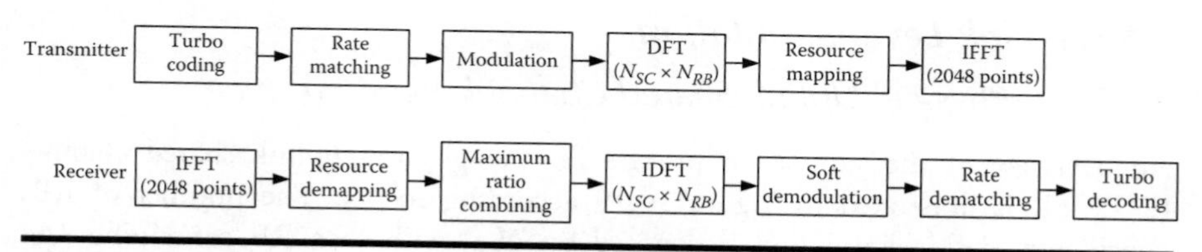

Figure 14.19 Overview of the PUSCH simulation process.

Table 14.2 Simulation Parameters for PUSCH

Parameters	*Values*
Transmission bandwidth	20 MHz
Subframe duration	0.5 ms
Subcarrier spacing	15 kHz
Sampling frequency	30.72 MHz (8 * 3.84 MHz)
FFT size	2048
CP length	(4.69/144)*6
(μs/samples)	(5.21/160)*1
Channel model	EPA5
Antenna configurations	1Tx × 2Rx
N_{RB}	10
MCS	QPSK, 1/2; 16QAM,1/2; 64QAM, 3/4

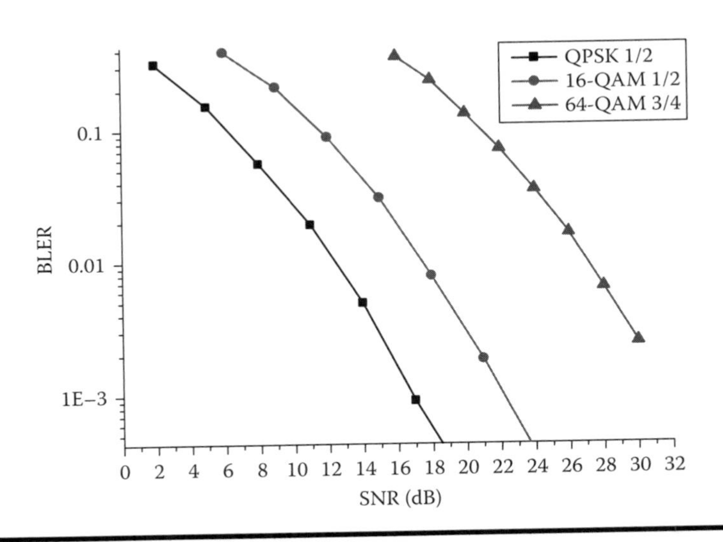

Figure 14.20 BLER performance for PUSCH with 1Tx × 2Rx antennas.

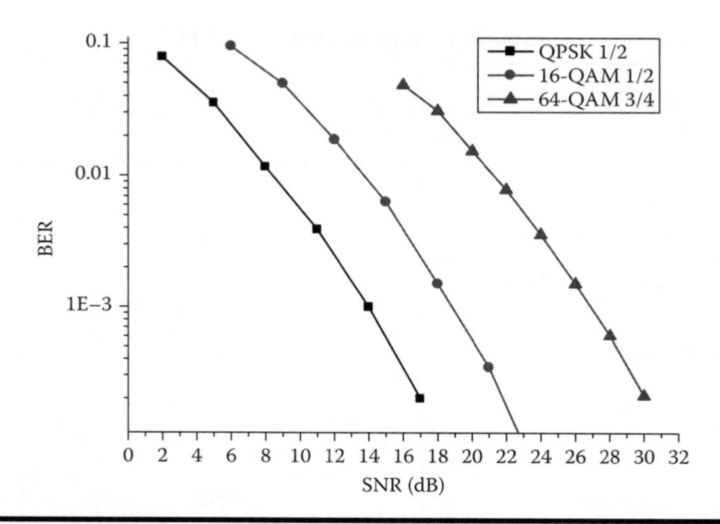

Figure 14.21 BER performance for PUSCH with 1Tx × 2Rx antennas.

Table 14.3 Simulation Parameters for PDSCH

Parameters	*Values*
Transmission bandwidth	20 MHz
Subframe duration	0.5 ms
Subcarrier spacing	15 kHz
Sampling frequency	30.72 MHz (8 * 3.84 MHz)
FFT size	2048
CP length (μs/samples)	(4.69/144) * 6 (5.21/160) * 1
Channel model	EPA5
Antenna configurations	8Tx × 2Rx
N_{RB}	10
MCS	QPSK, 1/2; 16QAM,1/2; 64QAM, 3/4

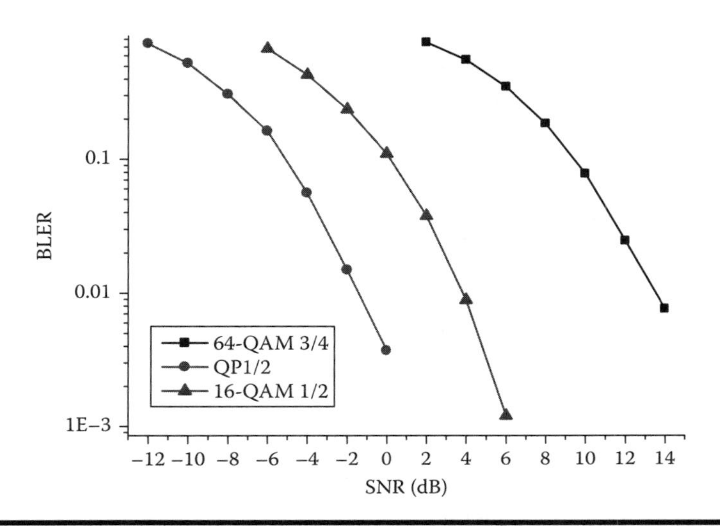

Figure 14.22 BLER performance for PDSCH with 8Tx × 2Rx antennas.

Figures 14.22 and 14.23, respectively, show BLER and BER performances of different modulation and coding schemes for PDSCH. For any given modulation and coding scheme, one can observe from Figures 14.20 and 14.22 that PDSCH requires a lower signal-to-noise ratio (SNR) to achieve a specific BLER than PUSCH. The superiority of PDSCH over PUSCH occurs mostly as a result of additional diversity gain provided by multiple transmit antennas.

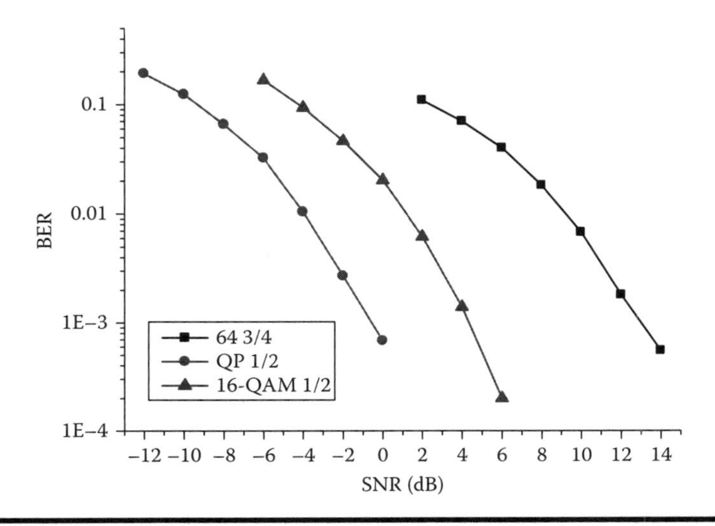

Figure 14.23 BER performance for PDSCH with 8Tx × 2Rx antennas.

Table 14.4 PUCCH Formats

PUCCH Format	*Modulation Scheme*	*Number of bits per Subframe, M_{bit}*
1	N/A	1
1a	BPSK	1
1b	QPSK	2
2	QPSK	20
2a	QPSK + BPSK	21
2b	QPSK + QPSK	22

14.5.1.3 Physical Uplink Control Channel (PUCCH)

The physical uplink control channel (PUCCH) supports multiple formats as shown in Table 14.4. Figures 14.24, 14.25, and 14.26 respectively, elaborate an overview of simulation process for each PUCCH formats. Take format 2a/2b for instance. A block of complex-valued symbols that are modulated from bits $b(0), \ldots, b(19)$ shall be multiplied with a length-12 cyclically shifted sequence and blockwise spread with a time-domain orthogonal sequence, while the single modulation symbol resulted from the modulation of $b(20), \ldots, b(M_{\text{bit}} - 1)$ is used in the generation of the reference signal. Afterward, such resulted modulated symbols are mapped to resources as defined in [4]. In uplink control signal transmission, a single antenna is assumed for UE, whereas multiple antennas perform MRC at eNB. Other simulation assumptions for PUCCH are shown in Table 14.5.

Performance curves in terms of BER and BLER versus SNR for PUCCH format 1/1a/1b and format 2/2a/2b are shown in Figures 14.27 and 14.28, respectively. Because there is only 1-bit for PUCCH format 1 and 1a in each subframe, their curves of BER and BLER overlap with each other in Figure 14.27. It can also be observed from Figure 14.27 that format 1 achieves the best BER/BLER performance, followed by format 1a, with format 1b as the worst of the three formats. In Figure 14.28, PUCCH formats 2/2a/2b demonstrate similar BER/BLER performance, due to their minor difference in the number of bits per subframe. Comparing Figure 14.28 with Figure 14.27 leads to the conclusion that PUCCH format 2/2a/2b yield inferior performance to PUCCH formats 1/1a/1b.

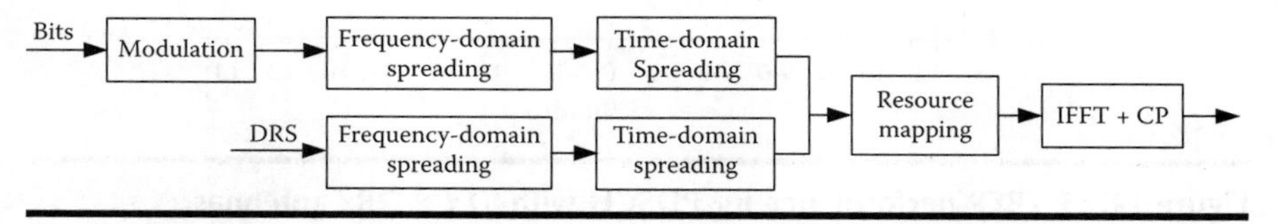

Figure 14.24 Overview of simulation process for PUCCH format 1/1a/1b.

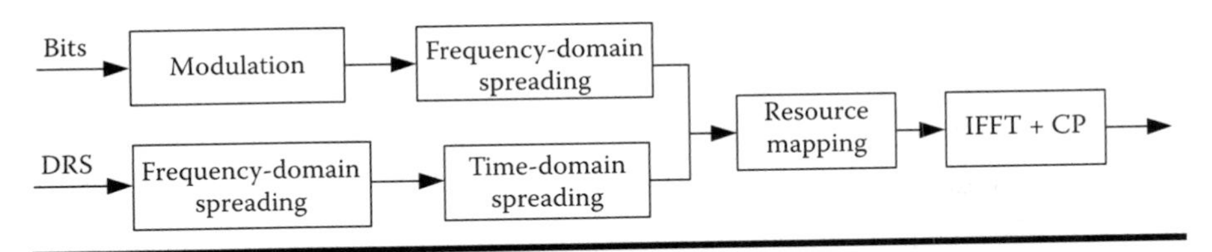

Figure 14.25 Overview of simulation process for PUCCH format 2.

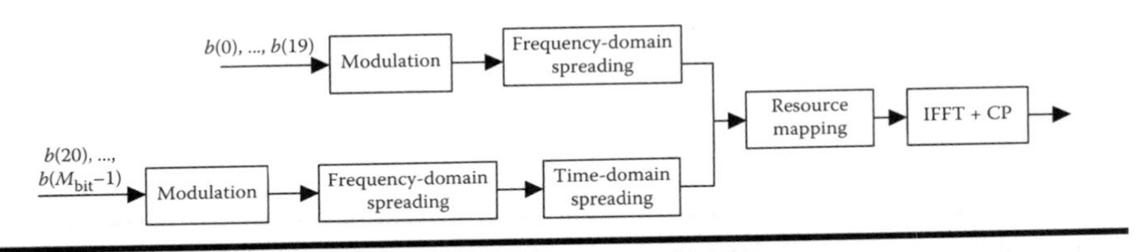

Figure 14.26 Overview of simulation process for PUCCH formats 2a/2b.

Table 14.5 Simulation Parameters for PUCCH

Parameters	*Values*
Transmission bandwidth	20 MHz
Subframe duration	0.5 ms
Subcarrier spacing	15 kHz
Sampling frequency	30.72 MHz (8 * 3.84 MHz)
FFT size	2048
CP length (μs/samples)	(4.69/144) * 6 (5.21/160) * 1
Channel model	EPA5
Antenna configurations	1Tx × 2Rx
N_{RB}	10
PUCCH formats	1/1a/1b/2/2a/2b

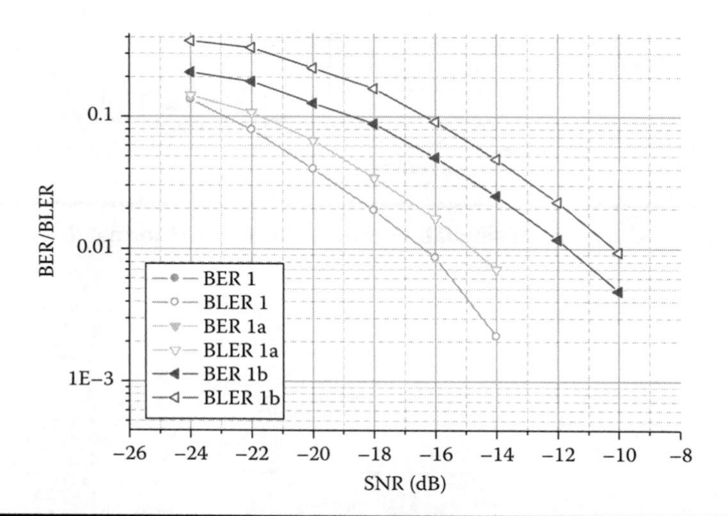

Figure 14.27 BER/BLER performance for PUCCH format 1/1a/1b with 1Tx × 2Rx antennas.

14.5.1.4 Physical Downlink Control Channel (PDCCH)

Simulation for physical downlink control channel (PDCCH) is implemented following the chart given in Figure 14.29. The number of bits transmitted on PDCCH is determined by various downlink control information (DCI) formats. Meanwhile, PDCCH supports four types of PDCCH formats that define the number of control

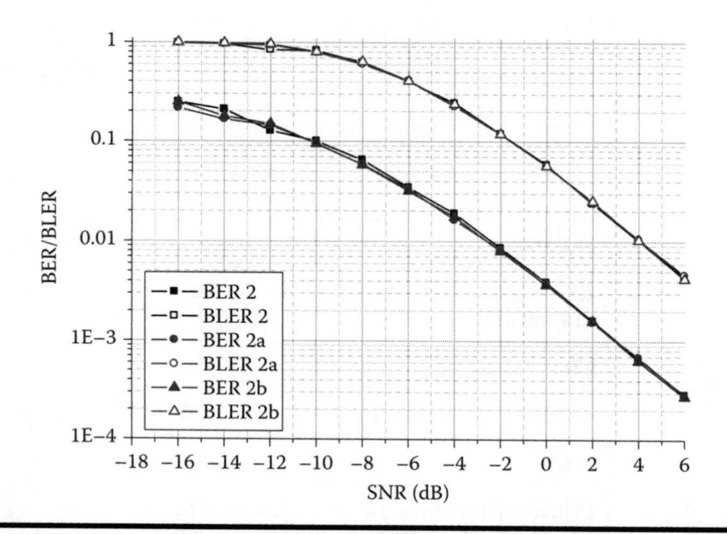

Figure 14.28 BER/BLER performance for PUCCH format 2/2a/2b with 1Tx × 2Rx antennas.

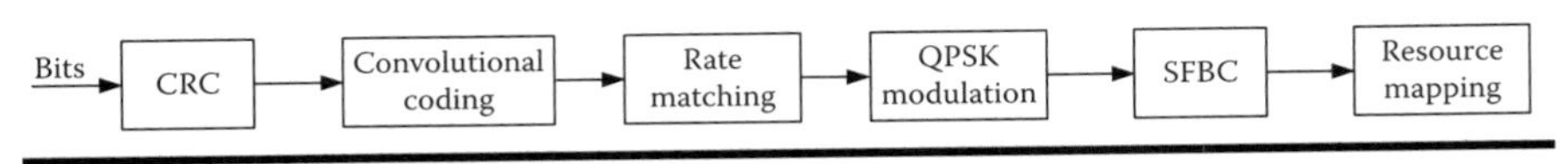

Figure 14.29 Overview of simulation process for PDCCH.

channel elements (CCEs) that could be used for DCI message transmission. Multiple antennas are assumed to be used in both transmitter and receiver. SFBC is performed before modulated symbols being mapped to physical resources. Table 14.6 gives other simulation assumptions for PDCCH.

In Figures 14.30 and 14.31, BER and BLER versus SNR curves are shown for PUCCH format 0 when one, two, four, and eight CCEs are used. Evidently, the performance of PDCCH is greatly enhanced as the number of CCEs used for transmission of a particular PDCCH increases, and the most distinctive performance gap exists between 1-CCE and 2-CCE curves at a high-SNR regime.

14.5.2 TD-LTE Link Budget

The link budget calculations estimate the maximum allowed signal attenuation, called path loss, between the UE and the eNB antenna. The maximum path loss allows the maximum cell range to be estimated with a suitable propagation model,

Table 14.6 Simulation Parameters for PDCCH

Parameters	*Values*
Transmission bandwidth	20 MHz
Subframe duration	0.5 ms
Sub-carrier spacing	15 kHz
Sampling frequency	30.72 MHz (8 * 3.84 MHz)
FFT size	2048
CP length (μs/samples)	(4.69/144) * 6 (5.21/160) * 1
Channel model	EPA5
Antenna configurations	2Tx × 2Rx
N_{RB}	10
DCI (format/number of bits)	Format 0/31

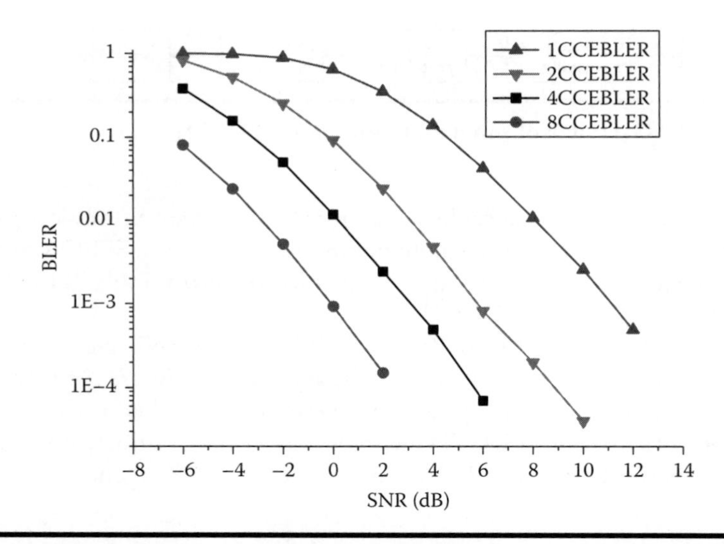

Figure 14.30 BLER performance for PDCCH DCI format 0 with 2Tx × 2Rx antennas.

such as COST231-Hata. The cell range gives the number of eNB sites required to cover the target geographical area. The link budget calculation can also be used to compare the relative coverage of the different systems.

Here we focus on the differences between link budgets for TDD and FDD modes. The differences relate mainly to the limited maximum UE transmit power.

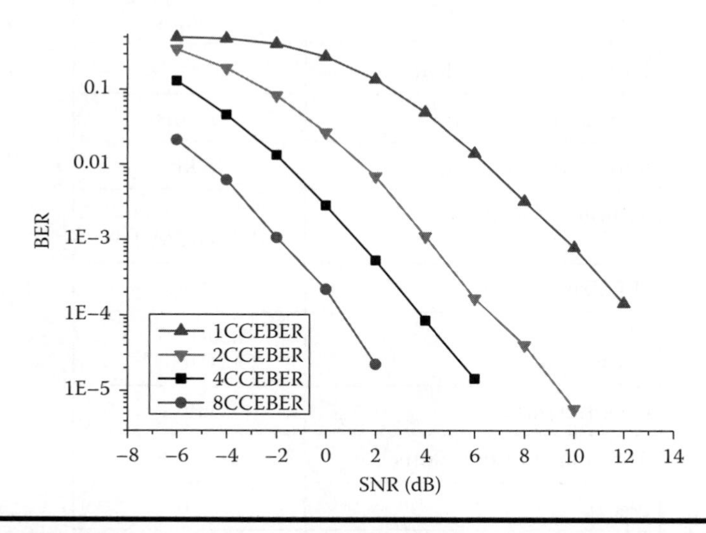

Figure 14.31 BER performance for PDCCH DCI format 0 with 2Tx × 2Rx antennas.

The TDD UE cannot transmit continuously as the transmission must be switched off during the downlink reception. The UE will thus need to transmit with a larger bandwidth and a lower power density to achieve a similar bit rate to an FDD system. The lower power density is because the UE transmitter is limited on total maximum power, not on power per hertz.

For a TDD mode link, budget results are for assuming downlink: uplink share is 2:2, a propagation model given by COST231-Hata work in frequency carriers in 2350 MHz.

$$\text{Propagation loss} = 37.8 + 34.8 \, log \, (\text{distance}) \tag{14.34}$$

with distance given in meters. Other parameters for the TD-LTE link budget for uplink are introduced in Table 14.9. More details on the link budget are given in Table 14.8, which provides references for the SINR targets used in the uplink link budget. For comparison, Tables 14.7 and 14.10 provide similar information for the downlink link budget.

Figure 14.32 shows coverage provided by the CQI and ACK/NACK information transported on the PUCCH, compared to the physical random access channel (PRACH). By comparing, the coverage of RACH sets the limit of UL control channel.

From a similar calculation, the coverage of an average border throughput of 500 kbps supported based on 10 RB (1TTI) PUSCH is about 700 m.

Figure 14.33 shows 90%-ile coverage (log-normal margin of 6.8 dB is used given 10 dB standard deviation). All control channels should be able to match but do not need to exceed P-BCH coverage.

The 36-bit and 48-bit PDCCH for transporting scheduling grants (and other grant types) require a power boost for full case 3 coverage, whereas the 24-bit PDCCH format (possibly for Paging, RACH, and P-BCH grants) does not. The 48-bit PDCCH is only for downlink scheduling grants. Coverage is extended by power boosting via puncturing (e.g., a CCE) or borrowing between PHICH, PDCCH, and PCFICH. A total of 12 resource elements (REs) can support eight PHICH where 6-dB boosting of a PHICH for full Case 3 coverage can be obtained by puncturing a single CCE or by borrowing power from other PHICH and PDCCH.

About PDSCH, the coverage of an average border throughout of 500 kbps supported based on 10 RB (1TTI) PDSCH is about 1200 m. With the same border throughout, PUSCH sets the limits to the traffic channel.

The above results show that Case 3 downlink control channel coverage can be achieved via power boosting using puncturing or borrowing from other PHICH and PDCCH. The coverage of other downlink and uplink control channels should be able to match but do not need to exceed P-BCH coverage.

PUCCH does not limit the UL coverage, and a RACH preamble repetition is required for optimized coverage. RACH coverage sets the limits to overall TD-LTE UL control channel coverage.

Traffic channel coverage is limited by PUSCH for a 500-kbps border throughout.

Table 14.7 Downlink Link Budget Parameters for TD-LTE

Simulation Conditions	
DL:UL	2:2
Antenna	82 beamforming
System bandwidth (MHz)	20M
No. of occupied RBs	10
Base Station EIRP	
Base Tx power (dBm)	46
Maximum power per UE (dBm) a	36
Transmission line loss (dB) b	2
Base Tx antenna gain (dBi) c	16
Base EIRP (dBm) A	$a - b + c$
Subscriber Sensitivity	
MS antenna gain (dBi) d	0
Thermal noise (kT) (dBm/Hz)	−174
MS noise power (dB) e	−112(10RB)
MS Noise figure (dB) f	5
SNR for MCS level (dB) g	Include MIMO gain
Subscriber Sensitivity B	$e + f + g - d$
System Margin	
Shadow fading margin (dB) h	6.8
Interference margin (dB) i	3–9
MS body loss (dB) j	0
Total system margin (dB) C	$h + i + j$
Maximum allowable pathloss	$A - B - C$

Table 14.8 Required UL SINR for Target Error Rate

DL Control Channel	*Required E_s/N_0 for Target ER*	*Reference*
ACK/NACK-PUCCH	$E_s/N_0 = -4.1$ dB for 0.1% BER	[14]
CQI-PUCCH	$E_s/N_0 = -10$ dB (1 bit) for 1% FER	[15]
	$E_s/N_0 = -7$ dB (5 bits) for 1% FER	
	$E_s/N_0 = -4$ dB (10 bits) for 1% FER	
RACH	$E_s/N_0 = -11.5$ dB for 1% PER	[16]

14.6 System Performance Evaluations

System level simulation is an effective method to evaluate system performance and analyze radio resource management (RRM) algorithms. The core feature of system level simulation lies in the fact that more attention has been put on the higher level of a system, while the specific signal processing procedure has been simplified. In a system level scenario, many UEs are distributed in a certain manner (usually uniformly) in the entire network, and establish communication links with its own serving eNB according to a specific access algorithm. Each eNB allocates the resource available to its UEs based on particular scheduling algorithms and each UE would calculate its received SINR and make a decision whether the received transport block can be decoded successfully. If the received SINR is greater than the target threshold, the transport block is more probable to be decoded inerrably, otherwise the failure probability becomes higher.

The features of system level simulation for TD-LTE system include the frame structure, the wraparound technique, the channel interface, the SINR mapping method and so on. These features would be discussed in detail as follows.

14.6.1 Frame Structure

As discussed in Section 14.1, the frame structure of TD-LTE is illustrated in [4], which is also known as frame structure type 2. The major feature of frame structure type 2 is its flexible configuration of special subframe and uplink-downlink ratio. Specifically, there are nine possible special subframe configurations and seven uplink-downlink configurations, and the details are listed in Tables 14.11 and 14.12. For Table 14.12, letter D means downlink subframe, letter S means special subframe, and letter U means uplink subrame.

Based on the frame structure illustrated above, a global timer is needed to control the network to work synchronously. In this section, uplink-downlink configuration 2 and special subframe configuration 8 are implemented as an instance. Deploying

Table 14.9 Uplink Link Budget Parameters for TD-LTE

Simulation Conditions	
DL:UL	2:2
Antenna	1*8
System bandwidth (MHz)	$20M$
No. of occupied RBs	10
UE Station EIRP	
UE Tx power (dBm) a	4623
UE Tx antenna gain (dBi) b	0
UE EIRP (dBm) A	$a + b$
Base Station Sensitivity	
BS Rx antenna gain (dBi) d	16
Transmission line loss (dB) e	2
Thermal noise (kT) (dBm/Hz)	−174
MS noise power (dB) f	$-112(10RB)$
MS noise figure (dB) g	5
SNR for MCS level (dB) h	Include MIMO Gain
Subscriber sensitivity B	$e + f + g + h - d$
System Margin	
Shadow-fading margin (dB) h	6.8
Interference margin (dB) i	$3\ to 9$
MS body loss (dB) j	0
Total system margin (dB) C	$h + i + j$
Maximum allowable path loss	$A - B - C$

Table 14.10 Required DL SINR for Target Error Target Error Rate

DL Control Channel	*Required E_s/N_0 for Target ER*	*Reference*
PCFICH	$E_s/N_0 = -2$ dB for 0.1% BER	[17]
P-BCH	$E_s/N_0 = -8.3$ dB for 1% BER	[18]
PHICH (12 REs per 8 PHICH)	$E_s/N_0 = -2.8$ dB for 0.1% BER	[19]
PDCCH (8 CCEs)	$E_s/N_0 = -8.3$ dB (48-bit) for 1% FER	[20]
	$E_s/N_0 = -5.3$ dB (36-bit)	
	$E_s/N_0 = -7.3$ dB (24-bit)	
PCH message	$E_s/N_0 = -8.3$ dB for 1% PER	

different frame configuration results in different uplink and downlink resources available and also has an impact on the CQI feedback mechanism.

14.6.2 Wraparound Technique

In system level simulations, cellular network modeling is limited to a certain finite number of cells with strict boundaries. If a certain UE is located in the boundary cells, the received interference from surrounding cells may be much less than that when the UE is positioned in a central cell because there is no interference received from the area outside of the boundaries. Besides, when mobility is taken into consideration, an UE may leave the network and lose its link to the network due to the limited boundaries. To avoid these problems, a wraparound technique has been implemented in the system level simulator.

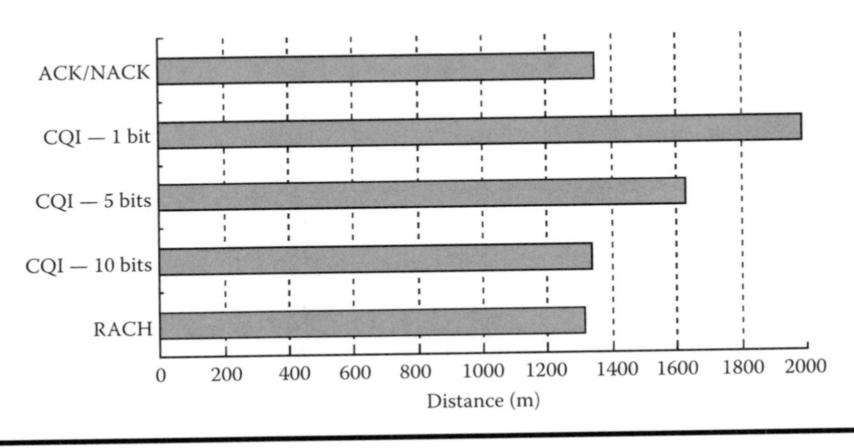

Figure 14.32 TD-LTE UL channel coverage.

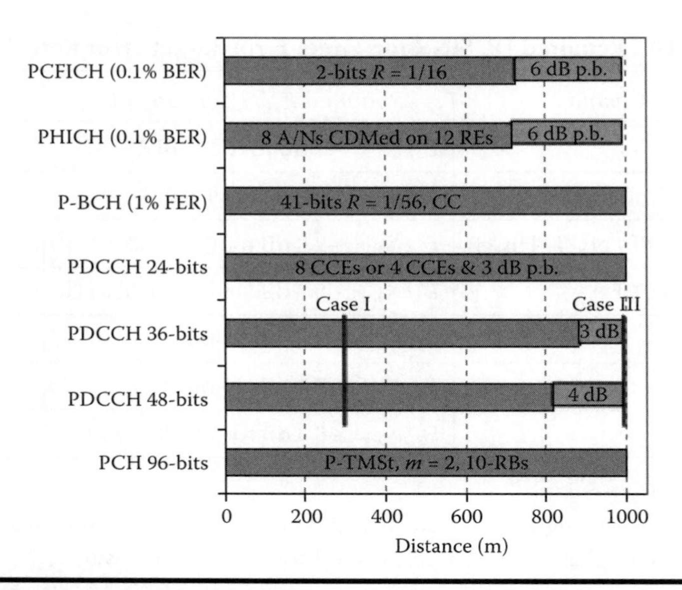

Figure 14.33 TD-LTE DL channel coverage.

Table 14.11 Configurations of Special Subframe (Unit: Symbol)

Special Subframe Configuration	*Normal Cyclic Prefix*			*Extended Cyclic Prefix*		
	DwPTS	*GP*	*UpPTS*	*DwPTS*	*GP*	*UpPTS*
0	3	10	1	3	8	1
1	9	4	1	8	3	1
2	10	3	1	9	2	1
3	11	2	1	10	1	1
4	12	1	1	3	7	2
5	3	9	2	8	2	2
6	9	3	2	9	1	2
7	10	2	2	—	—	—
8	11	1	2	—	—	—

Table 14.12 Uplink–Downlink Configurations

Uplink–Downlink Configuration	*Downlink-to-Uplink Switch-Point Periodicity*	*Subframe Number*									
		0	*1*	*2*	*3*	*4*	*5*	*6*	*7*	*8*	*9*
0	5 ms	D	S	U	U	U	D	S	U	U	U
1	5 ms	D	S	U	U	D	D	S	U	U	D
2	5 ms	D	S	U	D	D	D	S	U	D	D
3	10 ms	D	S	U	U	U	D	D	D	D	D
4	10 ms	D	S	U	U	D	D	D	D	D	D
5	10 ms	D	S	U	D	D	D	D	D	D	D
6	5 ms	D	S	U	U	U	D	S	U	U	D

The wraparound technique considered here is essentially a geometry mapping method to create an infinitely large network. To illustrate this technique briefly, the following terms are defined and used:

- **Real Cell:** The network entity exists in practice (i.e., the eNB and the areas where the UEs are located, as shown in the white area).
- **Virtual Cell:** The area where none of the real network entities exist, as shown in the color area surrounding the real cells.
- **Reference Cell:** A certain real cell where the calculation of the receiving signals is being carried out.
- **Mapping Cell:** S certain real or virtual cell that is mapped to from a particular reference cell according to the wraparound rule.
- **Real Coordinate:** The coordinate of a network entity in a real cell.
- **Mapping Coordinate:** The coordinate of a network entity in a mapping cell. When the mapping cell is the real cell itself, the mapping coordinate is the same as the real one.

Figure 14.34 illustrates the wraparound technique clearly. When an UE is located at the boundary cell (e.g., cell 7, through wraparound), a two-layer surround cells can be modeled by mapping some real cells to the virtual ones. The index of a virtual cell is composed of two parts, the one out of the bracket stands for the virtual index, whereas the one in the bracket denotes the mapped cell. Because the real network entities are only distributed in the real cells, when an UE is moving out of this area, it will be mapped to the other side of the area. For example, when an UE is moving from cells 9 to 22, then when the UE is about to cross the boundary, it will be mapped to the real cell 13 and enter cell 12. In this way, both the receiving and mobility problems can be solved.

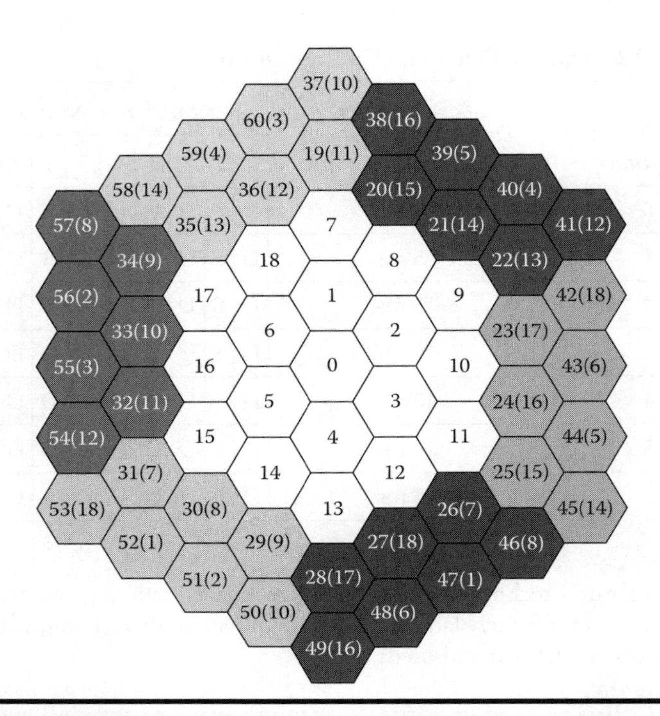

Figure 14.34 Illustration of the wraparound technique.

14.6.3 Channel Interface

Small-scale fading is very important in system level simulation because it has significant impact on the fast packet scheduling. In a TD-LTE system level simulator, extended spatial channel model (SCM-E) is used to model the fading in a small scale.

SCM-E is developed by the European WINNER project and it applies for 2 to 5 GHz frequency bands and supports bandwidths up to 100 MHz in three different outdoor environments [21]. As an enhanced version of SCM, the line-of-sight (LOS) option of SCM-E is extended to all scenarios. By adding short-term time-variability of some channel parameters within each drop, a time-varying feature is available in SCM-E.

Because in a system level simulation scenario, there are usually many UEs (e.g., 570) in the entire network. If the small-scale channel coefficients are generated in real time, the computation burden is tremendous, which would make it almost impossible to get an ideal result. To solve the problem, an off-line simulation method is proposed. We generate a lot of channel coefficients for each link and store them in a file. To simulate many eNB-UE transmissions, we also need to generate enough links. The more samples and links are available, the better the statistical feature would be obtained. During the simulation initialization process, a link is selected randomly for each eNB-UE link, and an index is generated randomly to indicate the reading entry of the channel coefficients. After reading the coefficients of each tap and their relative intervals, fast Fourier transform (FFT) is used to obtain the frequency response of the link and thus the coefficient of each subcarrier can be obtained.

The time-varying feature is modeled at system level. When the speed of each UE is given, the coherence time can be calculated and is used to control the updating interval of small-scale fading. For example, when the moving speed of each UE is 3 km/h, the updating interval is about 30 half-frames, which means in the 30 consecutive half-frames, the channel state of each link remains constant.

14.6.4 SINR Mapping Method

A key issue for accurate system level simulation is to be able to go from an instantaneous channel state to a corresponding block error probability (BLEP). Instead of directly finding the block error probability, it was proposed in [22] to use an effective SIR mapping (ESM) that maps the instantaneous channel state, for example, the set of subcarrier SNRs γ_k in case of OFDM into an instantaneous effective SNR γ_{eff}. The effective SNR is then used to find an estimate of the block error probability from basic AWGN link level performance.

For an ESM to be accurate it obviously has to fulfill the following approximate equivalence [23]:

$$\text{BLEP}(\{\gamma_k\}) \approx \text{BLEP}_{\text{AWGN}}(\gamma_{\text{eff}}) \tag{14.35}$$

where $\text{BLEP}(\{\gamma_k\})$ is the actual block error probability for the instantaneous channel state $\{\gamma_k\}$ and $\text{BLEP}_{\text{AWGN}}(\gamma_{\text{eff}})$ is the AWGN block error probability. It is important to understand that the above expression must be fulfilled for each instantaneous channel realization, or at least for almost all channel realizations, and not only "on average" for a given channel model. The detailed derivations of EESM is illustrated in [23], and generalized EESM can be expressed as:

$$\gamma_{\text{eff}} = -\beta \ln\left(\frac{1}{N}\sum_{i=1}^{N} e^{-\frac{\gamma_i}{\beta}}\right) \tag{14.36}$$

where N is the number of subcarriers, γ_k is the SNR on carrier k, and β is a parameter used to match the ESM to a specific combination of modulation scheme and coding rate. A suitable value for the parameter β for each modulation scheme and/or coding rate can be found from link level simulations, and some well-known values are listed in Table 14.13 [24].

14.6.5 Overhead Calculation

In order to make the transmission reliable, part of the resources is reserved as reference signals or for controlling. The existence of the overheads achieves a trade-off between efficiency and reliability. For example, the reference signals are commonly used for

Table 14.13 Values of the Parameter β

Modulation	*Code Rate*	*β Factor*
QPSK	1/3	1.49
QPSK	1/2	1.57
QPSK	2/3	1.69
QPSK	3/4	1.69
QPSK	4/5	1.65
16QAM	1/3	3.36
16QAM	1/2	4.56
16QAM	2/3	6.42
16QAM	3/4	7.33
16QAM	4/5	7.68
64QAM	1/3	9.21
64QAM	1/2	13.76
64QAM	2/3	20.57
64QAM	3/4	25.16
64QAM	4/5	28.38

channel measurement, and it is well known that the more resources are reserved as reference signals, the more accurate measurement results and more reliable communication can be expected. On the other side, however, if more resources are used for channel measurement, then less resources can be used for data transmission, which leads to the loss of efficiency.

In our simulator, the following control and reference signal overheads are assumed:

- **Downlink Reference Signals (RS):** The number of resources occupied by RS varies as the antenna configurations. For normal subframes, when single antenna is implemented, four REs out of one PRB (equivalent 84 REs) are reserved as RS, thus the overhead is 4.76%. When two antennas are implemented, the reserved REs increase to eight for every PRB, resulting in the overhead of 9.52%. When four antennas are deployed, 12 REs out of 84 REs are reserved as RS, thus the overhead grows to 14.29%. For a special subframe, the RS overhead can be calculated in a similar way, and the specific special subframe configuration should be considered jointly.

- **Physical Downlink Control Channel (PDCCH):** In general, the downlink control channels can be configured to occupy the first one, two, or three OFDM symbols in a subframe, extending over the entire system bandwidth. This flexibility allows the control channel overhead to be adjusted according to the particular system configuration and it varies between 7.14% and 21.43%.
- **Other Downlink Control Symbols:** Synchronization signal, physical broadcast channel (PBCH), physical control format indicator channel (PCFICH), and physical hybrid automatic repeat request indicator channel (PHICH). Because these channels occupy relatively fixed resources, the overhead varies as the system bandwidth changes. When the bandwidth is 20 MHz, the overhead is below 1%, whereas for 1.4 MHz, the overhead is approximately 9%.
- **Uplink Reference Signals and PUCCH:** Uplink reference signals take one out of seven symbols, resulting in an overhead of 14.29 and PUCCH slightly reduces the uplink data rate, thus it is not included in the overhead calculation.

14.6.6 System Performances Analysis

The main simulation parameters are listed in Table 14.14. The major difference between Cases 1 and 3 lies on the intersite distance (ISD). For Case 1, the ISD is 500 m, whereas for Case 3, the ISD is 1732 m. After dropped uniformly in the network, each UE calculates the reference signal receiving Power (rSRP) of each eNB and takes the strongest one as the serving eNB. After the access procedure, each eNB allocates resources to its UE based on the scheduling algorithms, and each UE would calculate the quality of the received signals and make the decision whether it can successfully decode the transport block. In each downlink subframe, each UE generates the CQI according to the measurements of the common RS and feeds it back to the eNB in the uplink subframe available. Multiple antenna techniques are also taken into consideration, and they include: 1Tx × 2Rx with MRC, 2Tx × 2Rx SFBC with MRC, 2Tx × 2Rx spatial multiplexing (SM) with MMSE, and closed-loop feedback.

Figure 14.35 shows the downlink geometry distribution of TD-LTE system level simulator. The geometry distribution is dependent on the geographical position of each UE, and it is reflected by the average SINR. The average SINR is calculated only based on the large- and middle-scale fading parameters such as the path loss fading, the shadow fading, the antenna gain, whereas the impact of the small-scale fading is not taken into consideration. Geometry distribution is very important in calibration among different simulators because different UE distribution would definitely result in different quality of receiving signals and thus make the results incomparable.

It is shown in Figures 14.36 and 14.37 and Tables 14.15 through 14.18 that the cell-average and cell-edge throughput for different scheduling algorithms and two cases. Here, three classical scheduling algorithms [i.e., max *C/I* (MCI), PF, and RR] are evaluated. MCI allocates the resources to the UE who has the best channel

Table 14.14 The Main Parameters of the TD-LTE System Level Simulator

Parameter	*Assumption*
Scenario	Case 1 & 3: 2G CF, 10M BW, speed 3 km/h
Cellular layout	Hexagonal grid, 19 cells, three sectors per cell
Load	Average 10 UE per sector
UE distribution	Users dropped uniformly in entire sector
Total eNB T_x power (P_{total})	46 dBm
BS antenna gain plus cable loss	14 dBi
UE antenna gain	0 dBi
Noise figure at relay	5 dB
Noise figure at UE	5 dB
Noise figure at UE	9 dB
Noise power spectral density of UE	−174 dBm/Hz
Distance-dependent path loss for macro to UE	$L = 128.1 + 37.6\log10(R)$, R in kilometers
Minimum distance between UE and cell	≥35 meters
Penetration loss	20 dB for eNB to UE
Antenna pattern (horizontal) (For three-sector cell sites with fixed antenna patterns)	$A(\theta) = -\min\left[12\left(\frac{\theta}{\theta_{3\,dB}}\right)^2, A_m\right]$ $\theta_{3\,dB} = 70^\circ$ $A_m = 25$ dB
Inter-cell interference modeling	Explicit modeling
Channel model	SCM-E
Traffic model	Full buffer
Number of antenna elements (BS, UE)	(1,1)/(1, 2)/(2,2)
Polarization	No
Scheduling algorithm	PF
Number of MCS candidates for link adaptation	15
Channel estimation error	Ideal estimation

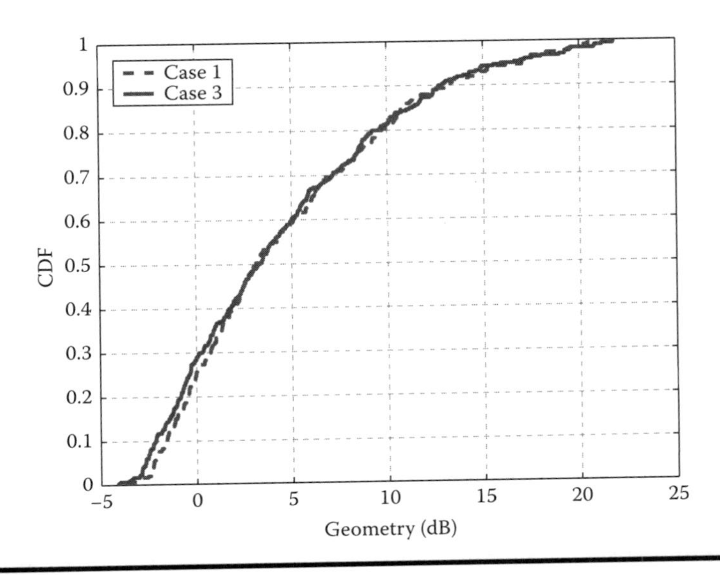

Figure 14.35 Downlink geometry distribution.

state, thus a better UE would obtain more resources than normal or worse UEs and multiuser diversity gain can be achieved. In an extreme situation, if an UE is very close to its serving eNB, then all the resources available may be allocated to the UE and other UEs may starve. This scheduling algorithm can achieve the highest average throughput, but the fairness among UEs can not be guaranteed at all. RR schedules

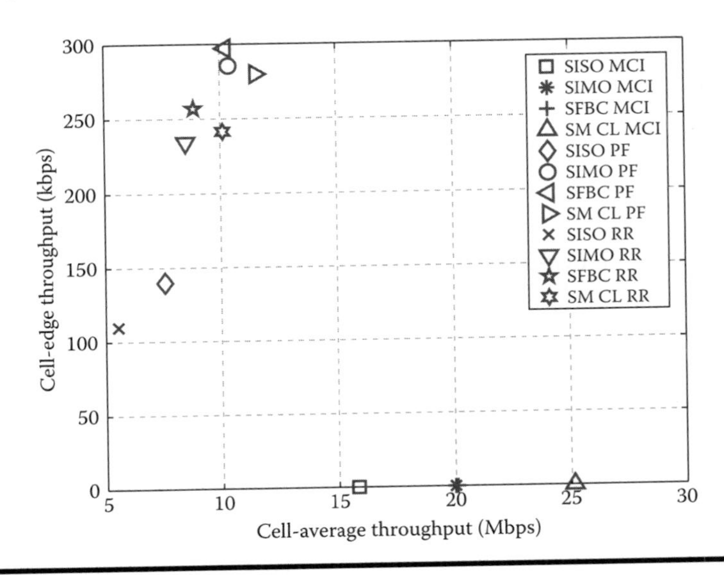

Figure 14.36 Downlink cell-average and cell-edge throughput of Case 1.

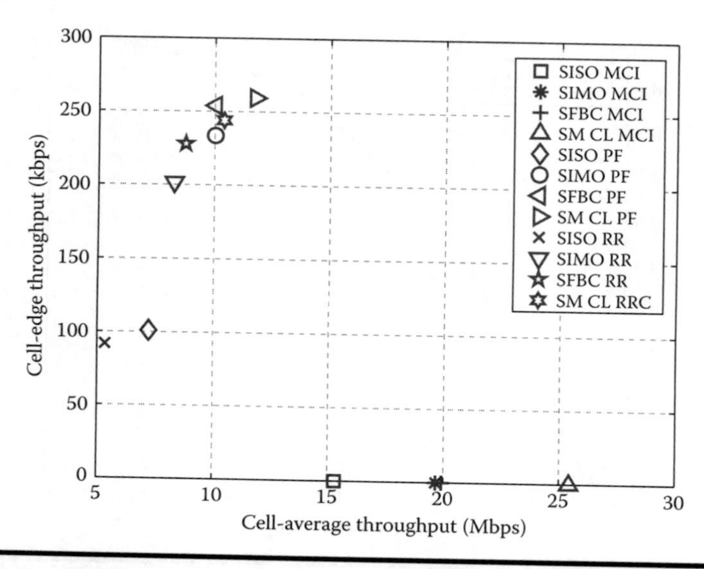

Figure 14.37 Downlink cell-average and cell-edge throughput of Case 3.

Table 14.15 Downlink Cell-Average Throughput (Mbps) of Case 1

Case 1	*Max C/I*	*PF*	*RR*
SISO	15.825	7.559	5.528
SIMO	19.992	10.381	8.509
SFBC	20.075	10.241	8.858
SM CL	25.132	11.533	10.109

Table 14.16 Downlink Cell-Edge Throughput (kbps) of Case 1

Case 1	*Max C/I*	*PF*	*RR*
SISO	0	139.8	109.7
SIMO	0	285.5	234.1
SFBC	0	297.3	256.8
SM CL	0	279.9	241.4

Table 14.17 Downlink Cell-Average Throughput (Mbps) of Case 3

Case 1	*Max C/I*	*PF*	*RR*
SISO	15.298	7.222	5.343
SIMO	19.669	10.025	8.268
SFBC	19.928	10.041	8.742
SM CL	25.43	11.755	10.404

each UE with equal probability, and this is realized through the utilization of list structure. In each cell, the eNB holds a list composed of all the UEs served by itself, and in each scheduling interval, the frequency resources are allocated one by one to each UE thus making each UE obtain the same amount of resources. RR is the fairest scheduling algorithm, where each UE can occupy the same amount of resources. Due to the lack of utilization of the channel state information, multiuser diversity gain is lost. PF is a tradeoff between the cell-average throughput and the fairness among UEs. Through the joint consideration of both the current channel state information and the situation of being served in the past several scheduling intervals, fairness and performance are obtained simultaneously. For each scheduling algorithm, it is clear to see that SM CL performs the best in average throughput while SIMO and SFBC perform better in edge throughput.

Figures 14.38 and 14.39 show the cumulative distribution of UE throughput obtained from TD-LTE system level simulator. It is defined that 5% fractile of the UE throughput is the cell-edge throughput, which means that 95% of the UEs in each cell have a higher throughput than this value. From the cumulative distribution results, we can also see that multiple antenna techniques can increase the system performance significantly, both in average throughput and edge throughput, as the curves of SIMO and MIMO are on the right side of the SISO curves in all the throughput range. In terms of SIMO and MIMO curves, SM CL performs better at high throughput regimes because they can achieve multiplexing gains effectively when the received SINR is high. Besides, SIMO MRC and SFBC MRC have higher

Table 14.18 Downlink Cell-Edge Throughput (kbps) of Case 3

Case 1	*Max C/I*	*PF*	*RR*
SISO	0	100.9	91.9
SIMO	0	233.4	201.4
SFBC	0	253.8	227.7
SM CL	0	259.4	243.9

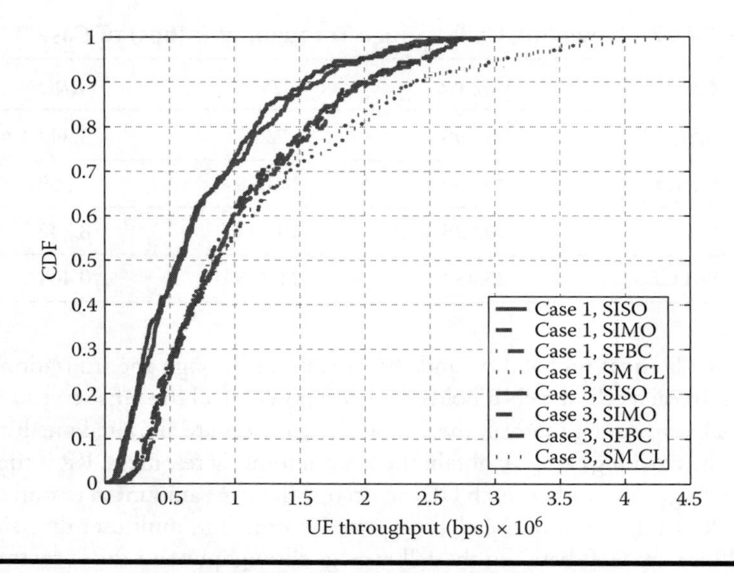

Figure 14.38 Downlink cumulative distribution of UE throughput (in a large scale).

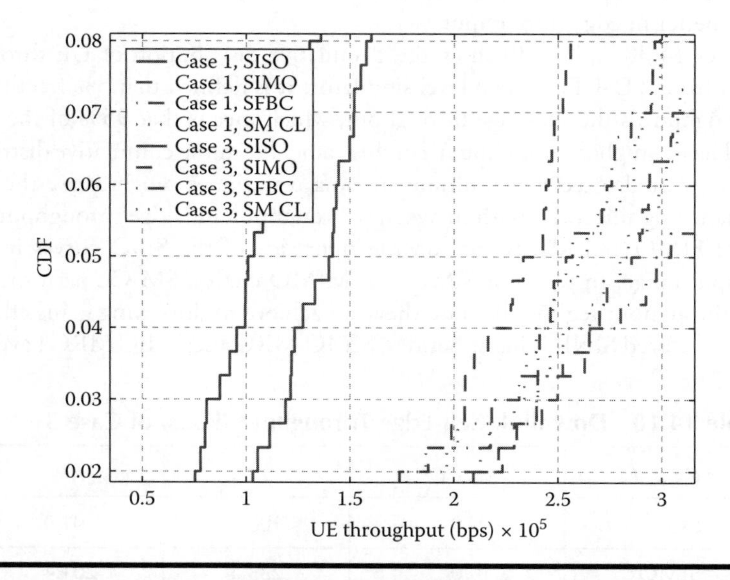

Figure 14.39 Downlink cumulative distribution of UE throughput (in a small scale).

cell-edge throughput because the spatial diversity gains they obtained can significantly improve the quality of the received signals.

14.7 Frequency Planning in TD-LTE

Frequency planning in TD-LTE systems is very important, which is not same as in CDMA systems, where the frequency reuse factor is 1, and the neighbor cells can use the same radio resources.

14.7.1 Frequency Reuse Factor(FRF) Deduction in OFDM/OFDMA Cellular Systems

In cellular systems based on OFDM/OFDMA technique, no intra-cell interference is assumed due to orthogonality of assigned resource blocks of different UEs in the same cell. Considering the classical deployment of hexagonal cells with frequency reuse factor N (N>1) as illustrated in Figure 14.40, and assuming that the inter-cell interference resulted from the first tier of six cochannel BSs is the major source of interference, the approximate forward link SINR of UE in the target cell is easy to obtain as:

$$SINR \approx \frac{1}{6}\left(\frac{r}{D}\right)^{-\alpha} \tag{14.37}$$

where r is the cell radius, D is the distance between two cochannel BSs, and α is a path loss exponent with a value range of ~3.5–5.5.

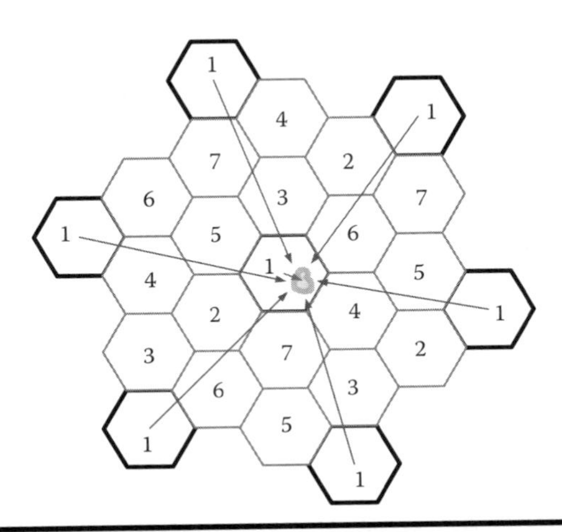

Figure 14.40 Scenario of first tier cochannel BSs with $N = 7$.

With a demodulation threshold $SINR_{th}$, that is, the minimum SINR requirement of the radio receiver, the following restriction (denoted as restriction A) must be satisfied in order to ensure the correct demodulation of the signal.

$$\frac{1}{6}\left(\frac{r}{D}\right)^{-\alpha} \geq SINR_{th} \tag{14.38}$$

Thus, we arrive at

$$D \geq \sqrt[\alpha]{6 \cdot SINR_{th}} \cdot r \tag{14.39}$$

In the N-cell reuse cluster, the relationship of distance D and cell radius r conforms to

$$D = \sqrt{3N}r \tag{14.40}$$

Substituting Equation 14.40 into Equation 14.39 yields Equation 14.41, which is further rewritten as Equation 14.42:

$$\sqrt{3N}r \geq \sqrt[\alpha]{6 \cdot SINR_{th}} \cdot r \tag{14.41}$$

$$N \geq \frac{1}{3}(6 \cdot SINR_{th})^{\frac{2}{\alpha}} \tag{14.42}$$

Until now, we can calculate the minimum size of cell cluster N according to Equation 14.42 with a given modulation threshold $SINR_{th}$ and a path loss exponent α.

In order to testify frequency reuse factor N obtained above, we further apply it to the worst-case SINR situation, which occurs when target UE locates at the cell edge and far from its serving BS. As illustrated in Figure 14.41, each two out of the

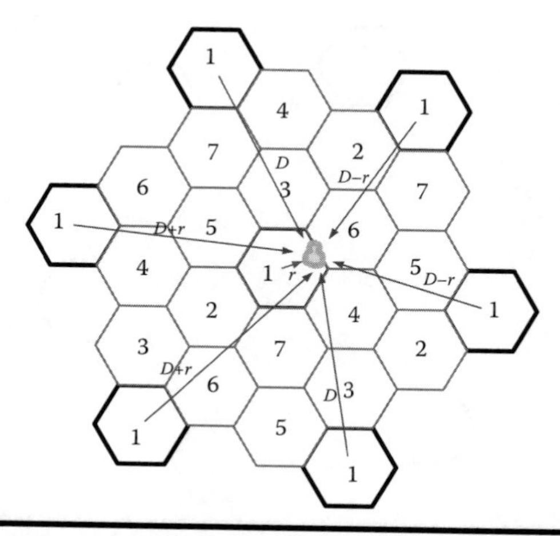

Figure 14.41 Worst-case SINR situation with $N = 7$.

six first-tier cochannel BSs are approximately located at distances of $D - r$, D, and $D + r$. Hence, the cell-edge SINR is [25]:

$$\begin{aligned} SINR_{\text{cell-edge}} &\approx \frac{1}{2}\frac{r^{-\alpha}}{(D+r)^{-\alpha} + D^{-\alpha} + (D-r)^{-\alpha}} \\ &= \frac{1}{2}\frac{1}{\left(\frac{D}{r}+1\right)^{-\alpha} + \left(\frac{D}{r}\right)^{-\alpha} + \left(\frac{D}{r}-1\right)^{-\alpha}} \end{aligned} \tag{14.43}$$

We now arrive at another restriction (denoted as restriction B), which is tighter than restriction A:

$$SINR_{\text{cell-edge}} \geq SINR_{th} \tag{14.44}$$

If restriction B comes into existence, it indicates that the minimum size of cell cluster N obtained from restriction B is essentially feasible. Otherwise, it is quite probable that the above frequency reuse factor N is too small to yield an acceptable performance due to serious inter-cell interference. In these circumstances, we need to enlarge N and check whether restriction B is satisfied or not. In practical network planning and deployment, an additional restriction C is often required for frequency reuse factor N as follows:

$$N = i^2 + ij + j^2 \qquad i, j \in Z \qquad ij \neq 0 \tag{14.45}$$

When taking into account restrictions A, B, and C simultaneously, frequency reuse factor N is determined by:

$$\begin{aligned} &N = arg \min_{i,j}\{i^2 + ij + j^2\} \\ &s.t.\ N \geq \frac{1}{3}(6 \cdot SINR_{th})^{\frac{2}{n}} \qquad SINR_{\text{cell-edge}} \geq SINR_{th} \\ &i, j \in Z \quad ij \neq 0 \end{aligned} \tag{14.46}$$

14.7.2 FRF of Downlink Control Channels in TD-LTE

The frequency reuse factor is often used to get an insight into a control channel's inherent robustness, which is extremely vital for the performance of the whole wireless system. Based on link level simulation discussed in Section 14.5.2, we are able to acquire a series of demodulation threshold $SINR_{th}$ necessary in the deduction of frequency reuse factor of downlink control channels in TD-LTE systems. We assume demodulation threshold $SINR_{th}$ as the specific SINR corresponding to 1% BLER on a SINR-BLER curve. When transmission bandwidth is 20 MHz, EPA5 channel model is utilized, $SINR_{th}$ of different downlink control channels of TD-LTE are listed in Table 14.19.

Table 14.19 Demodulation Threshold $SINR_{th}$ of TD-LTE Control Channels

Control Channels	*Configuration*			$SINR_{th}$*(dB)*
	Antenna (Tx × Rx)	*Channel Format*		
PDCCH	2 × 2	Format 0/1A	1 CCE	8.10
			2 CCE	3.12
			4 CCE	0.2
			8 CCE	−2.82
		Format 1	8 CCE	−1.84
		Format 1 B		−2.68
		Format 1 C		−4.37
		Format 2		−1.71
		Format 3/3A		−2.82
PCFICH		NA		−3.06

Using demodulation threshold $SINR_{th}$ given above, the frequency reuse factor for each control channel is further calculated and shown in Table 14.20 in compliance of FRF deduction process introduced in Section 14.5.7.1. Various path loss exponents of the $\alpha = 2$, 3, and 4 are considered respectively. We can observe there is minor difference between the performance of both downlink control channels of PDCCH and PCFICH. Both of downlink control channels concerned can only support a cell cluster size of or even greater than three under all path loss exponents, implying that desired cell cluster size as small as one may not be allowable for PDCCH and PCFICH, or other techniques for robustness enhancement are needed to achieve that. A step further for PDCCH, the number of CCEs used in DCI (downlink control information) transmission has greater impact on the performance of PDCCH than the DCI format itself. When the size of the CCEs for transmission is fixed (i.e., 8-CCE), different DCI formats achieve identical FRF with $N = 3$. If DCI format 0/1A is chosen, however, 1-CCE and 2-CCE demonstrate obvious inferiority to 4-CCE and 8-CCE. Therefore, transmission using 1-CCE and 2-CCE, which requires a high SINR for a target BLER, would be more appropriate for UEs distributed near serving BSs and far from interference sources, rather than for cell-edge UEs, who are the major victims of inter-cell interference.

Besides theoretical analysis present above, we can also obtain frequency reuse factor through system level simulation based on the Monte Carlo method (i.e., collecting

Table 14.20 Frequency Reuse Factors of TD-LTE Downlink Control Channels

Control Channels	*Configuration*			*SINR_{th} (dB)*	*Frequency Reuse Factor N*		
	Antenna (Tx × Rx)	*Channel Format*			$a = 2$	$a = 3$	$a = 4$
PDCCH	2 × 2	Format 0/1A	1 CCE	8.10	16	7	4
			2 CCE	3.12	7	3	3
			4 CCE	0.2	3	3	3
			8 CCE	−2.82	3	3	3
		Format 1	8 CCE	−1.84	3	3	3
		Format 1B		−2.68	3	3	3
		Format 1C		−4.37	3	3	3
		Format 2		−1.71	3	3	3
		Format 3/3A		−2.82	3	3	3
PCFICH	2 × 2	NA		−3.06	3	3	3

and analyzing SINR of UEs randomly distributed over a multicell system using a sufficient number of individual snapshots) to find out the minimum acceptable FRF for a target demodulation threshold. Simulation parameters for forward link are shown in Table 14.21. Simulation results for various frequency reuse factors are presented in terms of UEs SINR CDF curves plotted in Figure 14.42. SINR corresponding to 5% CDF UE is commonly regarded as the statistically SINR for UE located at cell edge, we thereby highlight 5% CDF SINR for each frequency reuse factor curve in the same figure. Such a 5% CDF SINR is denoted as $SINR^{N_i}_{\text{cell edge}}$, $N_i = 1, 3, 4, 7, \ldots$ for simplicity.

By comparing $SINR_{th}$ and $SINR^{N_i}_{\text{cell edge}}$, we are capable of finding the target minimum acceptable frequency reuse factor N_k such that $SINR^{N_k}_{\text{cell edge}} \leq SINR_{th} < SINR^{N_{k+1}}_{\text{cell edge}}$. Note that this method of finding target FRF holds true for both PDCCH and PCFICH. Table 14.22 shows comparison of frequency reuse factors

Table 14.21 Simulation Parameter for TD-LTE Forward Link

Parameters	*Values*
Cell Layout	19 macro cells, 3 sectors per site, with wrap-around for $N = 1$
	64 macro cells (only statistics in 16 central cells are collected), 3 sectors per site, without wrap-around for $N > 1$
Cell range	600 m
Carrier frequency	2.35 GHz
Bandwith	20 MHz
Bandwidth efficiency	90%
UE distribution	Randomly distributed throughout system area
System loading	Full buffer traffic
BS antenna pattern	$A(\theta) = -\min\left[12\left(\frac{\theta}{\theta_{3dB}}\right)^2, A_m\right]$ $\theta_{3dB} = 65°, A_m = 20$ dB
Antenna height above average rooftop level	15 m
Receiver antenna gain	0 dBi
Transmitter antenna gain	15 dBi
Maximum BS power	46 dBm
Propagation model	Vehicle model
Shadowing variance	10 dB
Noise power	−174 dBm/Hz
Noise figure	9 dB

for PDCCH and PCFICH obtained in two distinctive ways: one by theoretical analysis and the other by simulation. It is worthwhile to notice that these two methods yield results by and large the same, except for minor difference regarding PDCCH 2-CCE. Frequency reuse factors for other downlink control channels and uplink control channels can be obtained similarly.

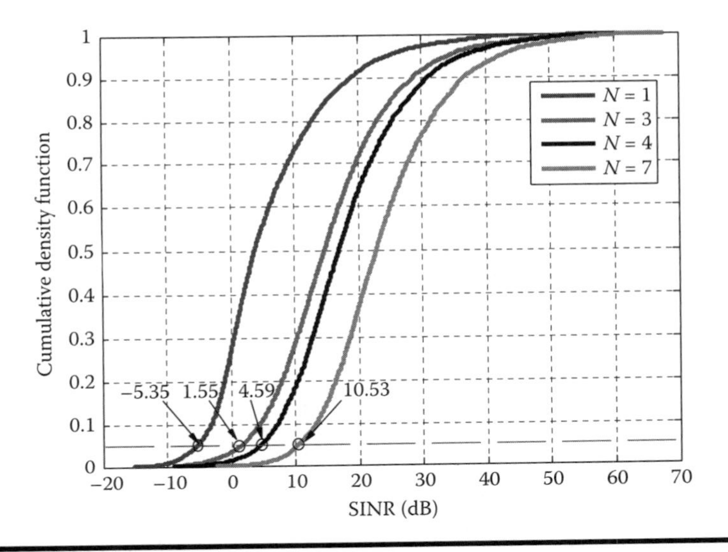

Figure 14.42 SINR versus CDF curves for different frequency reuse factors.

14.8 Performance Enhancement in TD-LTE

To improve cell-edge users' service quality, increase the system throughput and enlarge the cell coverage, the decode-and-forward (DF) relay station (RS) technique has been used in wireless cellular networks. When the DF-based RS is used in LTE systems, the TDD mode is more suitable than FDD because downlink and uplink resources can be assigned adaptively in a TD-LTE system.

Table 14.22 Theoretical and Simulation Value of FRF for TD-LTE Downlink Control Channels

	Configuration				*Frequency Reuse Factor N*	
Control Channels	*Antenna (Tx × Rx)*	*Channel Format*		*$SINR_{th}$ (dB)*	*Theoretical Value (α = 3)*	*Simulation Value*
PDCCH	2 × 2	Format 0/1A	1 CCE	8.10	7	7
			2 CCE	3.12	3	4
			4 CCE	0.2	3	3
			8 CCE	−2.82	3	3
PCFICH	2 × 2	NA		−3.06	3	3

14.8.1 Directional Relay in TD-LTE

A directional relay topology is shown in Figure 14.43, where eNB is marked as a black triangle and RS is denoted as a black rectangle. It is assumed that eNBs are fixed at cell centers and RSs are at each vertex of hexagonal cells. In the proposed DF-relay-based system model, each eNB is equipped with one omnidirectional antenna while each RS is equipped with three 120-degree directional antennas. Thus, three cells are served by one RS, and six RSs are configured in one cell. To avoid inter-cell interference between adjacent eNBs, the coverage of eNB is limited and the UEs at cell edges are only served by RSs [i.e., only UEs located in the inner loop of the cell (named as inner UE) are served by eNB, whereas RSs will serve UEs in the outer loop of

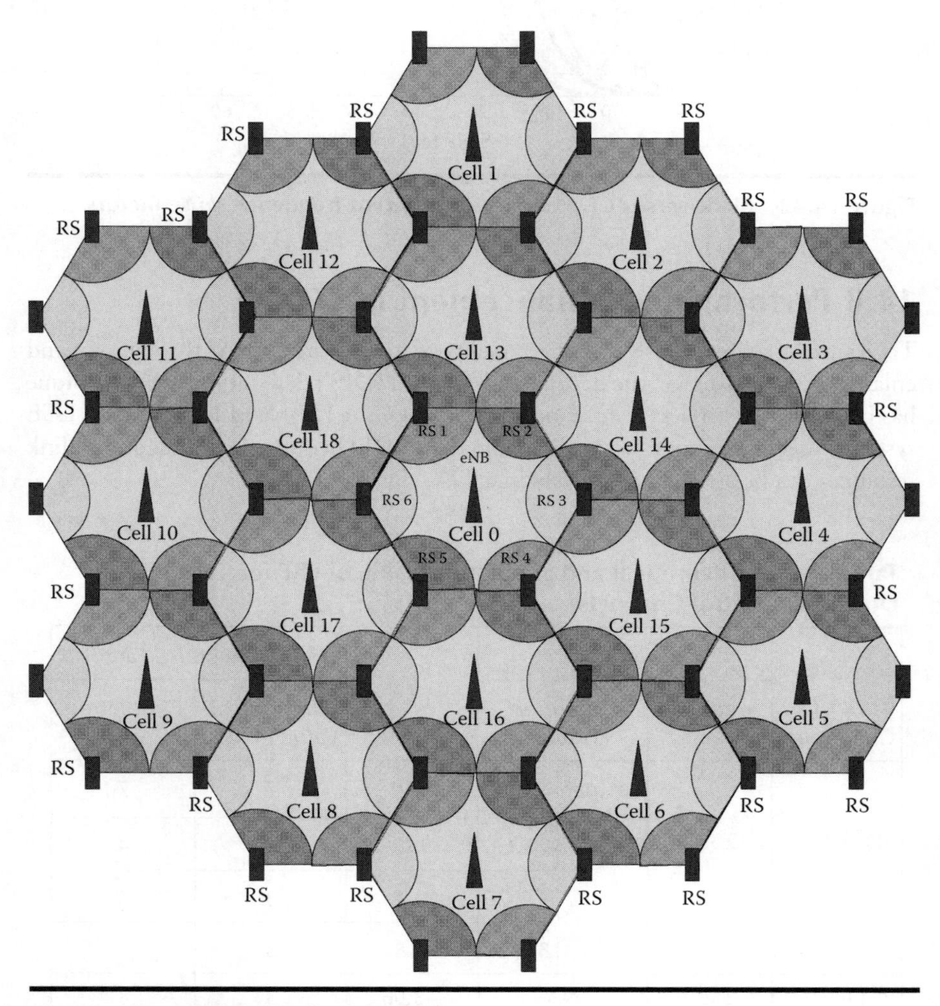

Figure 14.43 Directional relay topology.

the cell (named as outer UE)]. Considering huge interference between adjacent RSs, the directional antennas are deployed in RSs. Furthermore, to improve spectrum efficiency, the non-adjacent RSs will reuse the same radio resources. Both outer and inner UEs are administrated by the serving eNB.

To be fully compatible with TD-LTE systems, the frame structure for the proposed directional relay topology is shown in Figure 14.44, which is inherited from the configuration frame pattern for the TD-LTE systems [2]. For the configuration 2 frame pattern, each radio frame consists of two half-frames with a length $T_f = 153600 \cdot T_s = 5$ ms. Each half-frame consists of eight slots and three special fields, or DwPTS, GP, and UpPTS. To be compatible with the Type 1 frame structure (defined for FDD-LTE systems), two time slots are combined as one subframe (i.e., subframe i consists of slots $2i$ and $2i + 1$). Subframes 0, 5, and DwPTS are always reserved for downlink transmissions.

14.8.2 Performance Evaluation of Directional Relay

In this chapter, we investigate the downlink performance, and thus most subframes are reserved for downlink channels, where six subframes (marked in gray blocks) are reserved for downlink and two subframes (marked in dark gray blocks) are reserved for uplink in each frame. It is noted that there are two subframes for DwPTS, GP, and UpPTS. Every subframe has one long arrow and one short arrow, where the long one indicates the direct link between eNBs and inner UEs, while the short one represents the relay link between eNBs and RNs, or the access link between RNs and outer UEs. Since eNB and RS can communicate with their own UEs simultaneously, the RSs are non-transparent and categorized as Type 1. For demonstration simplicity, the inner UEs and outer UEs are denoted as the same UEs in Figure 14.44.

To reduce the interference, especially the interference from the adjacent RSs, the static interference coordination mechanism is adopted. Figure 14.45 shows the time-frequency resource allocation design for the direct and relay transmissions in downlink. For the right subfigure in Figure 14.45, the gray block resource is used for

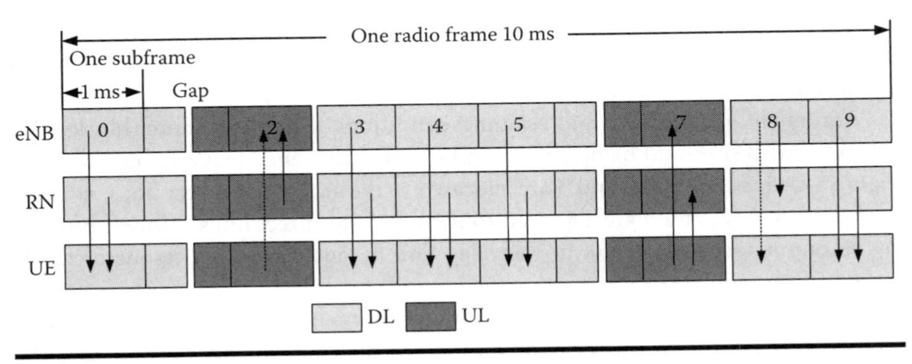

Figure 14.44 TD-LTE frame pattern for DF based relay protocol.

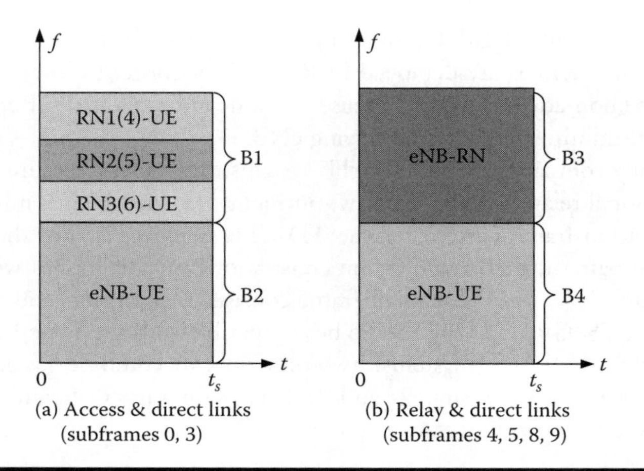

Figure 14.45 Radio resource reuse for access links.

the direct link (denoted as B4), and the dark gray block is for the relay link (denoted as B3). Since all radio resources are orthogonally allocated, there is no intra-cell interference in each subframe (i.e., there is no interference between different UEs, UE and RS, and different RSs located in the same cell).

For the left subfigure in Figure 14.45, the B1 resource block is used for the access link for the communications between RSs and UEs. If all RSs in the same cell are allocated with the same time-frequency resources, a huge interference may occur between RSs. Due to geometrical isolations, only RSs at the opposite positions will occupy the same time-frequency resource, which will degrade most RSs' interference and improve the transmission spectrum efficiency between eNB and RS. Therefore, if there are six RSs for each cell and the RS is numbered sequentially, the RS pairs such as 3 and 6, 2 and 5, and 1 and 4, can use the same radio resources. Consequently, radio resources for the access links between the RS and UE are divided into three parts, each of which is presented by a different shade in the left subfigure of Figure 14.45. Although radio resource reuse between the opposite RSs will help increase interference, the interference can be suppressed to a satisfactorily low level due to the geometrical isolation.

Accurately, according to radio channel conditions, the radio resource blocks (B1, B2, B3, and B4) should be optimized in both the time and frequency domains. In Figure 14.45, subframes 0 and 3 are allocated to the access and direct links, whereas subframes 4, 5, 8, and 9 are allocated to the relay and direct links. More subframes are recommended to allocate to the relay link because the radio resource can be reused at the access links.

The performances between the traditional nonrelay and the directional relay scenarios are simulated. In the simulations, the carrier frequency is set to 2 GHz, the frequency bandwidth is 10 MHz. The eNBs' transmission power is 46 dBm.

Table 14.23 Throughput Expectation

Throughput	*Theoretical (Mbps)*	*Simulation (Mbps)*
Nonrelay	9.11	12.87
Relay: inner UEs	20.84	22.43
Relay: outer UEs	11.08	12.74

The path loss factor is 2.5 for LOS (line of sight), and 4 for NLOS (nonline of sight). The noise power density is −174 dBm/Hz. The cell radius is 1000 m. The simulations generated 50,000 snapshots, each of which refers to a different and independent UE's location.

Table 14.23 shows the average throughput comparisons between the directional relay and nonrelay scenarios. When the directional relay protocol is utilized, the UEs in the inner loop of the cell have about two times better performance than that in the nonrelay scenario. Meanwhile, UEs in the outer loop of cell can achieve almost the same performance as the average throughput of the nonrelay scenario.

In a real network, SINR should be in a limited range due to the hardware restriction and the definition of the feedback information to RS or BS. Consequently, the maximum allowed SINR of UEs should be limited. The following simulations were performed under the assumption that the UE's SINR is not higher than 22 dB.

Table 14.24 shows the SINR comparison between the directional relay and the traditional nonrelay scenarios. It is noted that the SINR received at RSs is not limited. There is a relatively big difference between the theoretical analysis and the simulation results because the impact from the limited SINR is considered only for simulation.

Table 14.25 shows the average throughput comparison between the directional relay and traditional nonrelay scenarios, indicating the impacts from the maximal allowed SINR on the average throughput.

To take a fair comparison with the SINR distributions for different scenarios, the CDF curves of SINR obtained from the simulations are depicted in Figure 14.46. In the traditional nonrelay scenario, about 30% of UEs experience a SINR less than 0 dB, which means that they have to resort to other signal processing techniques to

Table 14.24 SINR Expectation (SINR is Restricted)

SINR	*Theoretical (dB)*	*Simulation (dB)*
Nonrelay: eNB → UE	11.10	8.67
Relay: eNB → UE	19.73	15.97
Relay: eNB → RS	36.50	36.50
Relay: RS → UE	17.20	14.73

Table 14.25 Throughput Expectation Value (SINR is Restricted)

Throughput	*Theoretical (Mbps)*	*Simulation (Mbps)*
Nonrelay	9.11	12.73
Relay: inner UEs	20.84	21.48
Relay: outer UEs	11.08	12.37

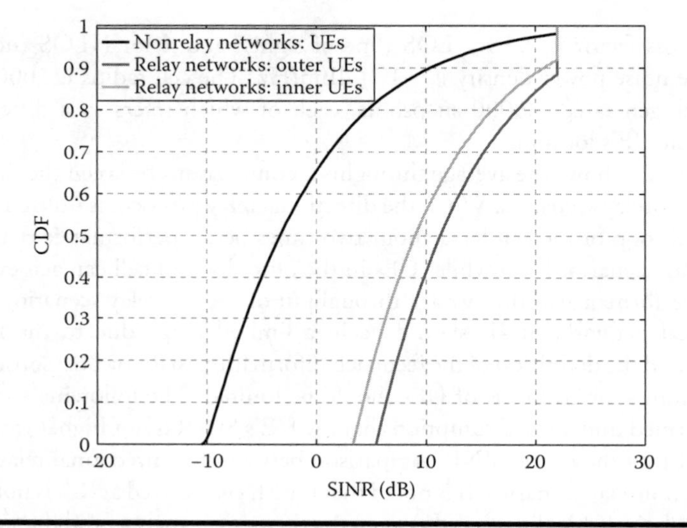

Figure 14.46 CDF of SINR in UEs when SINR is not bigger than 21 dB.

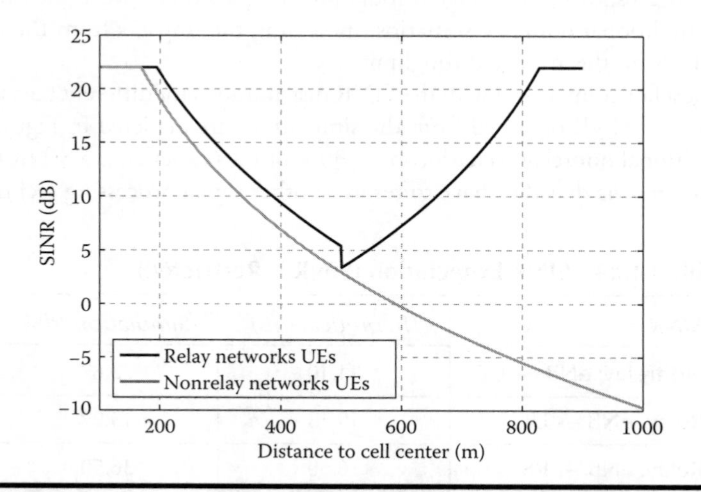

Figure 14.47 SINR profiles (maximum allowable SINR is 21 dB).

recover useful signals from overwhelming interference and noises. On the contrary, when utilizing the proposed directional relay scheme, both the outer and inner UEs will have a good transmission quality, where all SINRs are higher than 3 dB.

Figure 14.47 gives an explicit view of the geometrical SINR profile, where we assumed that an UE is moving from the cell center to the cell edge. The light gray curve represents SINRs for UEs in nonrelay networks, and the SINR decreases almost linearly with the increasing distance. However, for UEs in the proposed directional relay scenario, UEs experience a rising curve of SINR after they have gone across the eNB's serving boundary and into the served coverage by the RS. We can see that the SINR of cell-edge UEss can be improved remarkably with the assistance of directional relay nodes.

Acknowledgments

This work was supported in part by the China National Natural Science Foundation (grant 60602058) and the National Advanced Technologies R&D Programs (863 grant 2009AA01Z244).

References

[1] Hsiao-Hwa Chen, Chang-Xin Fan, and W.W., Lu, "China's perspectives on 3G mobile communications and beyond: TD-SCDMA technology." *Wireless Communications, IEEE* [see also *IEEE Personal Communications*]. vol. 9, iss. 2, April 2002, pp. 48–59.

[2] M. Peng and W. Wang, "A framework for investigating radio resource management algorithms in TD-SCDMA systems." *Communications Magazine, IEEE,* vol. 43, iss. 6, June 2005, pp. S12–S18.

[3] M. Peng and W. Wang, "Technologies and standards for TD-SCDMA evolutions to IMT-advanced," *Communications Magazine, IEEE,* vol. 47, iss. 12, Dec. 2009, pp. 50–58.

[4] 3GPP, 3rd generation partnership project; TS 36.211. "Evolved universal terrestrial radio access (E-UTRA); Physical channels and modulation (Release 8)," vol. 8.7.0, May 2009. Available: http://www.3gpp.org.

[5] Orange, China Mobile, KPN, NTT DoCoMo, Sprint,T-Mobile, Vodafone, Telecom Italia, and R1-070674. "LTE physical layer framework for performance verification," RAN1 48, St. Louis, MO, 12–16. Available: http://www.3gpp.org. February 2007

[6] S. Song, "On distribution of sums of n independent random variables subject to exponential distribution," *Journal of Liaoning Normal University* (natural science edition), vol. 4, April 1990.

[7] R. Jantti, and S. L., Kim, "Downlink resource management in the frequency domain for multicell OFCDM wireless networks," *IEEE Transactions in Vehicular Technology*, vol. 57, no. 5, pp. 3241–3246, September 2008.

[8] 3GPP Project Document R1-090133, "Beamforming enhancement in LTE-advanced," Huawei, CMCC, CATT, January 2009. [Online]. Available: http://www.3gpp.org.

[9] 3GPP Project Document R1-060291, "OFDMA Downlink inter-cell interference mitigation," February 2006. [Online]. Available: http://www.3gpp.org.

[10] 3GPP Project Document R1-050507, "Soft frequency reuse scheme for UTRAN LTE," May 2005. [Online]. Available: http://www.3gpp.org.
[11] 3GPP Project Document R1-050764, "Inter-cell interference handling for E-UTRA," August 2005. [Online]. Available: http://www.3gpp.org.
[12] S. Sesia, Issam Toufik, and Matthew Baker, *"LTE—The UMTS Long Term Evolution: From theory to Practice,"* John Wiley & Sons, Ltd, 2009.
[13] R. Kwan, C. Leung, and J. Zhang, "Resource allocation in an LTE cellular communications system," *in Proceeding of IEEE International Conference on Communications (ICC 2009)*, pp. 1–5, June 2009.
[14] 3GPP Project Document R1-071347, "Uplink ACK/NACK performance with and without reference signals," March 2007. [Online]. Available: http://www.3gpp.org.
[15] 3GPP Project Document R1-070778, "UCQI feedback overhead with CDM uplink control channel region," February 2007. [Online]. Available: http://www.3gpp.org.
[16] 3GPP Project Document R1-072174, "DL L1/L2 control channel coverage for E-UTRA," May 2007. [Online]. Available: http://www.3gpp.org.
[17] 3GPP Project Document R1-072692. "Coding and transmission of control channel format indicator," June 2007. [Online]. Available: http://www.3gpp.org.
[18] 3GPP Project Document R1-072665, "P-BCH Design," June 2007. [Online]. Available: http://www.3gpp.org.
[19] 3GPP Project Document R1-072689, "Downlink acknowledgement mapping to REs," June 2007. [Online]. Available: http://www.3gpp.org.
[20] 3GPP Project Document R1-072690, "Support of precoding for E-UTRA DL L1/L2 control channel," June 2007. [Online]. Available: http://www.3gpp.org.
[21] Baum, D. S.; Hansen, J.; Salo, J., "An interim channel model for beyond-3G systems: extending the 3GPP spatial channel model (SCM)," *IEEE Transactions in Vehicular Technology,* 2005, vol. 5, pp. 3132–3136, May 2005.
[22] 3GPP Project Document R1-030999, "Considerations on the system-performance evaluation of HSDP using OFDM modulation," October 2003. [Online]. Available: http://www.3gpp.org.
[23] 3GPP Project Document R1-031303, "System-level evaluation of OFDM-further considerations," November 2003. [Online]. Available: http://www.3gpp.org.
[24] 3GPP Project Document R1-061508, "System analysis comparing synchronous and asynchronous adaptive HARQ," May 2006. [Online]. Available: http://www.3gpp.org.
[25] G. L. Stuber, *Principles of mobile communication,* New York: Kluwer Academic Publishers, 2002.

Chapter 15

Planning and Optimization of Multihop Relaying Networks

Fernando Gordejuela-Sanchez and Jie Zhang

Contents

15.1 Introduction

To successfully compete with other existing and future wireless, cellular, and wireline services, wireless network designers need to fully consider the specific technical constraints that influence the whole design process. The number of combinations of network elements and parameters that can be configured during the design process [e.g., antenna tilt, azimuth, base station location, power, radio resource management (RRM) parameters], constitutes the solution space. The size of this space determines the degree of complexity of finding appropriate solutions. In wireless metropolitan area networks (WMAN) scenarios, and particularly in multihop environments, the number of options is high, so it is very unlikely that the optimal network configuration can be found using a manual method [1].

Multihop relaying networks involve a more complicated air interface than other kind of cellular networks and need to be rigorously analyzed during the process of network design to achieve an outstanding performance. The chapter explains the impact of new elements of the LTE access network architecture (relay stations) in the network design process. With the use of examples for capacity and coverage planning, the chapter highlights the complexity of the design process in this kind of network.

Base station location, antenna azimuth, and tilt optimization have been widely investigated as the main parts of advanced network planning and optimization tools, especially in the case of Universal Mobile Telecommunication System (UMTS) and carrier division multiplex access (CDMA)-based networks [1–5] and also mobile worldwide interoperability for microwave access (WiMAX) [6]. Few works specialized on LTE have been found. They focus on special cases [7] or self-healing/self-optimization capabilities of an already deployed network [8] and do not consider a precise formulation and computationally efficient methodology on multiobjective problems.

In relation to this topic, some approaches on technologies such as GSM suggest the use of radio frequency planning (RFP) in order to mitigate interference and enhance system performance [3, 9]. Similarly, orthogonal frequency division multiple access (OFDMA) supports reconfiguration of the subchannel usage with different frequency reuse schemes and subchannel allocation techniques in order to mitigate inter-cell interference [10]. Some of these strategies are proposed in the existing literature on multihop communications. However, they only consider fixed reuse patterns or static assignment of subchannels in regular or hexagonal scenarios [11–13] and do not describe a joint formulation with other parameters such as antenna tilt/azimuth.

Furthermore, most existing efforts have been focused on an access-oriented design [14], which may not be a good solution for the following reasons. Some of these designs may not be efficient for areas with a high number of users and with different service demands which often result in a bottleneck for the rest of the network. In addition, for many existing designs that are developed to gain basic access, the customer may not be able to obtain the desired quality service (QoS).

Also, access-oriented design may not be fair to the provider who develops the infrastructure because the service provider only earns an access fee, which is usually paid monthly and is relatively low compared to the deployment cost.

With the increasing need of supporting new services and applications for different scenarios or multiple objectives, the problem of wireless network architecture design becomes too large in scope to be handled efficiently with a single technique. This chapter presents a service-oriented optimization framework that provides a clear and comprehensive description of different options and solutions to achieve an optimal network configuration. The chapter first explains the impact of the LTE physical technology on the network architecture design. Then, an insight on the network planning and optimization process is given within a LTE context, and we give an economic perspective to the calibration of the factors in the cost function. Furthermore, we describe a way to deal with the trade-offs generated during the network architecture design.

15.2 Multihop Relaying Networks

15.2.1 Properties of Relay-Based Networks

The multihop relaying configuration aims to deal with challenging radio propagation characteristics and low CINR (carrier-to-interference and noise ratio) at the cell edge with the introduction of relay sectors (RSs). RSs forward the information from some mobile stations (MSs) to the legacy base sectors (BSs) that are connected to the core network. The BSs will provide resources to the RSs and other MS to which they are directly connected.

Figure 15.1 shows different usage scenarios for RSs. The RS is able to work in line of sight (LOS) and non-LOS propagation conditions and supports MS handover. Using RSs can help overcome the dependency on wired backbones and, therefore, can reduce the infrastructure costs. RSs can also extend the coverage beyond the BS range, increase the quality for indoor coverage, deal with shadowing in urban scenarios, balance the network load, help the terminals to save batteries, and may increase the overall system capacity, which can be particularly useful in hotspots. The RS can also provide broadband access in emergency situations or to locations in rural and developing areas where broadband is currently unavailable.

However, there are also some important challenges. A major concern for network designers is the management of interference [15], which can degrade the performance of the system considerably. In multihop networks, this is a critical issue, as adding RSs constitutes a new source of interference, and therefore its power, frequency, and antenna configuration must be carefully selected to avoid interference.

The RS normally works in half-duplex mode, which means the station does not receive and transmit using the same channel simultaneously. In addition, the RS can operate in two possible schemes, depending on how it processes the received

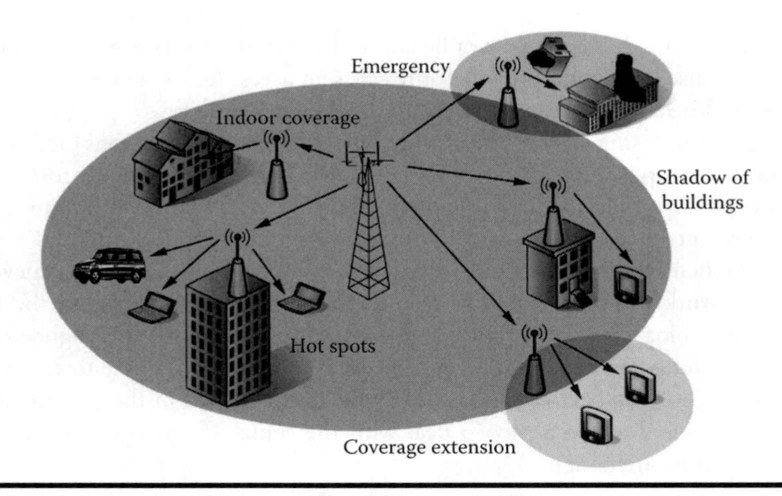

Figure 15.1 Example of usage scenarios for BS and RS.

signal: amplifying and forwarding (AF), where the RS just amplifies and retransmits symbols, or decoding and forwarding (DF), where the RS demodulates and decodes the received signal before retransmission.

The concept of cooperative relaying is based on one signal that can be received and forwarded by multiple stations to the same MS, which combines the signals received from the RSs, and possibly from the BS, in order to take advantage of channel diversity gain. The cooperation is considered as source diversity if identical signals are transmitted simultaneously in time and frequency by both BS and RSs, or transmit diversity if space-time codes are used.

15.2.2 Deployment of Relay-Based Networks

In general, the deployment of a relay-based network depends on a particular scenario. However, there are different strategies that can be adopted according to the service provider needs, illustrated in Figure 15.2. RSs can help to extend the BS coverage by placing them close to the edge of the coverage area. On the other hand, LTE supports a variety of modulation and coding schemes (MCS) [15], and the system can select the most robust schemes for poor radio propagation characteristics (i.e., low CINR) normally in regions close to the cell edge. The most robust MCSs generate low throughput levels and therefore the use of additional RSs in these areas can help to increase the CINR and also the throughput. Another aspect to take into account in a relay-based wireless network is network resilience, due to a higher probability of having a failure in a multihop communication. Different strategies can reflect diverse

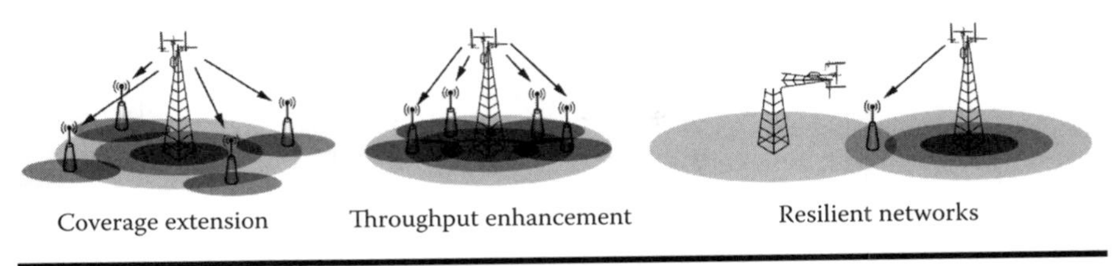

Figure 15.2 Scenarios for different strategies in BS and RS positions.

abilities of the network to recover from an event of having a problem in a station or a link failure.

15.3 Technology-Specific Features Affecting the Network Planning and Optimization Process

LTE describes the standardization work done by the Third Generation Partnership Project (3GPP) to define a new high-speed radio access method for mobile communication systems. LTE uses new multiple access schemes on the air interface: OFDMA in downlink and SC-FDMA (single-carrier frequency division multiple access) in the uplink. OFDMA is a physical technology that supports several key features necessary for delivering broadband services, for example, scalable channel bandwidths, high spectral efficiency, interference, and multipath tolerance due to subcarrier orthogonality and long symbols. SC-FDMA is a new single-carrier multiple access technique which has similar structure and performance as OFDMA. SC-FDMA uses single-carrier modulation and orthogonal frequency multiplexing using discrete fourier transform (DFT) spreading in the transmitter and frequency domain equalization in the receiver.

In both cases, subcarriers are grouped into resource blocks (RBs) of 12 adjacent subcarriers with an intersubcarrier spacing of 15 kHz [15]. Each RB has a time slot duration of 0.5 ms, which corresponds to six or seven symbols. Figure 15.3 shows the resource grid. The smallest resource unit that a scheduler can assign to a user is a scheduling block (SB), which consists of two consecutive RBs, spanning a subframe time duration of 1 ms. One LTE radio frame consist of 10 ms. All SBs belonging to a single user can be assigned to only one MCS in each scheduling period.

15.3.1 Interference Model

Establishing a reliable connection for the users in LTE is not as easy as just assigning enough transmission power for each SB. The reason is that there is a certain level of interference between users depending on the position and the number of BSs/RSs and

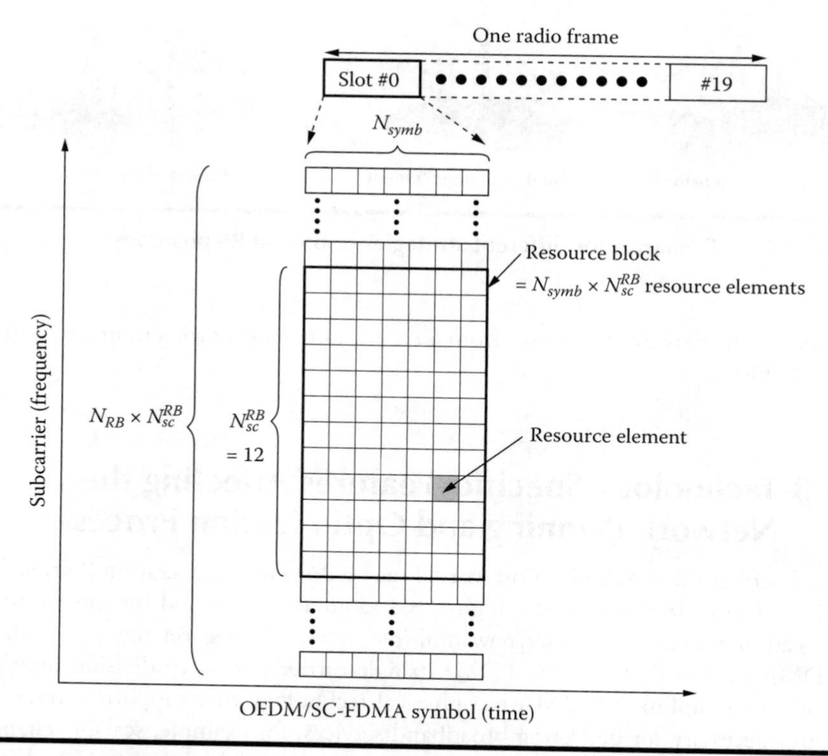

Figure 15.3 Resource grid.

the resources available for each of them. Within the same frequency band, some users interfere each other and some others do not. This particular matrix interference model of LTE should be taken into account when planning and optimizing a network.

In a system with a set of candidate sectors (that may include BSs and RSs): $S = \{S_1, S_i, S_j, \ldots, S_N\}$, with users associated to the active ones: $M_{S_i} = \{1, \ldots, m \ldots, M_i\}$, let us introduce the following notations:

- Set of subcarriers: $S_c = \{1, \ldots, x \ldots, X\}$.
- Set of slots: $S_l = \{1, \ldots, y \ldots, Y\}$.

Let us define a subchannel in LTE in a 12 consecutive subcarriers. As a result, we can define the following:

- Set of subchannels: $S_h = \{1, \ldots, k \ldots, K\}$ where $K = \lceil X/12 \rceil$
- Set of resources (scheduling blocks): $S_b = \{1, \ldots, r \ldots, R\}$ where $R = K \cdot \left(\frac{Y}{2}\right)$

The interference on a certain user (I_m) is the sum of the power received from other interfering base stations (j) that are using the same resources ($C_{m,j,r}$), as Equation 15.1 indicates. Depending on the channel conditions and the service requirements, users will require different number of resources and will be interfered differently, and thus diverse situations may be created. User m is assigned to sector S_i

$$I_m = \sum_{r \in Sb} \left(\sum_{j \in S, j \neq i} C_{m,j,r} \cdot u_{j,r} \right) \cdot v_{m,r} \tag{15.1}$$

where

$$v_{m,r} = \begin{cases} 1 & \text{if resource } r \text{ is used by user } m \\ 0 & \text{otherwise} \end{cases} \tag{15.2}$$

$$u_{j,r} = \begin{cases} 1 & \text{if resource } r \text{ is used in station } j \\ 0 & \text{otherwise} \end{cases} \tag{15.3}$$

In a multihop communication, relay links can be considered as a kind of user communication, where the above calculations also apply. The final QoS perceived by the MS will be determined by the weakest link. An example can be found in Section 15.4.3.

15.3.2 Cooperative Relay Networks

In relay topology, whenever the BS transmits, it is doing so in the downlink (DL). Whenever the MS transmits, it is doing so in the uplink (UL). The RS, however, must transmit and receive in both DL and UL. In order to accommodate these communications, the radio frame has to be split into several zones [16].

With the frame split, fewer slots will be available for the users of each zone, and therefore the efficiency decreases. However, in average, users will receive a stronger signal and therefore higher CINR levels. As a result, higher level MCSs may be used and thus fewer slots would be needed to achieve a certain throughput. In addition, the interference may decrease depending on how the separation of resources is done in the frame division.

When cooperative transmission is considered, transmissions instances are repeated in each cooperative path or frame zone (in both BS and RS), and therefore the pool of available resources is reduced due to redundancy [17]. The gains that the cooperative transmission may provide depend on the particular scenario and should overcome this drawback if taken into account.

It is accepted that, in most cases, more than three hops in the communication (two relay stations) is not practical due to the low radio resource efficiency and possible impairments in the communication such as delay [16]. Figure 15.4 shows

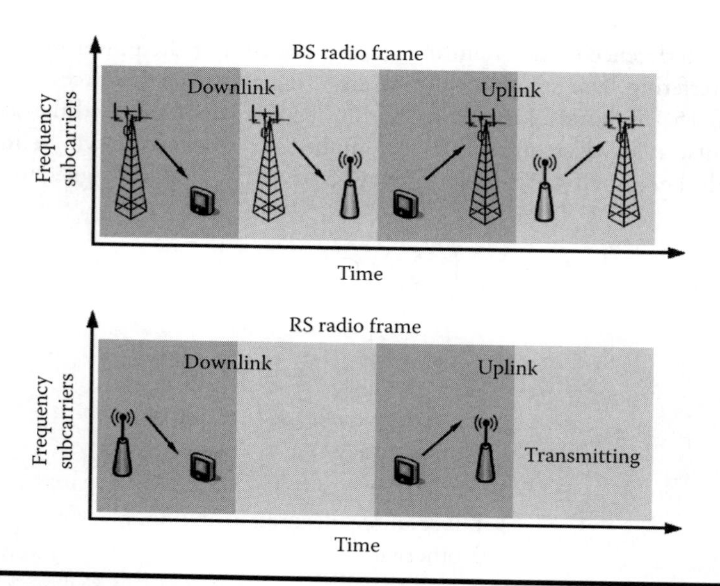

Figure 15.4 Time forwarding.

the frame structure for two hop communications and time domain forwarding. For more hops, the BS should reserve one zone to accommodate each relay link. The length of each zone of the frame will change to allow different MCSs, depending on the link quality and channel conditions.

Frequency domain forwarding and other duplex configurations are also possible [18], but in this chapter we focus on the described flexible scenario in Figure 15.4 [17]. In this configuration, some of the frame pieces are allocated to the access links and some others to the relay ones in order to avoid interference from the BS-MSs and RS-MSs communications (access links) to BS-RS (relay link). However, the designer still needs to deal with the interference between BS-MSs and RS-MSs. In this chapter, we propose the use of time domain forwarding and traditional frequency planning in the form of subchannel planning. The aim of frequency planning is to restrict the use of the subcarriers to a certain subset of subchannels in each BS/RS of the network. This subchannel restriction may involve another loss of cell resource utilization efficiency. However, the objective of this procedure is to reduce the interference and collisions, thus having higher CINR levels that should enhance the system performance.

In practice, the relays often operate in decode-and-forward mode because the MAC layer requires data contained in the header and subheaders to operate, and the physical (PHY) layer is generally unaware of the difference between the headers and the payload data. Therefore, the data may need several frames to be relayed by the RS. However, as mentioned, an amplify-and-forward operation would be allowed where the RS demodulates and deinterleaves its signal and then immediately interleaves and modulates it for transmission without decoding. In this case, the order of the RS frame zones shown in Figure 15.4 would swap.

15.4 Optimization Framework and Procedures

15.4.1 Parameter Reconfiguration

The first stage of network life is the actual definition of the network, also known as the preplanning phase [17]. In this phase, different criteria are defined, including description of the expected traffic, services, network topology and deployment scenarios, as well as coverage, throughput, system capacity requirements, relay type, and usage model.

The preplanning phase is followed by a complete network planning and optimization process which consists of other three main phases: dimensioning, detailed network planning and optimization, and operation and maintenance. Each of the three network planning and optimization phases needs to perform a process shown in Figure 15.5. The concept of planning involves finding the quality of the network performance for a given configuration, whereas optimization refers to finding the configuration for which the network quality is optimal. There are several methods and algorithms in the existing literature to perform the latter task that will be described in Section 15.5.

In the network-dimensioning phase, the first manual network adjustment based on data collection from a digital map and propagation model tuning can be done.

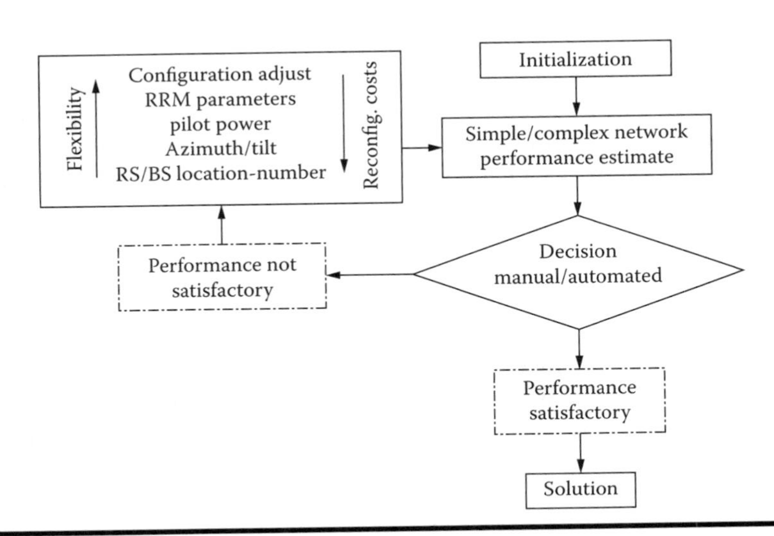

Figure 15.5 Network Planning and Optimization.

Pilot power, coverage range, and an estimate of the number of stations are calculated based on CINR predictions.

The following detailed network planning and optimization phase focus of this work, normally uses a tool for capacity calculation to predict the network performance of each configuration tested for different parameters with high reconfiguration costs (e.g., site location number, tilt, or azimuth), which is the case evaluated in this work. The tool should support traffic, propagation, RRM models, and other parameters. The objective of this tool is to calculate key performance indicators (KPI) that can represent the quality of a certain LTE network configuration.

The final operation and maintenance stage may configure parameters that can be easily reconfigured for testing, such as BS/RS transmit power or scheduler algorithm. Pilot pollution studies can be also carried out at this stage. Finally, further detailed network planning and optimization phases can be performed again after this stage in the event of a network expansion, for example, when more RSs are needed, or a malfunctioning of the network appears due to a bad network definition and dimensioning.

The optimization algorithm tests different configurations for different parameters (e.g., BS location from a set of candidate sites, antenna tilt, or azimuth) that need to be discrete, and the resolution chosen will determine the size of the solution space. The algorithm stops when the cost function based on some costs and a series of penalty functions based on different KPIs ($G = \{G_1, G_h, \ldots, G_H\}$) meet the needs of the service provider. Different criteria h can be considered in the cost function. For example, we propose infrastructure costs, user CINR (Cu), throughput (Tg), and RS reliability (Cr) in a four-objective minimization problem as in Equation 15.4.

$$\min[\lambda \cdot F_{\text{costs}} + F_{\text{pen}(Cu)} + F_{\text{pen}(Tg)} + F_{\text{pen}(Cr)}] \tag{15.4}$$

The first objective, F_{costs}, represents the aggregate of the installation and annual maintenance fee for the stations and no special penalty function is applied. The integer variable s_i in Equation 15.5 indicates whether sector S_i is selected in the solution. The weighting factor λ gives more or less importance to this objective, compared to the rest of the criteria.

$$F_{\text{costs}} = \sum_{i \in S} f_{\text{costs}}(S_i) \cdot s_i \tag{15.5}$$

Some other different criteria values are evaluated over all the users, with the use of penalty functions calibrated by the network designer. A binary variable $d_{m,h}$ in 0, 1 indicates if user m is evaluated by the cost function related to criterion h that has a value of $t_{m,h}$ (see Equation 15.6).

$$F_{\text{pen}(G_h)} = \sum_{i \in S} \sum_{m \in Ms_i} f_{\text{pen}(G_h)}(t_{m,h}) \cdot d_{m,h} \tag{15.6}$$

A threshold-based function avoids the excessive influence of very good or very bad users. A maximal penalty is applied when the value of the criterion $t_{m,h}$ from user m is smaller than a lower bound $T_{\min}$ and no penalty exists when it is higher than a threshold $T_{\max}$. In between these two values the function is linearly decreasing as Equation 15.7 and Figure 15.6 indicates.

$$f_{\text{pen}(G_h)}(t_{m,h}) = \begin{cases} 0 & \text{if } T_{m,h}^{(\max)} < t_{m,h} \\ f_{m,h}^{(\max)} \cdot \left[\dfrac{t_{m,h} - T_{m,h}^{(\max)}}{T_{m,h}^{(\min)} - T_{m,h}^{(\max)}} \right] & \text{if } T_{m,h}^{(\min)} < t_{m,h} \leq T_{m,h}^{(\max)} \\ f_{m,h}^{(\max)} & \text{if } t_{m,h} \leq T_{m,h}^{(\min)} \end{cases} \tag{15.7}$$

$F_{\text{pen}(Cu)}$ represents the wireless connection by penalizing the effective CINR perceived over all SBs of each user. Note that, different users can have the same "connectivity" by having similar average CINR over their set of SBs, but they may require different number of SBs and thus, different final throughput. Therefore, a separate indicator—$F_{\text{pen}(Tg)}$—is needed in the cost function, which ensures a certain throughput, and thus a quality of service (QoS) to the MSs with different services.

The RS reliability $F_{\text{pen}(Cr)}$ penalizes different degrees of network resiliency. A failure in the multihop communication is more likely to happen for weak relay links connections. Thus, CINR levels in these links can provide information to the service provider about RS reliability. Note that the performance in this criterion is penalized over the BS-RS link, and therefore over the subset of RSs. If we consider a set of active RSs attached to a sector i as a subset of S ($Rs_i = \{1, \ldots, w, \ldots, W_i\}$), a binary variable $d_{h'}^{w}$ in $\{0, 1\}$ (Equation 15.8) indicates whether RS_w is evaluated by the cost function related to criterion h', which may represent a subset of criteria related to RS

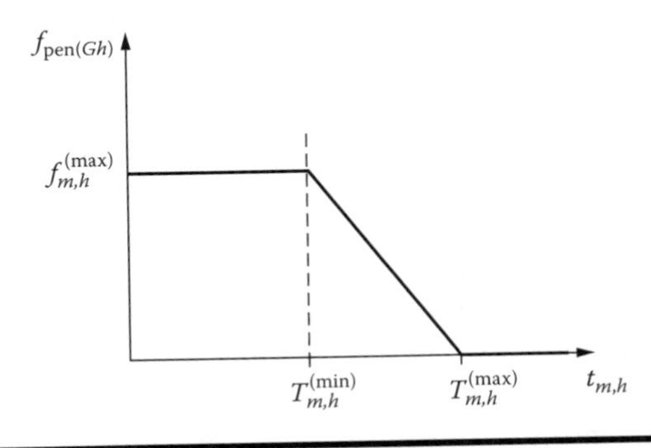

Figure 15.6 Threshold based penalty function for minimization.

resiliency ($Cr = \{Cr_1, Cr_{h'} \ldots, Cr_{H'}\}, Cr \subset G$), for example, when the CINR values of the relay link are penalized differently in UL or DL.

$$F_{\text{pen}(Cr)} = \sum_{h' \in Cr} F_{\text{pen}(Cr_{h'})} = \sum_{i \in S} \sum_{w \in Rs_i} \sum_{h' \in Cr} f_{\text{pen}(Cr_{h'})}[CINR(dB)_{w,h'}] \cdot d_{w,h'} \tag{15.8}$$

Different strategies can be set by the network designer by tuning the values (λ, $F_{\max}$, $T_{\min}$, $T_{\max}$) in order to give more or less preference to different objectives in the cost function (Equation 15.4). An economic perspective to the problem is taken in Section 15.6, where we show an example of deployment analysis.

The processes described in the detailed planning and optimization phase are considered off-line, as the optimization loop does not include the real network. Network simulation is used instead, as it would be unfeasible to iteratively reconfigure some parameters that need physical changes. The last operation and maintenance phase can make use of on-line optimization by using real measurement data and optimize parameters that can be reconfigured by software, such as RRM or handover parameters. In this case, the cost functions described in Equations 15.4 and 15.6 can also be used. The difference is that the KPIs to be penalized in such functions would correspond to real network data instead of the simulation procedure, as is explained in Section 15.4.3. The objective of the operation and maintenance phase is to respond to changes in traffic, architecture, or network demands, and further self-healing and self-optimization procedures can be applied (see [7,8] for more information).

15.4.2 Frequency Planning

There are several ways to perform frequency planning in this kind of networks. With the fixed subchannel allocation (FSA) procedure, the whole channel is divided in several segments, and each segment is assigned to each sector. This scheme simplifies the radio frequency planning design, as the operator only needs to assign segments to sectors. In the case of three sectors per site, this procedure mitigates inter-cell interference by reducing the probability of slots collision by a factor of 3. However, the sector capacity is also reduced by a factor of 3. This method is suitable for regular/hexagonal scenarios [19].

Dynamic frequency planning (DFP) in the form of subchannel planning has been recently been proposed for OFDMA in the existing literature (e.g., [19, 20]), where the use of a certain subset of subchannels are assigned to different sectors. Similarly, we propose a dynamic allocation of subchannels that can adapt to the environment conditions where loaded sectors may need to use more subchannels than neighbor sectors with low demand. In other words, DFP takes advantage of multiuser channel and traffic diversity to adjust the subchannel allocation. Basically, DFP can be divided into two categories: centralized DFP, where the subchannel allocation decision is made by a central controller; and distributed DFP, where the subchannel allocation decision is made by users or BSs independently.

In the centralized option, the procedure can be performed in the detailed planning and optimization phase as an inner optimization loop, where an optimal assignment of subchannels may be offered for each configuration tested (see Section 15.4.3), or in the operation and maintenance stage. For example, the subchannel usage can be reconfigured to react against the emergence of new hotspots, different service demands during the day, a sports event, or different traffic during seasons. The distributed scheme is only suitable in the operation and maintenance phase of an already deployed network, where self-configuration techniques can be applied [7].

In the centralized DFP scheme, system inter-cell interference can be characterized by a restriction Matrix W (see [9] and [19]), without the need of full simulation of RRM procedures. As a result, the optimization process becomes more efficient than the one described previously (Section 15.4.1). In this case, an optimization algorithm can also iteratively go through different possible solutions for subchannel assignments, aiming to minimize global interference, represented in the mentioned matrix. The difference with the optimization of other kind of parameters is that when different solutions of subchannel allocation are being tested by the algorithm, the best servers of MSs are not changing. Therefore, the interference relation between any two sectors is not changing for every solution tested. The MS service requirements and the MCS estimated from the channel quality indicator (CQI) can be taken into account to avoid full simulation. This is done in two stages, interference prediction and capacity demands, which are described as follows.

- The interference prediction stage characterizes the inter-cell interference of the network, represented by the matrix W of size $N \times N$, in which $w_{i,j}$ represents the inter-cell interference between sectors S_i, S_j.

  Let us introduce the concept of interference event (IE). There is an interference event between two sectors, S_i (server) and S_j (neighbor) over user m ($\text{IE}_{i,j,m} = 1$), if the power level of the carrier signal (from S_i to a served user m) is smaller than the power level of a neighboring interfering signal (from S_j to the same user) plus a given threshold. Cooperative communications between RS and the serving BS do not account for interference, as the MS must be able to identify and combine the different signals to obtain diversity gain.

  The threshold Thres is a protection margin against interference, and it is set empirically by the operator according to its planning targets. A small value in this threshold involves a more conservative approach by assuming that neighbor users are more likely to interfere. However, reasonable variations in this term does not significantly affect the optimal subchannel allocation, as it affects the whole system modeling. We have set the value to 12 dB, which shows a good performance in an LTE system. The following pseudocode shows the procedure:

```
for(i = 1;   i ≤ N;   i++)
    for(j = 1;   j ≤ N;   j++)
        for(m = 1;   m ≤ Ms_i;   m++)
            if(C_{m,i} < I_{m,j} + Thres(12 dB))    then
            {
                if(m ∈ Ms_i   &&   m ∈ Ms_j)    then    //Cooperative
                        communications continue
                IE_{i,j,m} = 1
            }
```

The system inter-cell interference is modeled by W, as Equation 15.9 indicates in terms of percentage of time that cells interfere with each other considering that the same subchannel is used, and taking into account the requested capacity by the MS Rc_m, which depends on the service used. This gives a service-oriented perspective to the problem, as users with more throughput demands will generally need more resources. The fact that some users will need more resources than other due to the channel conditions is still not considered in this estimation, as W models the interference statistically in a nonregular scenario. Note that the power level of the server, neighboring cells, and other information is reported using measurement reports within the LTE radio frame.

$$w_{i,j} = \frac{\sum_{m \in Ms_i} Ie_{i,j,m} \cdot Rc_m}{\sum_{m \in Ms_i} Rc_m} \tag{15.9}$$

- The optimal number of required subchannels D_i per sector S_i, is approximated in the capacity demand stage. We propose a subchannel allocation taking into account the traffic demands in the access zones in both BS and RS. The reason is that relay links are expected to be more reliable due to an optimal position of the RSs during the design process, and MSs communications are more likely to be affected by changes in the traffic density or channel conditions.

  Sectors may have different access zone lengths according the traffic requirements. Also, some sectors may not have associated RSs, and therefore MSs attached to this sector use the whole frame as an access zone and need fewer subchannels, D_i. This is represented in Figure 15.7 and should be taken into account in formulating the problem. Note that network time synchronization is necessary in OFDMA systems to minimize multiaccess interference, as well as for proper performance of handovers.

  The number of requested slots Rl_m for each user m of sector S_i is calculated from the requested capacity RC_m (kbps) and slot efficiency SE_m (kbps). The average symbol efficiency depends on the average modulation and coding selected by the users within the sector. It can be obtained from previous frames, or it can be derived from the following formula taking

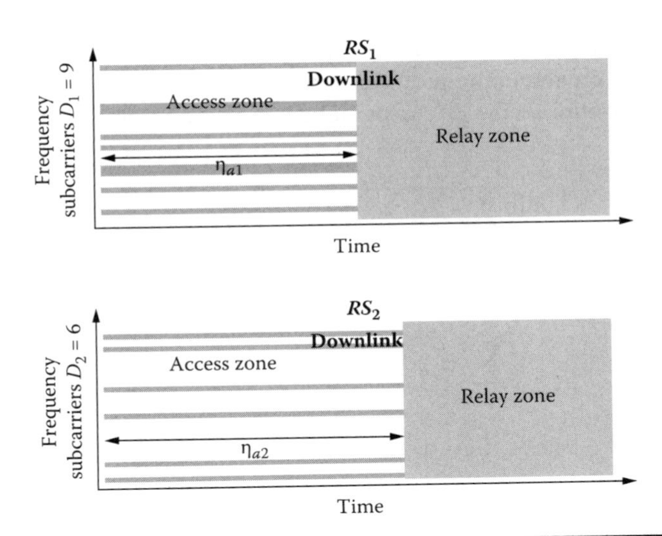

Figure 15.7 Example of frequency planning with access zone consideration.

into account network statistics:

$$RL_m = \frac{RC_m}{SE_m} \tag{15.10}$$

As Equation 15.11 indicates, SE_m depends on the number of subcarriers per SB η_s (12 in LTE), and the bearer efficiency $\eta_{B(CQI_m)}$ in terms of bits/subcarrier. The bearer efficiency depends on the MCS assigned to the user, and it is selected according to the CQI feedback from previous iterations. T_{frame} is the frame duration.

$$SE_m = \frac{\eta_s \cdot \eta_{B(CQI_m)}}{T_{\text{frame}}} \tag{15.11}$$

The requested subchannels D_i per sector are calculated by dividing sum of all RL_m over all MSs within the sector S_i—which includes cooperative communications—and the number of symbols in the access zone per subchannel η_{a_i}:

$$D_i = \frac{1}{\eta_{a_i}} \cdot \sum_{m \in Ms_i} RL_m \tag{15.12}$$

The factor $\eta_{c_{i,j}}$ represents the portion of access zone that is being interfered by the neighbor sector (see Equation 15.13).

$$\eta_{c_{i,j}} = \frac{\min(\eta_{a_i}, \eta_{a_j})}{\eta_{a_i}} \tag{15.13}$$

Once the inputs have been obtained, the optimization routine can be defined as a mixed integer programming problem, where the objective is to find the optimal solution that minimizes the given cost function representing the overall network interference:

$$\min \sum_{i \in S} \sum_{j \in S} \sum_{k \in Sh} w_{i,j} \cdot \frac{\eta_{c_{i,j}}}{D_i \cdot D_j} \cdot y_{i,j,k} \tag{15.14}$$

subject to:

$$\sum_{k \in Sh} x_{i,k} = D_i \qquad \forall i, k \tag{15.15}$$

$$x_{i,k} + x_{j,k} - 1 \leq y_{i,j,k} \qquad \forall i, j, k \tag{15.16}$$

$$y_{i,j,k} \geq 0 \qquad \forall i, j, k \tag{15.17}$$

$$x_{i,k} \in \{0, 1\} \qquad \forall i, k \tag{15.18}$$

where $x_{i,j}$ is a binary variable that indicates whether sector S_i uses subchannel k or not. Constraint (Equation 15.15) imposes that sector S_i must use D_i subchannels. Inequalities (Equations 15.16 and 15.18) force that if S_i and S_j use subchannel k, then $y_{i,j,k}$ is forced to equal 1, and $y_{i,j,k} = 0$ otherwise. Finally, the cost function is the sum of the interference between all pair of sectors S_i and S_j, taking into account all the frequencies k.

Since the capacity of the sectors is not considered when the restriction matrix W is built, the interference restrictions $w_{i,j}$ must be divided by the number of used subchannels D_i and D_j in both interfered sectors S_i, S_j, and thus, the bigger the number of subchannels per sector, the smaller the chance of interference.

15.4.3 System Level Simulation and Performance Evaluation Methods

As mentioned, the use of network simulation is needed in the detailed network planning and optimization phase. Monte Carlo simulation is often used in cellular network planning and optimization due to the high computational load that dynamic simulation would require in an iterative process that may need thousands of simulations. This kind of simulation is snapshot based and represents an instant of the network performance with fixed position of MSs. The simulator takes multiple Monte Carlo snapshots to statistically observe the network behavior.

The technology-specific integration of the optimization procedures with the simulation platform is shown in Figure 15.8 and described in this section. Within each Monte Carlo snapshot, the throughput is calculated iteratively until the performance of the system converges. Because the resources used by one user in one

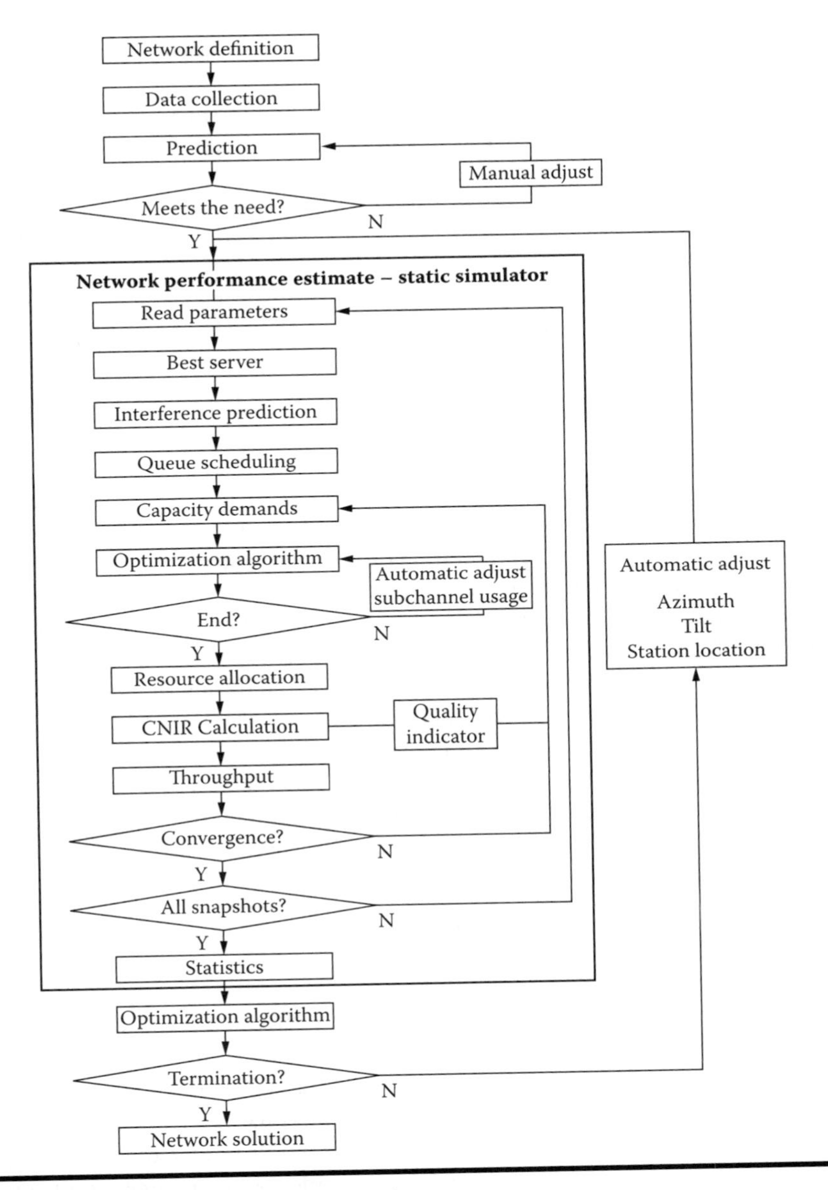

Figure 15.8 Optimization framework.

sector may interfere with other users, the system converges when it is stable in the selection of MCS for communication and assignment of resources. In order to fulfill the service requirements of the user, the CQI helps to set the adequate MCS for communication according to the performance obtained in the previous iteration. Link level simulation results are imported in the form of lookup tables (LUT) for throughput calculation, which should include effects of different wireless channels and effects like fading.

At each iteration of this process, the first automatic adjustment of the framework is done. A centralized DFP calculated by interference estimation is performed. The interference can be modeled with the restriction matrix described, which characterizes the possible interference between any two sectors. An optimization algorithm iteratively checks the estimated interference and goes through different possible solutions for subchannel assignments before the RRM process. As described in the previous section, the matrix is computed from the number of expected MSs in each sector as well as the MSs service requirements and the MCS estimated from the CQI.

In the outer loop, the optimization algorithm tests different configurations for different parameters that have high reconfiguration costs in a real network evaluation like site location and number, tilt, or azimuth. It stops when a cost function based on KPIs meets the needs of the service provider. With the use of an inner and outer optimization loop, the optimization procedure becomes an integrated two-step method, in which every tested configuration presents an optimal frequency planning.

The use of LUT is critical in network design in order to obtain an efficient network planning and optimization process. In the following, efficient capacity calculations that allow intensive system level simulations are introduced.

After the resource allocation is performed, CINR (and throughput) are calculated for each SB, r. A certain user, m, whose best server is S_i, is interfered in DL by other stations (S_j), only if S_i and S_j are using the same resource for transmission within a distance smaller than the system reuse distance. The final interference suffered by the user m will be the sum of all the interference rays coming from neighboring base stations (S_j). Equation 15.19 shows the case for DL communications.

$$I_{m,r}^{DL} = \sum_{j \in S,\, j \neq i} (P_{j,r} \cdot G_j \cdot L_j \cdot L_{j,m} \cdot G_m \cdot L_m) \cdot u_{j,r} \tag{15.19}$$

where m indicates the interfered user; i is the server base station; j are interfering stations; r is the studied resource; $P_{j,r}$ is the power applied by S_j in r; and $L_{j,m}$ is the path loss between user m and station j; G stands for antenna gain and L_j and L_m stand for feeding losses (a gain with a value smaller than 1). The integer variable $v_{j,r}$ indicates whether cell j is using slot r. Note that the power of all resources within a certain subchannel will be the same. Note that linear units must be used.

By using the interference model, the CINR of user m, can be calculated as follows:

$$CINR_{m,r} = \frac{C_{m,r}}{I_{m,r} + N_r} \tag{15.20}$$

where C and I are the power of the carrier and interference signals, respectively, and N is the noise power of the resource studied. The carrier signal power can be estimated by using:

$$C_{m,r}^{DL} = P_{i,r} \cdot G_i \cdot L_i \cdot L_{i,m} \cdot G_m \cdot L_m \quad (15.21)$$

The throughput of the sector is calculated, as shown in Equation 15.22, from the slot efficiency Se_m (see Equation 15.11), which depends on the bearer efficiency $\eta_{B(CQI_m)}$. Each bearer corresponds to a certain MCS (see Table 15.1). The bearer used for transmission is determined according to the CQI of the MS in a previous iteration, and it is represented by CINR measurements over the whole set of SBs on a user. Then, the LUT provides the block error rate (BLER) values, which is a function of the bearer B used and the CINR level of the studied resource r from the given set of available SBs. The $v_{m,r}$ represents the LTE matrixlike interference

Table 15.1 LTE Radio Access Bearers

Radio Access Bearer index	*Modulation*	*Coding*	*Bearer Efficiency η_B (Bits/Symbol)*	*CINR Threshold (dB)*
1	QPSK	0.0761719	0.1523	−6.5
2	QPSK	0.117188	0.2344	−4
3	QPSK	0.188477	0.377	−2.6
4	QPSK	0.300781	0.6016	−1
5	QPSK	0.438477	0.877	1
6	QPSK	0.587891	1.1758	3
7	16QAM	0.369141	1.4766	6.6
8	16QAM	0.478516	1.9141	10
9	16QAM	0.601563	2.4063	11.4
10	64QAM	0.455078	2.7305	11.8
11	64QAM	0.553711	3.3223	13
12	64QAM	0.650391	3.9023	13.8
13	64QAM	0.753906	4.5234	15.6
14	64QAM	0.852539	5.1152	16.8
15	64QAM	0.925781	5.5547	17.6

Table 15.2 System-Level Simulation Parameters

Parameter	*Value*
Fixed BSs	9
Candidate RSs	60
Channel bandwidth	10 MHz
Density wide area users	50 user/km^2
Density of hotspots users	70 user/km^2
Wide area users service	VoIP
Minimum VoIP throughput	12.2 Kbps
Maximum VoIP throughput	12.2 kbps
Hotspots users service	Data and VoIP
Minimum data throughput	64 Kbps
Maximum data throughput	128 Kbps
BS TX power	43 dBm
BS antenna gain	18 dBi
BS antenna pattern	90°
BS antenna height	40 m
BS antenna tilt	3
BS noise figure	4 dB
BS cable loss	3 dB
RS TX power	30 dBm
RS antenna gain	18 dBi
RS antenna pattern	90°
RS antenna height	30 m
RS noise figure	4 dB
RS cable loss	3 dB
MS Tx power	23 dBm
MS antenna pattern	Omnidirectional

(*Continued*)

Table 15.2 System-Level Simulation Parameters (*Continued*)

Parameter	*Value*
MS antenna height	1.5 m
MS noise figure	8 dB
MS cable loss	0 dB
Standard deviation of shadow fading	8 dB
Intrasite correlation of shadow fading	0.7
Intersite correlation of shadow fading	0.5
Snapshots	100
Path loss model	Ray launching
Scheduling algorithm	Round robin and best CINR

model by indicating if MS_m is using SB_r. Note that if sector S_i is not active (i.e., it is not selected in the solution), $M_i = 0$ and, therefore, $T_i = 0$.

$$T_m = \sum_{r \in Sb} [Se_m \cdot (1 - \mathrm{BLER}_{(B(CQI_m), \mathrm{CINR}_{m,r})})] \cdot v_{m,r} \tag{15.22}$$

In decode-and-forward multihop communications, the throughput is calculated separately in each link, as the signal is decoded and encoded again. The final value perceived by the user is determined by the weakest link. Therefore, in a two-hop communication, the final throughput is:

$$T_m = \min\left(T_m^{(S-R)}, T_m^{(R-D)}\right) \tag{15.23}$$

where $S - R$ represents the throughput in the source-relay link, and $R - D$ in the relay-destination link.

When cooperative communications are considered, the signal from other paths may contribute with some gain, which can be modeled according to the CINR values of the cooperative paths in additional LUTs. A common case is when a MS receives signal from both RS and BS. The LUT will relate the CINR of the BS-MS ($CINR_{m,r}^{S-D}$) link with a certain gain on the $CINR_{m,r}^{R-D}$ value.

In the following, an example of network design process has been performed over a rectangular area of the city of Munich (3.3 km × 2.3 km) using an LTE network. The scenario used is a nonregular network composed of 9 fixed and active BSs and 60 candidate RSs. In this area, an MS density of 50 MS/km^2 and two hot spots adding 20 MS/km^2, account for around 510 MSs per snapshot of the Monte Carlo routine. The rest of parameters are shown in Table 15.2.

Different configurations are searched by a simulated annealing algorithm in both the inner and outer optimization loop. The outer loop gets optimal values for antenna tilt and azimuth (in intervals of 10°), and RS position from a given set of candidate sites, with the optimal assignment of subchannels calculated in the inner loop. Figure 15.9 shows the pilot power and user status of a single snapshot in the solution found.

Hotspots and indoor coverage in the central area benefit from this configuration. However, there are a number of MSs with low CINR levels concentrated at the edge of the coverage areas, where they may have a poor channel quality. Note that the penalization is performed in different ranges and values and depends on the network designer criteria, as some objectives can be given more or less preference. Due to this

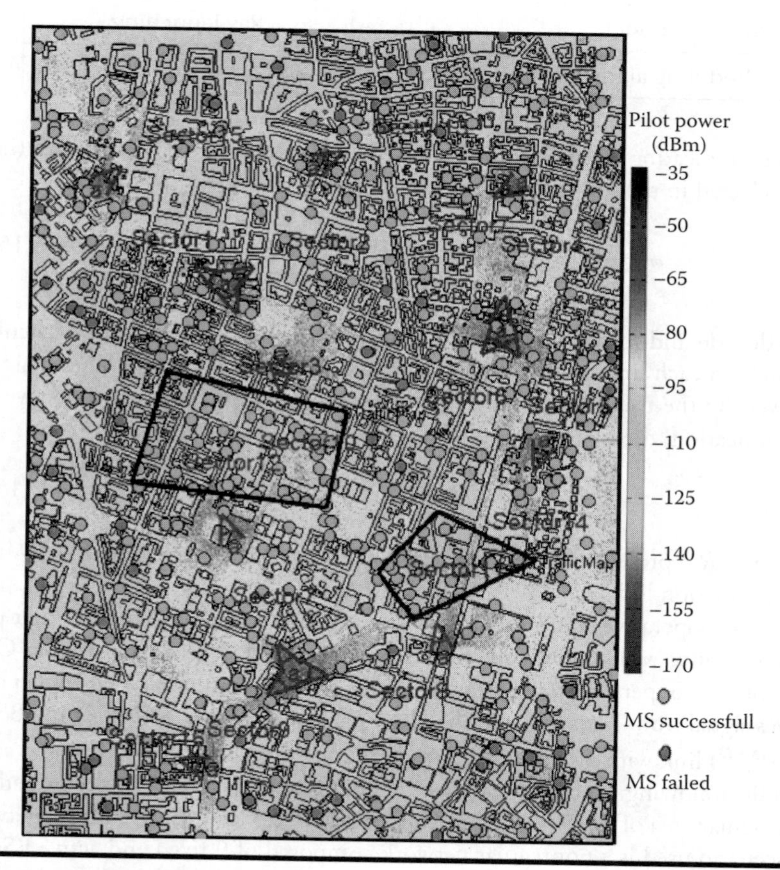

Figure 15.9 Optimal solution.

Table 15.3 Restriction Matrix and Demand Vectors

S_i	*1*	*2*	*3*	*4*	*5*	*6*	*7*	*8*	*9*	*10*	*11*	*12*	*14*	*15*	*17*	D_i
1	0	0	38	0	0	0	6	0	0	6	0	25	0	25	0	**21**
2	40	0	40	10	0	10	0	0	0	0	0	0	0	30	60	**8**
3	9	0	0	0	0	0	0	0	0	50	0	42	0	0	0	**11**
4	0	27	7	0	7	20	0	0	0	0	0	0	0	0	60	**9**
5	0	0	0	20	0	80	0	0	0	20	0	0	60	0	0	**7**
6	0	0	0	16	70	0	8	8	0	62	6	0	31	0	23	**11**
7	0	0	18	0	6	18	0	41	35	29	21	41	6	0	0	**28**
8	0	0	14	0	29	29	57	0	29	29	35	0	43	0	0	**29**
9	0	0	0	0	0	0	14	57	0	0	0	0	0	0	0	**11**
10	0	0	100	0	11	42	47	11	0	0	0	58	11	0	0	**24**
11	0	0	0	0	15	0	0	74	11	0	0	0	46	0	0	**41**
12	14	0	58	0	0	0	56	0	26	42	0	0	0	0	0	**46**
13	11	67	0	39	0	0	0	0	0	0	0	0	0	33	78	**13**
14	0	0	17	0	76	62	29	44	0	32	46	0	0	0	6	**47**
15	77	26	0	0	0	0	0	0	0	0	0	0	0	0	10	**40**
16	0	0	0	0	0	0	41	0	86	0	0	27	0	0	0	**18**
17	0	33	33	83	0	33	0	0	0	0	0	0	0	0	0	**7**

fact, diverse situations can be created. This solution is an example of throughput enhancement and network reliability, as RSs are well covered. Other strategies with different degrees of fairness for users or risky configurations for the operators can be set by tuning the values ($f^{(max)}_{m,h}$, $T^{(min)}_{m,h}$, $T^{(max)}_{m,h}$) in the penalty function.

The restriction matrix approximates the inter-cell interference between sectors of the network in terms of percentage of interfering users. An example can be seen in Table 15.3. This matrix characterizes the interference relationship between sectors, for example, the restriction is null when sectors are far away from each other. On the other hand, the interference is large between adjacent RS-BS, such as S_{10} and S_3 or S_{15} and S_1, as they reuse resources. Also, note that some relay sectors, such as S_{15}, S_{14}, and S_{12}, have the largest demand vector as they cover large areas or the hotspots.

We can also observe that the influence between two sectors is not reciprocal in most cases. The reason is that the number of MSs attached to several sectors is different. For example, S_2 is affected by 40% of the communications of S_1 but not the opposite, as S_1 demands are higher ($D_1 = 21$) than those for S_2 ($D_2 = 8$).

One snapshot of a solution obtained without the use of the inner optimization loop is shown in Figure 15.10. Instead, the frequency planning process has been performed after the optimization of tilt/azimuth and RS position. In this case, the throughput per user decreases specially in the MSs that are in the hotspots. The reason is that the tilt/azimuth and RS position set were calculated for a frequency reuse 1, and the subsequent automatic frequency planning has more difficulties to efficiently allocate more subchannels to sectors with large traffic demands due to the lower CINR levels with such antenna configuration.

Tilt/azimuth and RS position values have been set to avoid interference between sectors as much as possible. For example, compared to the optimal solution in

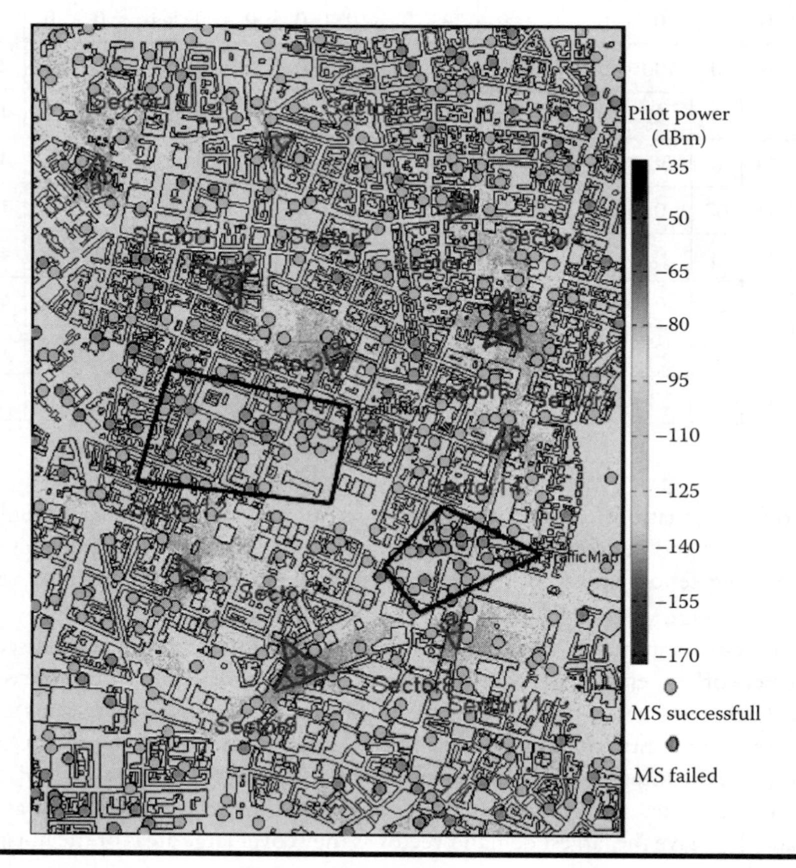

Figure 15.10 Solution without inner optimization loop.

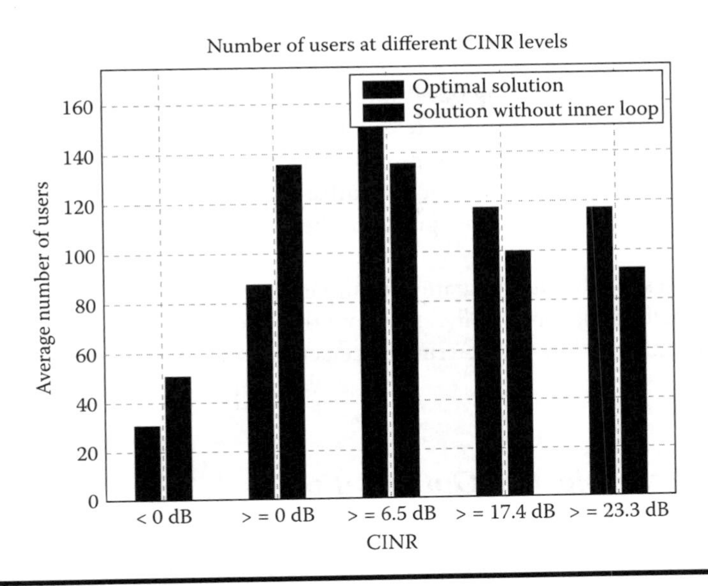

Figure 15.11 Number of users at different reference CINR.

Figure 15.10, it can be observed that sectors avoid pointing each other. Also, in some cases, RS positions are not set optimally for the relay link. As a result, the network is less resilient. Because the final throughput perceived by many MSs is the result of the communication of two links, the total throughput in this system is lower. The system throughput in this case is 853 Mbps, whereas in the optimal solution is 1125 Mbps.

Figure 15.11 shows the distribution of users at different CINR levels. We can observe that the main difference with respect to the solution without the inner loop is the lower number of users that cannot get the minimum MCS for communication or have low CINR levels that only allow low MCSs. This is due to the inefficient interference management. These users will need large bursts to fulfill their service requirements and the radio frame will be filled faster, thus leaving other users with the same situation in an unsuccessful state. This effect is critical in the two hotspots where traffic is higher.

15.5 Optimization Techniques

The network planning and optimization problem is hard as it is very difficult to find a theoretical optimum in polynomial time [1,6]. Different methods can be applied to solve the optimization problem presented in the form of the mixed integer program, minimizing the cost function.

15.5.1 Metaheuristics

Metaheuristics are algorithms based on a search within the solution space. They seem to be a common approach in related works to solve the network planning and optimization problem, as they can provide close to optimal solutions in reasonable time [1, 6]. Metaheuristics do not guarantee the optimality of the solution, but they often perform very well in practice when the number of variables becomes large.

The process of algorithm parameter tuning is customized for different problems and is normally done empirically. The time that the algorithm is running and the algorithm configuration will determine the performance of the whole optimization framework.

15.5.2 Multiobjective Optimization

The weighting of the different objectives in the cost function (Equation 15.4) may result in a difficult task during the network planning and optimization process, especially when the relay configuration is considered where more decisions need to be made and the different objectives incur in different trade-offs. For example, a final design with low infrastructure costs may have worse performance than another with more active stations.

The optimization problem can be solved by using the solutions of the Pareto front [21], where the network provider must select a posteriori the most appropriate ones according to some policies. The Pareto optimal solutions are called nondominated. Each of the obtained solutions will represent a certain optimal trade-off between the different factors in the cost function, as it is not possible to improve one of them without worsening the others. The objectives of the cost function should not be fully correlated in order to offer valuable information about the different solutions found. Figure 15.12 illustrates the solutions in an example of a two-objective minimization problem, where the stars represent the set of nondominated Pareto optimal solutions of rank 1. The rank of a solution n is defined by $Ra_n = 1 + So_n$, where So_n is the number of solutions by which n is dominated in the set of feasible solutions. To obtain the solutions of Pareto rank Ra_n, the solutions of rank $Ra_n - 1$ have to be removed.

The solutions are iteratively calculated by an optimization algorithm in three main stages:

- **Search Front Expansion:** The algorithm searches for neighbor solutions in the current search front. For example, a neighbor solution is obtained by removing/changing the position of one BS/RS or by changing the antenna azimuth/tilt.
- **Update of the Optimal Pareto Front:** The nondominated solutions of the neighborhood are selected.

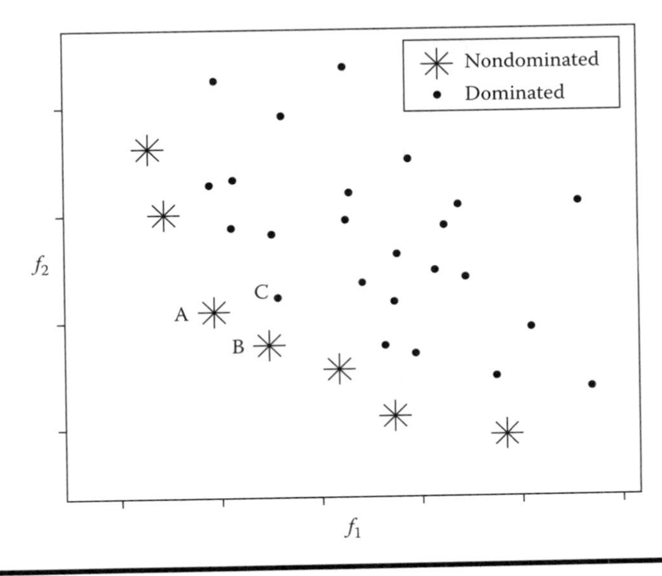

Figure 15.12 Example of a two- function minimization problem with dominated and nondominated solutions. Solution C has become dominated with the addition of solutions A and B.

- **Selection of the New Search Front:** This is done according to the optimization algorithm methodology.

Many algorithms can be adapted to the multicriteria optimization. In the following, a detailed description of the procedure for multiobjective TS is provided. The algorithm minimizes the previously defined criteria F_{costs}, $F_{\text{pen}(Cu)}$, $F_{\text{pen}(Tg)}$, and $F_{\text{pen}(Cr)}$ and try to improve the current search front $Sf_{u,n}$ composed of n solutions at every iteration u by calculating the neighboring solutions. A neighboring solution is a non-Tabu change of the state of the network by adding, removing, or moving a RS/BS or changing any other parameter of the network configuration. The total neighboring set is the aggregate of the neighborhoods of the n solutions of the search front, and it is stored in the short-term memory created by the algorithm. Similar to the standard algorithm, the Tabu list is maintained in long-term memory.

We sketch out this method in Figure 15.13. The optimal front Of is updated at the end of each iteration u by adding all the non-dominated solutions of the neighborhood of the search front. The previous solutions of the Of that have become dominated with the new ones (rank $Ra_n > 1$) have to be removed. The new search front $Sf_{u+1,n}$ is created randomly from previous solutions, thus introducing diversity in the search process [22]. In addition, the new solutions $n+1$ created from solution n are stored in the Tabu list, which is also updated at the end of each iteration u.

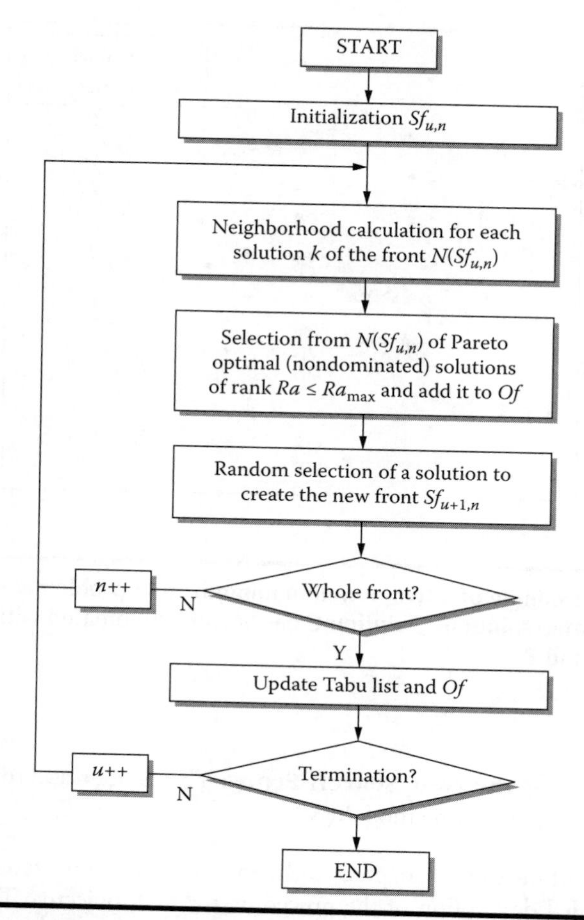

Figure 15.13 Multiobjective Tabu search algorithm.

The algorithm stops after a certain number of iterations defined by the user u_{max}, and the most representative solutions from the Pareto front are selected. This can be done manually by using graphical representation, or using some algorithms if the final set of non-dominated solutions is large, like the sharing function presented in [23]. The sharing function weights each solution with a proportional value to the density of its neighborhood in the Pareto front. Basically, the aim is to pick enough solutions to provide information about different trade-offs in the Pareto front. The maximum cost for each function should be taken into account by dismissing the solutions not fulfilling this requirement.

15.6 Planning Criteria for LTE—An Economic Perspective

A multicriteria and economic perspective can help the network provider to choose a suitable option for its strategies in complex optimization problems, such as the design of relay networks. An economic perspective can be given to the design activity as follows:

$$\text{Profit} = \text{ProfitEst} - \text{CostAdj} \tag{15.24}$$

In this model, some expected profits can be estimated during the network definition phase, and represented in ProfitEst in Equation 15.24. This factor includes ideal estimations for the profit in a business case, and it considers license acquisition costs, and installation and maintenance costs of the core network. The optimization process tries to minimize a cost adjustment factor, CostAdj, which includes access network costs (F_{costs}) and some penalty costs for the performance of a particular network configuration ($F_{pen(G_h)}$), (see Equation 15.25). These penalties balance the ideal estimations done in ProfitEst.

$$\min\left\{\text{CostAdj}\right\} = \min\left\{F_{\text{costs}} + \sum_{h\in G} F_{\text{pen}(G_h)}\right\} \tag{15.25}$$

Once the system has been dimensioned, the whole area under consideration can be divided into several regions ($Jz = \{1, \ldots, j, \ldots, J\}$), and each region j contains n_j candidate sites where BSs or RSs can be installed. A limited number of candidate sites are selected from a region to install a BS/RS. An installation cost is associated with each candidate site. With this simplified network scenario, the optimization procedure of the algorithm turns out to be more affordable. Significant solutions are not expected to be dismissed when using this method, as it sets an upper bound in the number of RSs/BSs installed and thus helps avoid testing solutions with high interference levels. The process of selecting the number and the appropriate areas corresponds to the first manual adjustment described in Section 15.4.1.

The problem now is to select candidate sites from each region to install RSs/BSs with a certain antenna configuration such that the traffic capacity and the number of covered MSs are maximized with the lowest installation cost and a resilient network. As we mentioned, the optimization procedure can be solved with the Pareto front method, which analyzes each objective separately and stops after a certain number of iterations [22].

The different criteria in the cost function can be grouped according to the main KPIs. Infrastructure costs, CINR (Cu), throughput (Tg), and RS resilience (Cr) are considered in Equation 15.26 in a four-objective minimization problem. Different subsets of criteria (Cu, Tg, $Cr \subset G$) related to the particular KPI nature will determine different constraints to the problem, for example, different throughput in services or different channel profiles, which are shown in Equations 15.27, 15.28, 15.29, 15.30 and described below.

$$\min \Bigg\{ \sum_{i \in S} f_{\text{costs}}(S_i) \cdot s_i, \quad \sum_{i \in S} \sum_{m \in Ms_i} \sum_{h' \in Cu} f_{\text{pen}(Cu_{h'})}[\text{CINR(dB)}_{m,h'}] \cdot d_{m,h'},$$

$$\sum_{i \in S} \sum_{m \in Ms_i} \sum_{h'' \in Tg} f_{\text{pen}(Tg_{h''})}[T(\text{kbps})_{m,h''}] \cdot d_{m,h''}, \qquad (15.26)$$

$$\sum_{i \in S} \sum_{w \in Rs_i} \sum_{h''' \in Cr} f_{\text{pen}(Cr_{h'''})}[\text{CINR(dB)}_{w,h'''}] \cdot d_{w,h'''} \Bigg\}$$

subject to:

$$\sum_{i \in S} s_i \leq J \cdot Sec_{\max} \qquad (15.27)$$

$$1 < \sum_{h' \in Cu} d_{m,h'} \leq 2 \qquad \forall m \qquad (15.28)$$

$$1 < \sum_{h'' \in Tg} d_{m,h''} \leq 2 \cdot Ser_{\max} \qquad \forall m \qquad (15.29)$$

$$\sum_{h''' \in Cr} d_{w,h'''} \leq 2 \qquad \forall w \qquad (15.30)$$

1. f_{costs} represents the aggregate of CapEx (capital expenditure) and OpEx (operational expenditure) in an LTE access network. This method represents a financial model over suitable return periods. The calculation of the infrastructure parameters of sites and sectors [$f_{\text{costs}}(S_i)$] is made from a set of economical parameters $P_c = \{P_1, \ldots, P_p, \ldots, P_P\}$ over a A return period from a business plan ($A \in An$, $\text{An} = \{1, \ldots, a, \ldots, A\}$). This is shown in Equation 15.31.

$$f_{\text{costs}}(S_i) = \sum_{p \in P_c} \sum_{a \in An} Bc_{a,p,i} \qquad (15.31)$$

where B_c indicates the value of the infrastructure economic parameter P_p at year a. Note that different sectors may use the same site.

One example for 5 year period is shown in Table 15.4. This economic study has been done for a typical RS/BS costs found in [6,24]. Other examples

Table 15.4 Example of Economic Parameters for Network Infrastructure Over a 5-Year Return Period

Year		1	2	3	4	5
Base site CapEx (k€)	Installation	10	8.8	7.7	6.8	6
	Establishment	0.75	0.75	0.75	0.75	0.75
Base site OpEx (k€)	Maintenance	1.5	1.33	1.18	1.06	0.95
Sector CapEx (k€)	Installation	5	4.35	3.78	3.29	2.87
	Establishment	0.07	0.07	0.07	0.07	0.07
Sector OpEx (k€)	Maintenance	0.25	0.22	0.19	0.17	0.14
	Rental	0.48	0.48	0.48	0.48	0.48

can be found in [25, 26]. The cost associated to RSs/BSs located in dense populated areas where more traffic is expected may be higher due to a more expensive renting fee or higher capacity equipment.

Constraint (Equation 15.27) indicates that a only limited number of candidate sectors ($Sec_{\max}$) may be selected from a region to install BSs/RSs.

2. $f_{\text{pen}(Cu_{h'})}$ represents the wireless connection by penalizing the effective CINR perceived by the MSs over all SBs. The thresholds $T_{m,h}^{(\min)}$ and $T_{m,h}^{(\max)}$ can be set to the CINR thresholds for the minimum and maximum MCS in the LUTs (Table 15.1). The assumptions about the expected users made by the network provider in the network definition, such as the user profile and speed, influence the final network design. Different MSs profiles may require different costs for the system due to the different channel conditions. Therefore, the network designer may penalize each of them differently (note that different LUTs will apply with different thresholds). As a result, a subset of criteria related to user CINR levels may apply for different situations ($Cu = \{Cu_1, Cu_{h'} \ldots, Cu_{H'}\}$, $Cu \subset G$). Table 15.5 shows the $T_{m,h}^{(\min)}$ and $T_{m,h}^{(\max)}$ thresholds for three different user profiles.

   Because, in a Monte Carlo simulation, each user will be in UL or DL (or both) and in certain channel conditions at each snapshot, constraint (Equation 15.28) indicates that a maximum of two penalty functions are applied per user. The maximum penalty $f_{m,h}^{(\max)}$ can be set in economic terms related to users. For example, in a user with low CINR, a penalty related to the user connection tariff and annual subscription may be applied (Table 15.6).

   In general, the value for the maximum penalty can be calculated as the aggregate of a subset of economic parameters for one user related to a certain

Table 15.5 CINR Thresholds

Modulation	*CINR (dB) Fixed user (Cu_1)*	*CINR (dB) Pedestrian user (Cu_2)*	*CINR (dB) 50 km/h user (Cu_3)*
QPSK1/2 [$T^{(min)}_{m,h}$]	−6.5	−4.3	−1.7
64QAM11/12 [$T^{(max)}_{m,h}$]	17.6	19.3	22.6

performance indicator G_h ($N' \subseteq N$, $N = \{N_1, \ldots, N_u, \ldots, N_U\}$) over an A return period (see Equation 15.32). With the appropriate selection of the elements of subset N', the network provider will be able to give more importance to certain performance indicators.

$$f^{(max)}_{m,h} = \sum_{u \in N'} \sum_{a \in An} Mc_{a,u,m} \tag{15.32}$$

where Mc indicates the value of the user economic parameter N_u at year a.

3. $f_{pen(Tg_{h'})}$ is the penalty function for the MSs throughput. $T^{(min)}_{m,h}$ and $T^{(max)}_{m,h}$ can be set to the maximum and minimum throughput request for each service. The use of this factor in the cost function provides a flexible and service-oriented network design, in which different business plans, based on certain services for some areas, can be applied. A subset of criteria related to the user throughput levels may apply for different services and situations (DL or UL) ($Tg = \{Tg_1, Tg_{h'}, \ldots, Tg_{H'}\}$, $Tg \subset G$). Table 15.7 shows the $T^{(min)}_{m,h}$ and $T^{(max)}_{m,h}$ thresholds for four different services. Constraint (Equation 15.29) indicates that the MS will use at least one service DL or UL, and a maximum of services, Ser_{max} (four in this case) in both UL and DL, which can be penalized in different ways.

The maximum penalty can be set to user economic parameters (e.g., a bad QoS will be reflected in losses in the Mb allowance and other charges

Table 15.6 Economic Parameters for the Users Over a 5 Year Return Period

Year	*1*	*2*	*3*	*4*	*5*
Annual subscription (€)	500	440	387	340	299
Connection tariff (€)	50	45	41	37	33
Mb allowance (€)	600	528	465	409	360
Other charges (€)	62	56	50	45	41

Table 15.7 Example of LTE Service Requirements Thresholds (kbps)

Name	Maximum DL throughput demand $T_{m,h}^{(max)}$	Maximum UL throughput demand $T_{m,h}^{(max)}$	Minimum DL throughput demand $T_{m,h}^{(min)}$	Minimum UL throughput demand $T_{m,h}^{(min)}$
FTP download (Tg_1)	1000	100	0	0
Video (Tg_2)	64	64	64	64
VoIP (Tg_3)	12.2	12.2	12.2	12.2
Web browsing (Tg_4)	128	64	64	32

related to services (see Table 15.6). Therefore, $f_{m,h}^{(max)}$ is also calculated as in Equation 15.32, and represents an estimation of the losses that the lack of certain QoS in a user mean for the total financial model. Besides the values presented in Table 15.6, the network designer may also penalize each service differently.

4. $f_{pen(Cr_{b'})}$ penalizes different degrees of network resiliency. In the event of having a failure in an RS, other RSs may cover the new empty area by extending the coverage of an alternative station. In addition, the RSs previously attached to the faulty station need to reconnect to a new one. These procedures incur in reconfiguration costs that are considered in this objective, and an example is shown in Table 15.8.

   During network planning, some failure events can be simulated for each configuration. A failure in the multihop communication is more likely to happen for weak relay links connections. Note in Equation 15.26 that the performance in this criterion is penalized over the BS-RS link, and therefore over the subset of RSs (as described in Section 15.4.1). Similarly to the Cu criterion (Table 15.5), and in the case that only BS configuration is being optimized, different RS profiles can be penalized differently for every channel, according to the LUTs provided, adjusting $T_{m,h}^{(min)}$ and $T_{m,h}^{(max)}$ to the CINR values to for the minimum and maximum MCS. Constraint (Equation 15.30) indicates that different penalties can be applied in both UL and DL.

Table 15.8 Example of Economic Parameters for RS Reconfiguration

Year		1	2	3	4	5
RS OpEx (k€)	Reconfiguration	0.25	0.22	0.19	0.17	0.15

In the following paragraphs, an example of multicriteria network design process has been performed over a rectangular area of the city of Munich in a relay network expansion scenario to contain 60 candidate RSs, and a previous BS infrastructure, which is kept fixed.

Different configurations are searched by a Tabu search algorithm in the inner and outer optimization loop. In the outer loop, a multiobjective version of the algorithm gets optimal values for RS position and number, antenna tilt and azimuth—in intervals of 10°—by using the method of the Pareto front. The Tabu list size is chosen empirically during the initial tuning. Economic parameters have been set to the ones presented in this section. The exception are the sites/sectors CapEx and OpEx, which may vary with the area (e.g., establishment fee or maintenance). Table 15.9 shows a final set of five representative solutions selected for the four-objective scenario shown in Figure 15.14.

In solution 1, the lower number of RSs results in low infrastructure costs and also provide low throughput penalties. Hotspots and indoor coverage in the central area benefit from this configuration. On the other hand, there are a number of MSs with low CINR levels concentrated at the edge of the coverage areas where they may have a poor channel and high interference. This solution can be a suitable option for service providers that plan a future expansion of the network according to real user demands.

In solution 2, some new users can gain connection to the network because of the addition of more RSs, and therefore some penalty values related to connection tariff or annual subscription are reduced. On the other hand, the throughput per user decreases, especially in the MSs that are in the two hotspots. Users that require services that are bandwidth consuming are penalized more and thus, this option would be suitable for network providers that prefer to ensure connectivity to as many users as possible in a wide area than providing services with high bit rates.

Solution 1 represents a throughput enhancement strategy, and also a reliable network, whereas solution 2 characterizes a coverage extension strategy. Other solutions

Table 15.9 Cost and Penalty Values of Five Representative Solutions of the Pareto Front

Solution	*Relay Sectors*	F_{costs} *(100k(€))*	$F_{\text{pen}(Cu)}$ *(100k(€))*	$F_{\text{pen}(Tg)}$ *(100k(€))*	$F_{\text{pen}(Cr)}$ *(100k(€))*
1	6	4.25	2.53	1.21	2.02
2	8	4.83	1.25	2.63	2.68
3	5	3.33	2.38	1.82	3.16
4	4	2.63	3.74	3.81	4.21
5	8	4.93	2.36	1.45	1.82

Figure 15.14 Two solutions from the Pareto front.

in Table 15.9 represent other trade-offs. For example solutions 3 and 4 can reduce infrastructure costs at the expense of more penalties in "connection" values or even more degraded performance in solution 4. Also, both solutions offer bad RS reliability values. Solution 5 shows the best network performance but with the highest infrastructure costs.

15.7 Conclusion and Open Issues

To achieve optimally performing and cost-efficient networks, the design of the problem should be simplified and translated into an optimization routine that considers technology-specific and economic factors. This chapter explained the impact of relays on the LTE network architecture and presented a service-oriented optimization framework that offers the service provider a comprehensive, detailed, and clear description of different scenarios during the network architecture design process so that the most suitable solution can be found.

Relay networks involve a more complicated air interface than other kind of cellular networks and need to be rigorously analyzed during the process of network design to achieve outstanding performance. Frequency planning and base station configuration are integrated in a framework that provides a wide perspective to the network designer.

The results of the LTE scenario analyzed in this chapter highlight the complexity of the design process, as the operator needs to estimate a number of parameters, but also shows the benefits of the described optimization framework, which can provide valuable help to the service provider even in medium-sized scenarios. The accuracy of the results relies on a network simulation tool and a previous economic study of several factors. Nevertheless, a multiobjective perspective can provide a good assistance when choosing a solution in any case by showing the trends of different configurations.

There are a few challenges in LTE network planning and optimization, in particular in multihop relaying networks, that need further investigation. Some work on extending the formulation shown to fully avoid the use of system level simulations—using the restriction matrix—would greatly speed up the whole process. This can possibly lead to solutions of better quality and analysis of larger and more complex scenarios.

In addition, little work has been done on LTE self-optimization procedures, which may be used in an on-line optimization stage once the network has been deployed. In this case, relay stations may need to change their power, frequency settings, and other RRM parameters in a dynamic manner, with the only information of their immediate neighborhood. This is of big importance in nomadic relay stations.

Note that the examples in this chapter focused on the two-hop case, as this may be the most common case of this kind of network. However, the formulation shown also applies to the multihop case, and the consideration of routing algorithms through the design of several hops together with other parameters may improve final design.

References

[1] A. Eisenblatter, H. F. Geerdes, T. Koch, A. Martin, and R. Wessly, "UMTS radio network evaluation and optimization beyond snapshots," *Mathematical Methods of Operations Research*, vol. 63, pp. 1–29, 2005.

[2] E. Amaldi, A. Capone, and F. Malucelli, "Planning UMTS base station location: Optimization models with power control and algorithms," *IEEE Transactions on Wireless Communications*, vol. 2, no. 5, September 2003.

[3] S. Hurley, "Planning effective cellular mobile radio networks," *IEEE Transactions on Wireless Communications*, vol. 51, no. 2, pp. 243–253, March 2002.

[4] I. Siomina, P. Vrbrand, and D. Yuan. "Automated optimization of service coverage and base station antenna configuration in UMTS networks," *IEEE Wireless Communications*, vol. 13. pp. 16–25, 2006.

[5] J. Laiho, A. Wacker, and T. Novosad, *Radio Network Planning and Optimisation for UMTS*. John Wiley and Sons, New York, 2002.

[6] F. Gordejuela-Sanchez and J. Zhang, "Practical design of IEEE 802.16e networks: A mathematical model and algorithms," *Global Communications Conference (GLOBECOM)*, 2008, pp. 1–5.

[7] H. Claussen, L.T.W. Ho, and L.G. Samuel, "Self-optimization of coverage for femtocell deployments," *Wireless Telecommunications Symposium*, 2008, pp. 278–285.

[8] S. Modarres Razavi and D. Yuan, "Performance improvement of LTE tracking area design: A re-optimization approach," *Proceedings of the 6th ACM International Workshop on Mobility Management and Wireless Access (MobiWac)*, 2008.

[9] A. Eisenblatter, "Frequency assignment in GSM networks: models, heuristics, and lower bounds," Ph.D. dissertation, Technische Universitt Berlin, Germany, 2001.
[10] F. Gordejuela-Sanchez, D. Lpez-Prez, and J. Zhang, "Frequency planning in IEEE 802.16j networks an optimisation framework and performance analysis," *IEEE Wireless Communications and Networking Conference (WCNC)*, April 2009, pp. 1–6.
[11] L. Guan, J. Zhang, J. Li, G. Liu, and P. Zhang, "Spectral efficient frequency allocation scheme in multihop cellular network," *66th IEEE Vehicular Technology Conference (VTC) Fall*, 2007, pp. 1446–1450.
[12] Y. Li, Q. Miao, M. Peng, and W. Wang, "Frequency allocation in two-hop cellular relaying networks," *18th Annual IEEE International Symposium on Personal, Indoor and Mobile Radio Communications (PIMRC)*, 2007, pp. 1–5.
[13] T. Liu, M. Rong, H. Shi, D. Yu, Y. Xue, and E. Schulz, "Reuse partitioning in fixed two-hop cellular relaying network," *IEEE Wireless Communications and Networking Conference (WCNC)*, 2006, vol. 1, pp. 177–182.
[14] M. Werner, P. Moberg, p. Skillermark, M. Naden, W. Warzanskyj, and I. Berberana, "Optimization of cellular network deployments with multiple access point types," *WWRF Meeting'20*, Ottawa, Canada, April 2008.
[15] "3GPP; Technical specification group radio access network; requirements for E-UTRA and E-UTRAN," 3GPP, March 2008, http://www.3gpp.org/ftp/Specs/html-info/36-series.htm.
[16] S. W. Peters and R. W. Heath Jr., "The future of WiMAX: Multi-hop relaying with IEEE 802.16j," *IEEE Communications Magazine*, vol. 1, no. 47, 2009.
[17] D. Soldani and S. Dixit, "Wireless relays for broadband access," *Communications Magazine, IEEE*, vol. 46, no. 3, pp. 58–66, 2008.
[18] IST-FIREWORKS Deliverable 2D1 "Cellular deployment concepts for relay-based systems," November 2007, http://fireworks.intranet.gr.
[19] D. Lopez-Perez, A. Juttner, and J. Zhang, "Dynamic frequency planning versus frequency re-use schemes in OFDMA networks," *IEEE Vehicular Technology Conference (VTC)*, April 2009.
[20] D. Lopez-Perez, A. Juttner, and J. Zhang, "Optimisation methods for dynamic frequency planning in OFDMA networks," at *IEEE Networks*, 2008.
[21] J. Yang, M. Aydin, J. Zhang, and C. Maple, "UMTS base station location planning: a mathematical model and heuristic optimization algorithms," *IET Communications*, vol. 1(5), pp. 1007–1014, 2007.
[22] K. Jaffres-Runser, J. Gorce, and S. Ubeda, "QoS constrained wireless LAN optimization within a multiobjective framework," *IEEE Wireless Communications*, vol. 13, no. 6, pp. 26–33, December 2006.
[23] Y. Collette and P. Siarry, *Multiobjective Optimisation. Principles and Case Studies*. Springer, Berlin, 2003.
[24] A. Hoikkanen, "Economics of 3G long-term evolution: The business case for the mobile operator" *International Conference on Wireless and Optical Communications Networks (WOCN)*, 2007, pp. 1–5.
[25] "Final report on link level and system level channel models," IST-2003- 507581 WINNER, D5.4 vol. 1.4, November 2005.
[26] B. Timus, "Studies on the viability of cellular multihop networks with fixed relays," Ph.D. dissertation, University of Stockholm, Sweden, 2009.

Chapter 16

LTE E-MBMS Capacity and Intersite Gains

Américo Correia, Rui Dinis, Nuno Souto, and João Silva

Contents

16.1 Introduction

The Third Generation Partnership Project (3GPP) has launched the study item evolved UMTS terrestrial radio access (UTRA) and UMTS terrestrial radio access network (UTRAN), which studies the means to achieve further substantial leaps in terms of service provisioning and cost reduction. The overall target of the long-term evolution (LTE) of 3G was to arrive at an evolved radio access technology that can provide service performance on parity with current fixed-line access. As it is generally assumed that there will be a convergence toward the use of Internet protocol (IP)-based protocols (i.e., all services in the future will be carried on top of IP), the focus of this evolution was on enhancements for packet-based services. 3GPP concluded the Release 8 of the evolved 3G radio access technology in 2008, with subsequent initial deployment in the 2009–2010 time frame. At this point, it is important to emphasize that this evolved RAN is an evolution of the current 3G networks, building on already made investments. The 3GPP community has been working on LTE, and various contributions were made to implement evolved MBMS in LTE [1].

Orthogonal frequency division multiplexing/orthogonal frequency division multiple access OFDM/OFDMA [2–4], used in the physical layer (downlink connection) of LTE is an attractive choice to meet requirements for high data rates, with correspondingly large transmission bandwidths and flexible spectrum allocation. OFDM also allows for a smooth migration from earlier radio access technologies and is known for achieving high performance in frequency-selective channels. Furthermore, it enables frequency domain adaptation, provides benefits in broadcast scenarios, and is well suited for multiple-input multiple-output (MIMO) processing.

The possibility to operate in vastly different spectrum allocations is essential. Different bandwidths are realized by varying the number of subcarriers used for transmission, whereas the subcarrier spacing remains unchanged. In this way, operation in spectrum allocations of 1.4, 3, 5, 10, 15, and 20 MHz, respectively, can be supported.

For MBMS support within a certain cell coverage area for a given coverage target, the modulation and coding scheme (MCS) of the MBMS transport channel typically has to be designed under worst-case assumptions. Except for cell-edge users that typically experience large inter-cell interference, users with better channel conditions (closer to the base station) could receive the same service with a better quality (e.g., video resolution), as their receiving signal-to-noise ratios (SNR) would allow usage of a higher-rate MCS. Hierarchical modulations [5–8], which have been specified for broadcast systems like digital video broadcast terrestrial (DVB-T) or MediaFLO, is one way of accounting for unequal receiving conditions. Here, a signal constellation like 16-QAM, with each symbol being represented by four bits, is interpreted in a sense that the first two bits belong to an underlying QPSK alphabet. This enables the use of two independent data streams with different sensitivity requirements. In the foregoing example, the so-called high-priority stream employs QPSK modulation and is designed to cover the whole service area. The low-priority stream requires the constellation to be demodulated as 16-QAM and provides an additional or refined service via the two additional bits. This may transport an additional MBMS channel with a different type of service, or an enhancement stream that, for example, leads to enhancing the resolution of the base stream. A design parameter that determines the constellation layout allows the control of the amount of distortion that the enhancements symbols add to the baseline constellation and can be used to control the ratio of coverage areas or service data rates. Theoretical evaluation of these types of modulations, where it is explicitly shown the dependence of the individual bit streams performance on the constellation design parameter, has been previously presented in [9, 10].

Specifically for broadcast and multicast transmissions in a mobile cellular network, depending on the communication link conditions, some receivers will have better SNR than others, and thus the capacity of the communication link for these users is higher. Hierarchical constellations and MIMO (spatial multiplexing [11, 12]) are methods able to offer multiresolution and take advantage of the different link capacities. In [13–15] these two forms of multiresolution methods (considering the WCDMA technology) have been evaluated. In OFDMA-based networks, the transmission of different fractions of the total set of subcarriers (chunks) depending on the position of the mobiles is another way to offer multiresolution. All of these methods are able to provide unequal bit error protection. In any case, there are two or more classes of bits with different error protection levels to which different streams of information can be mapped. Regardless of channel conditions, a given user always attempts to demodulate both types of bits—the most protected and the ones carrying additional resolution. Depending on its position inside the cell, more or less blocks with additional resolution will be correctly received by the mobile user. However, the basic quality will be always correctly received independently of the position of any user within the 95% coverage target.

For increased distance between terminals and base station, decreased bit rates are correctly received due to the decrease of SNR. Adaptive modulation and coding (AMC) is a technique that maximizes the total throughput for unicast transmissions. The decrease of SNR with distance is common to unicast or broadcast/multicast transmissions. However, for broadcast/multicast the same video content is transmitted, and AMC is not possible without personal uplink feedback. With the introduction of multiresolution techniques, maximization of the total throughput is the goal to achieve. There will be support for MBMS right from the first version of LTE specifications. However, specifications for E-MBMS are in the early stages. Two important scenarios have been identified for E-MBMS: one is single-cell broadcast, and the second is MBMS single-frequency network (MBSFN). MBSFN is a new feature that is being introduced in the LTE specification. MBSFN is envisaged for delivering services such as mobile TV using the LTE infrastructure and is expected to be a competitor to DVB-H-based TV broadcast. In MBSFN, the transmission happens from a time-synchronized set of enhanced-nodeBs (eNBs) using the same resource block. This enables over-the-air combining, thus improving the signal-to-interference plus noise ratio (SINR) significantly compared to non-SFN operation. The cyclic prefix (CP) used for MBSFN is slightly longer, and this enables the UE to combine transmissions from different eNBs, thus somewhat negating some of the advantages of SFN operation. There will be six symbols in a slot of 0.5 ms for MBSFN operation versus seven symbols in a slot of 0.5 ms for non-SFN operation.

System-level simulations for broadcast/multicast with multiresolution and different fractional frequency reuse for LTE are necessary to evaluate achievable capacity and intersite gains compared to single-resolution systems, WCDMA based. Taking the 95% coverage as reference the evaluation of the achievable capacity gain (number of transmitted mobile TV channels for WCDMA and LTE) is done [16]. The intersite distance gain is also evaluated, allowing for a substantial reduction in the number of cell sites when LTE replaces WCDMA. The scenario based on the use of single-frequency network (SFN), with the multimedia broadcast over SFN (MBSFN) channel, is also evaluated for 16-QAM/64-QAM hierarchical modulations and compared with the present MBMS network based on WCDMA.

16.2 Objectives and Requirements

The multimedia broadcast and multicast service (MBMS), introduced by 3GPP in Release 6 was intended to use network/radio resources efficiently (by transmitting data over a common radio channel), both in the core network and, most importantly, in the air interface of UTRAN, where the bottleneck is placed to a large group of users. MBMS included point-to-point (PtP) and point-to-multipoint (PtM) modes. The former allowed individual retransmissions but the latter did not. MBMS is targeting high (variable) bit rate services over a common channel. One of the most important properties of MBMS is resource sharing among many user equipments (UEs), meaning that many users should be able to listen to the same MBMS channel at the same time. Thus, power should be allocated to this MBMS channel for arbitrary UEs in the cell to receive MBMS service. PtM transmission does not employ feedback and therefore needs to be statically configured to provide desired coverage in the cell. The transmitted signal is lowest at the cell border, and therefore the PtM bearer can greatly benefit from exploiting also the signals from adjacent cells transmitting the same service (i.e., from soft combining). While in the 3GPP LTE specification, two types of evolved-MBMS transmission scenarios exist:

- Multicell transmission (MBSFN over an SFN) on a dedicated frequency layer or on a shared frequency layer
- Single-cell transmission (SC-PMP: Single cell point to multipoint) on a shared frequency layer

In 3GPP Release 6, the only specified transmission scenario for the MBMS transmission is SC-PMP. However, soft-diversity combining is possible as long as the delay between different base station transmissions of the same content allows macrodiversity.

Multicell transmission in an SFN area is a way to improve the spectral efficiency. Because all MBMS cells transmit the same MBMS session data, signals can be combined for a UE located at a cell boundary. Furthermore, the multicell transmission may be provided over a cell group that comprises cells that transmit the same service. In contrast, single-cell transmission covers only one cell or one eNode B. In addition, the concept of a dynamic MBSFN area is introduced where the MBMS transmission is switched off in some cells of the MBSFN area when a certain MBMS is not needed there. In some cases, the released resource can be reused for other MBMS or unicast services. The decision to turn off the MBSFN transmission in a cell is based on two factors:

- Local existence, which refers to the number of UEs that are both interested in the current MBMS and located in this cell
- Contribution for neighboring cells, which refers to the number of UEs that are both interested in the current MBMS and located in neighboring cells having the same MBSFN transmission. The cells of an MBSFN area contribute to the MBSFN transmission only if there are UEs that are interested in the particular service in this MBSFN area.

The introduction of hierarchical modulation in a broadcast cellular service like E-MBMS requires a scalable video codec as shown in Figure 16.1 [13, 14], where the base layer transmission provides the minimum quality, and one or more enhancement layers offer improved quality at increasing bit/frame rates and resolutions. Besides being a potential solution for content adaptation, scalable video schemes may also allow an efficient usage of radio resources in evolved MBMS (E-MBMS).

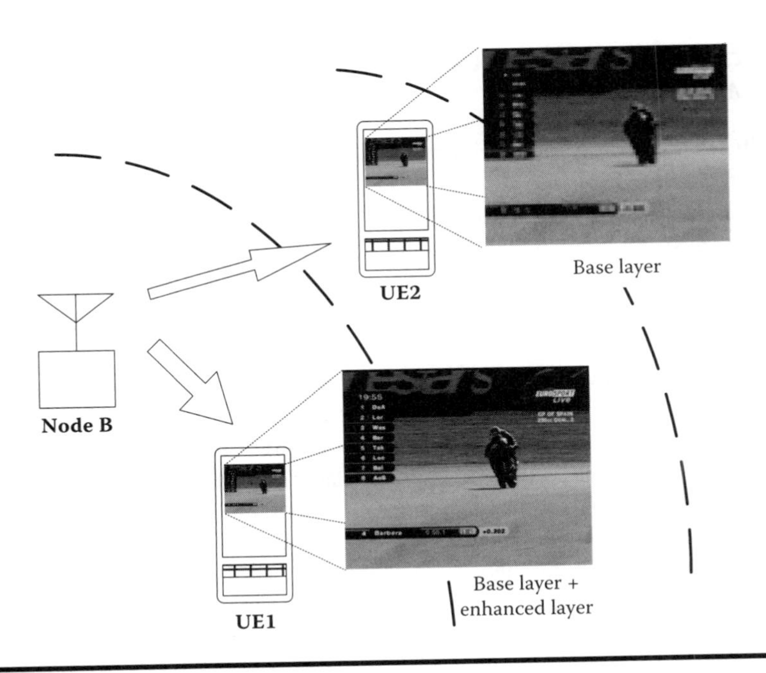

Figure 16.1 Scalable video transmission.

According to Release 6 of 3GPP, the single-resolution scheme corresponds to the transmission of QPSK with more than 95% coverage. The assignment of the fraction of the total transmission power reserved for MBMS has implications in the coverage and average throughput of the multiresolution based on the hierarchical 16-QAM scheme. The multicell interference distribution has also strong impact in the coverage and throughput. An interesting design parameter is the channel bit rate (and its coding rate) associated to the multiresolution scheme. An optimization of this parameter has also strong impact in achievable coverage and average throughputs.

Regardless of the channel conditions and user location, a given user always attempts to demodulate both the base layer and the enhancement layer carrying additional resolution. For good multiresolution design, the basic information will be always correctly received independently of the position of any user within the 95% coverage target. However, depending on its position inside the cell, more or fewer blocks with additional resolution will be correctly received by the mobile user.

The objective of this chapter is to design multiresolution schemes in the two different scenarios—MBSFN and SC-PMP with inter-cell interference without and with macrodiversity support—and to measure the corresponding multiresolution gain of total throughput compared to the reference total throughput of the single resolution scheme based on the QPSK transmission.

16.3 Evaluation Methodology and Simulation Assumptions

Typically, radio network simulations can be classified as either link level (radio link between the base station and the user terminal) or system level (several base stations with large number of mobile users). A single approach would be preferable, but the complexity of such a simulator (including everything from transmitted waveforms to multicell network) is far too high for the required simulation resolutions and simulation time. Therefore, separate but interconnected link and system level approaches are needed.

The link level simulator is needed for the system simulator to build a receiver model that can predict the receiver block error rate/bit error rate (BLER/BER) performance, taking into account channel estimation, interleaving, modulation, receiver structure, and decoding. The system level simulator is needed to model a system with a large number of mobiles and base stations and also to evaluate algorithms operating in such a system.

As the simulation is divided in two parts, an approach for linking the two simulators must be defined. Conventionally, the information obtained from the link level simulator is inserted into the system level simulator through the utilization of a specific performance parameter (BLER) corresponding to a specific SNR estimated in the terminal or base station. Figure 16.2 shows the interaction between the simulators.

16.3.1 Link-Level Simulator Design

The link level simulator (LLS) was developed in Matlab and took into account the specifications of 3GPP MBMS Release 7 [17] regarding the signal processing of transport and physical channels. It satisfied two essential requirements:

- Serve as reference for all the link level simulations with multiresolution and parameters estimation
- Serve as a platform to the different multiresolution improvements tested and quantified

A typical time interval of each link level simulation is 0.5 s (as shown in Table 16.1). The entire OFDMA signal processing at the transmitter was included in the LLS as well as in several different receiver structures. To achieve reliable channel estimation and data detection, a receiver capable of jointly performing these tasks through iterative processing is used. The structure of the iterative receiver is shown in Figure 16.3 (see also [18]). Clearly, the receiver structure for additive white Gaussian noise (AWGN) channel is less complex (only a few turbo-decoder iterations, and no channel estimation or channel equalization is required).

Multipath Rayleigh fading channels were considered in the simulator, as it comprises a more realistic scenario for evaluating hierarchical high-order QAM modulations due to their sensitivity to the channel parameters estimation. As indicated, the receiver structure is nonlinear, iterative, and includes channel parameters estimation for the analyzed multipath Rayleigh fading channels [19].

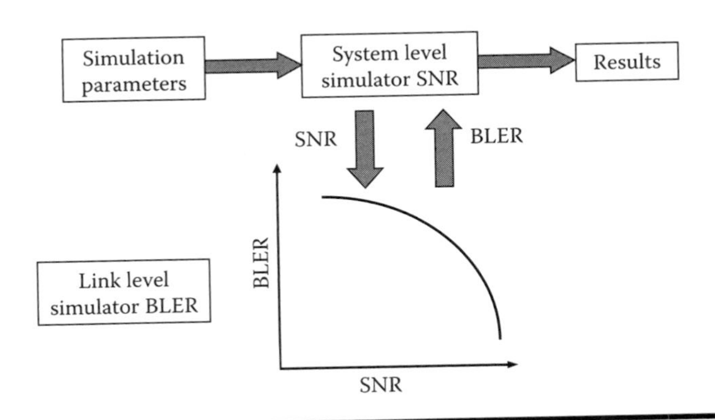

Figure 16.2 Interaction between link level simulator and system level simulator.

Table 16.1 Link and System Level Simulation Parameters for an Urban Macrocellular Scenario

Transmission bandwidth	10 MHz
Cyclic prefix size	72
FFT Size	1024
Carriers space (kHz)	15
Available bandwidth	9 MHz
Sample time (ns)	130
Max Tx power (dBm)/sector	46
Number of used subcarriers/sector	200
Number of used subcarriers/cell	600
Freq. reuse	1/3
Subframe duration (ms)	0.5
Interfering cells transmit with % of max power	90
Cellular layout	Hexagonal
Sectors	three sectors/cell
Number of cell sites	19
Antenna gain of the base station	17.5 dBi
Width of beam of the antenna at −3 dB	70 degrees
Front/back ratio of the antenna	20 dB
Antenna pattern radiation of the base station	Gaussian
Propagation model	Okumura-Hata
Downlink thermal noise	−100 dBm
Cable loss	3 dB
Fadeout standard deviation due to shadowing	10 dB

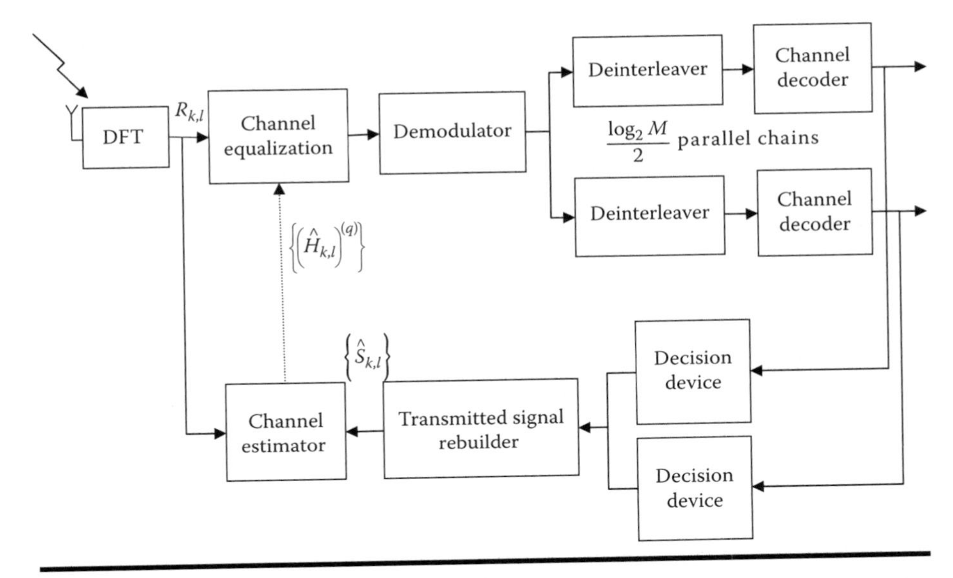

Figure 16.3 Iterative receiver structure.

16.3.2 Radio Access Network System Level Simulator

For the purpose of validating the work presented in this section, a system level simulator was developed in Java, using a discrete event-based philosophy, which captures the dynamic behavior of the RAN system. This dynamic behavior includes the user (e.g., mobility and variable traffic demands), radio interface, and RAN with some level of abstraction. The system level simulator (SLS) works at a subframe duration rate; a typical time interval of each simulation is 600 s. Table 16.1 shows the simulation parameters. It presents the parameters used in the link and system level simulations based on 3GPP documents [20–23].

The channel model used in the system level simulator considers three types of losses:

- Distance loss
- Shadowing loss
- Multipath fading loss (one value per 0.5 ms)

Model parameters depend on the environment. For distance loss, the Okumura-Hata model from the COST 231 project was used (see [24]). Shadowing is due to the existence of large obstacles like buildings and the movement of UEs in and out of the shadows. This is modeled through a process with a log-normal distribution and a correlation distance. The multipath fading employed in the system level simulator corresponds to the 3GPP channel models, where the ITU vehicular A (see [21] Annex B) and the MBSFN environments were chosen as references. The latter models were also used in the link level simulator but at a much higher rate. Vehicular A (with velocity $v = 30$ km/h) channel model was chosen because it is an important test channel in 3GPP specifications, furthermore, it allows direct comparison against previous system level simulations [16]. In OFDM systems, an important parameter is the maximum delay of the multipath profile and its relation with the duration of the time guard between OFDM symbols to avoid intersymbol interference. 3GPP has specified a short time guard with about 4.75 μs and a long time guard with 16.67 μs. The latter was considered in the model for achieving the results next presented, making the performance less sensitive to the chosen propagation channel. However, there is a reduction of the transmitted bit rates.

A uniform distribution of mobile users is generated at the beginning of each simulation. A typical number of users chosen for each simulation run was 20 per sector. Each mobile has random mobility with the specified speed of 30 km/h. Dynamic system level simulators like the one presented in this chapter are very accurate; the main limitation is the hypothetical urban macrocellular test scenario that is different from any real one.

Figure 16.4 illustrates the cellular layout (trisectorial antenna pattern) indicating the fractional frequency reuse of 1/3 considered in the system level simulations.

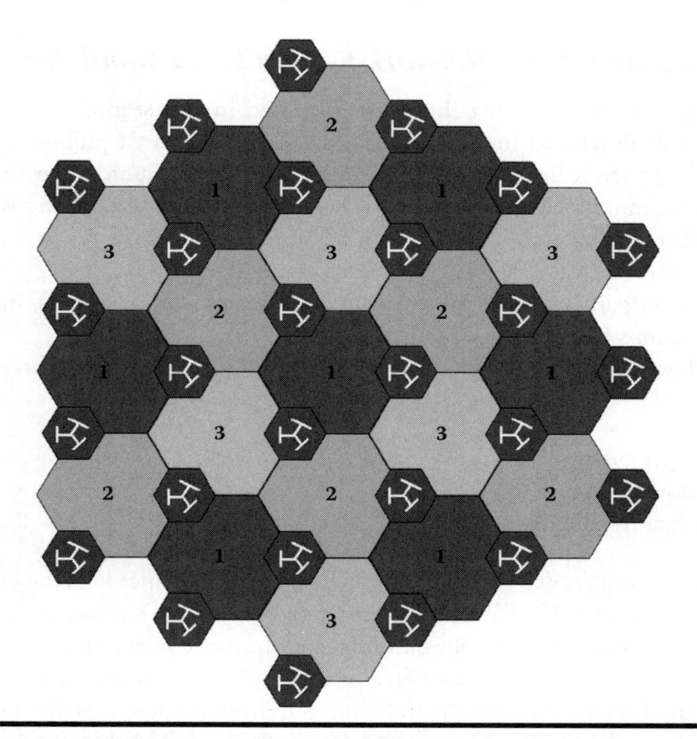

Figure 16.4 Cellular layout including the frequency reuse of 1/3 (gray shadows/numbers of the cells).

One third of the available bandwidth was used in each sector to reduce multicell interference. As indicated in Figure 16.4, the identification of the sources of multicell interference—the use of the same adjacent subcarriers (named physical resource blocks)—is given by the sectors with the same gray shadow/number: dark gray/one, average gray/two, or light gray/three.

For 16-QAM hierarchical constellations, two classes of bits with different error protection are used. The darkest gray around the antennas only indicates the approximate coverage of the weak bits blocks, whereas the other gray shadows indicate the coverage of the strong bits blocks.

In the analysis of the single-cell point-to-multipoint scenario (SC-PMP), there is one radio link between the mobile and the closest base station. It does not assume any time synchronism between the transmissions from different base stations with the same gray shadow resulting in interference from all cells without the same gray shadow. However, in the SC-PMP scenario with macrodiversity combining the two best radio links, it is assumed that there is time synchronization between the two closest base station sites with the same gray shadow. In this case, multicell interference is reduced because only the other base station sites with the same gray shadow remain asynchronous and capable of interference.

In the MBSFN scenario, there are at least three radio links between the mobile and the three closest base stations. Time synchronisation is assumed between the transmissions from the closest base stations with the same gray shadow, resulting in much less interference from the cellular environment. This results in macrodiversity combining of the three best radio links. In addition, the interfering base stations must be at least 5 km away from the reference base station considering a cyclic prefix (CP) of 16.67 s and a frequency of 2 GHz. Only distant base station sites are capable of introducing interference.

16.4 System Level Performance Results

To study the behavior of the proposed OFDM multiresolution schemes, several simulations were performed for 16-QAM hierarchical modulations. The 16-QAM hierarchical constellations are constructed using a main QPSK constellation where each symbol is, in fact, another QPSK constellation, shown in Figure 16.5.

The main parameter for defining one of these constellations is the ratio between d_1 and d_2 as shown in Figure 16.5:

$$\frac{d_1}{d_2} = k \tag{16.1}$$

where $0 < k \leq 0.5$.

For 16-QAM, two classes of bits with different error protection were used (for 64-QAM, three classes are used). Each information stream was encoded with a different

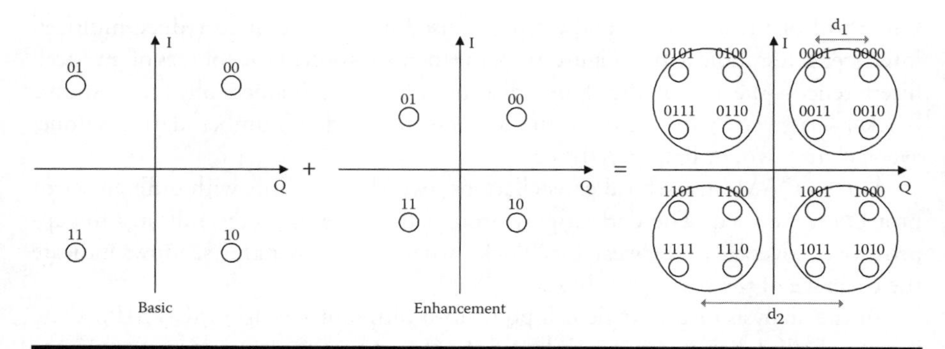

Figure 16.5 Signal constellation for 16-QAM hierarchical modulation.

block size, depending on the coding rate. Two different coding rates were considered: 1/2 and 3/4. This leads to block sizes of each information stream of 2400 and 3600 bits, respectively. This occupies a subframe 0.5 ms long. The total corresponding transmitted information bit rates per cell sector are 4800 kbps and 7200 kbps, respectively. A reference for measuring the capacity is the number of transmitted TV channels, each with a bit rate of 256 kbps. For the chosen fractional frequency reuse of 1/3, we get 18.75 and 28.125 TV channels, respectively. We want to compare the OFDM/OFDMA results directly with those obtained previously with the WCDMA technology [17]. All the parameters used for OFDM during these simulations were based on 3GPP documents [20–23].

About one-third of the total physical resource blocks (PRB) are transmitted in each sector. This corresponds to an instantly occupied bandwidth of 3.0 MHz, where we have considered an average of 16.67 PRBs per sector, each with 180 kHz of adjacent bandwidth (corresponding to 12 subcarriers with a frequency spacing of 15 kHz). The number of adjacent subcarriers in each PRB was chosen according to 3GPP specifications. With the dynamic allocation of the resources per sector, sectors 2 and 3 have 17 PRBs, and sector 1 has 16 PRBs in the first subframe duration. Sectors 1 and 3 have 17 PRBs, and sector 2 has 16 PRBs in the second subframe duration. Finally, sectors 1 and 2 have 17 PRBs, and sector 3 has 16 PRBs in the third subframe duration. On average, there are 16.67 PRBs per sector. We can conclude that the transmission of each TV channel with LTE technology requires less than one PRB for any analyzed coding rate.

16.4.1 BLER Results

In link level simulations, we have evaluated the hierarchical 16-QAM and 64-QAM with two different coding rates. In Figures 16.6 and 16.7 we consider the vehicular A propagation channel to be used in the SC-PMP scenario, and we present BLER versus E_s/N_o for the hierarchical 16-QAM and 64-QAM, respectively. In the legend,

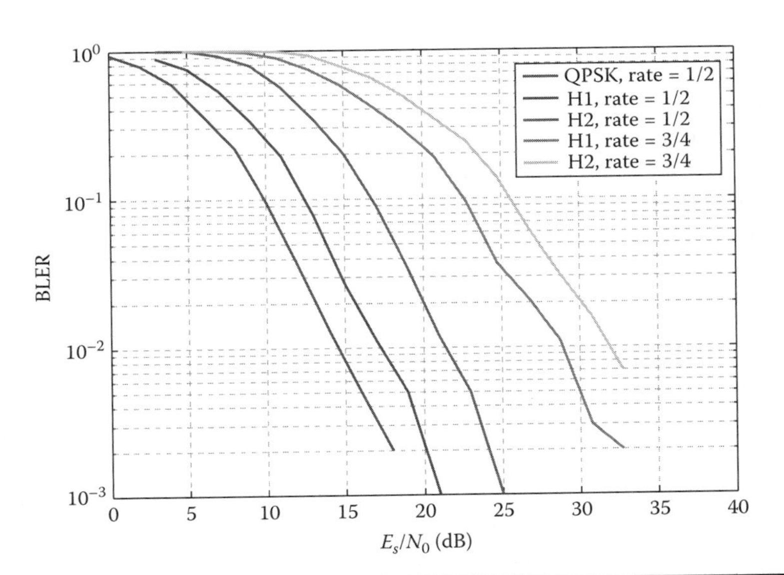

Figure 16.6 BLER vs. E_s/N_0 for hierarchical 16-QAM, VehA 30 km/h.

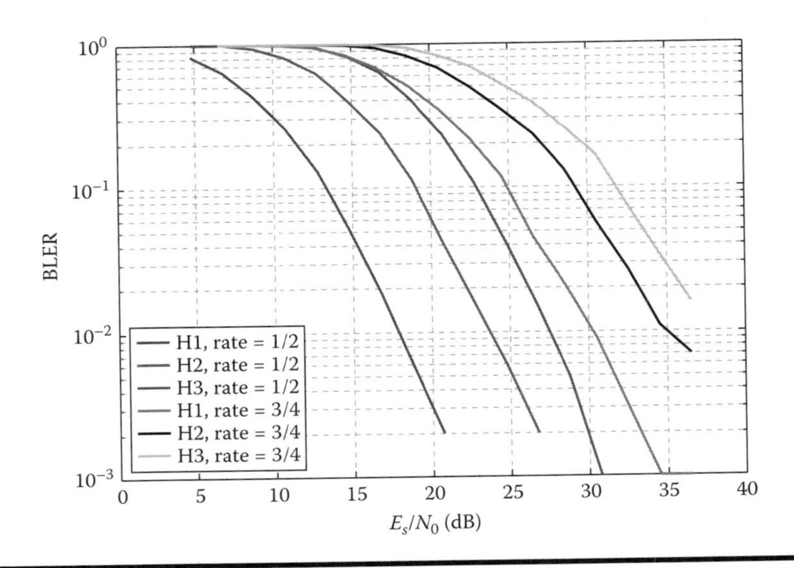

Figure 16.7 BLER vs. E_s/N_0 for hierarchical 64-QAM, VehA 30 km/h.

for 16-QAM, H1 means strong bit blocks and H2 weak bit blocks. For 64-QAM, H1 means strong bit blocks, H2 medium bit blocks, and H3 weak bit blocks. We have also included for comparison the QPSK performance for coding rate 1/2. We conclude that QPSK has the lowest E_s/N_0, consequently it will have the highest coverage. As expected, coding rate 1/2 provides a smaller E_s/N_0 compared to rate = 3/4, resulting in higher coverage. However, coding rate = 3/4 provides bit rates that are 1.5 higher than rate 1/2. There is a trade-off between bit rate (or throughput) and coverage. We will consider later on macrodiversity combining to increase the coverage (and throughput) at the cell borders.

When we compare Figures 16.6 and 16.7, we observe that due to the higher bit rates offered by 64-QAM, exactly 1.5 times the bit rates of 16-QAM, the corresponding E_s/N_0 are higher than 16-QAM, resulting in less coverage for 64-QAM. It seems that 64-QAM with coding rate 1/2 has a small E_s/N_0 advantage compared to 16-QAM rate 3/4 (both provide the same maximum bit rates). However, the sensitivity of 64-QAM to channel estimation errors is a feature that should not be forgotten, especially for hierarchical 64-QAM. The introduction of macrodiversity combining will increase the coverage of 16-QAM H2 blocks and H3 blocks (weak bits).

Figure 16.8 shows the BLER versus E_s/N_0 for the hierarchical 16-QAM in the MBSFN scenario with the MBSFN propagation channel. Comparison between Figures 16.6 and 16.8 indicate that the MBSFN channel due to the longer multipath power delay profile provides higher multipath diversity. This can be confirmed by a clear increase of the BLER performance of both coding rates compared to VehA. There is also a higher inherent intersymbol interference in the MBSFN channel, which is evident for rate 3/4 and weak bit blocks (H2). In spite of this, there is no significant loss in the BLER performance for rate 3/4 due to the redundancy of the channel coding. Recall that as the MBSFN scenario has lower inter-cell interference,

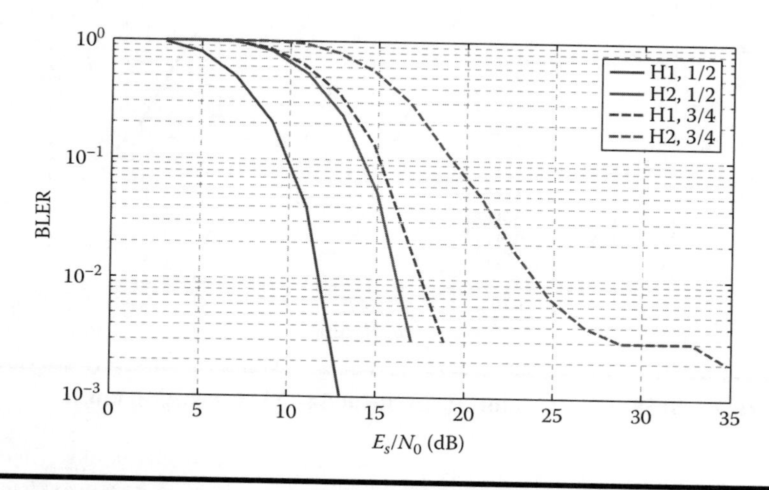

Figure 16.8 BLER vs. E_s/N_0 for hierarchical 16-QAM, MBSFN 30 km/h.

it is thus expected to compensate the lower bit rates resulting from the use of a longer guard time to avoid the effects of intersymbol interference.

16.4.2 Coverage Results

In system level simulations, mobile users receive strong and weak bits blocks transmitted from base stations. Each block undergoes small and large scale fading and multicell interference. In terms of coverage or throughput, the SNR of each block is computed taking into account all the above impairments. Based on the comparison between the reference SNR at a BLER of 1% and the evaluated SNR, it is decided whether the block is or not correctly received. This is done for all the transmitted blocks for all users in all sectors of the 19 cells, during typically 10 min.

Figure 16.9 presents the coverage versus the fraction of the total transmitted power (denoted as E_c/I_{or}), for SC-PMP scenario where there is interference only from one-third of the sectors due to the frequency reuse of 1/3 (see Figure 16.4). All interfering sites transmit with a maximum power of 90% according to the parameters indicated in Table 16.1. The cell radius is 750 m or 1500 m, and strong blocks (H1) are separated from weak blocks (H2) without including macrodiversity combining, denoted as 1RL, and also with macrodiversity combining the two best radio links (2RL). Recall that the basic scenario SC-PMP does not include macrodiversity. Multicell interference is 90% of the maximum transmitted power in each site. Previous coding rates were considered—rates 1/2 and 3/4, respectively. Recall that it is necessary to ensure the coverage of 95% for strong bit blocks (H1). The only case that never reaches the required coverage is when the cell radius has 1500 m

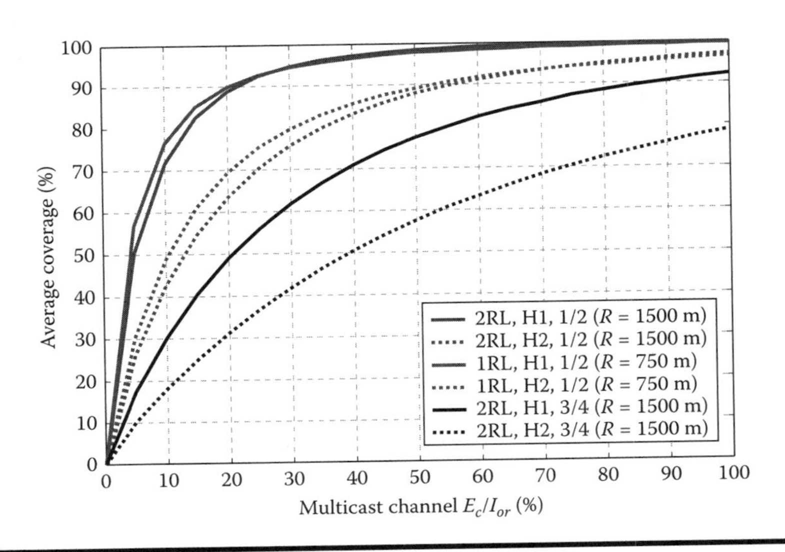

Figure 16.9 Average coverage (%) vs. E_c/I_{or}, of SC-PMP.

and the coding rate is 3/4, even with a macrodiversity of 2RL. For coding rate 1/2, it is equivalent to have 1RL and cell radius of 750 m or 2RL and cell radius of 1500 m. For $E_c/I_{or} = 50\%$ and rate 1/2, the coverage of H1 is 98%, and H2 coverage is around 89%. For the same E_c/I_{or}, but with rate 3/4, the coverage values of H1 and H2 are 77% and 57%, respectively. According to the coverage results of this figure, we can conclude that with a coding rate of 1/2, we can increase the cell radius from 750 m to 1500 m as long as we include macrodiversity combining of the two best radio links. However, for coding rate 3/4, we cannot double the cell radius and ensure the designed coverage, even if we add a macrodiversity of 2RL. We should choose between increasing the number of TV channels (or the TV channel bit rates) or increasing the cell radius.

In Figure 16.10 the coverage performance curves for MBSFN scenario, versus E_c/I_{or}, are presented for both cell radii of 750 m and 1500 m and should be compared to the corresponding results of Figure 16.9 for the SC-PMP scenario. As expected, there is a difference in the coverage between the two scenarios where MBSFN takes advantage of its lower inter-cell interference. The coverage values are above 95% even for small values of E_c/N_0, such as 25%; the only exception are the weak bit blocks (H2) of coding rate 3/4 and cell radius $R = 1500$ m. There is a coverage similarity between rate 1/2 with $R = 1500$ m and rate 3/4 with $R = 750$ m. This means that we can opt between increasing the coding rate (the average throughput) or increasing the coverage. When we increase both coverage

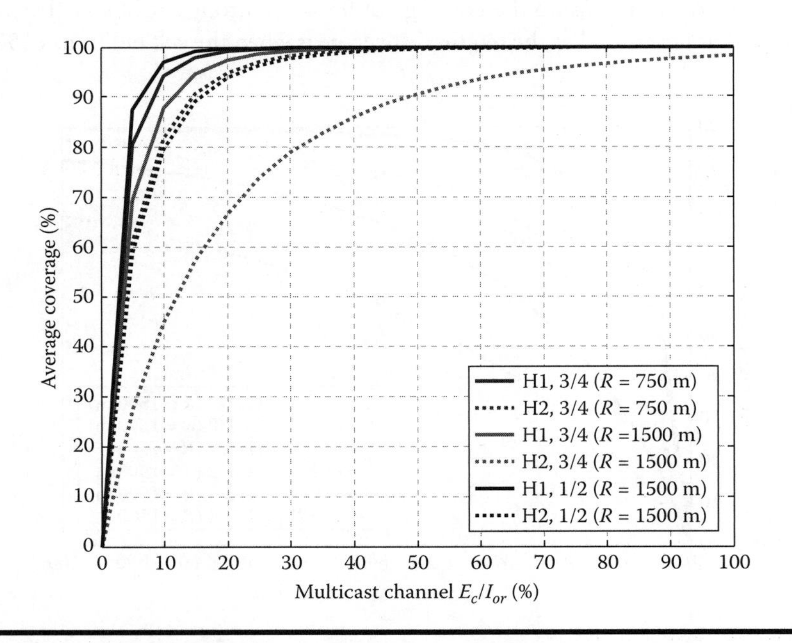

Figure 16.10 Average coverage (%) vs. E_c/I_{or}, of MBSFN.

and throughput at the same time, we observe a decrease of coverage, in particular, H2 blocks.

16.4.3 Throughput Results

Figure 16.11 presents the average throughput distribution as a function of the E_c/I_{or} for the SC-PMP scenario network with and without macrodiversity for both cell radii of 750 m and 1500 m. We observe a considerable gain in throughput when macrodiversity (2RL) is considered compared to the single radio link case. This is particularly true for the high coding rate of 3/4. For E_c/I_{or} above 50%, the average throughput for coding rate 3/4 is above 256 kbps, which is the maximum throughput for rate 1/2. However, not all UEs are able to achieve such high throughput, as users located at the cell borders never reach such high values of throughput.

Figure 16.12 considers the throughput distribution as function of the distance between UEs and BS for the $E_c/I_{or} = 90\%$, with and without macrodiversity for the same cell radius of 1500 m and different coding rates. For the chosen E_c/I_{or}, both 2RL and 1RL ensure the maximum throughput for users located near the base station. As the distance between UEs and BS increases the throughput of 1RL decreases significantly. However, the decrease in throughput is more obvious for rate 3/4 and when mobile users are at the cell bofrders. It is observed that with 2RL, only for rate 1/2, is the throughput almost independent of the distance. For the high coding rate of 3/4, a single radio link offers high throughput only for

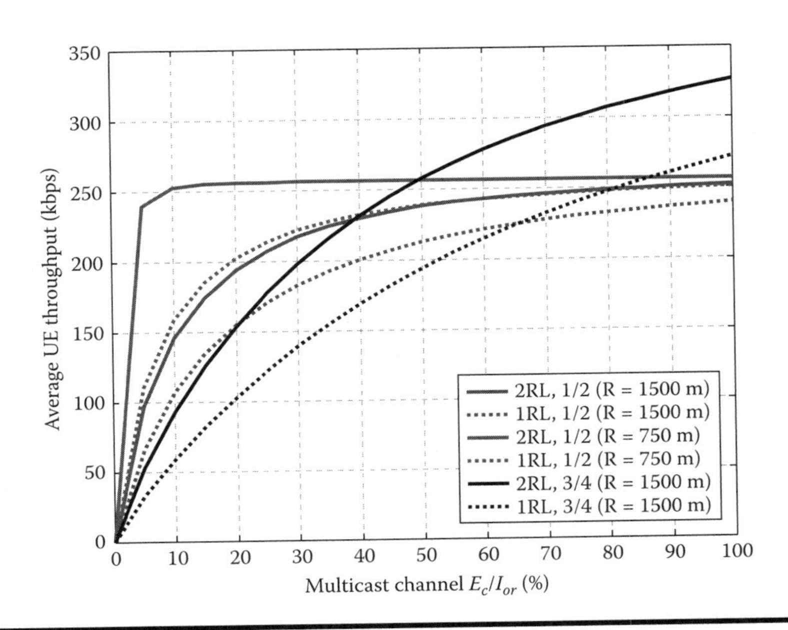

Figure 16.11 Average UEs throughput vs. E_c/I_{or}, SC-PMP.

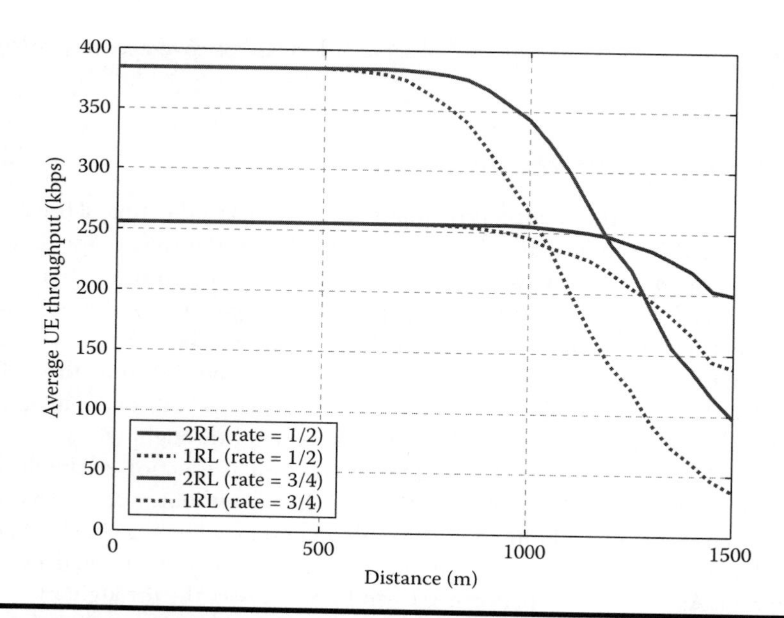

Figure 16.12 Throughput vs. distance for SC-PMP.

users close to the base station. This figure complements the previous one where the average throughput for the cell area was presented. When we consider that approximately three-fourths of users are located in the area between the cell border and the half radius of the cell, we realize the importance for the average throughput within in this area. For users located beyond 1200 m when there is 2RL, coding rate 1/2 provides higher throughput than rate 3/4. Again, we must choose between increasing the intersite distance (coverage) or increasing the number of TV channels (capacity).

Figure 16.13 presents the average throughput distribution as function of the E_c/I_{or} for the MBSFN scenario for both cell radii of 750 m and 1500 m. As expected, we observe that for high coding rate 3/4 the throughput performance is higher for $R = 750$ m compared to $R = 1500$ m due to the higher coverage of H2 blocks. However, for the E_c/I_{or} value of 90 (dedicated carrier to MBSFN) there is almost no difference between the two performance curves. The average throughput of 256 kbps, which is the maximum throughput for rate 1/2 is achievable for $E_c/I_{or} = 45\%$. This means that we can have two MBSFN carriers each transmitting at least 18 TV channels. The operator must always choose between increasing the capacity, $2 \times 18 = 36$ TV channels, keeping $R = 750$ m or increase the coverage to $R = 1500$ m with only 28 TV channels (see Table 16.2).

In Figure 16.14, the throughput distribution as function of the distance between UEs and BS is presented for the $E_c/I_{or} = 90\%$ and considering both scenarios. For the chosen E_c/I_{or}, both scenarios assure the maximum throughput for users

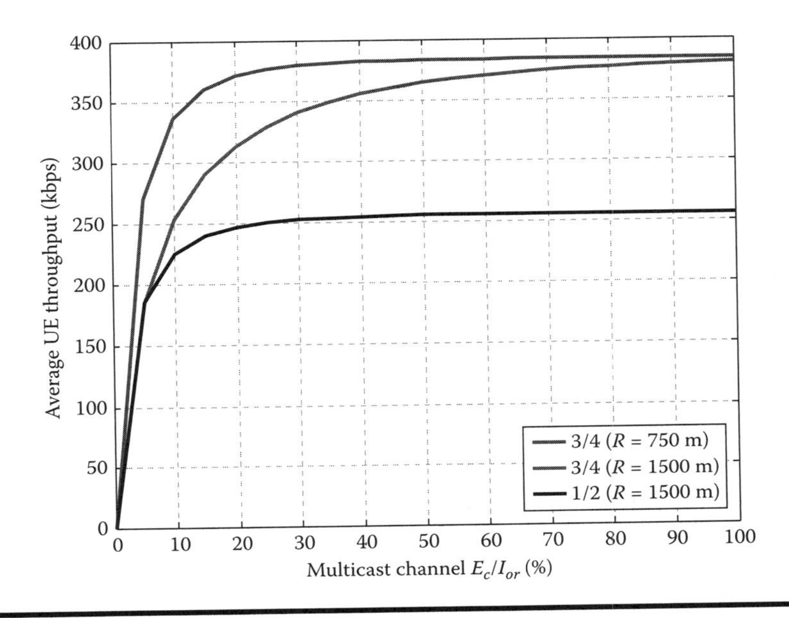

Figure 16.13 Average UEs throughput vs. E_c/I_{or} for MBSFN.

located near the base station. As the distance between UEs and BS increases, the throughput of SC (single cell with a macrodiversity of 2RL) decreases significantly. However, it is more obvious that there is more of a decrease in throughput for coding rate 3/4 than for rate 1/2. The SFN assures not only higher coverage than SC (single cell with 2RL) due to lower inter-cell interference but also higher throughput. For coding rate 3/4, the throughput gain of MBSFN compared SC-PMP, where we measure the gain, as the aggregate throughput in all cell areas (weighted by the distribution of users) under the two-throughput performance curves indicates a value close to 1.5 considering the use of the 16-QAM multiresolution scheme. Notice that for broadcasting mobile TV channels, it is also important to increase the intersite

Table 16.2 Capacity Values for MBSFN and SC-PMP Scenarios Using 16-QAM Hierarchical Multiresolution OFDMA (BW = 10 MHz)

QoS	*#TV Channels*	*Spectral Efficiency*	*ISD*	*Scenario*
256 kbps	18.75	0.48 bps/Hz/cell	1500 m	SC-PMP 1RL
256 kbps	18.75	0.48 bps/Hz/cell	3000 m	SC-PMP 2RL
256 kbps	28.125	0.72 bps/Hz/cell	3000 m	MBSFN
384 kbps	18.75	0.72 bps/Hz/cell	3000 m	MBSFN

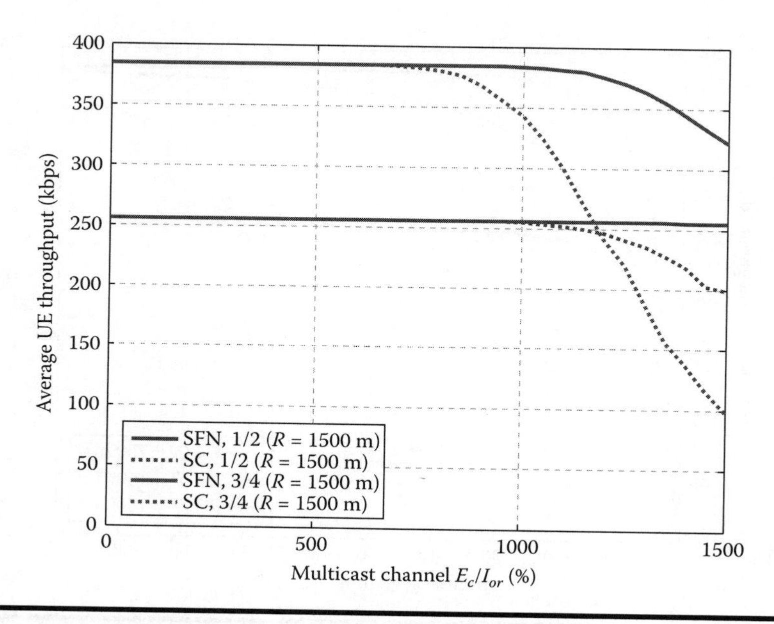

Figure 16.14 Throughput vs. distance.

distance (ISD) to 3000 m to reduce the number of sites and this is only easily ensured with MBSFN. Recall that as the intersite distance ISD $= 2R$ when we double R, we double the ISD allowing for a 50% reduction in the number of sites.

To obtain the 16-QAM multiresolution gain over the single resolution with QPSK used by MBMS with the WCDMA technology specified in Release 6, the aggregate throughput in all cell areas with multiresolution should be computed and divided by the single resolution aggregate throughput in the cell area. As the coverage of QPSK blocks becomes the same as strong bits blocks of hierarchical 16-QAM due to macrodiversity combining, comparing aggregate throughputs is based on different coverage of the weak bits blocks.

It is clear that the smallest throughput gain is achieved for coding rate = 1/2 (256 kbps). For this case, taking into account that the single resolution throughput of QPSK, which is 128 kbps, the throughput gain is 2. The highest throughput gain is achieved for coding rate = 3/4 (384 kbps) and the MBSFN scenario. For this case, the throughput gain is almost 3. However, for the SC-PMP scenario with macrodiversity the throughput gain stays close to 3/1.5 = 2 (see Figure 16.14).

In the MBSFN scenario due to the smaller inter-cell interference and substantial macrodiversity combining, in order to achieve higher multiresolution gains, it is suggested that both the channel bit rates from 256 kbps (i.e., the channel coding rate of 1/2) to 384 kbps corresponding to coding rate of 3/4 be increased; in addition, the ISD can also be extended from 1500 m to 3000 m. For the high channel bit rate of 384 kbps, the spectral efficiency achieved per cell sector considering the transmission

Table 16.3 Capacity Values for a QPSK Single-Resolution WCDMA Scheme for Bandwidth BW = 10 MHz

QoS	*#TV Channels*	*Spectral Efficiency*	*ISD*	*Scenario*
256 kbps	14	0.358 bps/Hz/cell	1000 m	SC-PMP 2RL (2×5MHz)

of 18.75 TV channels using all the 10 MHz bandwidth is 0.72 bps/Hz/cell. This value of spectral efficiency is valid for users at the cell border. The ISD associated to this spectral efficiency is 3000 m. Alternatively, 28.125 TV channels with 256 kbps could be transmitted at the same time as indicated in Table 16.2.

Table 16.3 shows the capacity of MBMS single resolution taking into account results for the standard MBMS normalized in Release 6 presented in [17] for the same SC-PMP scenario with the macrodiversity of two radio links.

The comparison between Tables 16.2 and 16.3 is not straightforward due to the different ISD used. However, it is possible to draw a capacity gain of at least 2 between hierarchical 16-QAM and QPSK (notice that the higher ISD is an advantage for broadcasting).

The ISD gain is related to the decrease in the number of sites due to a longer ISD. It has been said in this chapter that for the MBSFN scenario the cell radius can be increased (the ISD is the double the cell radius) from 750 m up to 1500 m without any decrease of average throughput as long as the fraction of the total transmitted power dedicated to E-MBMS reaches 90%. This corresponds to a reduction in the number of sites equal to 50%. An alternative is to use the same carrier for E-MBMS traffic and other unicast types of traffic without increasing the ISD.

16.5 Summary and Conclusions

We have evaluated the use of multiresolutions with hierarchical modulations for the E-MBMS to be standardized in the next release for the LTE-advanced. In this chapter, link and system level simulations of LTE multicellular networks considering broadcast/multicast transmissions using OFDM/OFDMA–based LTE technology were presented. The evaluation of the capacity, in terms of number of TV channels with given bit rates or total spectral efficiency and coverage, was presented.

Taking the 95% coverage as reference the evaluation of the achievable capacity gain (number of transmitted mobile TV channels for WCDMA and LTE) is next presented. The intersite distance gain is also presented allowing for a substantial reduction in the number of cell sites when LTE will replace WCDMA.

Scenarios based on the use of SFNs with the MBSFN channel and the SC-PMP network with the vehicular A channel were both evaluated for 16-QAM/64-QAM hierarchical modulations and compared with the present WCDMA-based MBMS network. In general, it can be stated that multiresolution is suitable for any of the analyzed scenarios MBSFN and SC-PMP. Indeed, it works fine in any single-cell scenario without macrodiversity combining or in multicells with macrodiversity.

In the SC-PMP scenario without macrodiversity (1RL), due to multiresolution, the channel bit rate of each TV channel (compared to single resolution provided by QPSK) for users can be increased close to the base station for the ISD equal to 1500 m. When macrodiversity (2RL) of the two best radio links is added, the multiresolution schemes become less sensitive to the used channel bit rates and it is possible to increase the channel coding rate, keeping the same ISD or to increase the ISD to 3000 m, keeping the channel bit rate. The operator must choose between the trade-off of increasing capacity or coverage (see Table 16.2); it is not possible to increase both at the same time.

16.6 Open Issues

Evolved MBMS is not a closed issue in the standardization carried out by 3GPP. It is not finalized if the hierarchical constellations already used by the DVB and MediaFLO standards will be chosen for LTE-advanced in the next release of E-MBMS. The combined use of 64-QAM hierarchical constellations and MIMO (spatial multiplexing) in the LTE-advanced as an additional flexible multiresolution scheme for the E-MBMS network is a topic that remains to be evaluated.

The scenario evaluated here was that MBSFN was based on a regular cellular grid. Real-life scenarios are not so uniform that will result in the reduction of the gain figures presented in this chapter. This is an issue that should be considered and evaluated in the near future when LTE deployment is carried out.

References

[1] 3GPP TR 25.905, version 7.2.0, Release 7 "Feasibility study on improvement of the multimedia broadcast multicast service (MBMS)." http://www.3gpp.org, January 2008.

[2] H. Sari, Y. Levy, and G. Karam, "An analysis of orthogonal frequency-division multiple access," *IEEE GLOBECOM'97*, November 1997.

[3] I. Koffman and V. Roman, "Broadband wireless access solutions based on OFDM access in IEEE 802.16," *IEEE Communications Magazine*, vol. 40, no. 4, pp. 96–103, April 2002.

[4] J. A. C. Bingham, "Multicarrier modulation for data transmission: An idea whose time has come," *IEEE Communications Magazine*, vol. 28, no. 5, pp. 5–14, May 1990.

[5] T. Cover, "Broadcast channels," *IEEE Transactions on Informational Theory*, vol. IT-18, pp. 2–14, January 1972.

[6] K. Ramchandran, A. Ortega, K. M. Uz, and M. Vetterli, "Multi-resolution broadcast for digital HDTV using joint source/channel coding," *IEEE Journal on Selected Areas in Communication*, vol. 11, January 1993.

[7] H. Jiang and P. A. Wilford, "A hierarchical modulation for upgrading digital," *IEEE Transactions on Broadcasting*, vol. 51, no. 2, pp. 223–229, June 2005.

[8] S. Wang, S. Kwon, and B. K. Yi, "On enhancing hierarchical modulation," *Proceedings of the IEEE International Symposium on Broadband Multimedia Systems and Broadcasting—BTS*, Las Vegas, NV, March 31–April 2, 2008.

[9] P. K. Vitthaladevuni and M.-S. Alouini, "A closed-form expression for the exact BER of generalized PAM and QAM constellations," *IEEE Transactions on Communications*, vol. 52, pp. 698–700, May 2004.

[10] N. Souto, F. Cercas, R. Dinis, and J. C. Silva, "On the BER performance of hierarchical M-QAM constellations with diversity and imperfect channel estimation," *IEEE Transactions on Communications*, vol. 55, no. 10, pp. 1852–1856, October 2007.

[11] G. Foschini, "Layered-space-time architecture for wireless communication in a fading environment when using multi-element antennas," *Bell Labs Technical Journal*, pp. 41–59, Autumn 1996.

[12] G. Foschini and M. Gans, "On limits of wireless communications in fading environments when using multiple antennas," *Wireless Personal Communications Journal*, vol. 6, pp. 315–335, March 1998.

[13] A. Soares, N. Souto, J. Silva, P. Eusbio, and A. Correia, "Effective radio resource management for MBMS in UMTS networks," *Wireless Personal Communications Journal*, vol. 42, no. 2, pp. 185–211, July 2007.

[14] A. Soares, J. Silva, F. Leito, A. Correia, and N. Souto, "MIMO based radio resource management for UMTS multicast broadcast multimedia services," *Wireless Personal Communications Journal*, vol. 42, Issue 2, pp. 225–246, July 2007.

[15] A. Correia, N. Souto, J. Silva, and A. Soares, Chapter 17, "Air interface enhancements for MBMS," *Handbook on Mobile Broadcasting*, Borko Furht and Syed Ahson, eds., CRC Press, Taylor and Francis, New York, 2008.

[16] A. Correia, J. Silva, N. Souto, L. Silva, A. Boal, and A. Soares, "Multi-resolution broadcast/multicast systems for MBMS," *IEEE Transactions on Broadcasting*, vol. 53, no. 1, pp. 224–234, March 2007.

[17] 3GPPP TR 25.814, version 7.1.0, Release 7, "Technical specification group radio Access Network; Physical Layers Aspects for Evolved (UTRA)," September 2006. http://www.3gpp.org

[18] N. Souto, A. Correia, R. Dinis, J. Silva, and L. Abreu, "Multiresolution MBMS Transmissions for MIMO UTRA LTE systems," *Proceedings of the IEEE International Symposium on Broadband Multimedia Systems and Broadcasting*—BTS, Las Vegas, NV, March 31–April 2, 2008.

[19] 3GPP, 25.101, version 6.2.0, Release 6, "User equipment radio transmission and reception (FDD)," June 2006, http://www.3gpp.org

[20] 3GPP TR 25.912, version 7.1.0, Release 7, "Feasibility study for evolved universal terrestrial radio access (UTRA) and universal terrestrial radio access network (UTRAN), http://www.3gpp.org

[21] 3GPP TR 25.892, version 6.0.0, Release 6, "Feasibility study for orthogonal frequency division multiplexing (OFDM) for UTRAN enhancement," September 2006, http://www.3gpp.org

[22] 3GPP TR 36.942, Release 8, "Evolved universal terrestrial radio access (E-UTRA); radio frequency (RF) system scenarios," December 2008, http://www.3gpp.org

[23] 3GPP, R1-070674, "LTE physical layer framework for performance verification," 3GPP TSG-RAN1#48, February 2007, http://www.3gpp.org

[24] E. Damosso, European Commission COST 231, *Digital Mobile Radio towards Future Generation Systems*, European Commission, 1999, Brussels.

Index

S

T

U

V